UNITEXT – La Matematica per il 3+2

Volume 74

For further volumes:
http://www.springer.com/series/5418

Alfredo Bermúdez · Dolores Gómez ·
Pilar Salgado

Mathematical Models and Numerical Simulation in Electromagnetism

Alfredo Bermúdez
Department of Applied Mathematics
Universidade de Santiago de Compostela
Spain

Dolores Gómez
Department of Applied Mathematics
Universidade de Santiago de Compostela
Spain

Pilar Salgado
Department of Applied Mathematics
Universidade de Santiago de Compostela
Spain

UNITEXT – La Matematica per il 3+2
ISSN 2038-5722   ISSN 2038-5757 (electronic)
ISBN 978-3-319-02948-1   ISBN 978-3-319-02949-8 (eBook)
DOI 10.1007/978-3-319-02949-8
Springer Cham Heidelberg New York Dordrecht London

Library of Congress Control Number: 2013949325

© Springer International Publishing Switzerland 2014
This work is subject to copyright. All rights are reserved by the Publisher, whether the whole or part of the material is concerned, specifically the rights of translation, reprinting, reuse of illustrations, recitation, broadcasting, reproduction on microfilms or in any other physical way, and transmission or information storage and retrieval, electronic adaptation, computer software, or by similar or dissimilar methodology now known or hereafter developed. Exempted from this legal reservation are brief excerpts in connection with reviews or scholarly analysis or material supplied specifically for the purpose of being entered and executed on a computer system, for exclusive use by the purchaser of the work. Duplication of this publication or parts thereof is permitted only under the provisions of the Copyright Law of the Publisher's location, in its current version, and permission for use must always be obtained from Springer. Permissions for use may be obtained through RightsLink at the Copyright Clearance Center. Violations are liable to prosecution under the respective Copyright Law.
The use of general descriptive names, registered names, trademarks, service marks, etc. in this publication does not imply, even in the absence of a specific statement, that such names are exempt from the relevant protective laws and regulations and therefore free for general use.
While the advice and information in this book are believed to be true and accurate at the date of publication, neither the authors nor the editors nor the publisher can accept any legal responsibility for any errors or omissions that may be made. The publisher makes no warranty, express or implied, with respect to the material contained herein.

Cover Design: Beatrice ẞ, Milano
Typesetting with LaTeX: PTP-Berlin, Protago TeX-Production GmbH, Germany (www.ptp-berlin.de)

Springer is a part of Springer Science+Business Media (www.springer.com)

To our students

Preface

This monograph has been developed as a basic support for a graduate course in electromagnetism oriented to numerical simulation. Thus, the emphasis is on the general equations rather that in solving analytically particular problems. The main goal is that the reader learns about the boundary-value problems of partial differential equations that should be solved, in general by using numerical methods, in order to perform computer simulation of electromagnetic processes. The book is more oriented to electrical engineering applications rather than to wave propagation problems related to telecommunication devices. Unlike most books on the subject, it goes from the general to the specific, namely, from the full Maxwell's equations to the particular cases of electrostatics, direct current, magnetostatics and eddy currents models. It also contains a first part devoted to electrical circuit theory based on ordinary differential equations. Apart from standard exercises related to analytical calculus, some others oriented to real-life applications are also included.

The book is structured in three parts. The first one is devoted to lumped parameter models so it includes an elementary circuit theory. It consists of three chapters. Chapter 1 is devoted to recall the harmonic oscillator. The use of complex functions to integrate the model is emphasized and the notion of impedance is introduced. In Chap. 2 the mechanical-electrical analogy is exploited to study series RLC circuits. Chapter 3 concerns more general linear circuits. They are modelled as digraphs leading to linear systems of integro-differential equations. Numerical solution is addressed both in the time and in the frequency domains.

The second part concerns distributed parameter models. In Chap. 4 Maxwell's equations in the empty space are recalled. Then they are reduced to vector wave equations in terms of either the electric field or the magnetic induction. The equations for the harmonic case are also obtained and then plane wave solutions are calculated. Chapter 5 includes several examples that can be solved by hand and Chap. 6 extends Maxwell's equations to material media. Concepts as electric polarization and magnetization are described and their influence in the electromagnetic field is analyzed. Next, the particular cases of electrostatics, direct current, magnetostatics and eddy currents are considered from Chaps. 7 to 10. The weak formulation of each problem most suitable to perform computer simulation of electromagnetic processes is chosen and detailed. Moreover, Chap. 10 includes a brief section on coupling distributed and lumped models. The last chapter of Part II is devoted to nonlinear magnetics. This is the subject of Chap. 11 where we recall the hysteresis behavior and its mathematical modelling following the Preisach methodology.

The third part of the book is devoted to numerical simulation of electromagnetic problems. The numerical analysis of Maxwell's equations has been extensively developed during the last decades by mathematicians and engineers. Thus we can mention the books by Alonso and Valli [4], Bossavit [23], Chari and Silvester [31], Jin [46], Monk [55], Silvester and Ferrari [64], just to list a representative sampling of books dealing with this subject.

Numerical methods can address situations that otherwise would become totally unmanageable given the complexity or impracticability of its analytical solution. Moreover, with a suitable post processing of the results, it is possible to qualitatively and quantitatively understand what is really happening in the system under study. Among the numerical methods found in the literature to approximate Maxwell's equations, the finite element method (FEM) is the most extended. Its main advantages are its geometric flexibility and the richness in theoretical mathematical tools useful to analyze the approximation of the problem.

In recent years, the proliferation of commercial software packages attests to the growing use of numerical simulation in industry. This is due, in large part, to its usefulness in the design process of products reducing the "trial and error" cycle by evaluating multiple designs in the computer.

Among the commercial programs for the simulation of electromagnetic problems, we can mention Flux 2D and 3D from Cedrat [69], Ansys Multiphysics [68], Magnet from Infolytica [71], Opera from Cobbham [73] or Comsol Multiphysics [70], to name the most used in low-frequency applications. Usually, due to their high cost, the use of these packages is mainly restricted to the research field. However, computer programs and packages are also increasingly being integrated into the graduate and undergraduate curriculum, serving as virtual laboratories to teach and reinforce concepts.

In this book we will use MaxFEM [72], a free software package developed in the Research Group of Mathematical Engineering from Universidade de Santiago de Compostela (USC), Spain.

MaxFEM is an open software for numerical simulation in electromagnetism based on the finite element method. It contains several applications covering electrostatics, direct current, magnetostatics and eddy currents in two and/or three dimensions and in Cartesian and/or cylindrical coordinates. In addition to the code itself, MaxFEM includes a user interface to facilitate its handling and it is fitted with an on-line help manual. It can be downloaded from http://sourceforge.net/projects/maxfem/.

By introducing these companion computational tools in the textbook, our goal is to strike a balance between the classical teaching of electromagnetism, which emphasizes its physical and mathematical foundations, and a numerical simulation approach which focuses on solving realistic problems using software packages.

With this goal in mind, and according to Parts I and II, we have structured Part III of the book in five chapters from 12 to 16 devoted, respectively, to electrical circuits, electrostatics, direct current, magnetostatics and eddy currents problems.

Finally, some basic and complementary subjects like graph theory, vector calculus or function spaces are included in appendices as well as Maxwell's equations in Lagrangian coordinates.

Most of the research in mathematical electromagnetism of the authors has been developed in collaboration with professor Rodolfo Rodríguez from the University of Concepción (Chile). Along many years we have enjoyed his intelligence, kindness and friendship. Some of our common former students, as Bibiana López and Pablo Venegas have carefully read several chapters of the manuscript or written computer programs now included in MaxFEM. We also express our gratitude to the rest of the team that has written this software: M. Carmen Muñiz, Francisco Pena, Marta Piñeiro, Víctor Sande and Rodrigo Valiña. Finally, we also acknowledge the excellent work by Marta Piñeiro and Víctor Sande for providing us with several examples and figures included in the book.

Santiago de Compostela and Lugo
July 2013

Alfredo Bermúdez
Dolores Gómez
Pilar Salgado

Contents

Part I Lumped Parameter Models: Electric Circuit Theory

1 **The harmonic oscillator** 3
   1.1 A mechanical system 3
   1.2 Using complex functions. Undamped oscillations 5
   1.3 Damped oscillations... 6
   1.4 Energy balance 8
   1.5 Forced oscillations 10
   1.6 Resonance 13
   1.7 The first order equation 14

2 **The series RLC circuit** 21
   2.1 Mathematical model 21
   2.2 Energy balance 23
   2.3 Harmonic source. Impedance. Resonance 23
   2.4 Active and reactive power 25
   2.5 The circuit without capacitor 26

3 **Linear electrical circuits** 33
   3.1 Mathematical model 33
       3.1.1 The incidence matrix. Properties 34
       3.1.2 The first Kirchhoff's law 35
       3.1.3 Constitutive laws 35
       3.1.4 The second Kirchhoff's law 37
   3.2 Numerical solution of the circuit equations 38
       3.2.1 Existence of solution to the discrete problem 39
       3.2.2 Generators without internal resistance 41
   3.3 The case of given potentials 43
   3.4 Harmonic regime. Impedance matrix 44

Part II Distributed Parameter Models

4 Maxwell's equations in free space 53
    4.1   An introduction to distributed models for wave propagation. The one-dimensional wave equation 53
        4.1.1   Energy conservation 55
        4.1.2   Harmonic solutions to the one-dimensional wave equation 56
    4.2   Charges and currents 56
    4.3   Electric and magnetic fields 57
    4.4   Maxwell's integral equations in free space 58
    4.5   Maxwell's equations in differential form in free space 58
    4.6   Wave equations for electric and magnetic fields 59
    4.7   Harmonic regime 60
    4.8   Behavior at infinity: the Silver-Müller conditions 61
    4.9   Plane waves in empty space 61
    4.10 Wave polarization 64

5 Some solutions of Maxwell's equations in free space 67
    5.1   Electrostatic fields 67
        5.1.1   Electric scalar potential 67
        5.1.2   Point charge **Q** at the origin. Coulomb's law 68
        5.1.3   Ball of radius r_0 with a uniform charge density ρ_V 69
        5.1.4   A long line charge with uniform line density ρ_l 70
        5.1.5   Infinite planar charge with uniform surface density ρ_S 72
    5.2   Magnetostatic fields 73
        5.2.1   Round toroidal core of circular cross section 74
        5.2.2   Infinitely long solenoid 74

6 Maxwell's equations in material regions 77
    6.1   A one-dimensional model 77
        6.1.1   Reflection and transmission of waves 77
        6.1.2   Damped waves 79
    6.2   Conductors and insulators 80
        6.2.1   Electrical conductivity. Ohm's law 80
        6.2.2   Electric polarization. Dielectrics 80
        6.2.3   The electric dipole 81
        6.2.4   The polarization vector 83
        6.2.5   Electric susceptibility and electric permittivity 84
        6.2.6   Electric Gauss' law for materials 84
    6.3   Magnetic materials 85
        6.3.1   The magnetic dipole 85
        6.3.2   The magnetization vector 90
        6.3.3   Magnetic susceptibility. Magnetic permeability 91
        6.3.4   Magnetic Gauss' law for materials 91
    6.4   Ampère's law for materials 92

	6.5	Maxwell's equations in differential form for materials	92
	6.6	Electric charge conservation equation	93
	6.7	Interface and boundary conditions	94
	6.7.1	Surface impedance boundary condition	95
	6.8	Electromagnetic energy	96
	6.8.1	Energy balance in a bounded domain	98
	6.9	Electromagnetic force	100
	6.10	Plane waves in an unbounded conductive region	105
	6.10.1	Harmonic regime	106
	6.10.2	Propagation constant. Attenuation constant. Phase constant	106
	6.10.3	Wavelength. Propagation velocity	107
	6.10.4	Skin depth	107
	6.10.5	Intrinsic wave impedance	108
	6.11	Electric classification of materials	108
	6.11.1	Conductive media	108
	6.11.2	Lossy dielectrics	109
	6.12	Microwave heating	111
	6.12.1	Electromagnetic model	111
	6.12.2	A simplified solution: Lambert's law	112
	6.12.3	Heating power	113
	6.13	Properties of materials: homogeneity, isotropy, linearity	113

7   Electrostatics 115
  7.1 The electrostatics model. Electrostatic potential 115
  7.2 Energy of the electrostatic field 117
  7.3 The electrostatics model in bounded domains 118
  7.4 Weak formulation of electrostatics problems 119
      7.4.1 Weak formulation of the 3D electrostatics model 119
      7.4.2 Weak formulation of the in-plane 2D electrostatics model . 120
      7.4.3 The axisymmetric case 121
  7.5 Electric field created by a set of charged conductors. Capacitance matrix 121
      7.5.1 An important case: the capacitor 125
      7.5.2 Energy 132

8   Direct current 139
  8.1 Maxwell's equations for direct current 139
      8.1.1 Weak formulation of the 3D direct current model 140
      8.1.2 Weak formulation of the in-plane 2D direct current model 140
      8.1.3 The axisymmetric case 141
  8.2 Conductance matrix. Resistance matrix 142
  8.3 Approximate resistance 146
  8.4 Lumped direct current circuits 147

9 Magnetostatics 151
- 9.1 Maxwell's equations for magnetostatics 151
- 9.2 Magnetic vector potential 152
- 9.3 A formulation in magnetic vector potential 155
- 9.4 A formulation in magnetic field 158
- 9.5 A formulation in terms of the reduced scalar magnetic potential ... 159
- 9.6 A formulation with two scalar potentials 159
- 9.7 Distributed magnetic circuits: reluctance matrix 163
- 9.8 Approximate reluctance 169
- 9.9 Lumped magnetic circuits 172
- 9.10 2D Magnetostatic model with **J** normal to the 2D section 174
- 9.11 Axisymmetric magnetostatic model with azimuthal current 176

10 The eddy currents model 183
- 10.1 The time-harmonic eddy currents model in a bounded domain 183
- 10.2 Magnetic vector potential/scalar electric potential formulation 187
- 10.3 A formulation in magnetic field 190
- 10.4 Magnetic field/scalar magnetic potential formulation 192
  - 10.4.1 Energy conservation 193
- 10.5 Impedance matrix 194
  - 10.5.1 Reduced impedances 196
- 10.6 2D eddy currents model with **J** normal to the section 200
- 10.7 2D eddy currents model with **J** lying on the section 203
- 10.8 Axisymmetric eddy currents model with azimuthal **J** 205
- 10.9 Axisymmetric eddy currents model with **J** lying on the meridional section 212
- 10.10 Coupling lumped and distributed models 213

11 An introduction to nonlinear magnetics. Hysteresis 217
- 11.1 Magnetic behavior of materials 217
  - 11.1.1 Magnetic hysteresis 218
  - 11.1.2 Energy balance 222
- 11.2 Mathematical modelling of magnetic hysteresis 224
  - 11.2.1 The classical Preisach model 224
  - 11.2.2 Geometric interpretation 225
  - 11.2.3 Identification problem 231
  - 11.2.4 Everett function 233
  - 11.2.5 Numerical computation of the electromagnetic field: the transient eddy currents model with hysteresis 239
- 11.3 Excess eddy current losses. The dynamic Preisach model 240

6.5	Maxwell's equations in differential form for materials		92
6.6	Electric charge conservation equation		93
6.7	Interface and boundary conditions		94
	6.7.1	Surface impedance boundary condition	95
6.8	Electromagnetic energy		96
	6.8.1	Energy balance in a bounded domain	98
6.9	Electromagnetic force		100
6.10	Plane waves in an unbounded conductive region		105
	6.10.1	Harmonic regime	106
	6.10.2	Propagation constant. Attenuation constant. Phase constant	106
	6.10.3	Wavelength. Propagation velocity	107
	6.10.4	Skin depth	107
	6.10.5	Intrinsic wave impedance	108
6.11	Electric classification of materials		108
	6.11.1	Conductive media	108
	6.11.2	Lossy dielectrics	109
6.12	Microwave heating		111
	6.12.1	Electromagnetic model	111
	6.12.2	A simplified solution: Lambert's law	112
	6.12.3	Heating power	113
6.13	Properties of materials: homogeneity, isotropy, linearity		113

7 Electrostatics ... 115

7.1	The electrostatics model. Electrostatic potential		115
7.2	Energy of the electrostatic field		117
7.3	The electrostatics model in bounded domains		118
7.4	Weak formulation of electrostatics problems		119
	7.4.1	Weak formulation of the 3D electrostatics model	119
	7.4.2	Weak formulation of the in-plane 2D electrostatics model	120
	7.4.3	The axisymmetric case	121
7.5	Electric field created by a set of charged conductors. Capacitance matrix		121
	7.5.1	An important case: the capacitor	125
	7.5.2	Energy	132

8 Direct current ... 139

8.1	Maxwell's equations for direct current		139
	8.1.1	Weak formulation of the 3D direct current model	140
	8.1.2	Weak formulation of the in-plane 2D direct current model	140
	8.1.3	The axisymmetric case	141
8.2	Conductance matrix. Resistance matrix		142
8.3	Approximate resistance		146
8.4	Lumped direct current circuits		147

9 Magnetostatics 151
9.1 Maxwell's equations for magnetostatics 151
9.2 Magnetic vector potential 152
9.3 A formulation in magnetic vector potential 155
9.4 A formulation in magnetic field 158
9.5 A formulation in terms of the reduced scalar magnetic potential ... 159
9.6 A formulation with two scalar potentials 159
9.7 Distributed magnetic circuits: reluctance matrix 163
9.8 Approximate reluctance 169
9.9 Lumped magnetic circuits 172
9.10 2D Magnetostatic model with **J** normal to the 2D section ... 174
9.11 Axisymmetric magnetostatic model with azimuthal current 176

10 The eddy currents model 183
10.1 The time-harmonic eddy currents model in a bounded domain 183
10.2 Magnetic vector potential/scalar electric potential formulation 187
10.3 A formulation in magnetic field 190
10.4 Magnetic field/scalar magnetic potential formulation 192
    10.4.1 Energy conservation 193
10.5 Impedance matrix 194
    10.5.1 Reduced impedances 196
10.6 2D eddy currents model with **J** normal to the section 200
10.7 2D eddy currents model with **J** lying on the section 203
10.8 Axisymmetric eddy currents model with azimuthal **J** 205
10.9 Axisymmetric eddy currents model with **J** lying on the meridional section 212
10.10 Coupling lumped and distributed models 213

11 An introduction to nonlinear magnetics. Hysteresis 217
11.1 Magnetic behavior of materials 217
    11.1.1 Magnetic hysteresis 218
    11.1.2 Energy balance 222
11.2 Mathematical modelling of magnetic hysteresis 224
    11.2.1 The classical Preisach model 224
    11.2.2 Geometric interpretation 225
    11.2.3 Identification problem 231
    11.2.4 Everett function 233
    11.2.5 Numerical computation of the electromagnetic field: the transient eddy currents model with hysteresis 239
11.3 Excess eddy current losses. The dynamic Preisach model 240

Part III Numerical Solution of Maxwell's Equations

12   Electrical circuits with MaxFEM 243
  12.1 Some examples with harmonic source voltage 243
    12.1.1 Magnetically coupled circuits. A simple transformer 243
    12.1.2 A circuit with three connected components 246
    12.1.3 Three-phase alternating current 250
  12.2 Some examples with transient source voltage 253
    12.2.1 Results for the linear voltage source 255
    12.2.2 Results for the sawtooth wave 256
    12.2.3 Results for the rectangular pulse train 257
    12.2.4 Results for the PWM signal 258
  12.3 Some other typical configurations 259
    12.3.1 Bridge circuit 259
    12.3.2 Star delta connection circuit 262

13   Electrostatics with MaxFEM 267
  13.1 Some classical problems with analytical solution 267
    13.1.1 Point charge at the origin 268
    13.1.2 Electric field of n point charges 269
    13.1.3 A uniformly charged spherical volume 271
    13.1.4 A sphere with a non-uniform volumetric charge density .. 273
    13.1.5 A uniformly charged infinite line 275
    13.1.6 A uniformly charged segment 276
    13.1.7 A uniformly charged infinite plane 280
    13.1.8 An infinitely long cylinder uniformly charged on its
            surface 283
    13.1.9 An infinitely long solid cylinder uniformly charged 284
    13.1.10 A uniformly charged circular disk 286
  13.2 Some problems arising in physical applications 288
    13.2.1 The coaxial cable 289
    13.2.2 A planar capacitor with parallel plates 292
    13.2.3 Planar capacitor with discontinuities in the dielectric 295

14   Direct current with MaxFEM 299
  14.1 Current traversing a copper bar 299
  14.2 An electrolytic cell for aluminium production 300

15   Magnetostatics with MaxFEM 307
  15.1 Problems with analytical solution 307
    15.1.1 An infinite cylinder carrying a static current intensity 307
    15.1.2 Magnetic field created by an infinite plane current 310
    15.1.3 Two infinite coaxial conductors carrying a static current .. 312
    15.1.4 An infinite cylinder surrounded by an infinitely thin coil .. 314

15.2 Problems arising in physical applications 318
    15.2.1 2D magnetostatic fields in an electromagnetic contactor .. 318
    15.2.2 A cylindrical electromagnet 319
    15.2.3 3D magnetostatic fields in a C-magnetic core 321

16 Eddy currents with MaxFEM 325
16.1 Problems with analytical solution 325
    16.1.1 An infinite cylinder carrying an alternating current 325
    16.1.2 Two infinite coaxial conductors carrying an alternating current.. 329
16.2 Problems arising in physical applications 332
    16.2.1 Metallurgical electrodes of an electric arc furnace 332
    16.2.2 A two-dimensional example related with non-destructive testing 336
    16.2.3 An axisymmetric induction heating furnace 338
    16.2.4 An induction furnace surrounded by an helical coil 343
    16.2.5 A plate over a coil 344

Appendix A. Elements of graph theory............................. 347
A.1 Definitions 347
A.2 Incidence matrix. Connected components 348
A.3 Cycle matrix. Fundamental cycles 351

Appendix B. Vector calculus 355
B.1 Vector and tensor algebra 355
    B.1.1 Vector space. Basis 355
    B.1.2 Inner product 356
    B.1.3 Tensors 357
    B.1.4 The affine space 363
B.2 Vector and tensor analysis 364
    B.2.1 Differential operators 364
    B.2.2 Curves and curvilinear integrals 368
    B.2.3 Gauss' and Green's formulas. Stokes' Theorem 370
    B.2.4 The area enclosed by a plane curve 371
    B.2.5 Change of variable in integrals 372
    B.2.6 Kinematics and transport theorems 372
    B.2.7 Localization theorem............................... 376
    B.2.8 Curvilinear coordinates 376

Appendix C. Function spaces for electromagnetism 387
C.1 Scalar, vector and tensor distributions. Differential operators 387
    C.1.1 Introduction 387
    C.1.2 The space of distributions............................ 388
C.2 Lebesgue and Sobolev spaces. Trace theorems 391
C.3 The spaces $\mathbf{H}(\mathrm{div}, \Omega)$ and $\mathbf{H}(\mathbf{curl}, \Omega)$........................ 393

C.4	The spaces $H_{00}^{1/2}(S)$ and $H_{00}^{-1/2}(S)$	394
C.5	Green's formulas	395
C.6	Weighted Sobolev spaces	396
C.7	Complements	397
	C.7.1 Distributions supported on a surface. Examples	397
C.8	Radial functions and distributions	401

Appendix D. Harmonic regime: average values 405

Appendix E. Linear nodal and edge finite elements 407
    E.1   Linear nodal finite elements 408
    E.2   Nédélec edge elements of first order 408

Appendix F. Maxwell's equations in Lagrangian coordinates 413
    F.1   Preliminary results and notation 413
    F.2   Transforming the Maxwell's equations 415
    F.3   Transforming the constitutive laws 418

References 421

Index 427

Part I
Lumped Parameter Models: Electric Circuit Theory

Lumped parameter models are zero-dimensional models, namely the involved physical magnitudes are independent of space variables. Thus, for transient phenomena they consist of systems of ordinary differential equations because physical magnitudes only depend on time, while for stationary cases they simply consist of system of numerical equations. In the framework of electromagnetism the study of lumped parameter models is the subject of the so-called electrical circuit theory.

1
The harmonic oscillator

We start this part devoted to lumped parameter models by studying a simple paradigmatic model in physics: the *harmonic oscillator*. In principle, it corresponds to a mechanical system consisting of a mass connected to a spring and a dashpot (see Fig. 1.1), but it turns out that the model for a series electrical circuit including a resistance, an inductor and a capacitor, the so-called *series RLC circuit*, is analogous. This chapter is devoted to the analysis of the harmonic oscillator. Then, in the next chapter we translate the results to a series RLC electrical circuit.

1.1 A mechanical system

Let us consider a mass-spring system vibrating with simple harmonic motion. We assume that the amount of stretch is proportional to the restoring force F and, in a first step, that the mass slides freely without loss of energy.

By combining Hooke's law, $F = -kx$, with Newton's second law, $F = ma = m\ddot{x}$, we get the model
$$m\ddot{x} = -kx,$$
where m denotes the mass, x the displacement of the mass from the equilibrium position and $\ddot{x} = d^2x/dt^2$ is the acceleration. The negative sign in Hooke's law reflects

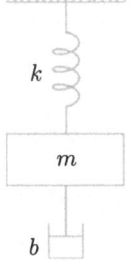

Fig. 1.1. A typical damped mass-spring system

A. Bermúdez, D. Gómez, P. Salgado: *Mathematical Models and Numerical Simulation in Electromagnetism.* UNITEXT – La Matematica per il 3+2 74
DOI 10.1007/978-3-319-02949-8_1, © Springer International Publishing Switzerland 2014

the opposing nature of the force. The positive constant k is called *stiffness* or *spring constant* and in the *International System of Units* (SI) it is measured in N/m.

By introducing

$$\omega_0 = \sqrt{\frac{k}{m}} \quad \text{(rad/s)},$$

the motion equation can be rewritten as

$$\ddot{x} + \omega_0^2 x = 0.$$

The general solution of this linear differential equation is a periodic function that has any of the two following forms:

$$x(t) = A\cos(\omega_0 t + \phi), \tag{1.1}$$

$$x(t) = B\cos(\omega_0 t) + C\sin(\omega_0 t), \tag{1.2}$$

where A, ϕ, B and C are free parameters. The *period* T is given by

$$T = \frac{2\pi}{\omega_0} = 2\pi\sqrt{\frac{m}{k}} \quad \text{(s)}.$$

The *frequency* is the number of periods per unit time, i.e.,

$$f = \frac{1}{T} = \frac{\omega_0}{2\pi} = \frac{1}{2\pi}\sqrt{\frac{k}{m}}$$

and the SI unit is *hertz* (Hz).

By taking derivatives in (1.1) we get the velocity and the acceleration which are given by

$$v(t) = -\omega_0 A\sin(\omega_0 t + \phi),$$

$$a(t) = -\omega_0^2 A\cos(\omega_0 t + \phi).$$

Moreover, the initial-value problem consists of finding a particular solution satisfying

$$x(0) = x_0,$$

$$\dot{x}(0) = v(0) = v_0.$$

Easy computations show that the corresponding values of A and ϕ are given by

$$A = \sqrt{x_0^2 + \left(\frac{v_0}{\omega_0}\right)^2}, \quad \phi = \tan^{-1}\left(\frac{-v_0}{\omega_0 x_0}\right).$$

In a similar way we can obtain B and C:

$$B = x_0, \quad C = \frac{v_0}{\omega_0}.$$

Parameters A and ϕ are related with B and C by

$$B = A\cos\phi, \quad C = -A\sin\phi,$$

$$A = \sqrt{B^2 + C^2}, \quad \phi = \tan^{-1}\left(\frac{-C}{B}\right).$$

1.2 Using complex functions. Undamped oscillations

In this section we introduce the use of complex-valued functions for solving linear ordinary differential equations like the one previously obtained for the harmonic oscillator.

In the rest of the chapter we will make use of some well-known equalities that have been summarized in Table 1.1.

Firstly, let us observe that, for any homogeneous linear ordinary differential equations with real coefficients, the complex-valued function $y(t) = y_1(t) + y_2(t)i$ is a complex solution if and only if the real functions $y_1(t)$ and $y_2(t)$ are solutions.

Then we seek a solution of problem (1.1) of the form

$$x(t) = \operatorname{Re}(\widetilde{A}e^{i\omega t}),$$

for some $\omega \in \mathbb{R}$ and $\widetilde{A} \in \mathbb{C}$.

For this purpose we replace x in (1.1) with $\widetilde{A}e^{i\omega t}$. We get,

$$-\omega^2 \widetilde{A} + \omega_0^2 \widetilde{A} = 0,$$

and then either $\widetilde{A} = 0$ (in which case we obtain the trivial solution) or $\omega = \omega_0$.

Let $\widetilde{A} = Ae^{i\phi}$. Then

$$x(t) = \operatorname{Re}(\widetilde{A}e^{i\omega_0 t}) = \operatorname{Re}(Ae^{i\phi}e^{i\omega_0 t}) = A\cos(\omega_0 t + \phi).$$

Thus $A = |\widetilde{A}|$ is the *amplitude* of the solution while $\phi = \arg(\widetilde{A})$ is its *phase*. $\widetilde{A}$ is called *complex amplitude* or *phasor*.

Table 1.1. Some useful identities

$\cos(x + \pi/2) = -\sin x,$
$\cos(x - \pi/2) = \sin x,$
$\sin(x + \pi/2) = \cos x,$
$\sin(x - \pi/2) = -\cos x,$
$e^{i\omega} = \cos\omega + i\sin\omega,$ if $\omega \in \mathbb{R}$ (Euler's formula).

1.3 Damped oscillations

We suppose there is a damping mechanism, as friction, that generates a force proportional to the velocity and opposite to the motion:

$$F_d = -b\dot{x},$$

where b is the *mechanical resistance* (measured in Ns/m in the SI). This force is added to the motion Eq. (1.1) which becomes

$$m\ddot{x} + b\dot{x} + kx = 0. \tag{1.3}$$

Dividing by m we obtain

$$\ddot{x} + 2\alpha\dot{x} + \omega_0^2 x = 0. \tag{1.4}$$

where $\alpha := b/(2m)$ and ω_0 has been introduced in (1.1). We look for a solution of the form $x(t) = \text{Re}(\widetilde{A}e^{\gamma t})$, $\widetilde{A}$ and γ being complex numbers. By replacing x with $\widetilde{A}e^{\gamma t}$ in (1.4) we get

$$(\gamma^2 + 2\alpha\gamma + \omega_0^2)\widetilde{A}e^{\gamma t} = 0.$$

In order to obtain a non-null solution we require,

$$\gamma^2 + 2\alpha\gamma + \omega_0^2 = 0,$$

or, equivalently,

$$\gamma = -\alpha \pm \sqrt{\alpha^2 - \omega_0^2}.$$

We distinguish three cases depending on the sign of $\alpha^2 - \omega_0^2$.

I. Case $\alpha < \omega_0$. It is the so-called *underdamped* case. We have,

$$\gamma = -\alpha \pm \sqrt{\omega_0^2 - \alpha^2}\, i = -\alpha \pm \omega_\alpha i,$$

where

$$\omega_\alpha = \sqrt{\omega_0^2 - \alpha^2}.$$

Then, the general solution of the motion equation is,

$$x(t) = Ae^{-\alpha t}\cos(\omega_\alpha t + \phi) = e^{-\alpha t}(B\cos(\omega_\alpha t) + C\sin(\omega_\alpha t)),$$

being $B = A\cos\phi$ and $C = -A\sin\phi$. The solution of the initial-value problem corresponding to the initial data $x(0) = x_0$, $\dot{x}(0) = v_0$ is given by

$$x(t) = e^{-\alpha t}\left(x_0 \cos(\omega_\alpha t) + \frac{v_0 + \alpha x_0}{\omega_\alpha}\sin(\omega_\alpha t)\right).$$

1.3 Damped oscillations

This motion is no longer periodic. Nevertheless, the time between zero crossings in the same direction remains constant and equal to

$$T_\alpha = \frac{2\pi}{\omega_\alpha},$$

which is defined as the *period* of the oscillation. The frequency f_α is given by

$$f_\alpha = \frac{1}{T_\alpha} = \frac{\omega_\alpha}{2\pi} = \frac{1}{2\pi}\sqrt{\omega_0^2 - \alpha^2} = \frac{1}{2\pi}\sqrt{\frac{k}{m} - \frac{b^2}{4m^2}}. \tag{1.5}$$

This number is usually called the *natural frequency* of the system. Of course, for $b = 0$ it coincides with (1.1). We notice that the oscillation frequency decreases because of the damping effect as it could be expected. We also notice that the amplitude decreases in time as $e^{-\alpha t}$ and the solution will oscillate between the two curves $-|A|e^{-\alpha t}$ and $|A|e^{-\alpha t}$ (see Fig. 1.2 d).

II. Case $\alpha = \omega_0$. In this case we said that the system is *critically damped*. We have that $\gamma = -\alpha$ is a double root of Eq. (1.3). Unlike the previous case, now the family of functions

$$x(t) = \mathrm{Re}(\widetilde{A}e^{\gamma t}) = A\cos\phi\, e^{-\alpha t}$$

only fills a one-dimensional space (there is only one free-parameter, namely, $A\cos\phi$), so we need to find another linearly independent solution. By replacing in Eq. (1.4), it is easy to see that function

$$x(t) = \mathrm{Re}(\widetilde{A}te^{\gamma t}),$$

with $\gamma = -\alpha$ is also a solution. Therefore, the general solution of the motion equation is

$$x(t) = (B + Ct)e^{-\alpha t}.$$

In particular, the displacement for $v_0 = 0$ is given by

$$x(t) = x_0(1 + \alpha t)e^{-\alpha t}.$$

The value of b corresponding to this case, that is,

$$b_c = 2\sqrt{km}$$

is called *critical resistance* or *critical damping coefficient*. In the general case, the ratio

$$\zeta := b/b_c = b/(2\sqrt{km})$$

is called *damping factor*.

III. Case $\alpha > \omega_0$. In this case, we said that the system is *overdamped*. Equation (1.3) has the roots
$$\gamma_\pm = -\alpha \pm \sqrt{\alpha^2 - \omega_0^2},$$
which satisfy $\gamma_- < \gamma_+ < 0$.

The two corresponding solutions are independent so the general solution of the motion equation is
$$x(t) = Be^{\gamma_+ t} + Ce^{\gamma_- t}.$$
The displacement for $v_0 = 0$ is given by
$$x(t) = \frac{x_0}{\gamma_+ - \gamma_-} \left(\gamma_+ e^{\gamma_- t} - \gamma_- e^{\gamma_+ t} \right).$$

This function is monotonically decreasing to 0 for $t > 0$.

When $\alpha \geq \omega_0$, that is, if the system is critically damped or overdamped, the solution does not oscillate: if we move the mass from its equilibrium position it asymptotically goes back to this position.

In the overdamped case, since $-\omega_0 < \gamma_+ < 0$, the system returns to the equilibrium position even more slowly that in the critically damped or the underdamped cases.

Remark 1.1. Notice that taking into account the definition of damping factor and using the natural frequency ω_0 of the system, Eq. (1.3) can be rewritten as
$$\ddot{x} + 2\zeta\omega_0\dot{x} + \omega_0^2 x = 0.$$

Again, this equation can be solved by replacing $x(t) = \mathrm{Re}(\widetilde{A}e^{\gamma t})$ in (1.1). We obtain the following expression for γ
$$\gamma = -\omega_0(\zeta \pm \sqrt{\zeta^2 - 1})$$
in terms of the values of the damping factor. Thus, if $\zeta < 1$ the system is underdamped, overdamped if $\zeta > 1$, critically damped if $\zeta = 1$ and undamped when $\zeta \to 0$.

The three different cases described above have been illustrated in Fig. 1.2 together with the undamped case. Notice that, qualitatively, the graphs for the overdamped and critically damped cases are similar.

1.4 Energy balance

Let us multiply the motion equation by velocity $\dot{x}$. We obtain
$$m\ddot{x}\dot{x} + b|\dot{x}|^2 + kx\dot{x} = 0$$
and then
$$\frac{1}{2}\frac{d}{dt}\left(m|\dot{x}|^2 + kx^2 \right) + b|\dot{x}|^2 = 0. \tag{1.6}$$

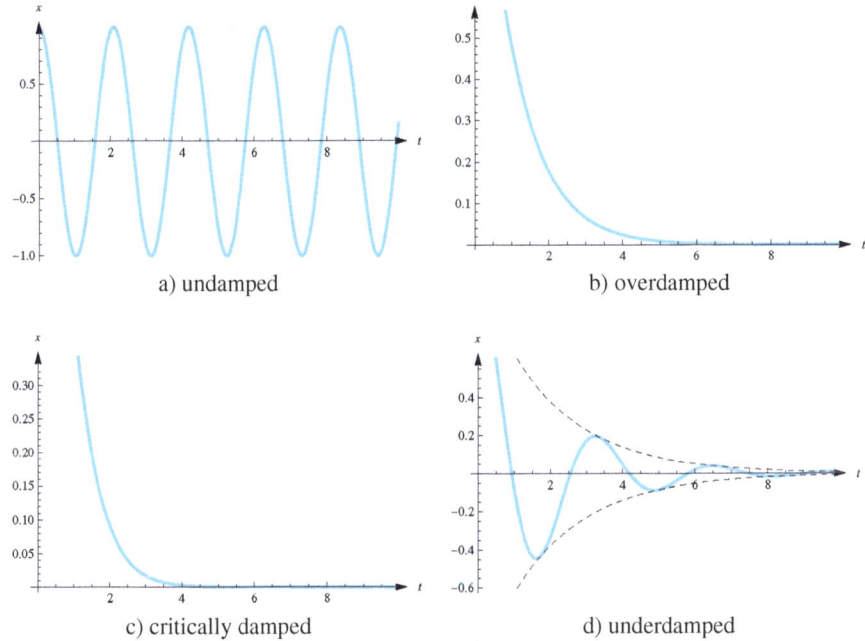

Fig. 1.2. Examples of different cases for unforced systems with zero-velocity initial condition

The expression

$$\mathcal{E}(t) = \frac{1}{2}m|\dot{x}(t)|^2 + \frac{1}{2}k|x(t)|^2,$$

is the total (mechanical) energy. The first term is the *kinetic energy* and the second one is the *elastic potential energy*. Let us justify the latter: in order to carry mass m from equilibrium point ($x = 0$) to position x, it is needed that an external force performs a work W against the spring. This work is stored as elastic potential energy. Let us calculate W: when the position is ξ, the spring exerts a force $F_s = -k\xi$, so the external force will be $F = k\xi$. Producing a small displacement, $d\xi$, the external force makes a work given by $dW = k\xi d\xi$. Thus, the total work will be

$$W = \int_0^x k\xi \, d\xi = \frac{1}{2}kx^2$$

and hence the potential energy corresponding to position x will be given by

$$\mathcal{E}_p(x) = \frac{1}{2}kx^2.$$

By integrating (1.6) from 0 to t we obtain

$$\mathcal{E}(t) + \int_0^t b|\dot{x}(s)|^2 \, ds = \mathcal{E}(0).$$

1.5 Forced oscillations

Let us assume there is a driving force f(t) that acts on the mass of the system in addition to the other forces that we have been considering. Then the motion equation becomes

$$m\ddot{x} + b\dot{x} + kx = f.$$

We show that, if f(t) is harmonic, then the solution consists of two parts: a *transient term* containing two arbitrary constants, which could be determined by two initial conditions, and a *stationary term* that depends only on f(t).

Let us assume that

$$f(t) = \mathbb{F}\cos(\omega t + \beta),$$

with $\omega > 0$, $\beta \in \mathbb{R}$ and $\mathbb{F} \in \mathbb{R}^+$. We notice that, if $\beta = -\pi/2$

$$f(t) = \mathbb{F}\sin(\omega t).$$

Let $\widetilde{\mathbb{F}} := \mathbb{F}e^{i\beta}$. Then,

$$f(t) = \mathrm{Re}(\widetilde{f}(t))$$

with $\widetilde{f}(t) = \widetilde{\mathbb{F}}e^{i\omega t}$. By replacing f with $\widetilde{f}$ and x with $\widetilde{A}e^{i\omega t}$ in (1.5) we obtain

$$\widetilde{A}e^{i\omega t}(-\omega^2 m + i\omega b + k) = \widetilde{\mathbb{F}}e^{i\omega t}, \tag{1.7}$$

from which it follows that

$$\widetilde{A} = \frac{\widetilde{\mathbb{F}}}{-\omega^2 m + i\omega b + k} = \frac{\widetilde{\mathbb{F}}/m}{\omega_0^2 - \omega^2 + 2\omega\alpha i}, \tag{1.8}$$

as far as the denominator is not null (we consider the case where it is null in Sect. 1.6). The corresponding solution is

$$x(t) = \mathrm{Re}\left(\widetilde{A}e^{i\omega t}\right) = \mathrm{Re}\left(\frac{\widetilde{\mathbb{F}}/m}{\omega_0^2 - \omega^2 + 2\omega\alpha i}e^{i\omega t}\right).$$

Let us denote by $\widetilde{x}(t)$ the complex displacement,

$$\widetilde{x}(t) = \widetilde{A}e^{i\omega t},$$

and by $\widetilde{v}(t)$ the corresponding complex velocity, i.e.,

$$\widetilde{v}(t) = i\omega \widetilde{A}e^{i\omega t}.$$

1.5 Forced oscillations

The *mechanical impedance* is defined by

$$\mathscr{L}(\omega) = \frac{\tilde{f}}{\tilde{v}} = \frac{\tilde{\mathbb{F}}e^{i\omega t}}{i\omega \tilde{\mathbb{A}}e^{i\omega t}} = \frac{\tilde{\mathbb{F}}}{i\omega \tilde{\mathbb{A}}} = \frac{\tilde{\mathbb{F}}}{i\omega \dfrac{\tilde{\mathbb{F}}}{-\omega^2 m + i\omega b + k}} = b + \left(\omega m - \frac{k}{\omega}\right)i$$

and it is a measure of how much the mass resists motion when subjected to a given force. The SI unit is Ns/m. It is a function of frequency ω of the applied force. At resonance frequencies (see Sect. 1.6), the mechanical impedance will be near to zero, meaning that less force is needed to cause the mass to move at a given velocity.

The imaginary part of the impedance

$$X(\omega) = \omega m - \frac{k}{\omega},$$

is called *mechanical reactance*.

Since

$$\tilde{\mathbb{A}} = \frac{\tilde{\mathbb{F}}}{i\omega \mathscr{L}(\omega)},$$

we can write the displacement in terms of $\mathscr{L}(\omega)$ as follows,

$$x(t) = \operatorname{Re}\left(\frac{\tilde{f}(t)}{i\omega \mathscr{L}(\omega)}\right). \tag{1.9}$$

Let us write $\mathscr{L}(\omega) = |\mathscr{L}(\omega)|e^{i\varphi(\omega)}$. Then

$$|\mathscr{L}(\omega)| = \sqrt{b^2 + X(\omega)^2} \quad \text{and}$$

$$\varphi(\omega) = \tan^{-1}\left(\frac{X(\omega)}{b}\right).$$

By replacing in (1.9) we get

$$x(t) = \frac{\mathbb{F}}{\omega|\mathscr{L}(\omega)|}\operatorname{Re}\left(\frac{e^{i(\omega t + \beta)}}{ie^{i\varphi(\omega)}}\right) = \frac{\mathbb{F}}{\omega|\mathscr{L}(\omega)|}\operatorname{Im}\left(\frac{e^{i(\omega t + \beta)}}{e^{i\varphi(\omega)}}\right)$$

$$= \frac{\mathbb{F}}{\omega|\mathscr{L}(\omega)|}\operatorname{Im}\left(e^{i(\omega t + \beta - \varphi(\omega))}\right) = \frac{\mathbb{F}}{\omega|\mathscr{L}(\omega)|}\sin(\omega t + \beta - \varphi(\omega)).$$

Since

$$\frac{\mathbb{F}}{\omega|\mathscr{L}(\omega)|}\sin(\omega t + \beta - \varphi(\omega)) = \frac{\mathbb{F}}{\omega|\mathscr{L}(\omega)|}\cos(\omega t + \beta - \varphi(\omega) - \frac{\pi}{2}),$$

we see that there is a phase difference between the force and the displacement of $\varphi(\omega) + \pi/2$.

Therefore, if $0 \leq \alpha < \omega_0$ and the denominator of (1.8) is not null, the general solution of the motion equation is

$$x(t) = Ae^{-\alpha t}\cos(\omega_a t + \phi) + \frac{\mathbb{F}}{\omega|\mathscr{Z}(\omega)|}\sin(\omega t + \beta - \varphi(\omega)). \qquad (1.10a)$$

Similarly, if $\alpha = \omega_0$ the general solution is

$$x(t) = (B + Ct)e^{-\alpha t} + \frac{\mathbb{F}}{\omega|\mathscr{Z}(\omega)|}\sin(\omega t + \beta - \varphi(\omega)), \qquad (1.10b)$$

and finally, if $\alpha > \omega_0$, the general solution is

$$x(t) = Be^{\gamma_+ t} + Ce^{\gamma_- t} + \frac{\mathbb{F}}{\omega|\mathscr{Z}(\omega)|}\sin(\omega t + \beta - \varphi(\omega)). \qquad (1.10c)$$

If $\alpha > 0$, that is, if there is some damping, the first term in the right-hand side of the general solution tends to zero when t goes to ∞, so this term is a transient one. Thus, for great values of t the displacement $x(t)$ will be approximately equal to the term

$$\frac{\mathbb{F}}{\omega|\mathscr{Z}(\omega)|}\sin(\omega t + \beta - \varphi(\omega)),$$

which is called *stationary solution*. The angular frequency of this forced oscillation is ω so it is equal to the frequency of the external forcing. The amplitude is

$$A = \frac{\mathbb{F}}{\omega|\mathscr{Z}(\omega)|} = \frac{\mathbb{F}}{\sqrt{(k - m\omega^2)^2 + \omega^2 b^2}}.$$

The units for A and $\mathbb{F}$ are different. The nondimensional parameter,

$$a := \frac{k}{\omega|\mathscr{Z}(\omega)|} = \frac{k}{\sqrt{(k - m\omega^2)^2 + \omega^2 b^2}}$$

is called *amplification factor*. We have

$$a = \frac{1}{\sqrt{\left(1 - (\frac{\omega}{\omega_0})^2\right)^2 + 4\zeta^2(\frac{\omega}{\omega_0})^2}}.$$

We notice that the value of a only depends on the damping factor ζ and the ratio between the frequency of the excitation force, ω, and the natural frequency of the free undamped vibration, ω_0. We also notice that, if $\alpha = 0$ then the first term in the right-hand side of (1.10a) does not tend to zero when t goes to ∞.

Moreover, if $\beta = -\pi/2$, that is, if $f(t) = \mathbb{F}\sin(\omega t)$ the stationary solution is given by

$$-\frac{\mathbb{F}}{\omega|\mathscr{L}(\omega)|}\cos(\omega t - \varphi(\omega)).$$

1.6 Resonance

This special case arises when $\mathscr{L}(\omega) = 0$ which is equivalent to

$$b = 0 \quad \text{and} \quad \omega m = \frac{k}{\omega}.$$

The latter means that $\omega = \sqrt{k/m} = \omega_0 = \omega_\alpha$.
In this situation, (1.10a) is no longer the general solution. In order to calculate it, we look for a particular solution of the form

$$x(t) = \operatorname{Re}(\widetilde{\mathbb{A}} t e^{i\omega_0 t}). \tag{1.11}$$

By replacing in the motion equation we get

$$m\widetilde{\mathbb{A}}(2i\omega_0 - \omega_0^2 t)e^{i\omega_0 t} + k\widetilde{\mathbb{A}} t e^{i\omega_0 t} = \widetilde{\mathbb{F}} e^{i\omega_0 t},$$

and then

$$2i\omega_0 m\widetilde{\mathbb{A}} + (-m\omega_0^2 + k)t\widetilde{\mathbb{A}} = \widetilde{\mathbb{F}}. \tag{1.12}$$

Since $m\omega_0^2 = k$, (1.12) yields

$$\widetilde{\mathbb{A}} = \frac{\widetilde{\mathbb{F}}}{2i\omega_0 m}.$$

Hence, a particular solution is given by

$$x(t) = \operatorname{Re}\left(\frac{\mathbb{F}}{2i\omega_0 m} t e^{i(\omega_0 t + \beta)}\right) = \frac{\mathbb{F} t}{2\omega_0 m}\sin(\omega_0 t + \beta).$$

This solution satisfies

$$x(0) = 0,$$

$$\dot{x}(0) = \frac{\mathbb{F}}{2\omega_0 m}\sin\beta.$$

The general solution is

$$x(t) = A\cos(\omega_0 t + \phi) + \frac{\mathbb{F} t}{2\omega_0 m}\sin(\omega_0 t + \beta),$$

while the particular solution of the initial-value problem

$$\begin{cases} m\ddot{x}(t) + kx(t) = \mathbb{F}\cos(\omega_0 t + \beta), \\ x(0) = x_0, \\ \dot{x}(0) = v_0, \end{cases}$$

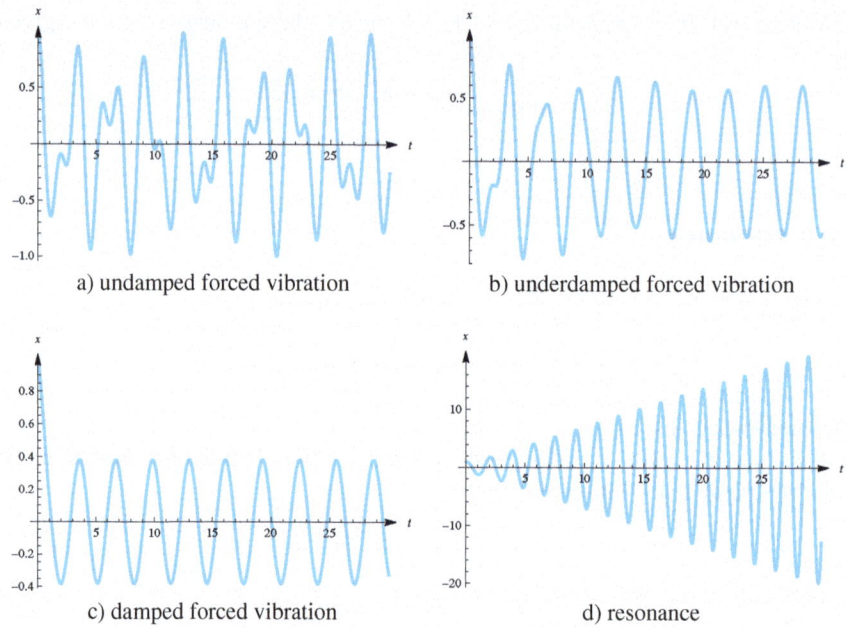

Fig. 1.3. Examples of different solutions for forced systems with zero-velocity initial condition

is
$$x(t) = \frac{\mathbb{F}t}{2\omega_0 m}\sin(\omega_0 t + \beta) + x_0\cos(\omega_0 t) + \frac{v_0 - \frac{\mathbb{F}}{2\omega_0 m}\sin\beta}{\omega_0}\sin(\omega_0 t).$$

We notice that the amplitude of the first term increases linearly with time.

In Fig. 1.3 we have illustrated different cases of forced systems. In particular, case d) illustrates the resonance phenomenon.

1.7 The first order equation

For the sake of completeness we recall the expression of the solution to the first order equation,
$$a\dot{u} + bu = 0.$$
The general solution is
$$u(t) = Ae^{-\frac{b}{a}t},$$
and then it is monotonically decreasing and tends to zero as $t \to \infty$, if $a, b > 0$.
The general solution to the non-homogeneous differential equation,
$$a\dot{u} + bu = \mathbb{F}\cos(\omega t + \beta).$$

is
$$u(t) = Ae^{-\frac{b}{a}t} + \frac{\mathbb{F}}{|\mathscr{L}(\omega)|}\cos(\omega t + \beta - \varphi(\omega))$$

where
$$\mathscr{L}(\omega) = b + i\omega a,$$

and $\varphi(\omega) = \arg \mathscr{L}(\omega)$. The corresponding stationary solution is

$$\frac{\mathbb{F}}{|\mathscr{L}(\omega)|}\cos(\omega t + \beta - \varphi(\omega)).$$

We notice that if the spring is absent, that is, if $k = 0$, then the displacement satisfies the differential equation,

$$m\ddot{x} + b\dot{x} = \mathbb{F}\cos(\omega t + \beta), \tag{1.13}$$

while the velocity is a solution to the first order equation,

$$m\dot{v} + bv = \mathbb{F}\cos(\omega t + \beta).$$

Hence,
$$v(t) = Ae^{-\frac{b}{m}t} + \frac{\mathbb{F}}{|\mathscr{L}(\omega)|}\cos(\omega t + \beta - \varphi(\omega)),$$

with $\mathscr{L}(\omega) = b + \omega mi$. By integrating we obtain the general solution to (1.13),

$$x(t) = D + Ee^{-\frac{b}{m}t} + \frac{\mathbb{F}}{\omega|\mathscr{L}(\omega)|}\sin(\omega t + \beta - \varphi(\omega)), \tag{1.14}$$

which depends on two constants D and E to be determined by adding two initial conditions.

Of course, this solution is coherent with the one for the second order equation previously obtained. Indeed, if $k = 0$ then $\omega_0 = 0$. Hence we are in the overdamped case ($\alpha > \omega_0$) with $\gamma_- = -2\alpha = -b/m$ and $\gamma_+ = 0$. Consequently the general solution (1.10c) becomes (1.14).

Problems

1.1. A spring is stretched 50 cm by a force of 4 N. A weight of 19.6 N is hung from the spring and pulled down 1 m below its equilibrium position. Determine its position x at any time t in the following cases:

(a) There is no air resistance. Compute the period, amplitude, and frequency of the motion.
(b) The air resistance is proportional to the velocity of the system with constant 8 Ns/m.
(c) The same as in case (b) but moreover the mass is subjected to an external force $f(t) = 80\sin(2t)$ N.

Solution: (a) $x(t) = \cos(2t)$, amplitude = 1m, period = π s, frequency = $1/\pi = 0.318$ cycles per second (b) $x(t) = e^{-t}(1+2t)$ (c) $x(t) = e^{-2t}(6+12t) - 5\cos(2t)$.

1.2. Consider a damped spring-mass system with a periodic external force given by $f(t) = \mathbb{F}\cos(\omega t)$. Compute initial conditions $x(0)$ and $\dot{x}(0)$ to obtain a stationary solution of the initial-value problem which models this system (express the initial conditions as a rational function of k, m, b, $\mathbb{F}$ and ω).

Solution: $x(0) = \dfrac{\mathbb{F}(k-m\omega^2)}{\omega^2 b^2 + (m\omega^2 - k^2)^2}$, $\dot{x}(0) = \dfrac{\mathbb{F}b\omega^2}{\omega^2 b^2 + (m\omega^2 - k)}$.

1.3. A linear spring-damped system is usually employed as a simple model to study a shock absorber mechanism in vehicles. In this model, the auto is a material point with mass m, and it is joined to a spring and a damper with respective constants k (N/m) and b (Ns/m). Let us denote by $y(t)$ the distance (in meters) from the mass center of the vehicle to a reference horizontal plane. Let a be the distance from the mass center to this plane in the equilibrium.

(a) Find the ordinary differential equation satisfied by $x(t) := y(t) - a$.
(b) Solve the initial-value problem that governs the motion of the vehicle assuming that the initial position and the initial velocity are x_0 and v_0, respectively. Find the conditions for which the system is underdamped, overdamped or critically damped.

Solution: (a) $m\ddot{x}(t) + b\dot{x}(t) + kx(t) = 0$ (b) If $b^2 < 4km$ the system is underdamped and $x(t) = e^{-\alpha t}[x_0 \cos(\omega_\alpha t) + \frac{v_0 + \alpha x_0}{\omega_\alpha} \sin(\omega_\alpha t)]$, with $\alpha = \frac{b}{2m}$, $\omega_0 = \sqrt{\frac{k}{m}}$ and $\omega_\alpha = \frac{\sqrt{4km-b^2}}{2m}$. If $b^2 > 4km$, the system is overdamped. In this case, the solution is given by $x(t) = \dfrac{m}{\sqrt{b^2-4km}}[(v_0 - \gamma_- x_0)e^{\gamma_+ t} + (\gamma_+ x_0 - v_0)e^{-\gamma_- t}]$ with $\gamma_\pm = -\alpha \pm \sqrt{\alpha^2 - \omega_0^2}$. If $b = 2\sqrt{km}$ the system is critically damped. Then $x(t) = x_0 + (v_0 + \alpha x_0)t$.

1.4. In a small factory, a comfort criterion for vehicles consists in assuming that the movement must be underdamped and, moreover, for the initial conditions $x(0) = x_0$ and $\dot{x}(0) = 0$ the time at which the mass first passes through its equilibrium position is $t = 0.05$ s. Find the value of k which satisfies this comfort criteria assuming that $m = 1.2 \, 10^3$ kg and $b = 10^4$ Ns/m.

Solution: $k = 1.397 \, 10^6$ N/m.

1.5. The stability criterion in the factory of Problem 1.4 requires that the vehicle is stable in rough roads. For simplicity, we will assume that the roughness is a sinusoidal function and the stability criterion requires that the maximum value of the stationary solution is 0.2 m when driving on roads with an amplitude of roughness equal to 0.1 m.
Let us assume, therefore, that the road surface is a function of the distance z traveled by the vehicle and that it is given by $s(z) = s_m \sin(2\pi z/D)$, where s_m is the amplitude and D is the distance between two consecutive peaks. Let us also suppose that

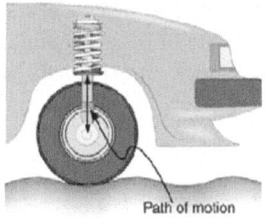

Fig. 1.4. Problem 1.5

the vehicle has a uniform velocity v. Then we have $\vartheta(t) := s(z(t)) = s_m \sin(2\pi vt/D)$. This movement forces the spring to compress and stretch periodically and therefore it responds with a force. Consequently, in the differential equation of motion a forcing term given by $f(t) = k s_m \sin(\frac{2\pi vt}{D})$ must be included.

Suspension designers must achieve an acceptable compromise between comfort criterion and stability criterion, so in what follows we will assume the value of k computed in Problem 1.4. Taking into account the previous considerations:

(a) Compute the frequency of the forcing term.
(b) Determine the stationary part $x_e(t)$ of the displacement.
(c) Compute the amplification factor a as a function of all the parameters involved in the model.
(d) Assuming that $s_m = 0.1$ m and $D = 20$ m, find the value of the velocity v satisfying the stability criterion, that is to say, the maximum displacement of the stationary solution is less or equal than 0.2 m. Which is the value of the corresponding amplification factor?

Hint: Notice that the dynamical study of the system is made not from the initial equilibrium position of the system but from its equilibrium position once the system has been charged and, as a consequence, the term concerning the weight of the vehicle does not appears explicitly in the model.

Solution: (a) The angular frequency is $\omega = 2\pi v/D$ and the circular frequency $f = v/D$. (b) $x(t) = -A\cos(\omega t - \varphi(\omega))$. (c) $a = k/\sqrt{(k - m(\frac{2\pi v}{D})^2)^2 + b^2(\frac{2\pi v}{D})^2}$.
(d) $0 \leq v \leq 285.40$ km/h which corresponds to $a \simeq 1$.

1.6. A spring with constant 20 N/m is attached to a 1 kg mass with friction constant 8 Ns/m. If the mass is initially at position 0.5 meter to the right of equilibrium and has an initial velocity of 1 m/s to the right, determine:
(a) When the mass will first return to the equilibrium position.
(b) The maximum displacement of the mass for $t > 0$.

Solution: (a) $t = \frac{1}{\sqrt{2}}(\pi + \arctan(-\frac{\sqrt{2}}{6})) \simeq 2.0578$ s (b) The position at any time t is given by $x(t) = \frac{\sqrt{19}}{2}e^{-4t}\cos(\sqrt{2}t - \varphi)$, with $\varphi = \arctan\frac{6}{\sqrt{2}}$. The maximum displacement is reached for $\dot{x}(t) = 0$, which corresponds to $t \simeq 0.07662$ s.

1 The harmonic oscillator

1.7. Often bumps are built into roads to discourage speeding. Let us suppose that the model of the vertical motion $x(t)$ of a car encountering a speed bump when it is moving at speed v (assuming its shock absorbers are broken down) is given by

$$m\ddot{x}(t) + kx(t) = \begin{cases} F_0 \cos(\pi vt/L) & \text{for } t < L/(2v), \\ 0 & \text{for } t \geq L/(2v), \end{cases}$$

where L is the width of the speed bump. Taking $m = k = 1$, $L = \pi$ and $F_0 = 1$ in appropriate units,

(a) Solve the initial-value problem assuming $x(0) = 0$, $\dot{x}(0) = v$.
(b) Thereby show that the formula for the oscillatory motion after the car has traversed the speed bump is $x(t) = A \sin t$, where A is a constant depending on v (see [56]).

Solution: $x(t) = \frac{\pi}{2} \sin t$ for $v = 1$, $x(t) = \frac{2v \cos(\pi/2v)}{v^2 - 1}$ for $v \neq 1$.

1.8. Consider a damped spring-mass with $m = 20$ kg, $k = 11520$ N/m. Let us suppose that the damping has a linear response with damping coefficient (or mechanical resistance) 39.2 Ns/m.

(a) Find the spring elongation if the system is driven by a constant force $f = 196$ N and the mass is at rest.
(b) Assume that the system is driven by a external force $f(t) = 196\cos(25t - \pi/4)$ in the SI units. Find the mechanical impedance, the stationary solution $x_e(t)$ of the displacement equation, its amplitude A and the amplification factor a.

Solution: (a) 0.01701 m (b) $\mathscr{Z}(\omega) = 39.2(1+i)$ Ns/m, $|\mathscr{Z}(\omega)| = 39.2\sqrt{2}$, $x_e(t) = 0.14142 \sin(25t - \pi/2) = -0.14142 \cos(25t)$, $A = 0.14142$ m, $a = 8.313$.

1.9. A stamping machine applies hammering forces on metal sheets by a die attached to the plunger. The plunger moves vertically up-and-down by a flywheel spinning at constant set speed. The constant rotational speed of the flywheel makes the impact force on the sheet metal, and therefore the supporting base, intermittent and cyclic. The heavy base on which the metal sheet is situated has a mass $m = 2000$ kg. The force acting on the base follows a function $f(t) = 2000 \sin(10t)$, where t is time in seconds. The base is supported by an elastic pad with an equivalent spring constant $k = 2\,10^5$ N/m. If the base is initially depressed down by an amount 0.1 m, solve the following questions:

(a) Determine the initial-value problem which solution is the position $x(t)$ of the base along time.
(b) Examine if this is a resonant vibration situation with the applied load.
(c) Should this be a resonant vibration, how long will take for the support to break at an elongation of 0.3 m?

Solution: (a) $2000\ddot{x}(t) + 2\,10^5 x(t) = 2000 \sin(10t)$, $x(0) = 0.1$, $\dot{x}(0) = 0$. (b) It is a resonance vibration situation as the natural frequency $\omega_0 = \sqrt{k/m} = \sqrt{2\,10^5/(2\,10^3)} = 10$ rad/s. (c) 8 s from the beginning of the resonant vibration.

1.10. A cylinder of radius 2 m is floating in a fluid of mass density 1000 kg/m³. If the cylinder is slightly depressed and then released, it oscillates in the vertical direction with an oscillation period of 1 s. Assuming that the viscous damping of the fluid and air can be neglected, compute the mass of the cylinder.

Hint: Use Archimedes' principle: an object that is completely or partially submerged in a fluid is acted on by an upward (buoyancy) force equal to the weight of the displaced fluid.

Solution: 3119 kg.

1.11. An airfoil is mounted in a wind tunnel for the purpose of studying the aerodynamic properties of the airfoil's shape. A simple model of this is illustrated in Fig. 1.5 as a rigid inertial body mounted on a rotational spring, fixed to the floor with a rigid support. Assume that the initial conditions are zero and that the driving frequency is such that $\omega_0^2 - \omega^2 > 0$.

(a) Write the motion equation.
(b) Find the maximum deflection for zero initial conditions.
(c) Find a design relationship for the spring stiffness k in terms of the rotational inertia, J, the magnitude of the applied moment, M_0, and the driving frequency, ω, that will keep the magnitude of the angular deflection $\theta(t)$ less than $5°$.

Solution: (a) $J\ddot{\theta}(t) + k\theta(t) = M_0 \cos(\omega t)$
(b) $\theta_{max} = \frac{(2M_0/J)}{\omega_0^2 - \omega^2}$ (c) $k > \frac{36J}{\pi}\left(\frac{2M_0}{J} + \frac{\pi\omega_0^2}{36}\right)$.

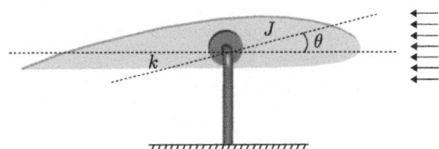

Fig. 1.5. Airfoil in a wind tunnel

2
The series RLC circuit

In this chapter we introduce a mathematical model for the series RLC circuit and then we apply the results obtained for the harmonic oscillator in the previous chapter to write its solution. An RLC circuit, apart from an electric generator, consists of three important electrical elements: a *capacitor* which stores energy in an electric field, an *inductor* that stores energy in a magnetic field, and a *resistor* that does not store energy but rather dissipates it as heat. Three physical parameters define the electric behavior of these elements: capacitance, inductance and resistance. We introduce important concepts as *electrical impedance, resonance, active power*, and *reactive power*.

2.1 Mathematical model

A similar model to the mechanical one described in the previous chapter is obtained for a series electrical circuit including an electric generator with *source voltage* E(t) (frequently referred as *electromotive force*), a capacitor with *capacitance* C, a resistor with *resistance* R and an inductor with *inductance* L (see Figs. 2.1 and 2.2).

In the SI, the units for these parameters are *henries* (H) for the inductance, *ohms* (Ω) for the resistance, *farads* (F) for the capacitance and *volts* (V) for the source voltage, together with multiples and sub-multiples.

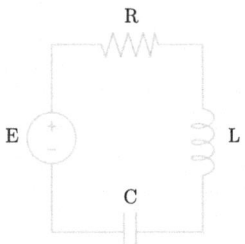

Fig. 2.1. A simple series RLC circuit

2 The series RLC circuit

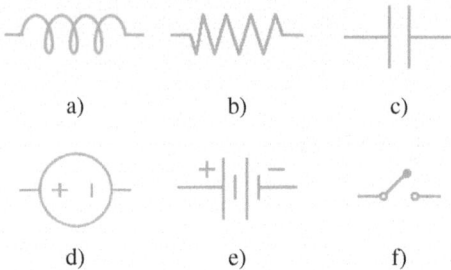

Fig. 2.2. Schematic representations of basic circuit elements employed along the book: a) inductor; b) resistance; c) capacitor; d) and e) source voltage; f) switch

If $I(t)$ denotes the *current intensity* flowing in the circuit at time t (in *amperes* (A)), the following differential equation holds:

$$L\ddot{I}(t) + R\dot{I}(t) + \frac{1}{C}I(t) = \dot{E}(t),$$

where L, R and C are all assumed to be known positive constants. Actually this equation is obtained by taking the time derivative of the following one which derives from Kirchhoff's laws (see Chap. 3):

$$L\dot{I}(t) + RI(t) + \frac{1}{C}Q(t) = E(t),$$

where

$$Q(t) := Q(0) + \int_0^t I(\tau)\,d\tau$$

is the *charge* (in *coulombs* (C)) of the capacitor at time t.
Since

$$I(t) = \dot{Q}(t) \text{ and } \dot{I}(t) = \ddot{Q}(t),$$

the above equation can also be written, in terms of the charge $Q(t)$, in the form

$$L\ddot{Q}(t) + R\dot{Q}(t) + \frac{1}{C}Q(t) = E(t).$$

The analogy with the mechanical system studied in the previous chapter is now clear. To complete the problem, initial data are given, namely, the initial charge of the capacitor $Q(0) = Q_0$ and the initial intensity in the circuit $\dot{Q}(t) = I(0) = I_0$.

If the alternative form (2.1) is used, we need to obtain the value of the first derivative of the intensity at initial time from (2.1). It is given by

$$\dot{I}(0) = \frac{1}{L}\left(-RI(0) - \frac{1}{C}Q(0) + E(0)\right) = \frac{1}{L}\left(-RI_0 - \frac{1}{C}Q_0 + E(0)\right).$$

Hence $\dot{I}(0)$ can be determined by the initial charge and current, which are physically measurable quantities.

2.2 Energy balance

In order to get an *energy balance principle*, let us multiply (2.1) by $I(t)$. We easily deduce

$$\frac{L}{2}\frac{d}{dt}I^2(t) + RI^2(t) + \frac{1}{2C}\frac{d}{dt}\left(Q(0) + \int_0^t I(\tau)\,d\tau\right)^2 = E(t)I(t).$$

Integrating in time,

$$\frac{L}{2}I^2(t) + R\int_0^t I^2(\tau)\,d\tau + \frac{1}{2C}\left(Q(0) + \int_0^t I(\tau)\,d\tau\right)^2 \quad (2.1)$$
$$= \frac{L}{2}I^2(0) + \frac{1}{2C}(Q(0))^2 + \int_0^t E(\tau)I(\tau)\,d\tau.$$

Terms $\frac{L}{2}I^2(t)$ and $\frac{1}{2C}(Q(0) + \int_0^t I(\tau)\,d\tau)^2$ represent the *magnetic energy* and the *electrostatic energy* stored at time t in the inductor and in the capacitor, respectively.

In its turn, term $R\int_0^t I^2(\tau)\,d\tau$ is the *dissipated* electromagnetic energy into heat from time 0 to time t (Joule heating), in the resistor.

Finally, term $\int_0^t E(\tau)I(\tau)\,d\tau$ is the energy supplied to the circuit by the source voltage E along the time interval $(0,t)$.

This energy balance is similar to the one to be obtained in Chap. 6 for distributed electromagnetic models (see (6.52)).

2.3 Harmonic source. Impedance. Resonance

Let us suppose that the source is *harmonic*, more precisely

$$E(t) = \text{Re}(\widetilde{\mathbb{E}}e^{i\omega t}), \text{ with } \widetilde{\mathbb{E}} \in \mathbb{C}. \quad (2.2)$$

If

$$\widetilde{\mathbb{E}} = \mathbb{E}e^{i\alpha},$$

this means

$$E(t) = \mathbb{E}\cos(\omega t + \alpha).$$

Then we make similar computations as for the mechanical system in Chap. 1. We try to find a solution of the form

$$I(t) = \text{Re}(\widetilde{\mathbb{I}}e^{i\omega t}),$$

with $\widetilde{\mathbb{I}}$ a complex number to be determined. For this purpose we replace this expression in Eq. (2.1). We obtain

$$-\omega^2 L\widetilde{\mathbb{I}} + i\omega R\widetilde{\mathbb{I}} + \frac{1}{C}\widetilde{\mathbb{I}} = i\omega\widetilde{\mathbb{E}}.$$

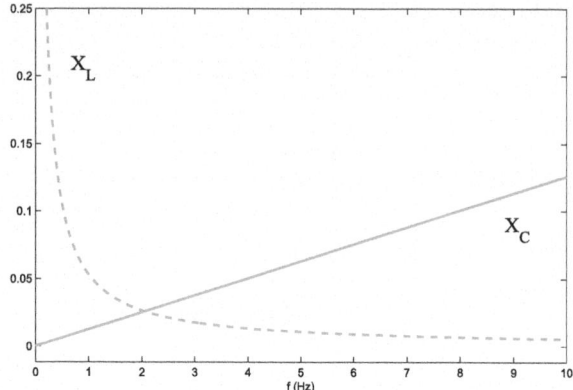

Fig. 2.3. Relationship between X_L and X_C as frequency increases

In this case, the *impedance* is defined by

$$\mathscr{Z}(\omega) := \frac{\widetilde{\mathbb{E}}}{\widetilde{\mathbb{I}}} = R + \left(\omega L - \frac{1}{\omega C}\right)i.$$

Impedance is a measure of the opposition that the circuit exhibits to the flow of a current when a voltage is applied. The real part is R, the *resistance* of the resistor. The imaginary part is called *reactance*:

$$X(\omega) = \omega L - \frac{1}{\omega C}. \tag{2.3}$$

The first term, $X_L(\omega) = \omega L$, is the *inductive reactance* and the second one, $X_C(\omega) = 1/(\omega C)$, is the *capacitive reactance* and are all measured in ohms (Ω). Since $\omega = 2\pi f$, for a given inductor, any increase in frequency f will cause a corresponding increase of inductive reactance, as deduced from (2.3). In contrast to the inductive reactance, the higher the frequency, the less the reactance for a given capacitor (see Fig. 2.3).

Finally, the *resonance frequency* is the one for which $X(\omega) = 0$, namely,

$$\omega_0 = \sqrt{\frac{1}{LC}}.$$

Remark 2.1. Notice that this frequency causes $X_C(\omega_0)$ and $X_L(\omega_0)$ to be equal and then we say the circuit is *resonant*. This can occur only at one frequency and this fact is the principle that enables tuned circuits in the radio receiver to select one particular frequency and reject all others (see Problem 2.10). Notice also that an increase in the value of either L or C, or both L and C, will lower the resonant frequency of a given circuit and, reciprocally, a decrease in some of these values will raise the resonant frequency.

Let us obtain a particular (harmonic) solution to (2.1). Firstly, we write the impedance and the complex source voltage in polar form,

$$\mathscr{Z}(\omega) = |\mathscr{Z}(\omega)|e^{i\varphi(\omega)}.$$

Then we get

$$I(t) = \frac{\mathbb{E}}{|\mathscr{Z}(\omega)|}\mathrm{Re}(e^{i\alpha}e^{-i\varphi(\omega)}e^{i\omega t}) = \frac{\mathbb{E}}{|\mathscr{Z}(\omega)|}\cos(\omega t + \alpha - \varphi(\omega)). \qquad (2.4)$$

Notice that the phase delay between the source voltage and the intensity is $\varphi(\omega)$.

Let us recall that the general solution of differential Eq. (2.1) is the sum of a particular solution, such as the one we have just obtained, and the general solution of the homogeneous equation that has been given in Sect. 1.3. We notice that if $R \neq 0$ the general solution of the homogeneous equation tends to 0 as $t \to \infty$. Hence, the general solution of Eq. (2.1) asymptotically converges to function (2.4), which is called the *stationary solution*.

2.4 Active and reactive power

Let us calculate the *average power* supplied to the circuit by the source voltage during a cycle. It is called the *active power* (or *real power*) and is given by (see Appendix D),

$$\mathscr{P}_a = \frac{\omega}{2\pi}\int_0^{\frac{2\pi}{\omega}} E(t)I(t)\,dt = \frac{1}{2}\mathrm{Re}(\widetilde{\mathbb{E}}\overline{\widetilde{\mathbb{I}}}) = \frac{1}{2}\mathrm{Re}(\frac{\widetilde{\mathbb{E}}\overline{\widetilde{\mathbb{E}}}}{\mathscr{Z}(\omega)})$$

$$= \frac{1}{2}\frac{\mathbb{E}^2}{|\mathscr{Z}(\omega)|}\cos\varphi(\omega) = \frac{1}{2}\mathbb{E}\mathbb{I}\cos\varphi(\omega),$$

and also by

$$\mathscr{P}_a = \frac{1}{2}\mathrm{Re}(\widetilde{\mathbb{E}}\overline{\widetilde{\mathbb{I}}}) = \frac{1}{2}\mathrm{Re}(\mathscr{Z}(\omega)\widetilde{\mathbb{I}}\overline{\widetilde{\mathbb{I}}}) = \frac{1}{2}\mathbb{I}^2|\mathscr{Z}(\omega)|\cos\varphi(\omega).$$

In its turn, the imaginary part,

$$\mathscr{P}_r = \frac{1}{2}\mathrm{Im}(\widetilde{\mathbb{E}}\overline{\widetilde{\mathbb{I}}}) = \frac{1}{2}\mathrm{Im}(\mathscr{Z}(\omega)\widetilde{\mathbb{I}}\overline{\widetilde{\mathbb{I}}}) = \frac{1}{2}\mathbb{I}^2\,\mathrm{Im}(\mathscr{Z}(\omega)) = \frac{1}{2}\mathbb{I}^2|\mathscr{Z}(\omega)|\sin\varphi(\omega),$$

is called *reactive power*. We notice that the reactive power corresponds to the first and third terms in the energy balance (2.1). This power only appears when there are inductors or capacitors in the circuit. In fact, this power is not expended by the circuit. It is the power needed to create the electric and magnetic fields of its components and fluctuates between them and the source of energy. The latter not only has to supply the energy consumed by the resistive elements, but it also has to generate the one that

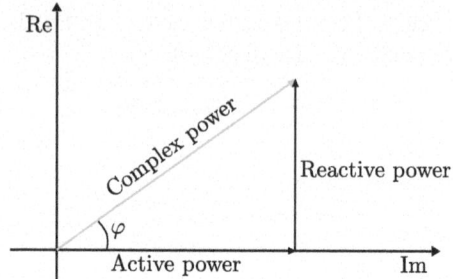

Fig. 2.4. Complex plane representation of the power

is being temporarily stored in the inductor and in the capacitor. This is an undesirable effect since electric companies charge extra for this "unused" energy.

Electrical loads consuming alternating current power consume both active and reactive power. In the case of a perfectly sinusoidal waveform, these quantities can be expressed as vectors that form a triangle as illustrated in Fig. 2.4. The modulus of the vector sum of active and reactive power is the *apparent power*.

The unit for all forms of power is the *watt* (W), but this unit is generally reserved for real power. The SI unit for reactive power is the *volt-ampere reactive* (var) and for apparent power is *volt-ampere* (va).

The *power factor* is the ratio of the active power that is used to do work and the apparent power that is supplied to the circuit. It is a dimensionless number. If both current and voltage are sinusoidal and in phase, the power factor is 1. If both are sinusoidal but not in phase, the power factor is the cosine of the phase angle φ.

A circuit with a low power factor will use higher currents to transfer a given quantity of active power than a circuit with a high power factor. This increases generation and transmission cost and also the losses in the circuit.

Let us remark that $\varphi(\omega) = 0$ if and only if the reactance $X(\omega) = 0$. Thus, if $\omega \neq \omega_0$ then the circuit produces a phase delay of $\varphi(\omega) = \arctan(X(\omega)/R)$ between the input source voltage and the intensity. This delay, in its turn, is responsible for the $\cos \varphi(\omega)$ factor in the formula of the active power. For a given frequency, angle $\varphi(\omega)$ can be reduced by an appropriate choice of the circuit parameters L and C. Usually, in most electrical engineering devices, parameters R and L are imposed so the only control variable is the capacitance C.

2.5 The circuit without capacitor

If there is no capacitor, the circuit equation becomes

$$L\dot{I}(t) + RI(t) = E(t).$$

For the harmonic excitation (2.2) we obtain the harmonic particular solution given by (2.4) but now the impedance is $\mathscr{Z}(\omega) = R + i\omega L$.

2.5 The circuit without capacitor

Concerning the transient solution, we notice that the general solution to the homogeneous ordinary differential equation

$$L\dot{I}(t) + RI(t) = 0$$

is

$$I(t) = Ae^{-\frac{R}{L}t},$$

A being a constant to be determined by the initial intensity. Hence, the general solution is

$$I(t) = Ae^{-\frac{R}{L}t} + \frac{\mathbb{E}}{|\mathscr{Z}(\omega)|} \cos(\omega t + \alpha - \varphi(\omega)).$$

If R is small then

$$\varphi(\omega) \approx \frac{\pi}{2}$$

and hence

$$I(t) \approx Ae^{-\frac{R}{L}t} + \frac{\mathbb{E}}{|\mathscr{Z}(\omega)|} \sin(\omega t + \alpha).$$

Taking (2.5) into account and depending on the values of α, we can deduce the following two important conclusions from the previous analysis:

1. If $E(t) = \mathbb{E}\cos(\omega t)$, that is, if $\alpha = 0$, we have

$$I(t) \approx Ae^{-\frac{R}{L}t} + \frac{\mathbb{E}}{|\mathscr{Z}(\omega)|} \sin(\omega t).$$

If $I(0) = 0$, then $A \approx 0$ and hence,

$$I(t) \approx \frac{\mathbb{E}}{|\mathscr{Z}(\omega)|} \sin(\omega t).$$

Thus, there is no transient part from the beginning.

2. If $E(t) = \mathbb{E}\sin(\omega t)$, that is, if $\alpha = -\pi/2$ we have,

$$I(t) \approx Ae^{-\frac{R}{L}t} + \frac{\mathbb{E}}{|\mathscr{Z}(\omega)|} \sin(\omega t - \pi/2) = Ae^{-\frac{R}{L}t} - \frac{\mathbb{E}}{|\mathscr{Z}(\omega)|} \cos(\omega t).$$

In this case, in order to eliminate the transient part of the solution (that is, in order for $A = 0$) we need to take as initial condition,

$$I(0) = -\frac{\mathbb{E}}{|\mathscr{Z}(\omega)|}.$$

On the contrary, if the equation is solved for $I(0) = 0$, the solution is

$$I(t) \approx \frac{\mathbb{E}}{|\mathscr{Z}(\omega)|}(e^{-\frac{R}{L}t} - \cos(\omega t)),$$

so it includes the transient part.

In Fig. 2.5 we have illustrated different responses of an RL circuit. In particular, pictures c) and d) illustrate the second case for different choices of the initial condition.

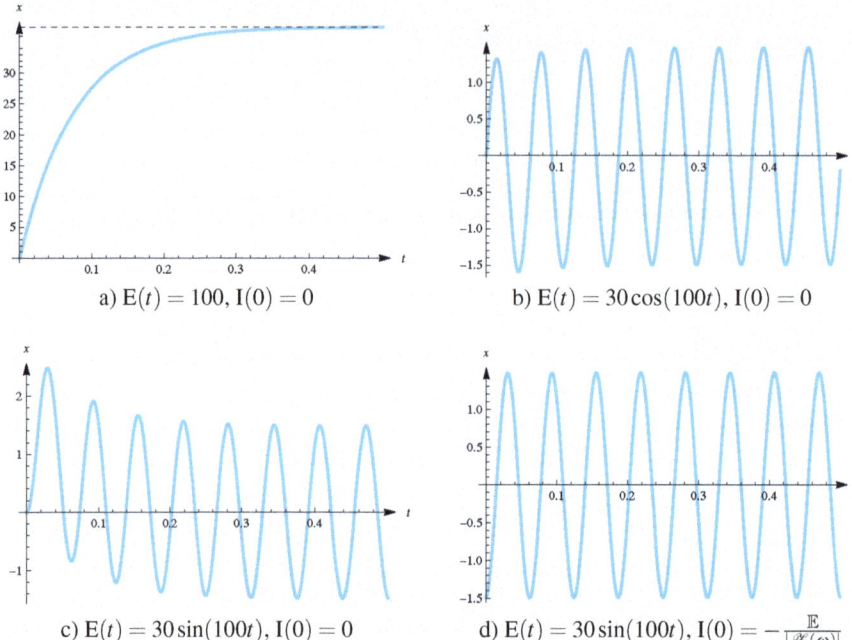

Fig. 2.5. Examples of different responses of an RL circuit with $R = 8/3 \, \Omega$, $L = 0.2$ H. a) Constant voltage source case. The solution tends to E/R; b) $E(t) = \mathbb{E}\cos(\omega t)$, ($\alpha = 0$). The transient part is negligible; c) $E(t) = \mathbb{E}\sin(\omega t)$, ($\alpha = -\pi/2$) with zero initial condition. A transient part can be observed; d) $E(t) = \mathbb{E}\sin(\omega t)$ with suitable initial condition. The solution is stationary from the beginning

---

Problems

2.1. A series RLC circuit with $L = 160 \cdot 10^{-3}$ H, $C = 99 \cdot 10^{-6}$ F, and $R = 68 \, \Omega$ is supplied with a sinusoidal voltage $E(t) = 40\sin(100t)$ V.

(a) Find the impedance of the circuit.
(b) Let $I(t) = I_0 \sin(\omega t - \phi)$ be the current at any instant in the circuit. Find I_0.
(c) What is the phase constant ϕ?

Solution: (a) $|\mathscr{Z}| = 109 \, \Omega$ (b) $I_0 = 0.367$ A (c) $\phi \simeq -0.89535$ rad.

2.2. An RLC circuit is connected to a current source which applies a voltage $E(t) = 100 e^{-5t}$ V. Assuming that $L = 0$ H, $R = 10 \, \Omega$, $C = 0.02$ F and there is no initial charge, compute:

(a) The charge and the current intensity as a function of time.
(b) The maximum value of the charge and the time needed to obtain such a value.

2.5 The circuit without capacitor

Solution: (a) $Q(t) = 10te^{-5t}$, $I(t) = 10e^{-5t}(1-5t)$ (b) The maximum value of the charge is 0.735 C and it is reached after 0.2 s.

2.3. Let us consider a simple LR circuit consisting of a resistor R, inductance L and source voltage $E(t) = E_0 \sin(\omega t)$, where $\omega > 0$. Let I_0 be the current in $t = 0$. Determine the expression for $I(t)$ as a function of time, identifying the transient and stationary terms.

Solution: $I(t) = \left(I_0 + \dfrac{E_0 \omega L}{R^2 + \omega^2 L^2}\right) e^{-\frac{R}{L}t} + \dfrac{E_0}{\sqrt{R^2 + \omega^2 L^2}} \sin(\omega t - \varphi(\omega))$.

2.4. A simple RLC electrical circuit consists of a capacitor with a capacitance of 0.02 F, a resistor with a resistance of 40 Ω and an inductor with an inductance of 8 H. The circuit is connected to a 24 volt battery. Initially there is no charge on the capacitor and no current in the circuit.

(a) Write a mathematical model which gives the charge on the capacitor for any time after the switch is closed.
(b) Find the charge on the capacitor and the current in the circuit after 1 second.

Solution: (a) $8\ddot{Q}(t) + 40\dot{Q}(t) + 50Q(t) = 24$, $Q(0) = 0$ C, $\dot{Q}(0) = 0$ A (b) $Q(t) = (-0.48 - 1.2t)e^{-2.5t} + 0.48$, $Q(1) = 0.3421$ C, $I(t) = 3te^{-2.5t}$, $I(1) = 0.2463$ A.

2.5. The RLC circuit in Problem 2.4 is now connected to an alternating current source which applies a voltage $E(t) = 100\cos(2t)$ (in volts). Assuming there is no initial charge on the capacitor or current in the circuit,

(a) Compute the angular frequency and the impedance of the circuit.
(b) Find a particular solution of the ordinary differential equation modelling this circuit in terms of Q by using the technique of search for solutions with complex values.
(c) Check that the amplitude of the solution computed in item (b) is $E_0/(i\omega \mathscr{L}(\omega))$ where E_0 is the amplitude of the source voltage.
(d) Find the charge on the capacitor for any time after the switch is closed.
(e) Find the charge on the capacitor and the current in the circuit after 1 second.
(f) Would a 5-ampere fuse have its capacity exceeded in this circuit?

Solution: (a) $\omega = 2$, $\mathscr{L}(\omega) = 40 - 9i\Omega$ (b)-(c) $Q(t) = \text{Re}(-\frac{50}{41}ie^{(2t-\phi)i})$ or, equivalently, $Q(t) = \frac{50}{41}\sin(2t - \phi)$ (d) $Q(t) = (-\frac{450}{1681} - \frac{5125}{1681}t)e^{-2.5t} + \frac{50}{41}\sin(2t - \phi)$ with $\phi = \arctan(-\frac{9}{40})$ (e) $Q(1) \simeq 0.698218$ C, $I(1) \simeq -1.046747$ A (f) No.

2.6. A series RLC circuit as the one sketched in Fig. 2.6 is supplied with a sinusoidal voltage $E(t) = E_0 \cos(\omega t)$. Assuming that R, L, E_0 and ω are known and that both switches are initially closed, find the following:

(a) The current traversing the circuit as a function of time.
(b) The current as a function of time after only switch s_1 is opened.
(c) The capacitance after switch s_2 is also opened, with the current and voltage in phase.

30 2 The series RLC circuit

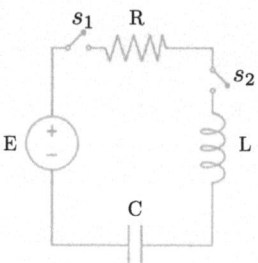

Fig. 2.6. Electric circuit for Problem 2.6

(d) The impedance of the circuit when both switches are open.
(e) The maximum energy stored in the capacitor during oscillations.
(f) The frequency that makes the inductive reactance one-half the capacitive reactance.

Solution: (a) $I(t) = \frac{E_0}{R} \cos(\omega t)$ (b) $I(t) = \frac{E_0}{\sqrt{R^2 + \omega^2 L^2}} \cos(\omega t + \arctan \frac{\omega L}{R})$
(c) $C = \frac{1}{\omega_0^2 L}$ (d) $\mathscr{Z} = R$ (e) $\mathfrak{E}_{max} = \frac{E_0^2 L}{2R^2}$ (f) $\omega = \frac{1}{\sqrt{2LC}}$.

2.7. *Unit step functions* or *Heaviside* functions are commonly used to represent switches in circuits. Let us recall the Heaviside function is defined as

$$u(t) = \begin{cases} 0 & \text{if } t < 0 \\ 1 & \text{if } t \geq 0 \end{cases}$$

It often represents a force, voltage, current, or signal that is turned on at time $t = 0$ and left on thereafter, assuming that it is possible to quickly change from "off" to "on" when the switch is thrown. Translations of unit step function are used to turn such functions on at times other than 0 (see Fig. 2.7). Multiplying any function $g(t)$ times $u(t-a)$ means that $g(t)$ is "turned off" before $t = a$ and "turned on" starting at $t = a$. Thus, we can think of any piecewise function as a combination of unit step functions that are being "turned on" and "turned off" at different points.

Consider a simple RLC electrical circuit consisting of a capacitor with a capacitance of 0.001 F, a resistor with a resistance of 110 Ω and an inductor with an inductance of 1 H. Initially there is no charge on the capacitor and no current in the circuit. The circuit is connected to a volt battery providing 90 V which is "on" at $t = 0$ and

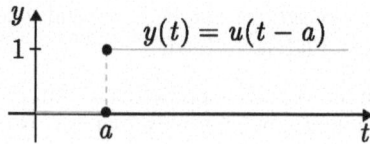

Fig. 2.7. Translated unit step function

turn off when $t = 1$, that is, $E(t) = 90 - 90u(t-1)$. Find the current resulting in the circuit.

Solution: $I(t) = \begin{cases} e^{-10t} - e^{-100t} & \text{if } t < 1, \\ (1-e^{10})e^{-10t} - (1-e^{100})e^{-100t} & \text{if } t > 1. \end{cases}$

2.8. Consider an electrical circuit analogous to that in Problem 2.7 with $L = 2$ H, $R = 1\,\Omega$ and $C = 0.5$ F. Compute the current $I(t)$ traversing the circuit assuming that the source voltage is given by $E(t) = u(t) + u(t-5)$.

Solution: $I(t) = \frac{2}{\sqrt{15}} e^{-t/4} \sin\left(\frac{\sqrt{15}}{4} t\right) u(t) + \frac{2}{\sqrt{15}} e^{-(t-5)/4} \sin\left(\frac{\sqrt{15}}{4}(t-5)\right) u(t-5)$.

2.9. A series RLC circuit is supplied with a source voltage which linearly increases from 0 to 100 volts in 10 seconds and then it is turned off. Compute the charge and the current traversing the circuit for the values $L = 1$ H, $R = 150\,\Omega$ and $C = 0.0002$ F.

Solution: $Q = u(t)(t - \frac{1}{5}) + \frac{1}{5} u(t) e^{-t} \left(\cos(3t) - \frac{4}{3}\sin(3t)\right)$, $I = u(t)\left(1 - e^{-t}\cos(3t) - \frac{1}{3} e^{-t}\sin(3t)\right) + u(t-10)e^{-(t-10)}\left(\cos(3t-30) - 33\sin(3t-30)\right)$.

2.10. The principle behind tuning a radio is the same as when an underdamped circuit is in resonance. In this case, the frequency of the transmission is fixed; for instance, an FM station may be broadcasting at a frequency of 95.5 MHz. If you tune a radio receiver to 95.5 MHz on your radio bandwave, the value of the capacitance C or inductance L change until you match the frequency of 95.5 MHz of the radio transmitter. This causes the receiver and the transmitter to vibrate and communicate at the same frequency or band wave; thus resonance is achieved allowing you to hear sound.

In other cases, to change from one band to another, you turn a switch. Each time you turn this switch, a different set of coils are connected into the circuit. The coils used to tune the receiver to the lowest band have the greatest number of turns. For each successive higher frequency band, the coils have fewer and fewer turns of wire. Thus the coils used with the highest frequency band have the least number of turns.

The ideal series-resonant circuit contains no resistance; it consists of only inductance and capacitance in series with the source voltage. Consider a LC series circuit associated to a radio.

(a) If f is the frequency of the broadcast, determine the value of L which makes the circuit resonates to a particular station.
(b) Determine how the value of L or C or both L and C influences in the resonant frequency of a given circuit.
(c) Assume that in this radio the switch is so worn that it provides only values of L between $0.5\,10^{-8}$ and $4\,10^{-8}$ H and values of C between 300 and 600 pF. Determine if it would be possible match the frequency of 95.5 MHz.

Solution: (a) Resonance is achieved when the angular frequency ω of the voltage source satisfies $\omega = 1/\sqrt{LC}$, or, equivalently $L = 1/((2\pi f)^2 C)$, where f is the frequency of the broadcast. Notice that any given combination of L and C can be resonant at only one frequency. (b) An increase in the value of either L or C, or both L and C, will lower the resonant frequency of a given circuit. Instead, a decrease in L or C, or both of them, will raise the resonant frequency. (c) Yes, it would be possible match frequencies between 32.48 and 129.94 MHz.

3
Linear electrical circuits

To finish this first part devoted to lumped models we deal with general linear electrical circuits, the RLC circuit studied in previous chapter being a particular case. We model the circuit as a directed graph where the edges are the electrical elements, namely, resistors, inductors, capacitors, magnetically coupled sub-circuits, and generators. Firstly, we write the equations for the charge conservation at the nodes and then the constitutive equation for each element. The unknowns are the electric potential at the nodes and the current intensities along the edges. This chapter intends to be a general simple introduction the subject. A more complete and deeper approach can be found, for instance in the references [1, 30]. Moreover, a brief summary of graph theory is given in Appendix A.

3.1 Mathematical model

As mentioned above we will model the circuit as a *graph*, which is a simple representation of the circuit using two basic elements: *nodes* (represented by points) and *edges* (line segments connecting two nodes). Let us suppose that the circuit consists of E electrical elements connected through N points.

In order to build a *directed graph* (or *digraph*) associated to the circuit we arbitrarily choose a positive sense for the electric current along each electrical element. This digraph, $\mathfrak{G}$, will have N nodes and E oriented edges. Each edge is defined by an ordered pair of nodes (see Fig. 3.1). We define the *connectivity matrix* of the graph, $\mathscr{M}$, as the $2 \times E$ matrix whose entries m_{1j} and m_{2j} are the ordered nodes of the j-th edge, for $j = 1,\ldots,E$. An edge for which $m_{1j} = m_{2j}$ is called a *loop*. We assume that $\mathfrak{G}$ has F connected components but some of them can be *magnetically coupled* (see Sect. 3.1.3).

3 Linear electrical circuits

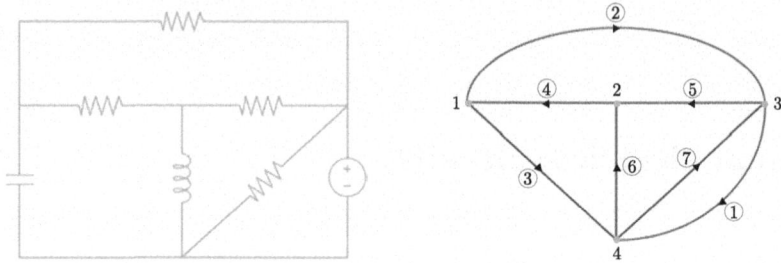

Fig. 3.1. Example of electric network and its graph

3.1.1 The incidence matrix. Properties

Let us recall that the *incidence matrix* of $\mathfrak{G}$ is the $N \times E$ matrix $\mathscr{A} = (a_{ij})$ defined by, for $j \in \{1,\ldots,E\}$,

$$a_{ij} = \begin{cases} 0 & \text{if node } i \text{ does not belong to edge } j, \\ -1 & \text{if } i = m_{1j}, \\ 1 & \text{if } i = m_{2j}. \end{cases}$$

Remark 3.1. Notice that each column of $\mathscr{A}$ has precisely a single $+1$ and a single -1.

In this paragraph we use elementary concepts and results from graph theory included in Appendix A to deduce some properties of matrix $\mathscr{A}$. Firstly, associated to the connected components let us introduce the linearly independent vectors $\mathbf{e}^i \in \mathbb{R}^N$, $i \in \{1,\ldots,F\}$, having all components null except for those corresponding to the nodes in the i-th connected component which are taken equal to one. It is straightforward to see that

$$\mathscr{A}^t \mathbf{e}^i = \mathbf{0}. \tag{3.1}$$

Therefore

$$\ker(\mathscr{A}^t) \supset \text{span}(\mathbf{e}^1,\ldots,\mathbf{e}^F).$$

Moreover, from Corollary A.1 in Appendix A we have

$$\text{rank}(\mathscr{A}) = N - F,$$

and, since

$$\dim \ker(\mathscr{A}^t) = N - \dim \text{im}(\mathscr{A}^t) = N - \dim \text{im}(\mathscr{A}) = F$$

we deduce

$$\ker(\mathscr{A}^t) = \text{span}(\mathbf{e}^1,\ldots,\mathbf{e}^F).$$

Moreover, from (3.1) we get

$$(\mathbf{e}^i)^t \mathscr{A} = \mathbf{0}, \quad i = 1,\ldots,F. \tag{3.2}$$

3.1.2 The first Kirchhoff's law

The first Kirchhoff's law establishes the charge conservation at the nodes of the circuit. Let us denote by $I_j(t)$ the current intensity along edge j. We notice that the sign of $I_j(t)$ is related to the orientation of this edge. Let us further assume that we introduce an external electric current of intensity $\psi_i(t)$ at node i and time t, for $i = 1, \ldots, N$ (we adopt the rule that $\psi_i(t)$ is positive if the current enters the circuit and negative, otherwise). Then we have

$$\sum_{j=1}^{E} a_{ij} I_j(t) = -\psi_i(t), \quad i = 1, \ldots, N.$$

3.1.3 Constitutive laws

Let us denote by $V_i(t)$, $i = 1, \ldots, N$ the electric potential at node i at time t. For each edge $j \in \{1, \ldots, E\}$, depending on the electrical element it contains, we have one of the following equations for the potential drop between its nodes:

- If edge j is a a *resistor*:

$$V_{m_{1j}}(t) - V_{m_{2j}}(t) = R_j I_j(t),$$

where R_j is the *resistance* (in Ω).
- If edge j is an *uncoupled inductor*:

$$V_{m_{1j}}(t) - V_{m_{2j}}(t) = L_j \frac{dI_j}{dt}(t),$$

where L_j is the *inductance* (in H).
- If edge j is a *capacitor*:

$$V_{m_{1j}}(t) - V_{m_{2j}}(t) = \frac{1}{C_j}\left(Q_j(0) + \int_0^t I_j(s)\, ds\right),$$

where C_j is the *capacitance* (in F) and $Q_j(0)$ the initial charge (in C).
- If the subset of edges $S = \{l_k : k = 1, \ldots, S\} \subset \{1, \ldots, E\}$ is a *magnetically coupled sub-circuit*:

$$\mathscr{L}^S \frac{d\vec{I}_S}{dt} = \Delta \vec{V}_S,$$

where $\mathscr{L}^S$ is a symmetric positive definite matrix of order S called *inductance matrix*,

$$\vec{I}_S = (I_{l_1}, \ldots, I_{l_S})^t$$

and

$$\Delta \vec{V}_S = (V_{m1l_1} - V_{m2l_1}, \ldots, V_{m1l_S} - V_{m2l_S})^t.$$

- If edge j is a *power generator*:

$$V_{m_{1j}}(t) - V_{m_{2j}}(t) = r_j I_j(t) - E_j(t),$$

where $E_j(t)$ is the *source voltage* (in V) and r_j is the *internal resistance*.

In order to write the above equations in a more compact form let us denote by $\vec{I}(t)$ and $\vec{V}(t)$ the vector functions of time:

$$\vec{I}(t) = (I_1(t), \ldots, I_E(t))^t, \quad \vec{V}(t) = (V_1(t), \ldots, V_N(t))^t.$$

Similarly,
$$\vec{E}(t) = (E_1(t), \ldots, E_E(t))^t$$
where $E_j(t) = 0$ if there is no a power source at edge j.

Moreover, let us denote by $\mathscr{D}$ the algebraic-differential linear operator defined on a vector of E functions of time, $\vec{I}(t)$, by:

- if edge j is a *resistor*:
$$\mathscr{D}(\vec{I})_j(t) := R_j I_j(t);$$

- if edge j is an uncoupled *inductor*:
$$\mathscr{D}(\vec{I})_j(t) := L_j \frac{dI_j}{dt}(t);$$

- if edge j is a *capacitor*:
$$\mathscr{D}(\vec{I})_j(t) := \frac{1}{C_j}\left(Q_j(0) + \int_0^t I_j(s)\,ds\right);$$

- if edge j is a *power generator*:
$$\mathscr{D}(\vec{I})_j(t) := r_j I_j(t);$$

- if the subset of edges $S = \{l_k : k = 1, \ldots, S\} \subset \{1, \ldots, E\}$ is a *magnetically coupled sub-circuit*:
$$\mathscr{D}(\vec{I})_S(t) := \mathscr{L}^S \frac{d\vec{I}_S}{dt}(t),$$

with $\mathscr{D}(\vec{I})_S := (\mathscr{D}(\vec{I})_{l_1}, \ldots, \mathscr{D}(\vec{I})_{l_S})$.

Then, given the source voltages $\vec{E}(t)$ and the input currents $\vec{\Psi}(t)$, the *circuit problem* consists in finding current intensities $\vec{I}(t)$ and potentials $\vec{V}(t)$ satisfying

$$\mathscr{D}(\vec{I})(t) + \mathscr{A}^t \vec{V}(t) = \vec{E}(t), \tag{3.3}$$

$$\mathscr{A}\vec{I}(t) = -\vec{\Psi}(t), \tag{3.4}$$

for all $t \in [0,T]$, where
$$\vec{\Psi}(t) = (\psi_1(t), \ldots, \psi_N(t))^t.$$

Remark 3.2. Let us multiply Eq. (3.4) by $(\mathbf{e}^i)^t$ on the left, $i = 1, \ldots, F$. From (3.2) we deduce,
$$0 = (\mathbf{e}^i)^t \vec{\Psi} = \sum_{j=1}^N (\mathbf{e}^i)_j \psi_j, \quad i = 1, \ldots, F.$$

Therefore, a necessary condition for problem (3.3)–(3.4) to have a solution is that the sum of the current intensities entering each connected component of the circuit is null.

Remark 3.3. Let us notice that operator $\mathscr{D}$ is block-diagonal. In fact, it is diagonal except for the blocks corresponding to magnetically coupled sub-circuits.

Remark 3.4. Magnetically coupled circuits refer to subsets of coils. They are physically separated but related due to the mutual induction phenomenon. A common example is a transformer. The diagonal terms of the inductance matrix $\mathscr{L}^S$ are called *self-inductances* while the non-diagonal terms are called *mutual inductances*.

3.1.4 The second Kirchhoff's law

For the sake of simplicity, in this section we assume the graph is connected. Then $F = 1$ and $\text{rank}(\mathscr{A}) = N - 1$. Along this section we will make use of several results in graph theory included in Appendix A.

The above Eqs. (3.3), (3.4) constitute a mathematical model for the circuit. From them we can deduce the second Kirchhoff's law for each cycle of the circuit. It states that the potential drop along any cycle is null. In principle not all the equations derived from all the cycles are linearly independent. One way of obtaining a set of linearly independent equations consists in using a tree and its corresponding cotree (see Appendix A). Indeed, by definition, if we add any edge of the cotree to the tree we generate a cycle. One can show that any tree has $N - 1$ edges and then its cotree has $E - N + 1$ edges. Furthermore, the $E - N + 1$ cycles obtained by adding to the tree any of the $E - N + 1$ edges of its cotree provide, via the second Kirchhoff's law, a set of linearly independent equations for the intensities. These equations and the N equations obtained from the first Kirchhoff's law at the nodes constitute a system of E algebraic-differential equations for the E intensities that is solvable. Thus, it constitutes an alternative model to the larger system (3.3)–(3.4).

The above arguments, based on well-know results from graph theory have their counterpart in linear algebra. Indeed, as we said above the rank of the incidence matrix $\mathscr{A} \in M_{N \times E}$ is $N - 1$ (see Sect. 3.1.1), and hence

$$\dim \ker(\mathscr{A}) = E - N + 1.$$

Let $\{\mathbf{u}_1, \ldots, \mathbf{u}_{E-N+1}\}$ be a basis of $\ker(\mathscr{A})$. We have $\mathscr{A} \mathbf{u}_k = \mathbf{0}\ \forall k = 1, \ldots, E - N + 1$. Then

$$\mathbf{u}_k^t \mathscr{A}^t = \mathbf{0} \quad \forall k = 1, \ldots, E - N + 1.$$

Let us multiply Eq. (3.3) on the left by $\mathbf{u}_k^t\ \forall k = 1, \ldots, E - N + 1$. We get

$$\mathbf{u}_k^t \mathscr{D}(\overrightarrow{\mathbf{I}})(t) = \mathbf{u}_k^t \overrightarrow{\mathbf{E}}(t) \quad \forall k = 1, \ldots, E - N + 1.$$

This is a linear system of $E - N + 1$ equations whose unknowns are exclusively the E intensities. Together with the $N - 1$ linearly independent equations in (3.4), they allow us to compute the intensities.

In Appendix A we give a method to determine a basis of $\ker(\mathscr{A})$.

3.2 Numerical solution of the circuit equations

Let us notice that some of the circuit equations are algebraic equations, others are ordinary differential equations and the rest are integral equations. For their numerical solution we propose a first order implicit discretization scheme. Let $0 = t_0 < t_1 < \cdots < t_M = T$ be a mesh of the time interval $[0,T]$ with equally spaced points, $t_m = m\Delta t$, $m = 0,\ldots,M$, being $\Delta t = T/M$.

Then we adopt the following approximations:

- A *backward formula* for the first derivative:

$$\frac{dI_j}{dt}(t_{m+1}) \simeq \frac{I_j(t_{m+1}) - I_j(t_m)}{\Delta t}.$$

- A *composed trapezoidal rule* for the integral:

$$\int_0^{t_{m+1}} I_j(s)\,ds \simeq \Delta t \left(\frac{1}{2} I_j(t_0) + \sum_{k=1}^{m} I_j(t_k) + \frac{1}{2} I_j(t_{m+1}) \right).$$

Let us denote by I_j^m y V_i^m the approximations of $I_j(t_m)$ and $V_i(t_m)$ ($j = 1,\ldots,E$, $i = 1,\ldots,N$, $m = 1,\ldots,M$), obtained as the solution of the following discretized problem:

$$\mathscr{D}_{\Delta t}\vec{I}^{m+1} + \mathscr{A}^t\vec{V}^{m+1} = \mathscr{C}^m(\vec{I}^0,\ldots,\vec{I}^m) + \vec{E}(t_{m+1}), \qquad (3.5)$$

$$\mathscr{A}\vec{I}^{m+1} = -\vec{\Psi}(t_{m+1}), \qquad (3.6)$$

where $\mathscr{D}_{\Delta t}$ is a block-diagonal matrix of order E defined by:

- Each *resistance, uncoupled inductor, capacitor or generator* yield a 1×1 diagonal block:

$$\mathscr{D}_{\Delta t, jj} = \begin{cases} R_j & \text{if edge } j \text{ is a resistance,} \\ \frac{L_j}{\Delta t} & \text{if edge } j \text{ is an uncoupled inductor,} \\ \frac{\Delta t}{2C_j} & \text{if edge } j \text{ is a capacitor,} \\ r_j & \text{if edge } j \text{ is a generator with internal resistance } r_j. \end{cases}$$

- Each *magnetically coupled sub-circuit* yields an $S \times S$ diagonal block:

$$\mathscr{D}_{\Delta t, S} = \frac{1}{\Delta t}\mathscr{L}^S.$$

Moreover, $\mathscr{C}^m$ denotes the column vector in $\mathbb{R}^E$ defined by

$$\mathscr{C}_j^m = \begin{cases} 0 & \text{if edge } j \text{ is a resistance,} \\ \frac{L_j}{\Delta t} I_j^m & \text{if edge } j \text{ is an inductor,} \\ -\frac{\Delta t}{C_j}\left(\frac{Q_j^0}{\Delta t} + \frac{1}{2}I_j^0 + \sum_{k=1}^{m} I_j^k\right) & \text{if edge } j \text{ is a capacitor,} \\ 0 & \text{if edge } j \text{ is a generator,} \end{cases}$$

where Q_j^0 and I_j^0 are the given charges and intensities at time $t = 0$. Moreover, the sub-vector corresponding to a magnetically coupled sub-circuit is

$$\mathscr{C}_S^m = \frac{1}{\Delta t}\mathscr{L}^S \vec{I}_S^m.$$

3.2.1 Existence of solution to the discrete problem

Let us assume that the inductance matrices $\mathscr{L}^S$ are positive definite for all magnetically coupled circuits and that the edges with generator have internal resistance (the case where there are generators without internal resistance will be consider below). Under these circumstances, the block-diagonal matrix $\mathscr{D}_{\Delta t}$ is invertible and, actually, positive definite. The same is true for its inverse so it must exist $\alpha > 0$ such that

$$\mathscr{D}_{\Delta t}^{-1} \mathbf{W} \cdot \mathbf{W} \geq \alpha \|\mathbf{W}\|^2.$$

In order to prove the existence of a solution to problem (3.5)–(3.6) let us determine the kernel of the whole block-matrix, namely,

$$\mathscr{B} = \begin{pmatrix} \mathscr{D}_{\Delta t} & \mathscr{A}^t \\ \mathscr{A} & 0 \end{pmatrix}.$$

Let $(\vec{I}, \vec{V}) \in \ker(\mathscr{B})$. This means that

$$\mathscr{D}_{\Delta t} \vec{I} + \mathscr{A}^t \vec{V} = \mathbf{0}, \quad (3.7)$$

$$\mathscr{A} \vec{I} = \mathbf{0}. \quad (3.8)$$

Let us make the scalar product of (3.7) with $\vec{I}$ and (3.8) with $\vec{V}$. We have

$$\mathscr{D}_{\Delta t} \vec{I} \cdot \vec{I} + \mathscr{A}^t \vec{V} \cdot \vec{I} = 0,$$
$$\mathscr{A} \vec{I} \cdot \vec{V} = 0.$$

By subtracting these equations and taking into account that $\mathscr{A}^t \vec{V} \cdot \vec{I} = \mathscr{A} \vec{I} \cdot \vec{V}$, we deduce

$$\mathscr{D}_{\Delta t} \vec{I} \cdot \vec{I} = 0,$$

and then $\vec{I} = \mathbf{0}$, because $\mathscr{D}_{\Delta t}$ is positive definite.

Now, let us replace this into (3.7); we deduce
$$\vec{V} \in \ker(\mathscr{A}^t) = \mathrm{span}(\mathbf{e}^1,\ldots,\mathbf{e}^F).$$
Hence, we have proved that
$$\ker(\mathscr{B}) = \{(\mathbf{0}, \vec{V}) \in \mathbb{R}^E \times \mathbb{R}^N : \vec{V} \in \mathrm{span}(\mathbf{e}^1,\ldots,\mathbf{e}^F)\}.$$
Since $\mathscr{B}$ is symmetric, we have
$$\mathrm{im}(\mathscr{B}) = \ker(\mathscr{B})^\perp.$$
In order to prove the existence of a solution for the discrete problem (3.5)–(3.6) we only need to show that the right-hand side belongs to $\mathrm{im}(\mathscr{B})$ which is equivalent to be orthogonal to vectors $(\mathbf{0}, \mathbf{e}^i)$, $i = 1,\ldots,F$. We have,
$$(\mathscr{C}^m(\vec{I}^0,\ldots,\vec{I}^m) + \vec{E}(t_{m+1}), -\vec{\Psi}(t_{m+1})) \cdot (\mathbf{0}, \mathbf{e}^i) = -\vec{\Psi}(t_{m+1}) \cdot \mathbf{e}^i.$$
Hence, we can affirm that, under the conditions
$$\vec{\Psi}(t) \cdot \mathbf{e}^i = 0 \; \forall t \in [0,T], \; i = 1,\ldots,F, \tag{3.9}$$
the discretized problem has, at least, a solution. In fact, it has many solutions, all corresponding to the same current intensities: if $(\vec{I}^1, \vec{V}^1)$ and $(\vec{I}^2, \vec{V}^2)$ are two solutions then $\vec{I}^1 = \vec{I}^2$ and $\vec{V}^1 - \vec{V}^2 \in \mathrm{span}(\mathbf{e}^1,\ldots,\mathbf{e}^F)$.

In practice, in order to define a single vector of potentials, one can impose the null value at one node of each connected component.

By so doing, the modified $\mathscr{B}$ is no longer singular but it is still symmetric. We notice that it is not positive definite, so the Cholesky method cannot be used for numerical solution of the linear system (3.5)–(3.6). Nevertheless, we can perform a Gauss' factorization of the form
$$\mathscr{B} = \mathscr{T}\mathscr{S}\mathscr{T}^t,$$
where $\mathscr{T}$ is a lower matrix with ones in the main diagonal and $\mathscr{S}$ is diagonal. Let us emphasize that this matrix is independent of m so it is convenient to make this factorization only once before the time-step cycle.

An interesting alternative to solve the discrete problem consists in using the *Schur complement* of the block-matrix $\mathscr{B}$: since $\mathscr{D}_{\Delta t}$ is invertible, from (3.5) we deduce
$$\vec{I}^{m+1} = \mathscr{D}_{\Delta t}^{-1}(\vec{E}(t_{m+1}) + \mathscr{C}^m(\vec{I}^0,\ldots,\vec{I}^m) - \mathscr{A}^t \vec{V}^{m+1}). \tag{3.10}$$
By replacing this expression in (3.6) we get
$$\mathscr{A}\mathscr{D}_{\Delta t}^{-1}(\vec{E}(t_{m+1}) + \mathscr{C}^m(\vec{I}^0,\ldots,\vec{I}^m) - \mathscr{A}^t \vec{V}^{m+1}) = -\vec{\Psi}(t_{m+1})$$
or, equivalently, the linear system of order N
$$\mathscr{A}\mathscr{D}_{\Delta t}^{-1}\mathscr{A}^t \vec{V}^{m+1} = \mathscr{A}\mathscr{D}_{\Delta t}^{-1}(\vec{E}(t_{m+1}) + \mathscr{C}^m(\vec{I}^0,\ldots,\vec{I}^m)) + \vec{\Psi}(t_{m+1}). \tag{3.11}$$
Matrix $\mathscr{S} = \mathscr{A}\mathscr{D}_{\Delta t}^{-1}\mathscr{A}^t$ is called the *Schur complement* of $\mathscr{B}$.

The above calculations show that if $(\vec{I}^{m+1}, \vec{V}^{m+1})$ is a solution of (3.5)–(3.6), then $\vec{V}^{m+1}$ is a solution of the reduced system (3.11). Conversely, if $\vec{V}^{m+1}$ is a solution of (3.11) then the couple made with $\overrightarrow{\mathbf{I}}^{m+1}$ given by (3.10) and $\overrightarrow{\mathbf{V}}^{m+1}$ is a solution to problem (3.5)–(3.6). Therefore, we can affirm that problem (3.11) has a unique solution up to an element in the linear space $\text{span}(\mathbf{e}^1, \ldots, \mathbf{e}^F)$.

Nevertheless, we will give a direct proof of this fact, by characterizing the kernel of matrix $\mathscr{S}$. More precisely, we will check that

$$\ker(\mathscr{S}) = \ker(\mathscr{A}^t) = \text{span}(\mathbf{e}^1, \ldots, \mathbf{e}^F).$$

Indeed, let us assume $\overrightarrow{\mathbf{U}} \in \ker(\mathscr{S})$. Then we have

$$\mathscr{A} \mathscr{D}_{\Delta t}^{-1} \mathscr{A}^t \overrightarrow{\mathbf{U}} = \mathbf{0},$$

from which it follows that

$$0 = \mathscr{A} \mathscr{D}_{\Delta t}^{-1} \mathscr{A}^t \overrightarrow{\mathbf{U}} \cdot \overrightarrow{\mathbf{U}} = \mathscr{D}_{\Delta t}^{-1} \mathscr{A}^t \overrightarrow{\mathbf{U}} \cdot \mathscr{A}^t \overrightarrow{\mathbf{U}} \geq 0,$$

and then $\mathscr{A}^t \overrightarrow{\mathbf{U}} = \mathbf{0}$, because $\mathscr{D}_{\Delta t}^{-1}$ is positive definite. Hence, $\overrightarrow{\mathbf{U}} \in \ker(\mathscr{A}^t)$. The reciprocal inclusion is obvious.

In order to prove the existence of a solution to the linear system (3.11) it is enough to see that the right-hand side belongs to image of $\mathscr{S}$. Since $\mathscr{S}$ is symmetric,

$$\text{im}(\mathscr{S}) = (\ker(\mathscr{S}))^\perp = \text{span}(\mathbf{e}^1, \ldots, \mathbf{e}^F).$$

We have

$$\mathbf{e}^i \cdot \left(\mathscr{A} \mathscr{D}_{\Delta t}^{-1} (\overrightarrow{\mathbf{E}}(t_{m+1}) + \mathscr{C}^m(\overrightarrow{\mathbf{I}}^0, \ldots, \overrightarrow{\mathbf{I}}^m)) + \overrightarrow{\Psi}(t_{m+1}) \right)$$

$$= \mathscr{A}^t \mathbf{e}^i \cdot \mathscr{D}_{\Delta t}^{-1} (\overrightarrow{\mathbf{E}}(t_{m+1}) + \mathscr{C}^m(\overrightarrow{\mathbf{I}}^0, \ldots, \overrightarrow{\mathbf{I}}^m)) + \mathbf{e}^i \cdot \overrightarrow{\Psi}(t_{m+1}) = 0, \quad i = 1, \ldots, F,$$

which completes the proof.

Again, the solution is not unique. Indeed, at each time step two solutions differ by a vector in the linear space $\text{span}(\mathbf{e}^1, \ldots, \mathbf{e}^F)$. In order to get a non-singular matrix we have to impose the potential at one node per connected component.

3.2.2 Generators without internal resistance

Let us assume that edge j is a generator without internal resistance ($r_j = 0$). Then matrix $\mathscr{D}_{\Delta t}$ is singular and the above results are no longer valid. In fact, we cannot guarantee the existence of a solution. A counterexample can be simply made by considering the case where two edges connecting the same couple of nodes are generators with different source voltages and without internal resistance. It is obvious that such a circuit is incompatible. In what follows we assume there are generators without internal resistance and try to determine the kernel of matrix $\mathscr{B}$. By analogous calculations to those made in Sect. 3.2.1 we deduce that if $(\overrightarrow{\mathbf{I}}, \overrightarrow{\mathbf{V}}) \in \ker(\mathscr{B})$ then $\overrightarrow{\mathbf{I}} \in \ker(\mathscr{D}_{\Delta t}) \cap \ker(\mathscr{A})$ and $\overrightarrow{\mathbf{V}} \in \ker(\mathscr{A}^t) = \text{span}(\mathbf{e}^1, \ldots, \mathbf{e}^F)$, and reciprocally.

In general we cannot affirm that $\ker(\mathscr{D}_{\Delta t}) \cap \ker(\mathscr{A}) = \{\mathbf{0}\}$, but if this occurred then we would have

$$\ker(\mathscr{B}) = \{(\mathbf{0}, \overrightarrow{\mathbf{V}}) \in \mathbb{R}^E \times \mathbb{R}^N : \overrightarrow{\mathbf{V}} \in \text{span}(\mathbf{e}^1, \ldots, \mathbf{e}^F)\}.$$

In particular, this is the case when there are no generators without internal resistance. This implies again that circuit model (3.5)–(3.6) has a linear manifold of solutions of dimension F as far as conditions (3.9) holds.

Moreover, if there are generators without internal resistance it will not be possible to compute the Schur complement as previously indicated because it requires matrix $\mathscr{D}_{\Delta t}$ to be invertible. This is why, in what follows, we introduce an alternative procedure to overcome this difficulty. For the sake of simplicity let us assume we can suppress the edges including resistance-free generators without changing the nodes of the graph.

Let j be an edge containing one of such generators. Let us assume for a moment that we know the current intensity along this edge, namely, I_j^{m+1}. In this case we can ignore edge j and consider this intensity as an external intensity entering/exiting the circuit trough the two nodes of j, i.e.,

$$\psi_{m_1 j}^{m+1} = -I_j^{m+1},$$

$$\psi_{m_2 j}^{m+1} = I_j^{m+1}.$$

Under this assumption we build system (3.11) as described above. Let us notice that, since we do not introduce edge j, matrix $\mathscr{D}_{\Delta t}$ is non-singular and we can follow the above procedure to obtain the Schur complement linear system.

Once the system is built the terms in the right-hand side $\psi_{m_1 j}^{m+1}$ and $\psi_{m_2 j}^{m+1}$ are moved to the left-hand side as I_j^{m+1} and $-I_j^{m+1}$, respectively. In this way, the resulting system has one unknown more than equations but it becomes equilibrated by adding the equation

$$-V_{m_2 j}^{m+1} + V_{m_1 j}^{m+1} = -E_j(t_{m+1}),$$

that has not been used yet because edge j has not been considered.

The new matrix has a new row and a new column but still keeps symmetry. Nevertheless, we emphasize that, for the reasons explained above, this linear system not always has a solution.

Let us summarize the above procedure:

1. Matrices $\mathscr{A}$ y $\mathscr{D}_{\Delta t}$ are built ignoring the edges having generators without internal resistance.
2. The Schur complement linear system is built:

$$\mathscr{A}\mathscr{D}_{\Delta t}^{-1}\mathscr{A}^t \vec{V}^{m+1} = \mathscr{A}\mathscr{D}_{\Delta t}^{-1}(\vec{E}(t_{m+1}) + \mathscr{C}^m(\vec{I}^0,\ldots,\vec{I}^m)) + \vec{\Psi}(t_{m+1}). \quad (3.12)$$

3. For each edge j representing a generator without internal resistance:
   a. Equation
   $$-V_{m_2 j}^{m+1} + V_{m_1 j}^{m+1} = -E_j(t_{m+1})$$

is added to the system (3.12) (this amounts to adding a new row to matrix $\mathscr{A}\mathscr{D}_{\Delta t}^{-1}\mathscr{A}^t$).

b. A column is added to the above matrix to keep symmetry (this means to add a new unknown: the current intensity along edge j).

In order to facilitate the computer implementation it is convenient to number the edges corresponding to generators without internal resistance at the end. In this way, their intensities, which are unknowns of the extended system, are added after vector $\vec{V}^{m+1}$.

3.3 The case of given potentials

In the previous sections we have assumed that the current intensities exchanged with the exterior by the circuit at the nodes are given (values ψ_i, $i = 1, \cdots, N$). Now we consider a more general case where in some nodes the potential is given rather than the current intensity. Let us denote by $\mathscr{B}_d$ (respectively, $\mathscr{B}_u$) the matrix that extracts from the vector of all potentials at the nodes the ones corresponding to those where the potentials (respectively, the current intensities) are given. Then, we have

$$\vec{V}(t) = \mathscr{B}_u^t \vec{V}^u(t) + \mathscr{B}_d^t \vec{V}^d(t)$$

and, similarly,

$$\vec{\Psi}(t) = \mathscr{B}_u^t \vec{\Psi}^d(t) + \mathscr{B}_d^t \vec{\Psi}^u(t),$$

where $\vec{V}^u$ (respectively, $\vec{\Psi}^u$) are the unknown potentials (respectively, current intensities) and $\vec{V}^d$ (respectively, $\vec{\Psi}^d$) are the given potentials (respectively, current intensities). By replacing the above expressions in (3.3)–(3.4) we obtain

$$\mathscr{D}(\vec{I})(t) + \mathscr{A}^t \mathscr{B}_u^t \vec{V}^u(t) = \vec{E}(t) - \mathscr{A}^t \mathscr{B}_d^t \vec{V}^d(t)$$
$$\mathscr{A}\vec{I}(t) \qquad + \mathscr{B}_d^t \vec{\Psi}^u(t) = -\mathscr{B}_u^t \vec{\Psi}^d(t),$$

for all $t \in [0,T]$. The unknowns are functions $\vec{I}$, $\vec{V}^u$ and $\vec{\Psi}^u$.

This system is not symmetric. In order to keep symmetry we can adopt the following alternative choice:

$$\mathscr{D}(\vec{I})(t) + \mathscr{A}^t \vec{V}(t) \qquad = \vec{E}(t), \qquad (3.13)$$

$$\mathscr{A}\vec{I}(t) \qquad + \mathscr{B}_d^t \vec{\Psi}^u(t) = -\mathscr{B}_u^t \vec{\Psi}^d(t), \qquad (3.14)$$

$$\mathscr{B}_d \vec{V}(t) \qquad = \vec{V}^d(t), \qquad (3.15)$$

where now the unknowns are functions $\vec{I}$, $\vec{V}$ and $\vec{\Psi}^u$.

3.4 Harmonic regime. Impedance matrix

In what follows we assume that the sources in the circuits are harmonic functions of time t. In this case, the model of the circuit can be reduced to a linear system of algebraic equations.

We forget about initial conditions. Since the circuit we are considering has a linear behavior, then all magnitudes are also harmonic functions of time, that is, they are of the form,
$$\vec{U}(t) = \text{Re}(\tilde{\mathbb{U}}e^{i\omega t}),$$
for some complex vector $\mathbb{U}$ called *complex amplitude* or *phasor*.

In a similar way as we did for the RLC circuit in Sect. 2.3, we replace $\vec{I}(t)$, $\vec{V}(t)$, $\vec{E}(t)$ and $\vec{\Psi}(t)$ in the system (3.3)–(3.4) by $\tilde{\mathbb{I}}e^{i\omega t}$, $\tilde{\mathbb{V}}e^{i\omega t}$, $\tilde{\mathbb{E}}e^{i\omega t}$ and $\tilde{\psi}e^{i\omega t}$, respectively, where $\tilde{\mathbb{I}}, \tilde{\mathbb{V}}, \tilde{\mathbb{E}}$ and $\tilde{\psi}$ are complex vectors. Since for $\vec{U}(t) = \mathbb{U}e^{i\omega t}$ we have
$$\frac{d\vec{U}}{dt}(t) = i\omega \mathbb{U} e^{i\omega t}$$
and, similarly,
$$\int \vec{U}(t)dt = \frac{1}{i\omega}\mathbb{U}e^{i\omega t},$$
we can easily deduce
$$\mathscr{D}(\omega)\tilde{\mathbb{I}} + \mathscr{A}^t\tilde{\mathbb{V}} = \tilde{\mathbb{E}}, \tag{3.16}$$
$$\mathscr{A}\tilde{\mathbb{I}} \qquad\qquad = -\tilde{\psi}, \tag{3.17}$$
where $\mathscr{D}(\omega)$ is the block-diagonal matrix built as follows:

- Each *resistance, uncoupled inductor, capacitor or generator* yields a 1×1 diagonal block:
$$\mathscr{D}(\omega)_{jj} = \begin{cases} R_j & \text{if edge } j \text{ is a resistance,} \\ i\omega L_j & \text{if edge } j \text{ is an uncoupled inductor,} \\ \frac{1}{i\omega C_j} & \text{if edge } j \text{ is a capacitor,} \\ r_j & \text{if edge } j \text{ is a generator with internal resistance } r_j. \end{cases}$$

- Each *magnetically coupled sub-circuit* yields an $S \times S$ diagonal block:
$$\mathscr{D}(\omega)_S = i\omega \mathscr{L}^S.$$

Let us notice that matrix $\mathscr{D}(\omega)$ is symmetric but not hermitian.

Assuming that $\ker(\mathscr{D}) \cap \ker(\mathscr{A}) = \{0\}$ (see Sect. 3.2.2) and $\tilde{\psi} \cdot \mathbf{e}^i = 0$, $i = 1,\ldots,F$, and using similar arguments as those in Sect. 3.2.1 we can prove the existence of a solution (in general not unique) to linear system (3.16)–(3.17). More

3.4 Harmonic regime. Impedance matrix

precisely, we can define an affine mapping

$$\Lambda(\omega) : \widetilde{\psi} \in \text{im}(\mathscr{A}) = \text{span}(\mathbf{e}^1, \ldots, \mathbf{e}^F)^\perp \subset \mathbb{C}^N \longrightarrow \widetilde{\mathbb{V}} \in \frac{\mathbb{C}^N}{\text{span}(\mathbf{e}^1, \ldots, \mathbf{e}^F)},$$

which is a bijection that can be written as

$$\Lambda(\omega)\widetilde{\psi} = \widetilde{\mathbb{V}}_{\widetilde{\mathbb{E}}} + \mathscr{L}(\omega)\widetilde{\psi},$$

where $\widetilde{\mathbb{V}}_{\widetilde{\mathbb{E}}}$ is the unique solution in $\mathbb{C}^N/\text{span}(\mathbf{e}^1, \ldots, \mathbf{e}^F)$ of the problem

$$\mathscr{D}(\omega)\widetilde{\mathbb{I}}_{\widetilde{\mathbb{E}}} + \mathscr{A}^t \widetilde{\mathbb{V}}_{\widetilde{\mathbb{E}}} = \widetilde{\mathbb{E}},$$

$$\mathscr{A}\widetilde{\mathbb{I}}_{\widetilde{\mathbb{E}}} = \mathbf{0},$$

and $\mathscr{L}(\omega)$ is an isomorphism called the *nodal impedance*. We notice that $\mathscr{L}(\omega)\widetilde{\psi} = \widetilde{\mathbb{V}}$, where $\widetilde{\mathbb{V}}$ is a solution of the complex linear system

$$\mathscr{D}(\omega)\widetilde{\mathbb{I}} + \mathscr{A}^t \widetilde{\mathbb{V}} = \mathbf{0}, \quad (3.18)$$

$$\mathscr{A}\widetilde{\mathbb{I}} = -\widetilde{\psi}. \quad (3.19)$$

The inverse of $\mathscr{L}(\omega)$ is called the *nodal admittance* and will be denoted by $\mathscr{Y}(\omega)$.

Assuming that $\mathscr{D}(\omega)$ is invertible (recall that this occurs if there are no resistance-free generators), from (3.18) we obtain

$$\widetilde{\mathbb{I}} = -\mathscr{D}(\omega)^{-1}\mathscr{A}^t \widetilde{\mathbb{V}},$$

and replacing in (3.19)

$$\mathscr{A}\mathscr{D}(\omega)^{-1}\mathscr{A}^t \widetilde{\mathbb{V}} = \widetilde{\psi}.$$

Therefore,

$$\mathscr{Y}(\omega) := \mathscr{A}\mathscr{D}(\omega)^{-1}\mathscr{A}^t.$$

Let us notice that $\mathscr{Y}(\omega)$ is an isomorphism:

$$\mathscr{Y}(\omega) : \mathbb{C}^N/\text{span}(\mathbf{e}^1, \ldots, \mathbf{e}^F) \longrightarrow \text{span}(\mathbf{e}^1, \ldots, \mathbf{e}^F)^\perp.$$

Moreover, it is symmetric but not hermitian. The same is true for $\mathscr{L}(\omega)$.

In practice, the circuit is connected to the exterior through a few nodes, let say P nodes, called *ports*. Let us assume that these ports belong to only one of the connected components of the circuit, the first one for simplicity, and that their numbers are $\{k_i : i = 1, \ldots, P\}$.

It is easy to see that there is an affine mapping giving the vector of potentials at these ports from the vector of input current intensities entering the circuit through them. The matrix associated to this affine mapping is called the *reduced impedance matrix* for the above set of ports.

In order to define this affine mapping let us denote by $\mathscr{P}$ the $(P \times N)$-matrix that extracts the P components corresponding to these ports from a vector in $\mathbb{C}^N$, namely, $\mathscr{P}_{ij} = \delta_{jk_i}$. For $\widetilde{\mathbb{F}} \in \mathbb{C}^P$ let us consider the linear system

$$\mathscr{D}(\omega)\widetilde{\mathbb{I}} + \mathscr{A}^t\widetilde{\mathbb{V}} = \widetilde{\mathbb{E}}, \tag{3.20}$$

$$\mathscr{A}\widetilde{\mathbb{I}} \qquad\qquad = \mathscr{P}^t\widetilde{\mathbb{F}}. \tag{3.21}$$

If $\mathscr{P}^t\widetilde{\mathbb{F}} \cdot \mathbf{e}^1 = \widetilde{\mathbb{F}} \cdot \mathscr{P}\mathbf{e}^1 = 0$, i.e., if $\sum_{i=1}^{P}\widetilde{\mathbb{F}}_i = 0$ (recall that all of the ports belong to the first connected component of the circuit), then this system has a solution $(\widetilde{\mathbb{I}}, \widetilde{\mathbb{V}})$. Furthermore, the intensities are unique and the potentials are unique up to a constant in each connected component. In particular, the potentials corresponding to the ports, $\mathscr{P}\widetilde{\mathbb{V}}$, are unique up to a constant.

Let us define the affine mapping

$$\Lambda_{\mathrm{P}}(\omega) : \widetilde{\mathbb{F}} \in \{\widetilde{\mathbb{F}} \in \mathbb{C}^P : \sum_{i=1}^{P}\mathbb{F}_i = 0\} = \operatorname{span}(\mathscr{P}\mathbf{e}^1)^\perp \longrightarrow \mathscr{P}\widetilde{\mathbb{V}} \in \mathbb{C}^P/\operatorname{span}(\mathscr{P}\mathbf{e}^1).$$

We have

$$\Lambda_{\mathrm{P}}(\omega)\widetilde{\mathbb{F}} = \mathscr{P}\widetilde{\mathbb{V}}_{\widetilde{\mathbb{E}}} + \mathscr{P}\mathscr{Z}(\omega)\mathscr{P}^t\widetilde{\mathbb{F}}.$$

Hence, the reduced impedance matrix is the matrix associated to the restriction of the linear mapping

$$\mathscr{Z}_{\mathrm{P}}(\omega) = \mathscr{P}\mathscr{Z}(\omega)\mathscr{P}^t$$

to the space $\operatorname{span}(\mathscr{P}\mathbf{e}^1)^\perp$. For its computation, we can proceed as follows:

1. We solve linear system (3.20)–(3.21) for $\widetilde{\mathbb{E}} = \mathbf{0}$, $(\widetilde{\mathbb{V}})_{k_P} = 0$, null potential at one node per connected component different from the first one, and $\widetilde{\mathbb{F}}$ successively equal to $\widetilde{\mathbb{F}}^i$, $i = 1, \ldots, P-1$, with $(\widetilde{\mathbb{F}}^i)_j = \delta_{ij}$, $j = 1, \ldots, P-1$ and $(\widetilde{\mathbb{F}}^i)_P = -1$. We call $\mathscr{R}$ the matrix whose columns are these solutions.

2. We extract from matrix $\mathscr{R}$ the $P-1$ rows $\{k_j, j = 1, \ldots, P-1\}$. The obtained square matrix, of order $P-1$, is the searched reduced impedance matrix.

Problems

3.1. Write the incidence matrix associated to the circuit represented in Fig. 3.1 assuming that the numbering of the edges and the nodes correspond, respectively, with those inside the circles and the points of the digraph.

Solution: $\mathscr{A} = \begin{pmatrix} 0 & -1 & -1 & 1 & 0 & 0 & 0 \\ 0 & 0 & 0 & -1 & 1 & 1 & 0 \\ -1 & 1 & 0 & 0 & -1 & 0 & 1 \\ 1 & 0 & 1 & 0 & 0 & -1 & -1 \end{pmatrix}.$

3.4 Harmonic regime. Impedance matrix   47

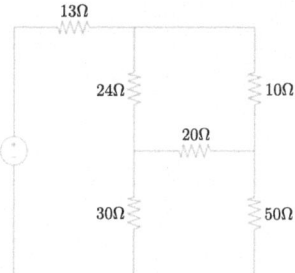

Fig. 3.2. Circuit for Problem 3.2

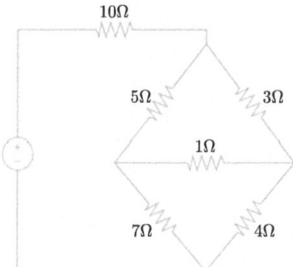

Fig. 3.3. Circuit for Problem 3.3

3.2. Compute the current intensity traversing the circuit in Fig. 3.2, considering that the power generator has no internal resistance and supplies an harmonic electromotive force with complex amplitude 100 V and frequency 1 Hz.

Solution: $I_1(t) = I_7(t) = 2.50\cos(2\pi t)$, $I_2(t) = 1.50\cos(2\pi t)$, $I_3(t) = \cos(2\pi t)$, $I_4(t) = 0.45\cos(2\pi t + \pi)$, $I_5(t) = 1.05\cos(2\pi t)$, $I_6(t) = 1.45\cos(2\pi t)$.

3.3. Compute the current intensity traversing the circuit in Fig. 3.3, considering that the power generator has no internal resistance and supplies an harmonic electromotive force with complex amplitude 15 V and frequency 1 Hz.

Solution: $I_1(t) = 1.04\cos(2\pi t)$, $I_2(t) = 0.39\cos(2\pi t)$, $I_3(t) = 0.65\cos(2\pi t)$, $I_4(t) = 0.01\cos(2\pi t)$, $I_5(t) = 0.38\cos(2\pi t)$, $I_6(t) = 0.66\cos(2\pi t)$, $I_7(t) = 1.04\cos(2\pi t)$.

3.4. Compute the current intensity traversing the circuit in Fig. 3.4, considering that the power generator has no internal resistance and supplies an harmonic electromotive force with complex amplitude 120 V and frequency 1 Hz.

Solution: $I_1(t) = 8.56\cos(2\pi t)$, $I_2(t) = 6.67\cos(2\pi t + \pi)$, $I_3(t) = 1.89\cos(2\pi t + \pi)$, $I_4(t) = 4.22\cos(2\pi t + \pi)$, $I_5(t) = 4.33\cos(2\pi t)$, $I_6(t) = 2.33\cos(2\pi t + \pi)$.

3.5. Compute the current intensity traversing the circuit in Fig. 3.5, considering that all power generators have no internal resistance and that all of them supply a harmonic electromotive force with frequency 1 Hz. The complex amplitudes are 12, 6, 9, 10 and 4 V respectively. Furthermore, in the leftmost node enters a current of

48   3 Linear electrical circuits

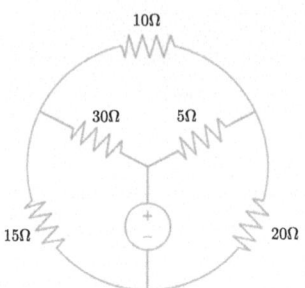

Fig. 3.4. Circuit for Problem 3.4

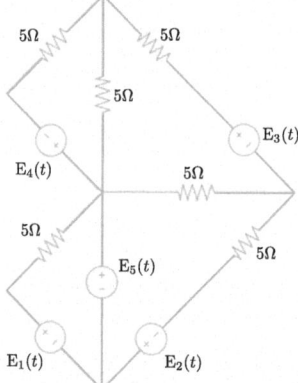

Fig. 3.5. Circuit for Problem 3.5

complex amplitude 2 A and in the node situated in the south end of the circuit exits a current of complex amplitude 1 A, both having a frequency of 1 Hz.

Solution: $I_1(t) = 0.90\cos(2\pi t + \pi)$, $I_2(t) = I_3(t) = 2.50\cos(2\pi t)$, $I_4(t) = I_5(t) = 1.84\cos(2\pi t)$, $I_8(t) = I_9(t) = 1.39\cos(2\pi t)$, $I_6(t) = I_7(t) = 1.39\cos(2\pi t)$, $I_{10}(t) = 0.29\cos(2\pi t + \pi)$, $I_{11}(t) = 0.16\cos(2\pi t + \pi)$, $I_{12}(t) = 0.34\cos(2\pi t)$ (see Sect. 12.3.2).

3.6. Consider the electrical circuit shown in Fig. 3.6. There are two state variables, I_1 and I_2, representing the currents passing through each of the two inductors, and one state variable, V, representing the voltage on the capacitor. If we assume zero initial conditions for the state variables and circuit parameter values $L_1 = 32$ H, $L_2 = 12$ H, $C = 43$ F, and $R = 1$ Ω find the output voltage across the resistor due to a harmonic input $E(t) = \sin(\omega t)$ when:

(a) $\omega = 1/2$.
(b) $\omega = 2$.

For each value of ω, plot the graphs of $V(t)$ and $E(t)$ as a function of t on the same set of coordinates and explain the meaning of the graphs.

3.4 Harmonic regime. Impedance matrix

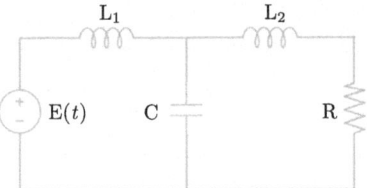

Fig. 3.6. Circuit for Problem 3.6

Solution: With the given parameter values, the circuit acts as a low-pass filter. At the lower frequency the amplitude of the steady-state response is approximately equal to 1, whereas at the higher frequency the steady-state response is greatly attenuated.

(a) $V(t) = \dfrac{1}{5e^t} - \dfrac{48\cos\left(\frac{t}{2}\right)}{65} + \dfrac{46\sin\left(\frac{t}{2}\right)}{65} + \dfrac{7\cos\left(\frac{\sqrt{3}t}{2}\right)}{13\,e^{t/2}} + \dfrac{\sqrt{3}\sin\left(\frac{\sqrt{3}t}{2}\right)}{13\,e^{t/2}}.$

(b) $V(t) = \dfrac{544\sqrt{3}\,e^{-\frac{t}{2}}\sin\left(\frac{\sqrt{3}t}{2}\right) - 476\,e^{-\frac{t}{2}}\cos\left(\frac{\sqrt{3}t}{2}\right) - 6\cos(4t) - 143\sin(4t) + 482}{4097}.$

Part II
Distributed Parameter Models

In this case the physical magnitudes are distributed, namely they also depend on the position in the affine space. As a consequence, the models are partial differential equation systems. These systems are deduced from the general mathematical model for electromagnetism, namely, the Maxwell's equations, making use of the specific assumptions of electrostatics, direct current, magnetostatics and eddy currents.

4
Maxwell's equations in free space

Electromagnetic field theory is a discipline concerned with the study of charges, at rest and in motion, producing currents, and electric and magnetic fields. The mathematical model relating all these fields is the Maxwell's equations system. As we will see, Maxwell's equations are wave equations. They can be reduced to vector equations similar to the standard wave equation arising in acoustics. However, electromagnetic waves may propagate in the empty space while mechanical waves need a material media. The latter are particular solutions of the conservation equations in thermomechanics. In solid we refer to mechanical vibrations and in fluids, besides the important example of the above mentioned acoustic waves, there are also gravity waves or capillary waves, among others. In all cases, an important feature of waves is its finite velocity of propagation.

For the sake of introducing the subject in a simple framework, we begin by recalling some results from a distributed scalar model of wave propagation: the equation of the vibrating string.

4.1 An introduction to distributed models for wave propagation. The one-dimensional wave equation

Let us consider the initial-value problem

$$\rho \frac{\partial^2 u}{\partial t^2} - K \frac{\partial^2 u}{\partial x^2} = 0, \ x \in \mathbb{R}, \ t > 0, \tag{4.1}$$

$$u(x,0) = u_0(x), \ x \in \mathbb{R}, \tag{4.2}$$

$$\frac{\partial u}{\partial t}(x,0) = u_1(x), \ x \in \mathbb{R}. \tag{4.3}$$

We recall that this problem is a model for a *vibrating string* in which case $\rho = \rho_l$ (kg/m) is its *linear density* and $K = T$ (N) is its *tension*. But it is also a one-dimensional version of the *Lamé-Navier* equations (see, for instance, [41]) for the elastic vibrations of a solid with *mass density* ρ (kg/m^3) and *Lamé's coefficients*

A. Bermúdez, D. Gómez, P. Salgado: *Mathematical Models and Numerical Simulation in Electromagnetism*. UNITEXT – La Matematica per il 3+2 74
DOI 10.1007/978-3-319-02949-8_4, © Springer International Publishing Switzerland 2014

4 Maxwell's equations in free space

λ (N/m^2) and μ (N/m^2). In this case, only longitudinal waves, the so-called *pressure waves*, are included in the model. Moreover, $K = 2\mu + \lambda$ and the *propagation velocity* is

$$c = \sqrt{\frac{2\mu + \lambda}{\rho}}.$$

Problem (4.1)–(4.3) is also a one-dimensional model for the vibrations of an elastic fluid (see [41]), i.e., for one-dimensional *acoustics*. In this case, $K = \rho c^2$, c being the *speed of sound* in the fluid. This simplified model is extensively used in, for instance, architectural acoustics.

In what follows we will obtain the *d'Alembert's solution* (4.1)–(4.3). For the sake of simplicity in notations we divide the first equation by ρ and denote

$$c := \frac{K}{\rho}.$$

By doing the change of variables,

$$\xi = x - ct,$$
$$\eta = x + ct,$$

we have,

$$\frac{\partial \widehat{u}}{\partial \xi} = -\frac{1}{2c}\left(\frac{\partial u}{\partial t} - c\frac{\partial u}{\partial x}\right), \tag{4.4}$$

$$\frac{\partial \widehat{u}}{\partial \eta} = \frac{1}{2c}\left(\frac{\partial u}{\partial t} + c\frac{\partial u}{\partial x}\right), \tag{4.5}$$

where $\widehat{u}(\xi,\eta) = u(x(\xi,\eta),t(\xi,\eta))$.
Since

$$\frac{\partial^2 u}{\partial t^2} - c^2\frac{\partial^2 u}{\partial x^2} = \left(\frac{\partial}{\partial t} + c\frac{\partial}{\partial x}\right)\left(\frac{\partial u}{\partial t} + c\frac{\partial u}{\partial x}\right), \tag{4.6}$$

the equation (4.1) becomes

$$\frac{\partial^2 \widehat{u}}{\partial \xi \partial \eta} = 0, \tag{4.7}$$

the general solution of which is

$$\widehat{u}(\xi,\eta) = f(\xi) + g(\eta), \tag{4.8}$$

for any continuous functions f and γ_{-}. This leads to the following general solution of the original problem:

$$u(x,t) = f(x - ct) + g(x + ct). \tag{4.9}$$

By imposing the initial conditions we easily obtain

$$g(x) = \frac{1}{2}\left\{u_0(x) + \frac{1}{c}\int_0^x u_1(s)ds\right\} + \frac{A}{2}, \tag{4.10}$$

$$f(x) = \frac{1}{2}\left\{u_0(x) - \frac{1}{c}\int_0^x u_1(s)ds\right\} - \frac{A}{2}, \tag{4.11}$$

from which the well known *d'Alembert's formula* is deduced,

$$u(x,t) = \frac{1}{2}\{u_0(x+ct) + u_0(x-ct)\} + \frac{1}{2c}\int_{x-ct}^{x+ct} u_1(s)\,\mathrm{d}s. \qquad (4.12)$$

It is easy to see that, if

$$\mathrm{supp}\,(u_0) \cup \mathrm{supp}\,(u_1) \subset [a,b],$$

then

$$\mathrm{supp}\,(u(\cdot,t)) \subset [a-ct, b+ct].$$

This means that signals propagate at finite velocity c.

4.1.1 Energy conservation

From the d'Alembert's solution (4.12) it is straightforward to obtain the following equality:

$$\frac{1}{2}\int_{\mathbb{R}} \rho |u(x,t)|^2 \,\mathrm{d}x + \frac{1}{2}\int_{\mathbb{R}} K \left|\frac{\partial u}{\partial x}(x,t)\right|^2 \mathrm{d}x$$
$$= \frac{1}{2}\int_{\mathbb{R}} \rho |u_0(x)|^2 \,\mathrm{d}x + \frac{1}{2}\int_{\mathbb{R}} K \left|\frac{\partial u_1}{\partial x}(x)\right|^2 \mathrm{d}x. \qquad (4.13)$$

In the case of the vibrating string, the first terms of the two members represent the *kinetic energy* at times t and 0, respectively, while the second ones give the *potential (elastic) energy*. The above equality means that the whole mechanical energy is conserved along the time.

In more general situations, the solution of the wave equation cannot be obtained in a closed form. However, we can still get a conservation principle like (4.13) by making a priori formal computations as follows. Let us multiply Eq. (4.1) by $\frac{\partial u}{\partial t}$ and integrate in $\mathbb{R}$. We obtain

$$\int_{\mathbb{R}} \rho \frac{\partial^2 u}{\partial t^2}(x,t) \frac{\partial u}{\partial t}(x,t)\,\mathrm{d}x - \int_{\mathbb{R}} K \frac{\partial^2 u}{\partial x^2}(x,t) \frac{\partial u}{\partial t}(x,t)\,\mathrm{d}x = 0.$$

By making integration by parts using the fact that the solution has compact support, we deduce

$$\int_{\mathbb{R}} \frac{1}{2}\rho \frac{\partial}{\partial t}\left|\frac{\partial u}{\partial t}(x,t)\right|^2 \mathrm{d}x + \int_{\mathbb{R}} \frac{1}{2}K \frac{\partial}{\partial t}\left|\frac{\partial u}{\partial x}(x,t)\right|^2 \mathrm{d}x = 0$$

and then

$$\frac{\mathrm{d}}{\mathrm{d}t}\int_{\mathbb{R}} \frac{1}{2}\rho \left|\frac{\partial u}{\partial t}(x,t)\right|^2 \mathrm{d}x + \frac{\mathrm{d}}{\mathrm{d}t}\int_{\mathbb{R}} \frac{1}{2}K \left|\frac{\partial u}{\partial x}(x,t)\right|^2 \mathrm{d}x = 0.$$

Equality (4.13) is now obtained by integrating in time from 0 to t.

4.1.2 Harmonic solutions to the one-dimensional wave equation

Let us consider again a homogeneous media in the whole domain. We seek harmonic solutions of the wave equation, i.e., functions of the following form:

$$u(x,t) = \text{Re}(\widetilde{\mathbb{U}}(x)e^{i\omega t}), \quad (4.14)$$

where the real number ω is the *angular frequency* and $\widetilde{\mathbb{U}}$ is a complex-valued function.

By replacing $u(x,t)$ with $\widetilde{\mathbb{U}}(x)e^{i\omega t}$ in (4.1)–(4.3) we get

$$-\omega^2 \widetilde{\mathbb{U}}(x)e^{i\omega t} - c^2 \frac{d^2 \widetilde{\mathbb{U}}}{dx}(x)e^{i\omega t} = 0,$$

and then $\widetilde{\mathbb{U}}$ is solution of the one-dimensional *Helmholtz's equation*:

$$-\omega^2 \widetilde{\mathbb{U}} - c^2 \frac{d^2 \widetilde{\mathbb{U}}}{dx^2} = 0. \quad (4.15)$$

The complex function $\widetilde{\mathbb{U}}(x) = e^{ikx}$ is a solution to this equation if and only if $k = \pm \omega/c$.

Thus the general solution is

$$\widetilde{\mathbb{U}}(x) = Be^{i\beta x} + Ce^{-i\beta x}, \quad (4.16)$$

where $\beta := \omega/c$, and B and C are two complex numbers.

Let $B = \mathbb{B}e^{i\varphi}$ and $C = \mathbb{C}e^{i\psi}$ be the polar representations of B and C, respectively. By replacing into (4.16) and then into (4.14) we get the following solution of the wave equation:

$$u(x,t) = \mathbb{B}\cos(\omega t + \beta x + \varphi) + \mathbb{C}\cos(\omega t - \beta x + \psi). \quad (4.17)$$

The first term represents a wave travelling to the left at velocity $c = \frac{\omega}{\beta}$ (*regressive wave*), while the second term represents a wave travelling to the right at the same velocity (*progressive wave*). Real numbers $\mathbb{B}$ and $\mathbb{C}$, and φ and ψ are their respective *amplitudes* and *phases*, while β is the *wave number*. We notice that the latter can be considered as the *spatial frequency* of the wave.

4.2 Charges and currents

Charges are represented by scalar density fields. These densities can be volume, surface or line densities. From a mathematical point of view, they will be distributions (see Appendix C) supported in a volume, a surface or a line, respectively. We will also consider point charges that will be represented by Dirac measures.

Suppose there are electric charges of volumetric density $\rho_V(x,t)$ in a region of the affine space $\mathscr{E}$ and that charge at point x moves with velocity $\mathbf{v}(x,t)$ at time t. We define the *current density* field $\mathbf{J}$ by

$$\mathbf{J}(x,t) := \rho_V(x,t)\mathbf{v}(x,t). \tag{4.18}$$

Notice that the current density is also the charge flux density.
Thus, for a surface S, the surface integral

$$\mathscr{I}_S = \int_S \mathbf{J} \cdot \mathbf{n}\, dA \tag{4.19}$$

represents the charge flux through S which is also called the *current intensity* through S.

In the International System of Units, the charge is measured in *coulomb* (C) and the current density in $C/sm^2 = A/m^2$, where coulomb over second is *ampere* (A). Hence, the SI unit for current intensity is ampere (A).

4.3 Electric and magnetic fields

Electric and magnetic fields are fundamentally fields of force that originate from electric charges. Electric charges at rest relative to an observation point give rise to an *electric field* at this point. It is static, i.e., time independent. The relative motion of charges provides an additional force called *magnetic force*. If charges are moving at constant velocities relative to the observation point, this added magnetic field is static (time independent). Accelerated motions produce both time-varying electric and magnetic fields.

The symbol for the *electric field intensity* is vector $\mathbf{E}$. It is a force per unit charge so in the SI the unit is newton over coulomb (N/C). In its turn, the magnetic field is represented by a vector field $\mathbf{B}$ called *magnetic flux density*. Its SI unit is weber per square meter (Wb/m^2), also called *tesla* (T). The presence of $\mathbf{E}$ and $\mathbf{B}$ at a point in the space may be physically detected by means of a charge Q moving at velocity $\mathbf{v}$. The force acting on this charge is given by the *Lorentz's force law*,

$$\mathbf{F} = Q(\mathbf{E} + \mathbf{v} \times \mathbf{B}). \tag{4.20}$$

In the case of a charge distribution with (volumetric) charge density ρ_V, the density of force of the electromagnetic field is

$$\mathbf{f} = \rho_V(\mathbf{E} + \mathbf{v} \times \mathbf{B}) = \rho_V \mathbf{E} + \mathbf{J} \times \mathbf{B}. \tag{4.21}$$

4.4 Maxwell's integral equations in free space

The connection of electric and magnetic fields to their charge and current sources is provided by Maxwell's equations. Their *integral forms* in free space are

$$\Gamma = \partial\Omega \qquad \int_\Gamma \varepsilon_0 \mathbf{E} \cdot \mathbf{n}\, dA = \mathcal{Q}_\Omega, \qquad \text{(C)} \qquad (4.22)$$

$$\Gamma = \partial\Omega \qquad \int_\Gamma \mathbf{B} \cdot \mathbf{n}\, dA = 0, \qquad \text{(Wb)} \qquad (4.23)$$

$$l \text{ border of } S \qquad \int_l \mathbf{E} \cdot d\mathbf{l} = -\frac{d}{dt}\int_S \mathbf{B}\cdot\mathbf{n}\, dA, \qquad \text{(V)} \qquad (4.24)$$

$$l \text{ border of } S \qquad \int_l \frac{\mathbf{B}}{\mu_0}\cdot d\mathbf{l} = \mathcal{I}_S + \frac{d}{dt}\int_S \varepsilon_0 \mathbf{E}\cdot\mathbf{n}\, dA, \qquad \text{(A)} \qquad (4.25)$$

where $\mathcal{Q}_\Omega$ and $\mathcal{I}_S$ denotes, respectively, the charge contained in volume Ω and the current intensity through surface S. Symbol V denotes *volt*, i.e. joule over coulomb. Equations (4.22) and (4.23) are the *Gauss' laws* for electric charge and magnetic field, respectively. Equation (4.24) is *Faraday's law* and Eq. (4.25) is *Ampère's law*.

Constant ε_0 is called *electric permittivity of the free space*. Its value is approximately $\varepsilon_0 \simeq 10^{-9}/36\pi$ in SI units, i.e., in C^2/Nm^2 or F/m, where we recall F is farad. Constant μ_0 is called *magnetic permeability* of the free space. Its value is $\mu_0 = 4\pi 10^{-7}$ in SI units, i.e., in H/m, where H is *henry*.

4.5 Maxwell's equations in differential form in free space

In this section we deduce the local or *differential forms* of Maxwell's equations from the integral ones given above.

For the sake of simplicity, let us assume that charges and currents are volumetric. This means that $\mathcal{Q}_\Omega$ and $\mathcal{I}_S$ are given by

$$\mathcal{Q}_\Omega = \int_\Omega \rho_V\, dV, \qquad (4.26)$$

$$\mathcal{I}_S = \int_S \mathbf{J}\cdot\mathbf{n}\, dA, \qquad (4.27)$$

where ρ_V is the *volumetric charge density* and $\mathbf{J}$ is the current density. Since the integral forms of Maxwell's equations must hold for any volume Ω (see (4.22) and (4.23)) and any surface S (see (4.24) and (4.25)), we can deduce local forms which are partial differential equations.

Indeed, by using the Gauss' theorem in (4.22) and (4.23) we obtain

$$\int_\Omega (\varepsilon_0 \operatorname{div} \mathbf{E} - \rho_V)\, dV = 0,$$

$$\int_\Omega \operatorname{div} \mathbf{B}\, dV = 0,$$

and then
$$\varepsilon_0 \operatorname{div} \mathbf{E} = \rho_V, \tag{4.28}$$
$$\operatorname{div} \mathbf{B} = 0. \tag{4.29}$$

Now, transforming (4.24) and (4.25) by Stokes' theorem, we get
$$\int_S [\operatorname{curl} \mathbf{E} \cdot \mathbf{n} + \frac{\partial \mathbf{B}}{\partial t} \cdot \mathbf{n}] \, dA = 0,$$
and hence
$$\frac{\partial \mathbf{B}}{\partial t} + \operatorname{curl} \mathbf{E} = 0. \tag{4.30}$$

Similarly,
$$\int_S \left[\frac{1}{\mu_0} \operatorname{curl} \mathbf{B} \cdot \mathbf{n} - \mathbf{J} \cdot \mathbf{n} - \varepsilon_0 \frac{\partial \mathbf{E}}{\partial t} \cdot \mathbf{n} \right] dA = 0,$$
and then
$$\varepsilon_0 \frac{\partial \mathbf{E}}{\partial t} - \frac{1}{\mu_0} \operatorname{curl} \mathbf{B} = -\mathbf{J}. \tag{4.31}$$

4.6 Wave equations for electric and magnetic fields

We come back to Maxwell's equations in the time domain. Our goal is to eliminate one of the fields $\mathbf{B}$ or $\mathbf{E}$ and to obtain a wave equation for the other field.

In order to eliminate $\mathbf{B}$ we take the **curl** of both sides of (4.30) to get
$$\operatorname{curl} \operatorname{curl} \mathbf{E} = -\frac{\partial}{\partial t}(\operatorname{curl} \mathbf{B}). \tag{4.32}$$

Next, we obtain **curl B** from Eq. (4.31) and replace it in (4.32). We deduce
$$\operatorname{curl} \operatorname{curl} \mathbf{E} + \mu_0 \varepsilon_0 \frac{\partial^2 \mathbf{E}}{\partial t^2} = -\mu_0 \frac{\partial \mathbf{J}}{\partial t}, \tag{4.33}$$
which is a non-homogeneous vector wave equation.

Now, let us eliminate $\mathbf{E}$. For this purpose we take the **curl** of Eq. (4.31) to get
$$\varepsilon_0 \frac{\partial \operatorname{curl} \mathbf{E}}{\partial t} - \frac{1}{\mu_0} \operatorname{curl} \operatorname{curl} \mathbf{B} = -\operatorname{curl} \mathbf{J}.$$

Then we replace **curl E** by $-\frac{\partial \mathbf{B}}{\partial t}$, as obtained from (4.30). We deduce
$$\varepsilon_0 \frac{\partial^2 \mathbf{B}}{\partial t^2} - \frac{1}{\mu_0} \operatorname{curl} \operatorname{curl} \mathbf{B} = -\operatorname{curl} \mathbf{J},$$
or, equivalently,
$$\operatorname{curl} \operatorname{curl} \mathbf{B} + \mu_0 \varepsilon_0 \frac{\partial^2 \mathbf{B}}{\partial t^2} = \mu_0 \operatorname{curl} \mathbf{J}, \tag{4.34}$$
which is a vector wave equation similar to (4.33).

Since we have,
$$-\Delta \mathbf{u} = \operatorname{curl} \operatorname{curl} \mathbf{u} - \operatorname{grad}(\operatorname{div} \mathbf{u}),$$

for any smooth vector field, then

$$\mathbf{curl\,curl\,B} = -\Delta\mathbf{B}, \qquad (4.35)$$

while

$$\mathbf{curl\,curl\,E} = -\Delta\mathbf{E} + \mathrm{grad}\left(\frac{\rho_V}{\varepsilon_0}\right). \qquad (4.36)$$

By using (4.35) and (4.36) in (4.34) and (4.33), respectively, we obtain the vector wave equations

$$\mu_0\varepsilon_0\frac{\partial^2\mathbf{B}}{\partial t^2} - \Delta\mathbf{B} = \mu_0\,\mathbf{curl\,J}, \qquad (4.37)$$

$$\mu_0\varepsilon_0\frac{\partial^2\mathbf{E}}{\partial t^2} - \Delta\mathbf{E} = \mu_0\frac{\partial\mathbf{J}}{\partial t} - \frac{1}{\varepsilon_0}\,\mathrm{grad}\,\rho_V. \qquad (4.38)$$

4.7 Harmonic regime

Time harmonic fields **E** and **B** are generated whenever their sources (i.e., charges and currents) have densities varying sinusoidally in time. More generally, let us assume that ρ_V and **J** are of the form,

$$\rho_V(x,t) = \mathrm{Re}(e^{i\omega t}\widehat{\rho}_V(x)),$$

$$\mathbf{J}(x,t) = \mathrm{Re}(e^{i\omega t}\widehat{\mathbf{J}}(x)),$$

where $\widehat{\rho}_V$ and $\widehat{\mathbf{J}}$ are time-independent complex-valued fields.

Then the solutions of Maxwell's equations also have this form, namely,

$$\mathbf{E}(x,t) = \mathrm{Re}(e^{i\omega t}\widehat{\mathbf{E}}(x)),$$

$$\mathbf{B}(x,t) = \mathrm{Re}(e^{i\omega t}\widehat{\mathbf{B}}(x)),$$

where, again, $\widehat{\mathbf{E}}$ and $\widehat{\mathbf{B}}$ are time-independent complex-valued fields.

By replacing the above expressions in the differential form of Maxwell's equations we obtain

$$\mathrm{div}(\varepsilon_0\widehat{\mathbf{E}}) = \widehat{\rho}_V, \qquad (4.39)$$

$$\mathrm{div}\,\widehat{\mathbf{B}} = 0, \qquad (4.40)$$

$$\mathbf{curl}\,\widehat{\mathbf{E}} = -i\omega\widehat{\mathbf{B}}, \qquad (4.41)$$

$$\mathbf{curl}\left(\frac{\widehat{\mathbf{B}}}{\mu_0}\right) = \widehat{\mathbf{J}} + i\omega\varepsilon_0\widehat{\mathbf{E}}. \qquad (4.42)$$

These equations are called the *harmonic Maxwell's equations*. We notice that all fields are complex-valued. However, they do not depend on the time variable.

In the time-harmonic regime, Eqs. (4.37) and (4.38) become, respectively,

$$-\mu_0\varepsilon_0\omega^2\widehat{\mathbf{B}} - \Delta\widehat{\mathbf{B}} = \mu_0\,\mathbf{curl}\,\widehat{\mathbf{J}},$$

$$-\mu_0\varepsilon_0\omega^2\widehat{\mathbf{E}} - \Delta\widehat{\mathbf{E}} = \mu_0 i\omega\widehat{\mathbf{J}} - \frac{1}{\varepsilon_0}\mathrm{grad}\,\widehat{\rho}_V.$$

These equations can also be directly obtained from (4.39)–(4.42).

4.8 Behavior at infinity: the Silver-Müller conditions

Let us suppose that sources $\widehat{\mathbf{J}}$ and $\widehat{\rho}_V$ have compact support. Then the radiation conditions for the harmonic Maxwell's equations are the following:

$$\left|\widehat{\mathbf{E}}(x)\right| = O\!\left(\frac{1}{r(x)}\right),$$

$$\left|\widehat{\mathbf{B}}(x)\right| = O\!\left(\frac{1}{r(x)}\right),$$

$$\left|\sqrt{\varepsilon_0}\widehat{\mathbf{E}}(x) - \frac{1}{\sqrt{\mu_0}}\widehat{\mathbf{B}}(x)\times\frac{\mathbf{r}(x)}{r(x)}\right| = o\!\left(\frac{1}{r(x)}\right),$$

$$\left|\sqrt{\varepsilon_0}\widehat{\mathbf{E}}(x)\times\frac{\mathbf{r}(x)}{r(x)} + \frac{1}{\sqrt{\mu_0}}\widehat{\mathbf{B}}(x)\right| = o\!\left(\frac{1}{r(x)}\right),$$

for $r(x) \to \infty$. They are called the *Silver-Müller conditions*.

4.9 Plane waves in empty space

In what follows we will look for some particular solutions of the harmonic Maxwell's equations in the case where charge and current densities are everywhere zero in the space, that is, fields ρ_V and $\mathbf{J}$ (and then $\widehat{\rho}_V$ and $\widehat{\mathbf{J}}$) are identically zero.

Let us look for plane waves, i.e., for solutions of the form,

$$\widehat{\mathbf{E}}(x) = \widehat{\mathbf{E}}_0 e^{i\mathbf{k}\cdot\mathbf{r}(x)}, \tag{4.43}$$

$$\widehat{\mathbf{B}}(x) = \widehat{\mathbf{B}}_0 e^{i\mathbf{k}\cdot\mathbf{r}(x)}, \tag{4.44}$$

where $\mathbf{k}$, the *wave vector*, $\widehat{\mathbf{E}}_0$ and $\widehat{\mathbf{B}}_0$ are three complex vectors. Let us recall that for

$$\mathbf{U}(x) = \mathbf{U}_0 e^{i\mathbf{k}\cdot\mathbf{r}(x)}$$

we have (see Appendix B):

- $\text{div}\,\mathbf{U} = i\mathbf{U}_0 \cdot \mathbf{k} e^{i\mathbf{k}\cdot\mathbf{r}}$,
- $\mathbf{curl}\,\mathbf{U} = i\mathbf{k} \times \mathbf{U}_0 e^{i\mathbf{k}\cdot\mathbf{r}}$,
- $\mathbf{curl}\,\mathbf{curl}\,\mathbf{U} = -\mathbf{k} \times (\mathbf{k} \times \mathbf{U}_0) e^{i\mathbf{k}\cdot\mathbf{r}}$.

By replacing (4.43) and (4.44) in (4.39)-(4.42) with $\widehat{\rho}_V = 0$ and $\widehat{\mathbf{J}} = \mathbf{0}$, we obtain

$$i\varepsilon_0 \widehat{\mathbf{E}}_0 \cdot \mathbf{k} e^{i\mathbf{k}\cdot\mathbf{r}} = 0,$$

$$i\widehat{\mathbf{B}}_0 \cdot \mathbf{k} e^{i\mathbf{k}\cdot\mathbf{r}} = 0,$$

$$i\mathbf{k} \times \widehat{\mathbf{E}}_0 e^{i\mathbf{k}\cdot\mathbf{r}} = -i\omega \widehat{\mathbf{B}}_0 e^{i\mathbf{k}\cdot\mathbf{r}},$$

$$i\frac{1}{\mu_0} \mathbf{k} \times \widehat{\mathbf{B}}_0 e^{i\mathbf{k}\cdot\mathbf{r}} = i\omega\varepsilon_0 \widehat{\mathbf{E}}_0 e^{i\mathbf{k}\cdot\mathbf{r}},$$

from which it follows that

$$\widehat{\mathbf{E}}_0 \cdot \mathbf{k} = 0, \qquad \widehat{\mathbf{B}}_0 \cdot \mathbf{k} = 0.$$

These equations mean that the oscillation direction is orthogonal to the propagation direction. In other words, electromagnetic plane waves are *transversal*. Furthermore,

$$\widehat{\mathbf{B}}_0 = -\frac{1}{\omega} \mathbf{k} \times \widehat{\mathbf{E}}_0,$$

$$\widehat{\mathbf{E}}_0 = \frac{1}{\varepsilon_0 \mu_0 \omega} \mathbf{k} \times \widehat{\mathbf{B}}_0.$$

Hence, vectors $\mathbf{k}$, $\widehat{\mathbf{B}}_0$ and $\widehat{\mathbf{E}}_0$ have to be mutually orthogonal and, moreover,

$$|\mathbf{k}|^2 = \varepsilon_0 \mu_0 \omega^2.$$

Let us take $\mathbf{k} = k\mathbf{e}_3$. Then we consider two cases having a similar treatment:

a) $\widehat{\mathbf{E}}_0 \in <\mathbf{e}_1>$, $\widehat{\mathbf{B}}_0 \in <\mathbf{e}_2>$,
b) $\widehat{\mathbf{E}}_0 \in <\mathbf{e}_2>$, $\widehat{\mathbf{B}}_0 \in <\mathbf{e}_1>$.

Case a) Let $\widehat{\mathbf{E}}_0 = \widehat{E}_m \mathbf{e}_1$ and $\widehat{\mathbf{B}}_0 = \widehat{B}_m \mathbf{e}_2$. We have

$$\widehat{\mathbf{E}}(x) = \widehat{E}_m e^{ikx_3} \mathbf{e}_1,$$

$$\widehat{\mathbf{B}}(x) = \widehat{B}_m e^{ikx_3} \mathbf{e}_2,$$

with $\widehat{B}_m = -(k/\omega)\widehat{E}_m$.

4.9 Plane waves in empty space

Since $|\mathbf{k}| = |k| = \omega\sqrt{\varepsilon_0\mu_0}$, we have as general solution

$$\widehat{\mathbf{E}}(x) = \left[\widehat{E}_m^+ e^{-i\beta_0 x_3} + \widehat{E}_m^- e^{i\beta_0 x_3}\right]\mathbf{e}_1, \tag{4.45}$$

$$\widehat{\mathbf{B}}(x) = \left[\sqrt{\mu_0\varepsilon_0}\,\widehat{E}_m^+ e^{-i\beta_0 x_3} - \sqrt{\mu_0\varepsilon_0}\,\widehat{E}_m^- e^{i\beta_0 x_3}\right]\mathbf{e}_2, \tag{4.46}$$

where $\beta_0 := \omega\sqrt{\mu_0\varepsilon_0}$ is the *phase constant*.

The corresponding solution of Maxwell's equations is

$$\mathbf{E}(x,t) = \mathrm{Re}\left(e^{i\omega t}\widehat{\mathbf{E}}(x)\right) = \left[E_m^+\cos(\omega t - \beta_0 x_3 + \phi^+) + E_m^-\cos(\omega t + \beta_0 x_3 + \phi^-)\right]\mathbf{e}_1,$$

$$\mathbf{B}(x,t) = \left[\left(\sqrt{\mu_0\varepsilon_0}E_m^+\right)\cos(\omega t - \beta_0 x_3 + \phi^+) - \left(\sqrt{\mu_0\varepsilon_0}E_m^+\right)\cos(\omega t + \beta_0 x_3 + \phi^-)\right]\mathbf{e}_2,$$

where $\widehat{E}_m^+ = E_m^+ e^{i\phi^+}$ and $\widehat{E}_m^- = E_m^- e^{i\phi^-}$.

From these expressions we deduce that signals propagate in the $\pm\mathbf{e}_3$-directions at the *phase velocity*,

$$c = \frac{\omega}{\beta_0} = \frac{1}{\sqrt{\varepsilon_0\mu_0}} \simeq 3.10^8 \text{ m/s}.$$

For fixed t, the distance in the $\mathbf{e}_3$-direction separating two nearest points having 2π rad of phase shift is called the *wavelength* and denoted by λ. We have

$$(\omega t - \beta_0 x_3 + \phi^+) - (\omega t - \beta_0(x_3 + \lambda) + \phi^+) = 2\pi$$

and hence,

$$\beta_0\lambda = 2\pi.$$

Thus

$$\lambda = \frac{2\pi}{\beta_0} = \frac{2\pi}{\omega\sqrt{\varepsilon_0\mu_0}} = \frac{c}{f},$$

$f = \frac{\omega}{2\pi}$ being the *frequency*.

At this time we introduce the *magnetic intensity* field $\mathbf{H}$ as follows,

$$\mathbf{H} = \frac{\mathbf{B}}{\mu_0} \text{ (A/m)}.$$

The ratio

$$\frac{\widehat{E}_m e^{ikx_3}}{\frac{1}{\mu_0}\widehat{B}_m e^{ikx_3}} = \frac{\widehat{E}_m e^{ikx_3}}{-\frac{1}{\mu_0}(k/\omega)\widehat{E}_m e^{ikx_3}} = -\frac{\omega}{k}\mu_0$$

does not depend on x_3. For a wave travelling in the positive x_3-direction ($k = -\beta_0$) its value is

$$-\frac{\omega}{k}\mu_0 = \frac{\omega}{\beta_0}\mu_0 = c\mu_0 = \sqrt{\frac{\mu_0}{\varepsilon_0}} = \eta_0 \simeq 120\pi \ (\Omega),$$

while for a wave travelling in the negative x_3-direction ($k = \beta_0$) it is

$$-\frac{\omega}{k}\mu_0 = -\frac{\omega}{\beta_0}\mu_0 = -\eta_0.$$

Constant η_0 is called the *intrinsic wave impedance* for empty space.

The fact that η_0 is real means that electric and magnetic fields of plane waves in empty space are in phase with one another.

Case b) It is analogous to case a).

4.10 Wave polarization

The vector orientation of an electromagnetic wave in space is called *polarization*. It is usually described with reference to its electric field direction. For instance, the x_3-travelling uniform plane wave corresponding to fields (cf. Case a) in Sect. 4.9)

$$\mathbf{E}^{x_1}(x_3,t) = E^+_{mx_1}\cos(\omega t - \beta_0 x_3 + \phi^+)\mathbf{e}_1, \tag{4.47}$$

$$\mathbf{B}^{x_2}(x_3,t) = \sqrt{\mu_0\varepsilon_0}E^+_{mx_1}\cos(\omega t - \beta_0 x_3 + \phi^+)\mathbf{e}_2, \tag{4.48}$$

is said to be polarized in the x_1-direction (or simply x_1-polarized). Similarly, the plane wave with fields (cf. Case b))

$$\mathbf{E}^{x_2}(x_3,t) = E^+_{mx_2}\cos(\omega t - \beta_0 x_3 + \phi^+)\mathbf{e}_2 \tag{4.49}$$

$$\mathbf{B}^{x_1}(x_3,t) = \sqrt{\mu_0\varepsilon_0}E^+_{mx_2}\cos(\omega t - \beta_0 x_3 + \phi^+)\mathbf{e}_1 \tag{4.50}$$

is x_2-polarized. These waves are said *linearly* polarized, because the electric field vector in any plane orthogonal to x_3 describes straight-line paths as time pases.

Since Maxwell's equations are linear, a vector superposition of the above two linearly polarized plane waves will also provide a solution. However, this solution needs not to be linearly polarized, depending on the phase condition between the x_1 and the x_2-polarized electric field components. For example, for

$$\mathbf{E}^{x_1}(x_3,t) = E_{mx_1}\cos(\omega t - \beta_0 x_3)\mathbf{e}_1$$

$$\mathbf{E}^{x_2}(x_3,t) = E_{mx_2}\cos(\omega t - \beta_0 x_3)\mathbf{e}_2$$

propagating in phase and at the same frequency along the x_3-axis, their sum will produce a linearly polarized field $\mathbf{E}$. Indeed, the straight line described by the electric field vector can be found by inserting $x_3 = 0$ into the E^{x_1} and E^{x_2} expressions and making their ratio to eliminate ωt, namely,

$$\mathrm{E}^{x_2} = \frac{E_{mx_2}}{E_{mx_1}}\mathrm{E}^{x_1},$$

which is the equation of a straight line in the plane with E^{x_1} and E^{x_2} as the variables.

4.10 Wave polarization

On the other hand, if the two component fields were 90° out of the phase, more precisely, if

$$\mathbf{E}^{x_1}(x_3,t) = E_{mx_1}\cos(\omega t - \beta_0 x_3)\mathbf{e}_1$$
$$\mathbf{E}^{x_2}(x_3,t) = E_{mx_2}\cos(\omega t - \beta_0 x_3 + \pi/2)\mathbf{e}_2$$

the sum $\mathbf{E} = a_{x_1}\mathbf{E}^{x_1} + a_{x_2}\mathbf{E}^{x_2}$ would produce the spiraling locus of the $\mathbf{E}$ vector about the x_3-axis. In the fixed $x_3 = 0$ plane, the component fields are

$$\mathbf{E}^{x_1}(0,t) = E_{mx_1}\cos(\omega t)\mathbf{e}_1,$$

$$\mathbf{E}^{x_2}(0,t) = E_{mx_2}\cos(\omega t + \pi/2)\mathbf{e}_2 = -E_{mx_2}\sin(\omega t)\mathbf{e}_2 = -E_{mx_2}\sqrt{1-\cos^2(\omega t)}\mathbf{e}_2.$$

Eliminating ωt from these equation yields the locus of $\mathbf{E}$ in the $(\mathbf{E}^{x_1},\mathbf{E}^{x_2})$ plane:

$$\frac{E^{x_1 2}}{E_{mx_1}^2} + \frac{E^{x_2 2}}{E_{mx_2}^2} = 1. \tag{4.51}$$

This is the equation of an ellipse with principal axes of half lengths E_{mx_1} and E_{mx_2}. Thus, the tip of the total $\mathbf{E}$ vector describes an elliptical locus in any fixed x_3-plane as the wave moves, indicating the elliptical polarization of the wave. It is also evident that a circular polarization of the $\mathbf{E}$ vector would occur if $E_{mx_1} = E_{mx_2}$ in (4.51).

One may show that if the 90° phase condition between $\mathbf{E}^{x_1}$ and $\mathbf{E}^{x_2}$ were replaced by a general angle θ, then the polarization locus would be the ellipse

$$\frac{E^{x_1 2}}{E_{mx_1}^2} + \frac{E^{x_2 2}}{E_{mx_2}^2} - \frac{2\cos\theta}{E_{mx_1}E_{mx_2}}E^{x_1}E^{x_2} - \sin^2\theta = 0,$$

whose major axis is tilted, depending on the choice of θ.

Wave polarization is of practical importance in, for instance, radio communication because the power extracted by a receiving antenna from the arriving wave depends on the polarization of the latter.

5
Some solutions of Maxwell's equations in free space

In this chapter we solve Maxwell's equations in some particular situations. They correspond to special geometries and assumptions on the time dependence of charges and currents. Thus, we consider classical examples in electrostatics and magnetostatics. By choosing suitable sets Ω and Γ in the integral form of Maxwell's equations we will be able to obtain the electromagnetic fields.

5.1 Electrostatic fields

In this section, we solve some classical examples in electrostatics, i.e., in the case where there are only charges at rest. Hence, the magnetic field is null and we have to determine the electric field.

5.1.1 Electric scalar potential

To begin with, let us remark that, in electrostatics, Faraday's law becomes,

$$\int_l \mathbf{E} \cdot \mathbf{dl} = 0, \tag{5.1}$$

for all closed line l. Then one can show that there exists a scalar field, V, called *electric potential*, such that $\mathbf{E} = -\operatorname{grad} V$. Furthermore V is unique up to a constant. Field V can be defined as follows:

Let p be an arbitrary fixed point in $\mathscr{E}$. For any point $x \in \mathscr{E}$ let (see Sect. B.2.2 to recall the meaning of this integral)

$$V(x) := -\int_{l(p,x)} \mathbf{E} \cdot \mathbf{dl}, \tag{5.2}$$

where l is any curve joining p to x (see Fig. 5.1). We notice that V is well-defined because it does not depend on the particular curve $l(p,x)$, thanks to (5.1). Moreover, it is not difficult to prove that $\mathbf{E} = -\operatorname{grad} V$.

A. Bermúdez, D. Gómez, P. Salgado: *Mathematical Models and Numerical Simulation in Electromagnetism*. UNITEXT – La Matematica per il 3+2 74
DOI 10.1007/978-3-319-02949-8_5, © Springer International Publishing Switzerland 2014

5.1.2 Point charge Q at the origin. Coulomb's law

Let us obtain the electric field and the electric potential created by a charge Q (C) located at origin o. Let x be a point located at distance $r(x)$ from Q. Equation (4.22) yields

$$\int_{\partial B} \varepsilon_0 \mathbf{E} \cdot \mathbf{n} \, dA = Q,$$

for any ball B centered at the origin, where ∂B denotes its boundary.

For symmetry reasons, the electric field has to be of the form $\mathbf{E}(x) = E_r(r(x))\mathbf{e}_r(x)$, with

$$\mathbf{e}_r(x) = \frac{\mathbf{r}(x)}{r(x)},$$

where $\mathbf{r}(x)$ is the position vector of point x with respect to the origin o and $r(x) = |\mathbf{r}(x)|$. By replacing in (4.22) we get

$$\varepsilon_0 E_r(r) 4\pi r^2 = Q$$

and hence,

$$\mathbf{E}(x) = \frac{1}{4\pi\varepsilon_0} \frac{Q}{r^3(x)} \mathbf{r}(x). \tag{5.3}$$

The force exerted by this field on a charge Q' located at point x is given by

$$\mathbf{F}(x) = Q'\mathbf{E}(x) = \frac{1}{4\pi\varepsilon_0} \frac{QQ'}{r^3(x)} \mathbf{r}(x) = \frac{1}{4\pi\varepsilon_0} \frac{QQ'}{r^2(x)} \mathbf{e}_r(x),$$

which is the well-known *Coulomb's law* on electrostatic forces. According to this law, the force acting between two static electric charges is central, inverse-square, and proportional to the product of the charges.

In what follows we will compute the electric potential. By choosing l as the curve shown in Fig. 5.1, from (5.2) and (5.3), we get

$$V(x) = -\int_{l(p,x)} \frac{Q}{4\pi\varepsilon_0 r^2} \mathbf{e}_r \cdot d\mathbf{l}. \tag{5.4}$$

Let c be a parametrization of curve $l(p,x)$ in spherical coordinates (see Fig. B.4), namely,

$$\mathbf{c}(s) = (\mathrm{r}(s), \phi(s), \theta(s)), \ s \in [0,1].$$

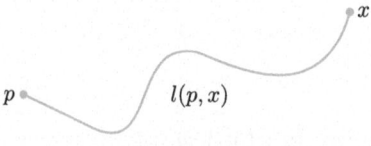

Fig. 5.1. Curve $l(p,x)$

5.1 Electrostatic fields

If we use the spherical coordinates to express the vector line element we have (see (B.59) in Appendix B)

$$d\mathbf{l} = dr\,\mathbf{e}_r + r\sin\theta\,d\phi\,\mathbf{e}_\phi + r\,d\theta\,\mathbf{e}_\theta$$

and then

$$d\mathbf{l}(c(s)) = \left(r'(s)\,\mathbf{e}_r + r(s)\sin\theta(s)\phi'(s)\,\mathbf{e}_\phi + r(s)\theta'(s)\,\mathbf{e}_\theta\right)ds.$$

As in this case the electric field is radial, only the radial component of the previous expression contributes to the integral (5.4) and then it reads

$$V(x) = -\int_0^1 \frac{Q}{4\pi\varepsilon_0 r^2(s)} r'(s)\,ds = -\int_{r(0)=r(p)}^{r(1)=r(x)} \frac{Q}{4\pi\varepsilon_0 r^2}\,dr$$

$$= \frac{Q}{4\pi\varepsilon_0}\left[\frac{1}{r(x)} - \frac{1}{r(p)}\right].$$

Usually, when the charge has a bounded support, we take p at infinity and hence

$$V(x) = \frac{Q}{4\pi\varepsilon_0 r(x)}. \tag{5.5}$$

Remark 5.1. We recall that $V \in W^{1,1}_{loc}(\mathscr{E})$ (see Sect. C.8). Furthermore it is the unique solution of the problem

$$-\Delta V = \frac{Q}{\varepsilon_0}\delta_0,$$

$$\lim_{r(x)\to\infty} V(x) = 0.$$

We notice that the first equation can be rewritten as

$$\text{div}\,\mathbf{E} = \frac{Q}{\varepsilon_0}\delta_0,$$

which is nothing but the local form of Gauss' law for electric charge.

5.1.3 Ball of radius r_0 with a uniform charge density ρ_V

Let us compute the electric field due to a ball of radius r_0 centered at the origin with a uniform volumetric charge density ρ_V (C/m^3) and total charge Q (C). In a similar way as in the previous example, the electric field can be obtained by applying the Gauss' law. Indeed, let x be a point with $r(x) \geq r_0$ and B a ball centered at the origin with radius $r(x)$. From (4.22) we deduce

$$\int_{\partial B} \varepsilon_0 \mathbf{E}\cdot\mathbf{n}\,dA = \rho_V \frac{4}{3}\pi r_0^3.$$

By symmetry, **E** must be of the form $\mathbf{E}(x) = E_r(r(x))\mathbf{e}_r(x)$. Hence,

$$4\pi r^2 \varepsilon_0 E_r(r) = \rho_V \frac{4}{3}\pi r_0^3,$$

from which it follows that

$$\mathbf{E}(x) = \frac{\rho_V r_0^3}{3\varepsilon_0 r^3(x)}\mathbf{r}(x) = \frac{\rho_V r_0^3}{3\varepsilon_0 r^2(x)}\mathbf{e}_r(x), \quad \text{for } r \geq r_0. \tag{5.6}$$

Now, let x be a point with $r(x) < r_0$ and B a ball centered at the origin with radius $r(x)$. Then Eq. (4.22) yields

$$4\pi r^2 \varepsilon_0 E_r(r) = \rho_V \frac{4}{3}\pi r^3$$

and hence,

$$\mathbf{E}(x) = \frac{\rho_V}{3\varepsilon_0}\mathbf{r}(x) = \frac{\rho_V r(x)}{3\varepsilon_0}\mathbf{e}_r(x), \quad \text{for } r(x) < r_0. \tag{5.7}$$

Notice that the electric field outside the ball is identical to that of a point charge $Q = 4/3(\rho_V \pi r_0^3)$ located at the origin.

Similar computations to those made in the case of the point charge lead us to the value of the electric potential for points x outside the ball, which is given by

$$V(x) = \frac{\rho_V r_0^3}{3\varepsilon_0 r(x)} \quad \text{if } r(x) \geq r_0. \tag{5.8}$$

However, for points x inside the ball the electric field is not the same as the field outside and hence one has to calculate the potential difference between a point x_0 on the surface of the ball and the point inside. Thus, for $r(x) < r_0$, we have

$$V(x) = -\int_{l(p,x_0)} \mathbf{E} \cdot d\mathbf{l} - \int_{l(x_0,x)} \mathbf{E} \cdot d\mathbf{l}$$

$$= \frac{\rho_V r_0^2}{3\varepsilon_0} + \frac{\rho_V}{6\varepsilon_0}(r_0^2 - r^2(x))$$

$$= \frac{\rho_V}{6\varepsilon_0}(3r_0^2 - r^2(x)). \tag{5.9}$$

5.1.4 A long line charge with uniform line density ρ_l

Let us consider an infinitely long, straight line in free space (see Fig. 5.2) with a uniform line charge density ρ_l (C/m). From symmetry, **E** must be radially directed and only dependent on the radial coordinate ρ, i.e.,

$$\mathbf{E}(x) = E_\rho(\rho(x))\mathbf{e}_\rho(x).$$

To obtain an expression for the electric field in a point x at a distance $\rho(x)$ from the line charge, let us consider a cylinder Ω, coaxial with this line, of radius $\rho(x)$ and

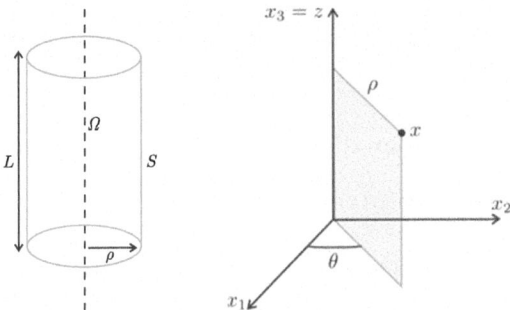

Fig. 5.2. Cylinder coaxial with the infinite charged line (left) and cylindrical coordinate system (right)

length L as shown in Fig. 5.2. From (4.22), we get

$$\int_{\partial\Omega} \varepsilon_0 E_\rho(\rho) \mathbf{e}_\rho \cdot \mathbf{n} dA = \rho_l L.$$

thanks to Gauss' law for electric charge.

Since $\mathbf{e}_\rho$ is orthogonal to $\mathbf{n}$ on the two bases of the cylinder and $\mathbf{n} = \mathbf{e}_\rho$ on its lateral surface S, we deduce

$$2\pi\rho L \varepsilon_0 E_\rho(\rho) = \rho_l L$$

and then,

$$\mathbf{E}(x) = \frac{\rho_l}{2\pi\varepsilon_0 \rho(x)} \mathbf{e}_\rho(x). \tag{5.10}$$

Notice that the electric field has the same magnitude at every point of the lateral boundary of the cylinder and that its length does not appear in this formula so we could choose a cylinder with arbitrary length.

Again, to compute the electric potential, let l be a curve as shown in Fig. 5.1. From (5.2) and (5.10), we get

$$V(x) = -\int_{l(p,x)} \frac{\rho_l}{2\pi\varepsilon_0 \rho(x)} \mathbf{e}_\rho(x) \cdot d\mathbf{l}. \tag{5.11}$$

Now let us consider c a parametrization of curve $l(p,x)$ in cylindrical coordinates (see Fig. 5.2), namely,

$$c(s) = (\rho(s), \theta(s), z(s)), \ s \in [0,1].$$

If we use cylindrical coordinates to express the vector line element we have (see (B.52) in Appendix B)

$$d\mathbf{l} = d\rho \, \mathbf{e}_\rho + \rho \, d\theta \, \mathbf{e}_\theta + dz \, \mathbf{e}_z.$$

We have

$$d\mathbf{l}(c(s)) = (\rho'(s) \mathbf{e}_\rho + \rho \, \theta'(s) \mathbf{e}_\theta + z'(s) \mathbf{e}_z) ds.$$

5 Some solutions of Maxwell's equations in free space

In this case only the azimuthal component of the previous expression contributes to the integral (5.11) which, as a consequence, reads

$$V(x) = -\int_0^1 \frac{\rho_l}{2\pi\varepsilon_0 \rho(x)} \rho'(s) ds = -\int_{\rho(0)=\rho(p)}^{\rho(1)=\rho(x)} \frac{\rho_l}{2\pi\varepsilon_0 \rho(x)} d\rho$$

$$= -\frac{\rho_l}{2\pi\varepsilon_0}[\ln(\rho(x)) - \ln(\rho(p))].$$

Notice that, for an infinite charged line, taking the potential reference point p at infinity is not a logical choice, since the local values of the potential would be infinite. This also occurs in some other academic problems for which one must use a zero potential reference value at a finite distance from the origin (see [38, 61]). In the present case, it it is convenient to take the zero of the potential to be at unit distance from the line, i.e, at $\rho(p) = 1$. Hence

$$V(x) = \frac{\rho_l}{2\pi\varepsilon_0}\ln\left(\frac{1}{\rho(x)}\right). \tag{5.12}$$

5.1.5 Infinite planar charge with uniform surface density ρ_S

Let us suppose the charged plane is the $x_2 x_3$-coordinate plane to be called S. Let ρ_S (C/m^2) be the (uniform) surface charge density of the plane. By symmetry, we expect the electric field on both sides of the plane to be a function of x_1 only, directed normal to the plane, and to point away from/towards the plane depending on whether ρ_S is positive/negative. That is, $\mathbf{E}(x) = E_1(x_1)\mathbf{e}_1$ if $x_1 > 0$ and $\mathbf{E}(x) = -E_1(-x_1)\mathbf{e}_1$ if $x_1 < 0$.

In order to apply Gauss' law, let us take Ω as a parallelepiped extending equally on both sides of the planar charge at distance a. Let us denote by S^+ and S^- the two faces parallel to the charged plane (see Fig. 5.3). From Gauss' law,

$$\int_{S^+} \varepsilon_0 E_1(x_1)\mathbf{e}_1 \cdot \mathbf{e}_1 + \int_{S^-} \varepsilon_0 E_1(-x_1)(-\mathbf{e}_1) \cdot (-\mathbf{e}_1) = bc\rho_S$$

and then,

$$2bc\varepsilon_0 E_1(a) = bc\rho_S,$$

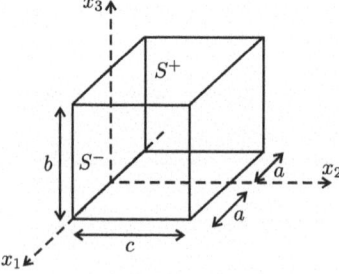

Fig. 5.3. Parallelepiped

which yields

$$\mathbf{E}(x) = \begin{cases} \dfrac{\rho_S}{2\varepsilon_0}\mathbf{e}_1 & \text{if } x_1 > 0, \\ -\dfrac{\rho_S}{2\varepsilon_0}\mathbf{e}_1 & \text{if } x_1 < 0. \end{cases} \qquad (5.13)$$

Notice that the electric field is uniform and does not depend on x_1.

From (5.2) and (5.13) we can easily deduce the expression for the electric potential

$$V(x) = -\frac{\rho_S}{2\varepsilon_0}|x_1|. \qquad (5.14)$$

In this case, a good reference point for the zero of the potential is the origin.

Remark 5.2. Let us notice that the normal component of the above electric field is discontinuous across S. Indeed, let us denote by $+$ the half space $x_1 > 0$ and by $-$ the half space $x_1 < 0$. Then their respective outward unit vectors normal to the plane boundary $x_1 = 0$ are $\mathbf{n}^+ = -\mathbf{e}_1$ and $\mathbf{n}^- = \mathbf{e}_1$. Thus we have,

$$[\mathbf{E}\cdot\mathbf{n}]_S := \mathbf{E}^+\cdot\mathbf{n}^+ + \mathbf{E}^-\cdot\mathbf{n}^- = \frac{\rho_S}{2\varepsilon_0}\mathbf{e}_1\cdot(-\mathbf{e}_1) + \left(-\frac{\rho_S}{2\varepsilon_0}\mathbf{e}_1\cdot\mathbf{e}_1\right) = -\frac{\rho_S}{\varepsilon_0}.$$

Moreover, let us compute the divergence of $\mathbf{E}$ in the distributional sense. For any $\varphi \in \mathscr{D}(\mathscr{E})$ we have

$$\langle \operatorname{div}\mathbf{E}, \varphi \rangle = -\int_{\mathscr{E}} \mathbf{E}\cdot\operatorname{grad}\varphi \, dV = -\int_{\operatorname{supp}\varphi} \mathbf{E}\cdot\operatorname{grad}\varphi \, dV$$

$$= -\int_{\operatorname{supp}\varphi \cap \{x_1>0\}} \mathbf{E}\cdot\operatorname{grad}\varphi \, dV - \int_{\operatorname{supp}\varphi \cap \{x_1<0\}} \mathbf{E}\cdot\operatorname{grad}\varphi \, dV$$

$$= -\int_S [\mathbf{E}\cdot\mathbf{n}]\varphi \, dA = \frac{1}{\varepsilon_0}\langle T_{\rho_S}, \varphi \rangle,$$

where we have used a Green's formula and T_{ρ_S} is the distribution defined by the surface charge ρ_S as follows:

$$\langle T_{\rho_S}, \varphi \rangle = \int_S \rho_S \varphi \, dA.$$

Hence, $\varepsilon_0 \operatorname{div}\mathbf{E} = T_{\rho_S}$ in the sense of distributions.

5.2 Magnetostatic fields

Magnetostatics refers to the static magnetic field created by static, i.e., time independent currents. We notice that, for time independent fields, Ampère's law becomes,

$$\int_l \frac{\mathbf{B}}{\mu_0}\cdot d\mathbf{l} = \int_S \mathbf{J}\cdot\mathbf{n}\, dA = \mathscr{I}_S. \qquad (5.15)$$

Moreover, we recall that $\mathbf{B}$ has to be divergence-free (Gauss' magnetic law).

74   5 Some solutions of Maxwell's equations in free space

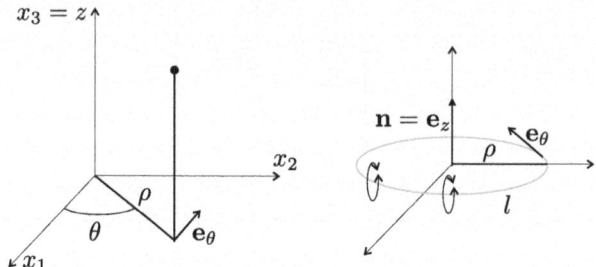

Fig. 5.4. Curve *l* used in calculus

5.2.1 Round toroidal core of circular cross section

Let us consider a n-turn, closely round toroidal core of circular cross section carrying a static current I (A). We will compute the magnetostatic field inside and outside the coil.

Since $\mathbf{J} = J_\rho \mathbf{e}_\rho + J_z \mathbf{e}_z$ (see Fig. 5.4), $\mathbf{B}$ must be of the form

$$\mathbf{B}(x) = B_\theta(\rho(x), z(x))\mathbf{e}_\theta(x),$$

for symmetry reasons and the Ampère's law in differential form, $\mathbf{curl\,B} = \mu_0 \mathbf{J}$.

Let *l* be a circumference of radius ρ inside the torus as in Fig. 5.4 and *S* be the corresponding circle. Since ρ and z are constant on *l*, we have,

$$\int_l B_\theta \mathbf{e}_\theta \cdot \mathbf{dl} = \int_l B_\theta \mathbf{e}_\theta \cdot \mathbf{e}_\theta dl = \mu_0 n I,$$

by Ampère's law, and then

$$2\pi\rho B_\theta(\rho,z) = \mu_0 n I,$$

which yields

$$\mathbf{B}(x) = \frac{\mu_0 n I}{2\pi\rho(x)}\mathbf{e}_\theta(x),$$

inside the torus.

Similar computations show that outside the torus the magnetic flux is zero. Indeed, it is enough to take a circumference $\rho = \bar{\rho}, z = \bar{z}$, outside the torus and a surface outside the torus having this circumference as its border, and to apply Ampère's law in integral form.

5.2.2 Infinitely long solenoid

Let us consider an infinitely long solenoid where the coil carries a static intensity I with n/d turns per unit length.

In this case $\mathbf{J}(x) = J_\theta(\rho(x))\mathbf{e}_\theta$ and

$$\mathbf{B}(x) = B_\rho(\rho(x))\mathbf{e}_\rho(x) + B_z(\rho(x))\mathbf{e}_z.$$

However, the existing symmetry requires that either $B_\rho \equiv 0$ or $B_z \equiv 0$.

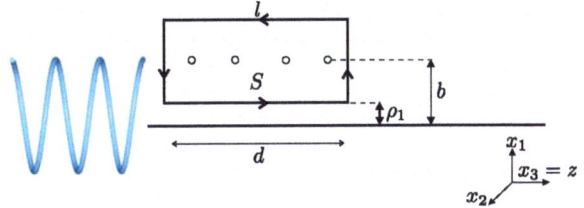

Fig. 5.5. Infinitely long solenoid

Since

$$\mathbf{curl\,u} = \left(\frac{1}{\rho}\frac{\partial u_z}{\partial \theta} - \frac{\partial u_\theta}{\partial z}\right)\mathbf{e}_\rho + \left(-\frac{\partial u_z}{\partial \rho} + \frac{\partial u_\rho}{\partial z}\right)\mathbf{e}_\theta + \left(\frac{1}{\rho}\frac{\partial}{\partial \rho}(\rho u_\theta) - \frac{1}{\rho}\frac{\partial u_\rho}{\partial \theta}\right)\mathbf{e}_z,$$

if $B_z \equiv 0$ then

$$\mathbf{curl\,B} = \mathbf{0},$$

which is a contradiction. Then $B_\rho = 0$ and $B_z \neq 0$ in which case,

$$\mathbf{B}(x) = B_z(\rho(x))\mathbf{e}_z.$$

By applying the Ampère's law to surface S in Fig. 5.5 we get

$$\int_l B_z(\rho)\mathbf{e}_z \cdot \mathbf{dl} = \int_l B_z(\rho)\mathbf{e}_z \cdot \mathbf{e}_z dl = \mu_0 nI.$$

The integral on the left-hand side is

$$\int_0^d B_z(\rho_1)\mathrm{d}z = B_z(\rho_1)d$$

because, as in the previous example, $\mathbf{B}$ outside the solenoid must be zero. Notice that the long solenoid can be considered as a torus the radius of which becomes infinite. Hence,

$$B_z(\rho_1)d = \mu_0 nI$$

and then

$$B_z(\rho_1) = \frac{\mu_0 nI}{d}$$

which is independent of ρ_1 if $\rho_1 < b$.

Therefore,

$$\mathbf{B}(x) = \frac{\mu_0 nI}{d}\mathbf{e}_z$$

inside the solenoid, i.e., for $\rho(x) < b$.

One can show that the following equality holds on the surface of the solenoid:

$$\mathbf{B}^+ \times \mathbf{n}^+ + \mathbf{B}^- \times \mathbf{n}^- = \mu_0 \frac{nI}{d}\mathbf{e}_\theta.$$

6
Maxwell's equations in material regions

In this chapter we rewrite the Maxwell's equations for the case where the propagation of the electromagnetic field takes place inside material media rather than in the empty space. With respect to the Maxwell's equations in the empty space, the presence of materials may lead to the following features:

- Different electric permittivity and magnetic permeability depending on each particular material.
- The appearance of dissipation mechanisms transforming electromagnetic energy into heat as those related to electrical resistance.

The former leads to a new phenomenon: *reflection*, while the latter yields a new term in the equations involving the first time derivative of the field also leading to another new phenomenon: *dissipation*. In order to get insight on these equations we will first consider a one-dimensional case as in Chap. 4.

6.1 A one-dimensional model

We first analyze wave reflection at the interface of two media.

6.1.1 Reflection and transmission of waves

Actually, Eq. (4.1) is only valid in homogeneous media. Otherwise, it has to be written in the form

$$\rho(x)\frac{\partial^2 u}{\partial t^2} - \frac{\partial}{\partial x}(K(x)\frac{\partial^2 u}{\partial x}) = 0.$$

Let us consider waves propagating in two infinite one-dimensional media in contact. More precisely, let us assume that functions ρ and K are of the following form:

$$\rho(x) = \begin{cases} \rho_1 & \text{if } x < 0, \\ \rho_2 & \text{if } x > 0. \end{cases} \qquad K(x) = \begin{cases} K_1 & \text{if } x < 0, \\ K_2 & \text{if } x > 0. \end{cases}$$

A. Bermúdez, D. Gómez, P. Salgado: *Mathematical Models and Numerical Simulation in Electromagnetism*. UNITEXT – La Matematica per il 3+2 74
DOI 10.1007/978-3-319-02949-8_6, © Springer International Publishing Switzerland 2014

6 Maxwell's equations in material regions

Then, the problem to be solved is

$$\rho_1 \frac{\partial^2 u}{\partial t^2} - K_1 \frac{\partial^2 u}{\partial x^2} \quad \text{if } x < 0,$$

$$\rho_2 \frac{\partial^2 u}{\partial t^2} - K_2 \frac{\partial^2 u}{\partial x^2} \quad \text{if } x > 0,$$

together with the transmission conditions,

$$u(0^-) = u(0^+),$$

$$K_1 \frac{\partial u}{\partial x}(0^-) = K_2 \frac{\partial u}{\partial x}(0^+).$$

Let us consider the initial-value problem corresponding to $u(x,0) = u_0(x)$, being u_0 a continuous function having compact support in $\mathbb{R}^-$ and $\frac{\partial u}{\partial t}(x,0) = 0$.

Some tedious computations lead to the following expression for the solution:

$$u(x,t) = \frac{1}{2} u_0(x + c_1 t) + \frac{1}{2} u_0(x - c_1 t) + \frac{1}{2} \mathscr{R} u_0(-(x + c_1 t)), \quad \text{if } x \leq 0, \qquad (6.1)$$

$$u(x,t) = \frac{1}{2} \mathscr{T} u_0(\frac{c_1}{c_2}(x - c_2 t)), \quad \text{if } x \geq 0, \qquad (6.2)$$

where

$$c_i = \sqrt{\frac{K_i}{\rho_i}}, \ i = 1, 2,$$

are the propagation velocity of signals in media 1 and 2, respectively, and

$$\mathscr{R} = \frac{\mathscr{Z}_1 - \mathscr{Z}_2}{\mathscr{Z}_1 + \mathscr{Z}_2}$$

and

$$\mathscr{T} = \frac{2\mathscr{Z}_1}{\mathscr{Z}_1 + \mathscr{Z}_2}$$

are the *reflection* and *transmission* coefficients, respectively.

In the above expressions, constants $\mathscr{Z}_i$, $i = 1, 2$, are called the *intrinsic impedances* of media 1 and 2, respectively. They are given by

$$\mathscr{Z}_i = \frac{K_i}{c_i} = \sqrt{\rho_i K_i}, \ i = 1, 2.$$

The third term on the right-hand side of (6.1) is the *reflected wave*, while the term on the right-hand side of (6.2) is the *transmitted wave*.

6.1.2 Damped waves

Let us introduce a *damping* mechanism in the one-dimensional wave equation (4.1) introduced in Sect. 4.1. It will be modelled by a term of the form

$$R\frac{\partial u}{\partial t}. \tag{6.3}$$

Let us notice that many dissipation effects in continuum mechanics depends on the velocity and hence are modelled by similar terms to (6.3).
Then the wave equation becomes

$$\rho\frac{\partial^2 u}{\partial t^2} + R\frac{\partial u}{\partial t} - K\frac{\partial^2 u}{\partial x^2} = 0. \tag{6.4}$$

Again, we look for a solution of the form

$$u(x,t) = \mathrm{Re}(\widetilde{U}(x)e^{i\omega t}), \tag{6.5}$$

ω being a real number.
By replacing the above expression for $u(x,t)$ in (6.4) we get

$$-\rho\omega^2\widetilde{U}(x)e^{i\omega t} + Ri\omega\widetilde{U}(x)e^{i\omega t} - K\frac{d^2\widetilde{U}}{dx^2}e^{i\omega t} = 0, \tag{6.6}$$

and then $\widetilde{U}$ must be a solution to the *Helmholtz's equation*,

$$-(\rho\omega^2 - iR\omega)\widetilde{U} - K\frac{d^2\widetilde{U}}{dx^2} = 0. \tag{6.7}$$

The general solution to this equation is

$$\widetilde{U}(x) = Be^{\gamma x} + Ce^{-\gamma x}, \tag{6.8}$$

where γ and $-\gamma$ are the two complex roots of the algebraic equation

$$K\gamma^2 + (\rho\omega^2 - iR\omega) = 0. \tag{6.9}$$

Let $\gamma = \alpha + \beta i$. Complex number γ is the *propagation constant*, while $k = \gamma/i = -i\gamma$ is the complex *wave number*. Real numbers α and β are called the *attenuation constant* and the *phase constant*, respectively. They are given by

$$\alpha = \frac{\omega}{\sqrt{2K}}\left[-\rho + \sqrt{\rho^2 + (\frac{R}{\omega})^2}\right]^{1/2}$$

$$\beta = \frac{\omega}{\sqrt{2K}}\left[\rho + \sqrt{\rho^2 + (\frac{R}{\omega})^2}\right]^{1/2}.$$

We notice that, if $R = 0$ (no damping), then $\gamma = i\omega\sqrt{\rho/K} = i\omega/c$ and the complex wave number becomes the wave number of undamped waves, namely, $\beta = \omega/c$.

By replacing (6.8) in (6.5) we get the following harmonic plane wave solution of the 1D damped model:

$$u(x,t) = \mathbb{B}e^{\alpha x}\cos(\omega t + \beta x + \varphi) + \mathbb{C}e^{-\alpha x}\cos(\omega t - \beta x + \psi). \tag{6.10}$$

We notice that, as the wave propagates in the 1D space, its amplitude exponentially decays. A similar behavior will be observed in Sect. 6.10 for the propagation of electromagnetic waves in electric conductors.

6.2 Conductors and insulators

In terms of their charge conduction properties materials can be classified as *insulators*, which possesses essentially no free electrons to provide currents under an electric field, and *conductors*, in which free, outer orbit electrons are readily available to produce a conduction current when an electric field is present.

6.2.1 Electrical conductivity. Ohm's law

Many conductors exhibit a linear dependence of **J** on the applied electric field **E**, namely,

$$\mathbf{J} = \sigma \mathbf{E}. \tag{6.11}$$

This is the *Ohm's law*. Parameter σ is called *electrical conductivity* of the material. The SI unit for σ is $(\Omega m)^{-1}$. The inverse of ohm is called *siemens* (S) so the SI unit for σ is also S/m.

Electrical conductivity depends on temperature. It is of the order of 10^8 (S/m) for the best conductors at room temperature to 10^{-16} (S/m) for the best insulators.

6.2.2 Electric polarization. Dielectrics

Insulators are incapable of carrying appreciable conduction currents when they are placed in a moderate external electric field. However, their molecules can undergo slight relative shift of positive and negative electric charges in opposite directions due to the Lorentz electric field forces. Such displacements are usually only a fraction of a molecular diameter in the material, but the sheer numbers of particles involved may cause a significant change in the electric fields from its value in the absence of the dielectric material. This phenomenon is called *electric polarization*. Insulators that can be polarized under the action of an applied magnetic field are called *dielectrics*.

In order to determine the macroscopic effect of polarization we start by introducing the so-called electric dipole and computing the electric field produced by it.

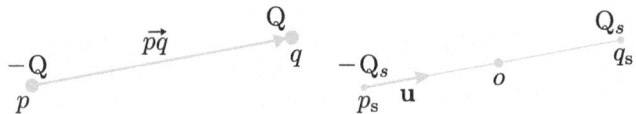

Fig. 6.1. Vector $\vec{pq}$ **Fig. 6.2.** Electric dipole

6.2.3 The electric dipole

Let us consider two charges of opposite sign, Q and $-Q$, located at points q and p in the affine space, respectively (see Fig. 6.1). We call *moment* the vector,

$$\mathbf{m} = Q\,\vec{pq} \quad (Cm).$$

Now let $\mathbf{u}$ be a unit vector and o any point in the affine space. For $s > 0$ we consider two charges $-Q_s$ and Q_s located at points $p = o - s\mathbf{u}$ and $q = o + s\mathbf{u}$ (see Fig. 6.2). Their momentum is given by

$$\mathbf{m}_s = 2Q_s s \mathbf{u}.$$

Let us take Q_s such that $2Q_s s = m$, m being a given constant. The limit when s goes to zero is called *electric dipole*.

6.2.3.1 Electric field created by an electric dipole

Our goal is to obtain the electric potential and then the electric field created by a dipole. For this purpose let us recall (see Remark 5.1) that, for $s > 0$, the electric potential created by charges $-Q_s$ and Q_s located at points p_s and q_s, respectively, is given as the solution of the Poisson's problem

$$-\Delta V_s = \frac{1}{\varepsilon_0}[-Q_s \delta_{p_s} + Q_s \delta_{q_s}], \qquad (6.12)$$

$$\lim_{r(x) \to \infty} V_s(x) = 0. \qquad (6.13)$$

Let us also recall that this Poisson's equation must be understood in the sense of distributions. Let us determine the limit of the distribution on the right-hand side as $s \to 0$.

For $\psi \in \mathscr{D}(\mathscr{E})$, we have

$$\langle -Q_s \delta_{p_s} + Q_s \delta_{q_s}, \psi \rangle = -Q_s \psi(o - s\mathbf{u}) + Q_s \psi(o + s\mathbf{u}) = \frac{m}{2s}(\psi(o + s\mathbf{u}) - \psi(o - s\mathbf{u})),$$

and passing to the limit we get

$$\lim_{s \to 0} \langle -Q_s \delta_{p_s} + Q_s \delta_{q_s}, \psi \rangle = m\frac{\partial \psi}{\partial \mathbf{u}}(o) = m\mathbf{u} \cdot \mathrm{grad}\,\psi(o).$$

It is quite natural to call *dipole moment* to vector $\mathbf{m} = m\mathbf{u}$.

Moreover, from the definition of distributional derivatives we deduce,

$$\lim_{s \to 0}\{-Q_s\delta(p_s) + Q_s\delta(q_s)\} = -\text{div}(\mathbf{m}\delta_o),$$

where $\mathbf{m}\delta_o$ is the vector distribution defined by $\langle \mathbf{m}\delta_o, \varphi \rangle := \mathbf{m} \cdot \varphi(o)$.
Indeed,

$$-\langle \text{div}(\mathbf{m}\delta_o), \psi \rangle = \langle \mathbf{m}\delta_o, \text{grad}\,\psi \rangle = \mathbf{m} \cdot \text{grad}\,\psi(o).$$

On the other hand, let us recall that the solution to the above Poisson's problem for the two charges $-Q_s$ and Q_s located at points p_s and q_s is given by (see (vii) in Sect. C.8):

$$V_s(x) = \frac{1}{4\pi\varepsilon_0}\left[\frac{-Q_s}{|x-o+su|} + \frac{Q_s}{|x-o-su|}\right]$$

$$= \frac{1}{4\pi\varepsilon_0}\left[\frac{-Q_s}{r(x+su)} + \frac{Q_s}{r(x-su)}\right].$$

It is straightforward to show that

$$\lim_{s \to 0} V_s = V \text{ in } L^1_{\text{loc}}(\mathcal{E}),$$

where

$$V(x) = -\frac{1}{4\pi\varepsilon_0}\mathbf{m}\,\text{grad}\left(\frac{1}{r(x)}\right)\cdot \mathbf{u} = \frac{1}{4\pi\varepsilon_0}\frac{\mathbf{m}\cdot\mathbf{r}(x)}{r^3(x)}.$$

Hence, V is the solution of the following Poisson's problem:

$$-\Delta V = -\frac{1}{\varepsilon_0}\text{div}(\mathbf{m}\delta_o),$$

$$\lim_{r(x)\to\infty} V(x) = 0.$$

Moreover, except for the origin o, the electric field created by the dipole is

$$\mathbf{E} = -\text{grad}\,V = -\frac{1}{4\pi\varepsilon_0}\text{grad}(\frac{\mathbf{m}\cdot\mathbf{r}}{r^3}) = -\frac{1}{4\pi\varepsilon_0}\text{grad}(\frac{\mathbf{r}}{r^3})^t\mathbf{m}$$

$$= -\frac{1}{4\pi\varepsilon_0}(\frac{1}{r^3}\text{grad}\,\mathbf{r} + \mathbf{r}\otimes\text{grad}\frac{1}{r^3})^t\mathbf{m} = \frac{1}{4\pi\varepsilon_0}(-\frac{1}{r^3}\mathbf{m} + \frac{3}{r^5}\mathbf{r}\cdot\mathbf{mr}),$$

where we have used equalities (B.20) and (B.19) and the fact that $\text{grad}\,\mathbf{r} = \mathbf{I}$.

6.2.3.2 Force and moment on an electric dipole

We compute the force and the moment exerted by an electric field on an electric dipole. The force is

$$\mathbf{F} = \lim_{s \to 0,\, 2sQ_s=m} \{-Q_s \mathbf{E}(o-s\mathbf{u}) + Q_s \mathbf{E}(o+s\mathbf{u})\}$$

$$= m \lim_{s \to 0} \frac{\mathbf{E}(o+s\mathbf{u}) - \mathbf{E}(o-s\mathbf{u})}{2s} = m \operatorname{\mathbf{grad}} \mathbf{E}(o)\mathbf{u} = \operatorname{\mathbf{grad}} \mathbf{E}(o)\mathbf{m}.$$

The moment with respect to point o is

$$\mathbf{M}_o = \lim_{s \to 0}(-Q_s \mathbf{r}(o-s\mathbf{u}) \times \mathbf{E}(o-s\mathbf{u}) + Q_s \mathbf{r}(o+s\mathbf{u}) \times \mathbf{E}(o+s\mathbf{u}))$$

$$= m \lim_{s \to 0} \frac{1}{2s}(s\mathbf{u} \times \mathbf{E}(o-s\mathbf{u}) + s\mathbf{u} \times \mathbf{E}(o+s\mathbf{u}))$$

$$= m\mathbf{u} \times \mathbf{E}(o) = \mathbf{m} \times \mathbf{E}(o).$$

6.2.4 The polarization vector

Let us come back to the study of electric polarization. Under the action of an electric field, a microscopic distribution of electric dipoles arises in the dielectric. The volumetric density of these dipoles is denoted by **P** and called *density of dipole moment* or *polarization vector*. Let us recall that the moment of an electric dipole **m** has units Cm. Hence **P** has units C/m^2.

In its turn, **P** creates an electric field the potential of which is the solution of the Poisson's problem (6.12)–(6.13),

$$-\Delta V_\mathbf{P} = -\frac{1}{\varepsilon_0} \operatorname{div} \mathbf{P}.$$

Hence, the electric field produced by the polarization is equivalent to the one created by the charge density

$$\rho_\mathbf{P} = -\operatorname{div} \mathbf{P} \; C/m^3.$$

Let us assume that a free charge density ρ_V exists in the polarized material. According to the Gauss' law for electric charge,

$$\operatorname{div} \varepsilon_0 \mathbf{E} = \rho_V + \rho_\mathbf{P} = \rho_V - \operatorname{div} \mathbf{P}, \tag{6.14}$$

and then,

$$\operatorname{div}(\varepsilon_0 \mathbf{E} + \mathbf{P}) = \rho_V.$$

The vector field

$$\mathbf{D} = \varepsilon_0 \mathbf{E} + \mathbf{P}, \tag{6.15}$$

is called *electric displacement*.

In terms of **D**, the Gauss' law for electric charge (6.14) becomes

$$\operatorname{div} \mathbf{D} = \rho_V. \tag{6.16}$$

6.2.5 Electric susceptibility and electric permittivity

Experiments reveal that many dielectric substances are essentially linear, meaning that **P** is proportional to the applied electric field, **E**. For such materials,

$$\mathbf{P} = \chi_e \varepsilon_0 \mathbf{E}, \tag{6.17}$$

where parameter χ_e is called *electric susceptibility* of the dielectric. The factor ε_0 is retained to make χ_e dimensionless.

According to the definition of **D** we have

$$\mathbf{D} = (1 + \chi_e)\varepsilon_0 \mathbf{E}. \tag{6.18}$$

It is usual to denote $1 + \chi_e$ by the dimensionless symbol

$$\varepsilon_r = 1 + \chi_e, \tag{6.19}$$

which is called *relative (electric) permittivity* of the region.

The *electric permittivity* of the material is defined by

$$\varepsilon = \varepsilon_r \varepsilon_0 = (1 + \chi_e)\varepsilon_0 \quad (\text{F/m}). \tag{6.20}$$

Thus $\mathbf{D} = \varepsilon \mathbf{E}$.

6.2.6 Electric Gauss' law for materials

It is

$$\text{div}\,\mathbf{D} = \rho_V, \tag{6.21}$$

in differential form, and in integral form

$$\int_\Gamma \mathbf{D} \cdot \mathbf{n}\,dA = \mathcal{Q}_\Omega, \tag{6.22}$$

where Γ is the boundary of any regular volume Ω and $\mathcal{Q}_\Omega$ is the total charge enclosed in Ω.

According to the previous discussion an electric field produces, in a dielectric material, an electric polarization **P** leading to an effect similar to an electric charge $\rho_\mathbf{P} = -\text{div}\,\mathbf{P}$.

If the electric field changes with time, field **P**, and then field $\rho_\mathbf{P}$, also varies with time. The time variation of $\rho_\mathbf{P}$ yields an electric current, called *polarization current*, the density of which, $\mathbf{J}_\mathbf{P}$, will be determined in the sequel.

For this purpose we notice that, by taking the divergence operator of Ampère's law we obtain,

$$\varepsilon_0 \frac{\partial \,\text{div}\,\mathbf{E}}{\partial t} = -\text{div}\,\mathbf{J},$$

and using the Gauss' law for electric charge (6.16),

$$\frac{\partial \rho_V}{\partial t} = -\operatorname{div} \mathbf{J}. \tag{6.23}$$

This equation is a local form of the *charge conservation principle*.

In a similar way, the polarization charge density and its corresponding polarization current density would satisfy

$$\operatorname{div} \mathbf{J_P} = -\frac{\partial \rho_P}{\partial t} = \operatorname{div} \frac{\partial \mathbf{P}}{\partial t}.$$

This equality leads to take as polarization current density,

$$\mathbf{J_P} = \frac{\partial \mathbf{P}}{\partial t}. \tag{6.24}$$

6.3 Magnetic materials

Magnetic materials are those that exhibit *magnetic polarization*, also called *magnetization*, under the presence of magnetic fields. Furthermore, certain materials can produce magnetic fields in the absence of external magnetic fields. They are called *permanent magnets*.

The development of magnetization in this section parallels that for electric polarization in the previous one. Just as the polarization density was used to represent the effect of electric dipoles on the electric field intensity, the magnetization density to be introduced below will account for the contributions of magnetic dipoles to the magnetic field intensity.

6.3.1 The magnetic dipole

Magnetization is due to the propensity of the atomic constituents of matter to behave as magnetic dipoles. The magnetic properties of a material are attributed to the tendency for the microscopic currents circulating on an atomic scale within the substance to align with an applied magnetic field **B**.

Three types of currents are associated with the atomic structure: those attributed to orbiting electrons, and those associated with electron spin and with nuclear spin. Each of these phenomena are equivalent to the circulation of a current I about a small closed path bounding a small surface S. This current defines a magnetic moment that creates a magnetic field.

Let meas(S) be the area of S. As meas(S) goes to zero and current I goes to infinity in such a way that the magnetic moment is constant, we get the so-called *magnetic dipole*. In what follows we will obtain a distribution modelling the magnetic dipole and then we will compute the created magnetic field.

Let us consider the small plane circular path c shown in Fig. 6.3 along which an electric current of intensity I is circulating.

Let S be the circle enclosed by c and **n** the unit normal vector, related by the right-hand rule to the direction of the current as in Stokes' theorem.

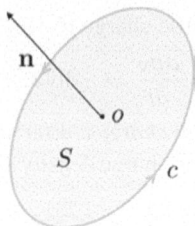

Fig. 6.3. Surface S and curve c

Let us define the *magnetic moment* of this current by

$$\mathbf{m} = I\,\text{meas}(S)\mathbf{n} = m\mathbf{n} \quad (\text{Am}^2). \tag{6.25}$$

We call *magnetic dipole* of amount $\mathbf{m}$ the limit of the above circuit when area $\text{meas}(S)$ goes to zero and, simultaneously, current I goes to infinity in such a way that the magnetic moment is constant and equal to $\mathbf{m}$. Our next goal is to write a distribution representing a magnetic dipole.

6.3.1.1 Magnetic field created by a magnetic dipole

Firstly, we consider the circuit for finite I and $\text{meas}(S)$. The current density can be modelled by the vector distribution

$$\mathbf{R} : \vec{\mathscr{D}}(\mathscr{E}) \longrightarrow \mathbb{R},$$

defined by

$$\langle \mathbf{R}, \varphi \rangle := \int_c I\varphi \cdot d\mathbf{l}, \tag{6.26}$$

for $\varphi \in \vec{\mathscr{D}}(\mathscr{E})$.

According to the Ampère's law, this current creates a magnetic field which, in the free space, satisfies,

$$\mathbf{curl}\,\mathbf{B} = \mu_0 \mathbf{R},$$

in the sense of distributions, i.e.,

$$\int_{\mathscr{E}} \mathbf{B} \cdot \mathbf{curl}\,\varphi \, dA = \mu_0 I \int_c \varphi \cdot d\mathbf{l} \quad \forall \varphi \in \vec{\mathscr{D}}(\mathscr{E}).$$

Now we pass to the limit in the right-hand side of this equality as $\text{meas}(S) \to 0$ and $I \to \infty$ with $I\,\text{meas}(S) = m$. By using Stokes' theorem we get

$$I\int_c \varphi \cdot d\mathbf{l} = I\int_S \mathbf{curl}\,\varphi \cdot \mathbf{n}\,dA = \frac{m}{\text{meas}(S)} \int_S \mathbf{curl}\,\varphi \cdot \mathbf{n}\,dA,$$

6.3 Magnetic materials

and then

$$\lim_{\substack{\text{meas}(S)\to 0 \\ I\to\infty \\ I\text{meas}(S)=m}} I\int_C \varphi\cdot d\mathbf{l} = \lim_{\text{meas}(S)\to 0} m\frac{1}{\text{meas}(S)}\int_S \text{curl}\,\varphi\cdot \mathbf{n}\,dA$$

$$= m\,\text{curl}\,\varphi(o)\cdot\mathbf{n} = \mathbf{m}\cdot\text{curl}\,\varphi(o),$$

where we have used the Localization theorem. Thus we have

$$\lim_{\substack{\text{meas}(S)\to 0 \\ I\to\infty \\ I\text{meas}(S)=m}} \mathbf{R} = \mathbf{S} := \text{curl}(\mathbf{m}\delta_o),$$

where $\mathbf{m}\delta_o$ is the vector distribution defined by $\langle \mathbf{m}\delta_o, \Psi\rangle := \mathbf{m}\cdot\Psi(o)$. Indeed,

$$\langle \text{curl}(\mathbf{m}\delta_o), \varphi\rangle := \langle \mathbf{m}\delta_o, \text{curl}\,\varphi\rangle = \mathbf{m}\cdot\text{curl}\,\varphi(o).$$

Thus, the magnetic field created by a magnetic dipole of amount $\mathbf{m}$ satisfies the equation,

$$\text{curl}\,\mathbf{B} = \mu_0\,\text{curl}(\mathbf{m}\delta_o), \tag{6.27}$$

in the sense of distributions, as well us $\text{div}\,\mathbf{B} = 0$.

Hence, the effect of a magnetic dipole of magnetic moment $\mathbf{m}$ is similar to that of the current density $\mathbf{J_m} = \text{curl}(\mathbf{m}\delta_o)$.

Since $\mathbf{B}$ is divergence-free we deduce the existence of a magnetic vector potential $\mathbf{A}$ such that $\mathbf{B} = \text{curl}\,\mathbf{A}$. For the purpose of uniqueness we choose $\mathbf{A}$ satisfying the *Coulomb's gauge*, i.e., $\text{div}\,\mathbf{A} = 0$. In terms of $\mathbf{A}$ Eq. (6.27) becomes

$$\text{curl}\,\text{curl}\,\mathbf{A} = \mu_0\,\text{curl}(\mathbf{m}\delta_o).$$

By using (B.30), it can be rewritten as

$$-\Delta\mathbf{A} = \mu_0\,\text{curl}(\mathbf{m}\delta_o).$$

Now by using the fundamental solution of Poisson's equation we can get the solution to this problem by convolution,

$$\langle \mathbf{T_A}, \Psi\rangle = \langle \mu_0\,\text{curl}(\mathbf{m}\delta_o), \int_{\mathcal{E}} \frac{\Psi(z)}{4\pi|x-z|}dz\rangle$$

$$= \mu_0\mathbf{m}\cdot\text{curl}_x\int_{\mathcal{E}} \frac{\Psi(z)}{4\pi|x-z|}dz\,|_{x=o} = \mu_0\mathbf{m}\cdot\int_{\mathcal{E}} \text{curl}_x\left(\frac{\Psi(z)}{4\pi|x-z|}\right)dz\,|_{x=o}$$

$$= \frac{\mu_0}{4\pi}\mathbf{m}\cdot\int_{\mathcal{E}} \nabla_x(\frac{1}{|x-z|})\,|_{x=o}\times\Psi(z)dz = \frac{\mu_0}{4\pi}\mathbf{m}\cdot\int_{\mathcal{E}} -\frac{x-z}{|x-z|^3}\,|_{x=o}\times\Psi(z)dz$$

$$= \frac{\mu_0}{4\pi}\mathbf{m}\cdot\int_{\mathcal{E}} \frac{\mathbf{r}(z)}{r^3(z)}\times\Psi(z)dz = \frac{\mu_0}{4\pi}\int_{\mathcal{E}} \Psi(z)\cdot\mathbf{m}\times\frac{\mathbf{r}(z)}{r^3(z)}dz,$$

6 Maxwell's equations in material regions

where we have used (B.26). Then we have

$$\mathbf{A} = \frac{\mu_0}{4\pi}\mathbf{m} \times \frac{\mathbf{r}}{r^3},$$

in the sense of distributions.

Except for the origin o, we have

$$\mathbf{B} = \mathrm{curl}\,\mathbf{A} = \frac{\mu_0}{4\pi}\mathrm{curl}(\mathbf{m} \times \frac{\mathbf{r}}{r^3}) = \frac{\mu_0}{4\pi}(\mathbf{m}\,\mathrm{div}(\frac{\mathbf{r}}{r^3}) - \mathrm{grad}\,(\frac{\mathbf{r}}{r^3})\mathbf{m})$$

$$= \frac{\mu_0}{4\pi}(-\frac{1}{r^3}\mathrm{grad}\,\mathbf{r} - \mathbf{r} \otimes \mathrm{grad}\,\frac{1}{r^3})\mathbf{m} = \frac{\mu_0}{4\pi}(-\frac{1}{r^3}\mathrm{grad}\,\mathbf{r} + 3\frac{1}{r^5}\mathbf{r} \otimes \mathbf{r})\mathbf{m}$$

$$= \frac{\mu_0}{4\pi}(-\frac{1}{r^3}\mathbf{m} + 3\frac{1}{r^5}\mathbf{r} \cdot \mathbf{m}\mathbf{r}),$$

where we have used (B.27), (B.18) and the fact that, for $x \neq o$ (see C.8),

$$\mathrm{div}(\frac{\mathbf{r}}{r^3}) = 0.$$

For instance, if $\mathbf{m} = m\mathbf{e}_3$ then

$$\mathbf{B}(x) = \frac{\mu_0}{4\pi}m\left(\frac{2\cos\phi(x)}{r^3(x)}\mathbf{e}_r + \frac{\sin\phi(x)}{r^3(x)}\mathbf{e}_\phi\right).$$

6.3.1.2 Force and moment of a magnetic field on a magnetic dipole

They are given, respectively, by

$$\mathbf{F} = [\mathrm{grad}\,\mathbf{B}(o)]^t\,\mathbf{m}, \qquad (6.28)$$

and

$$\mathbf{M}_o = \mathbf{m} \times \mathbf{B}(o). \qquad (6.29)$$

Indeed, according to Lorentz's law,

$$\mathbf{F} = \lim_{\substack{\mathrm{meas}(S) \to 0 \\ I \to \infty \\ I\mathrm{meas}(S) = m}} \int_C I\tau \times \mathbf{B}dl = \lim_{\substack{\mathrm{meas}(S) \to 0 \\ I \to \infty \\ I\mathrm{meas}(S) = m}} -I\int_C \mathbf{B} \times dl. \qquad (6.30)$$

In order to compute this limit we use the following formula of vector calculus:

$$\int_C \mathbf{w} \times dl = \int_S \mathrm{div}\,\mathbf{w}\mathbf{n}\,dA - \int_S \mathrm{grad}\,\mathbf{w}^t\mathbf{n}\,dA. \qquad (6.31)$$

To prove this formula we take any constant vector $\mathbf{a}$ and make the scalar product with the left-hand side of (6.31). We get

$$\mathbf{a} \cdot \int_C \mathbf{w} \times dl = \int_C \mathbf{a} \cdot \mathbf{w} \times dl = \int_C \mathbf{a} \times \mathbf{w} \cdot dl = \int_S \mathrm{curl}(\mathbf{a} \times \mathbf{w}) \cdot \mathbf{n}dA.$$

But,

$$\mathrm{curl}(\mathbf{v} \times \mathbf{w}) = \mathrm{grad}\,\mathbf{v}\mathbf{w} - \mathbf{w}\,\mathrm{div}\,\mathbf{v} + \mathbf{v}\,\mathrm{div}\,\mathbf{w} - \mathrm{grad}\,\mathbf{w}\mathbf{v},$$

and then,
$$\operatorname{curl}(\mathbf{a} \times \mathbf{w}) = \mathbf{a} \operatorname{div} \mathbf{w} - \operatorname{grad} \mathbf{w}\mathbf{a},$$
because **a** is constant.

By using this equality we get
$$\mathbf{a} \cdot \int_C \mathbf{w} \times d\mathbf{l} = \int_S \operatorname{div} \mathbf{w} \mathbf{a} \cdot \mathbf{n} \, dA - \int_S \operatorname{grad} \mathbf{w} \mathbf{a} \cdot \mathbf{n} \, dA = \mathbf{a} \cdot \int_S \operatorname{div} \mathbf{w} \mathbf{n} \, dA - \mathbf{a} \cdot \int_S \operatorname{grad} \mathbf{w}^t \mathbf{n} \, dA, \tag{6.32}$$
which finishes the proof.

By using formula (6.32) we obtain
$$-I \int_C \mathbf{B} \times d\mathbf{l} = -I \left(\int_S \operatorname{div} \mathbf{B} \mathbf{n} \, dA - \int_S \operatorname{grad} \mathbf{B}^t \mathbf{n} \, dA \right) = I \int_S \operatorname{grad} \mathbf{B}^t \mathbf{n} \, dA,$$
because $\operatorname{div} \mathbf{B} = 0$.

Hence,
$$\mathbf{F} = \lim_{\substack{\operatorname{meas}(S) \to 0 \\ I \to \infty \\ I \operatorname{meas}(S) = m}} -I \int_C \mathbf{B} \times d\mathbf{l} = \lim_{\operatorname{meas}(S) \to 0} \frac{m}{\operatorname{meas}(S)} \int_S \operatorname{grad} \mathbf{B}^t \mathbf{n} \, dA = m \operatorname{grad} \mathbf{B}^t(o) \mathbf{n}$$
$$= \operatorname{grad} \mathbf{B}^t(o) \mathbf{m}.$$
by using the Localization theorem.

Now, let us compute the moment of the magnetic force with respect to the position, o, of the magnetic dipole. We have,
$$\mathbf{M}_o = \lim_{\substack{\operatorname{meas}(S) \to 0 \\ I \to \infty \\ I \operatorname{meas}(S) = m}} I \int_C \mathbf{r} \times (\boldsymbol{\tau} \times \mathbf{B}) d\mathbf{l}$$
$$= \lim_{\operatorname{meas}(S) \to 0} \frac{m}{\operatorname{meas}(S)} \int_C (\mathbf{r} \cdot \mathbf{B} \boldsymbol{\tau} - \mathbf{r} \cdot \boldsymbol{\tau} \mathbf{B}) d\mathbf{l} = \lim_{\operatorname{meas}(S) \to 0} \frac{m}{\operatorname{meas}(S)} \int_C \mathbf{r} \cdot \mathbf{B} d\mathbf{l}$$
because $\mathbf{a} \times (\mathbf{b} \times \mathbf{c}) = \mathbf{a} \cdot \mathbf{c} \mathbf{b} - \mathbf{a} \cdot \mathbf{b} \mathbf{c}$ and $\mathbf{r} \perp \boldsymbol{\tau}$.

Let us recall the equality
$$\int_C \varphi d\mathbf{l} = -\int_S \operatorname{grad} \varphi \times \mathbf{n} \, dA \tag{6.33}$$
which can be proved from Stokes' theorem,
$$\int_C \mathbf{w} \cdot d\mathbf{l} = \int_S \operatorname{curl} \mathbf{w} \cdot \mathbf{n} \, dA, \tag{6.34}$$
by taking $\mathbf{w} = \varphi \mathbf{a}$, **a** being any vector in $\mathscr{V}$. Indeed, firstly we have
$$\mathbf{a} \cdot \int_C \varphi d\mathbf{l} = \int_S \operatorname{curl}(\varphi \mathbf{a}) \cdot \mathbf{n} \, dA.$$

Secondly,
$$\mathbf{curl}(\varphi \mathbf{a}) = \mathrm{grad}\,\varphi \times \mathbf{a} + \varphi\,\mathbf{curl}\,\mathbf{a} = \mathrm{grad}\,\varphi \times \mathbf{a},$$
because **a** is constant.

Hence,
$$\mathbf{a} \cdot \int_C \varphi \mathrm{d}\mathbf{l} = \int_S \mathrm{grad}\,\varphi \times \mathbf{a} \cdot \mathbf{n}\,\mathrm{d}A = \mathbf{a} \cdot \int_S \mathbf{n} \times \mathrm{grad}\,\varphi \mathrm{d}A,$$
which yields the equality.

By using (6.33) we get,
$$\mathbf{M}_o = \lim_{\mathrm{meas}(S) \to 0} \frac{m}{\mathrm{meas}(S)} \int_C \mathbf{r} \cdot \mathbf{B}\,\mathrm{d}\mathbf{l} = -m \lim_{\mathrm{meas}(S) \to 0} \frac{1}{\mathrm{meas}(S)} \int_S \mathrm{grad}(\mathbf{r} \cdot \mathbf{B}) \times \mathbf{n}\,\mathrm{d}A.$$

But,
$$\mathrm{grad}\,(\mathbf{r} \cdot \mathbf{B}) = \mathrm{grad}\,\mathbf{B}^t \mathbf{r} + \nabla \mathbf{r}^t \mathbf{B} = \mathrm{grad}\,\mathbf{B}^t \mathbf{r} + \mathbf{B},$$

and then,
$$\mathbf{M}_o = -m \lim_{\mathrm{meas}(S) \to 0} \frac{1}{\mathrm{meas}(S)} \int_S \left[\mathrm{grad}\,\mathbf{B}^t \mathbf{r} + \mathbf{B}\right) \times \mathbf{n}\right] \mathrm{d}A$$
$$= -m\left(\mathrm{grad}\,\mathbf{B}^t(o)\mathbf{r}(o) + \mathbf{B}(o)\right) \times \mathbf{n} = -m\mathbf{B}(o) \times \mathbf{n} = \mathbf{m} \times \mathbf{B}(o),$$
because, $\mathbf{r}(o) = \mathbf{0}$.

6.3.2 The magnetization vector

Let us come back to the study of magnetization.

Magnetic materials are those that exhibit magnetic polarization when they are subjected to an applied magnetic field. The magnetization phenomenon is represented by the alignment of the magnetic dipoles of the material with the applied magnetic field.

A magnetic material is presented by a number of magnetic dipoles and thus by many magnetic moments. In the absence of an applied magnetic field the magnetic dipoles and their corresponding electric loops are oriented in a random way so that, on a macroscopic scale, the vector sum of the magnetic moments is equal to zero.

When the magnetic material is subjected to an applied magnetic field represented by the magnetic flux density **B**, then a torque is exerted on the magnetic dipole given by $\mathbf{m} \times \mathbf{B}$. As a consequence, it tends to align in the direction of **B**. Thus, the resultant magnetic field at every point in the material would be larger than its corresponding value at the same point when the material is absent.

The macroscopic density of magnetic dipoles is represented by a vector field **M** called *magnetization density* or simply *magnetization*. Its SI unit is (Am^2/m^3= A/m).

The magnetic field created by **M** must satisfy,
$$\mathbf{curl}\,\mathbf{B_M} = \mu_0\,\mathbf{curl}\,\mathbf{M}.$$

In other words, the effect of a magnetic polarization represented by a *magnetization vector* **M** is similar to that of the current density,

$$\mathbf{J_M} = \operatorname{curl} \mathbf{M}.$$

Let us introduce the *magnetic intensity* field **H** defined by

$$\mathbf{H} = \frac{\mathbf{B}}{\mu_0} - \mathbf{M} \quad (\text{A/m}). \tag{6.35}$$

The field given by the product $\mu_0 \mathbf{M}$ is called *intensity of magnetization* (also referred to as *magnetic polarization*) and the SI unit is tesla (T).

6.3.3 Magnetic susceptibility. Magnetic permeability

We restrict ourselves to linear magnetic materials (an introduction to nonlinear magnetic materials including hysteresis can be found in Chap. 11). This means that **M** is proportional to **H**:

$$\mathbf{M} = \chi_m \mathbf{H}, \tag{6.36}$$

where the dimensionless parameter χ_m is called *magnetic susceptibility* of the material.

Inserting this equality in the definition of **H** we have

$$\mathbf{H} = \frac{\mathbf{B}}{\mu_0} - \chi_m \mathbf{H},$$

and then

$$\mathbf{B} = (1 + \chi_m)\mu_0 \mathbf{H}.$$

The parameter

$$\mu_r = 1 + \chi_m \tag{6.37}$$

is called the *relative (magnetic) permeability* of the material. Further, the *(magnetic) permeability* is defined by

$$\mu = \mu_r \mu_0, \tag{6.38}$$

so, finally,

$$\mathbf{B} = \mu \mathbf{H}. \tag{6.39}$$

6.3.4 Magnetic Gauss' law for materials

Recall that the electric Gauss' law for materials was developed by adding the effect of electric polarization charge density to the corresponding law for free-space.

The magnetic Gauss' law for materials can be developed analogously. However, no additional term is required in this case because no free magnetic charges exist physically in any known material. Thus **B** remains divergence-free in materials, that is,

$$\operatorname{div} \mathbf{B} = 0,$$

6.4 Ampère's law for materials

In free-space the **curl** of $\mathbf{B}/\mu_0$ has been expressed as the sum of a conduction current density $\mathbf{J}$ plus a displacement current density $\partial/\partial t\,(\varepsilon_0\mathbf{E})$ at any point, i.e.,

$$\mathbf{curl}\left(\frac{\mathbf{B}}{\mu_0}\right) = \mathbf{J} + \frac{\partial}{\partial t}(\varepsilon_0\mathbf{E}).$$

For materials, two additional types of currents may occur: $\mathbf{J_P} = \partial\mathbf{P}/\partial t$ and $\mathbf{J_M} = \mathbf{curl}\,\mathbf{M}$. They arise from electric and magnetic polarization effects, respectively. Adding these currents together leads to a revision of the Ampère's law for a material region, namely,

$$\mathbf{curl}\left(\frac{\mathbf{B}}{\mu_0}\right) = \mathbf{J} + \frac{\partial}{\partial t}(\varepsilon_0\mathbf{E}) + \frac{\partial\mathbf{P}}{\partial t} + \mathbf{curl}\,\mathbf{M}.$$

Grouping the **curl** terms on one hand and the time derivative terms on the other hand, we obtain

$$\mathbf{curl}\left(\frac{\mathbf{B}}{\mu_0} - \mathbf{M}\right) = \mathbf{J} + \frac{\partial}{\partial t}(\varepsilon_0\mathbf{E} + \mathbf{P}).$$

Recalling that

$$\mathbf{H} := \frac{\mathbf{B}}{\mu_0} - \mathbf{M} \quad \text{and} \quad \mathbf{D} := \varepsilon_0\mathbf{E} + \mathbf{P},$$

we finally get

$$\frac{\partial\mathbf{D}}{\partial t} - \mathbf{curl}\,\mathbf{H} = -\mathbf{J}, \qquad (6.40)$$

which is Ampère's law for materials.

6.5 Maxwell's equations in differential form for materials

We summarize the Maxwell's equations describing the propagation of the electromagnetic fields in materials:

$$\text{div}\,\mathbf{D} = \rho_V,$$

$$\text{div}\,\mathbf{B} = 0,$$

$$\frac{\partial\mathbf{D}}{\partial t} - \mathbf{curl}\,\mathbf{H} = -\mathbf{J},$$

$$\frac{\partial \mathbf{B}}{\partial t} + \operatorname{curl} \mathbf{E} = \mathbf{0},$$

$$\mathbf{D} = \varepsilon \mathbf{E},$$

$$\mathbf{B} = \mu \mathbf{H},$$

$$\mathbf{J} = \sigma \mathbf{E}.$$

6.6 Electric charge conservation equation

From Maxwell's equations we can obtain a relationship between charge and current densities. By applying the divergence operator to the Ampère's law we get

$$\frac{\partial \operatorname{div} \mathbf{D}}{\partial t} + \operatorname{div} \mathbf{J} = 0,$$

because $\operatorname{div} \mathbf{curl} \equiv 0$. Since $\operatorname{div} \mathbf{D} = \rho_V$, this equation yields

$$\frac{\partial \rho_V}{\partial t} + \operatorname{div} \mathbf{J} = 0.$$

This equality is nothing but a local form of the conservation of the electric charge. Indeed, let us consider a volume $\Omega \subset \mathscr{E}$. The total charge enclosed by Ω is given by

$$\mathscr{Q}_\Omega(t) = \int_\Omega \rho_V(x,t) dV \quad (C).$$

The time rate of change of $\mathscr{Q}_\Omega$ within Ω, namely,

$$\frac{d\mathscr{Q}_\Omega}{dt} = \int_\Omega \frac{\partial \rho_V}{\partial t}(x,t) dV,$$

is a measure of the current flowing into the set Ω. Moreover, the current flowing out of Ω through its boundary Γ is

$$\mathscr{I}_\Gamma = \int_\Gamma \mathbf{J} \cdot \mathbf{n} \, dA.$$

The postulate that electric charge is neither created nor destroyed yields,

$$-\frac{d\mathscr{Q}_\Omega}{dt} = \mathscr{I}_\Gamma,$$

that is,

$$\int_\Omega \frac{\partial \rho_V}{\partial t} dV + \int_\Gamma \mathbf{J} \cdot \mathbf{n} \, dA = 0.$$

By using the Gauss' theorem in the second integral we get,

$$\int_\Omega \left(\frac{\partial \rho_V}{\partial t} + \operatorname{div} \mathbf{J} \right) dV = 0,$$

and, since Ω is any volume in $\mathscr{E}$, this implies

$$\frac{\partial \rho_V}{\partial t} + \operatorname{div} \mathbf{J} = 0 \quad (\text{A/m}^3),$$

by the Localization theorem.

We notice that, for time independent field problems,

$$\operatorname{div} \mathbf{J} = 0.$$

This is the case of *direct currents* (see Chapt. 8).

6.7 Interface and boundary conditions

We assume that both the magnetic and the electric fields have locally finite energy, i.e., they belong to the space $\mathbf{L}^2_{\text{loc}}(\mathscr{E})$. Since $\operatorname{div} \mathbf{B} = 0$ and $\partial \mathbf{B}/\partial t + \operatorname{\mathbf{curl}} \mathbf{E} = \mathbf{0}$, we also have $\mathbf{B} \in \mathbf{H}_{\text{loc}}(\operatorname{div}, \mathscr{E})$ and $\mathbf{E} \in \mathbf{H}_{\text{loc}}(\operatorname{\mathbf{curl}}, \mathscr{E})$, at any time. This implies that on any surface S in the affine space $\mathscr{E}$, the normal component of $\mathbf{B}$ and the tangential component of $\mathbf{E}$ are continuous; more precisely,

$$\mathbf{B}^+ \cdot \mathbf{n}^+ + \mathbf{B}^- \cdot \mathbf{n}^- = 0, \tag{6.41}$$

and

$$\mathbf{E}^+ \times \mathbf{n}^+ + \mathbf{E}^- \times \mathbf{n}^- = 0. \tag{6.42}$$

Let us further assume that ρ_V and $\mathbf{J}$ are distributions of charge and current defined by functions in the spaces, say for instance, $\mathbf{L}^2_{\text{loc}}(\mathscr{E})$ and $\mathbf{L}^2_{\text{loc}}(\mathscr{E})$, respectively. Then $\mathbf{D} \in \mathbf{H}_{\text{loc}}(\operatorname{div}, \mathscr{E})$ and $\mathbf{H} \in \mathbf{H}_{\text{loc}}(\operatorname{\mathbf{curl}}, \mathscr{E})$ which means, in particular, that through any surface S we have (see Remark 5.2),

$$\mathbf{D}^+ \cdot \mathbf{n}^+ + \mathbf{D}^- \cdot \mathbf{n}^- = 0, \tag{6.43}$$

and

$$\mathbf{H}^+ \times \mathbf{n}^+ + \mathbf{H}^- \times \mathbf{n}^- = 0. \tag{6.44}$$

Moreover, if we have a surface charge density ρ_S, supported on a surface S, then

$$\operatorname{div} \mathbf{D} = T_{\rho_S},$$

in the sense of distributions, with

$$\langle T_{\rho_S}, \psi \rangle := \int_S \rho_S \psi \, dA,$$

and we have

$$\mathbf{D}^+ \cdot \mathbf{n}^+ + \mathbf{D}^- \cdot \mathbf{n}^- = -\rho_S. \tag{6.45}$$

6.7 Interface and boundary conditions

Similarly, if there is a surface current of density $\mathbf{J}_S$ on S, then the tangential component of $\mathbf{H}$ jumps across S. More precisely,

$$\mathbf{H}^+ \times \mathbf{n}^+ + \mathbf{H}^- \times \mathbf{n}^- = \mathbf{J}_S.$$

An important case corresponds to a surface S separating two different materials:

i) One of the regions, let say, the "$-$" region, is a perfect conductor ($\sigma \to \infty$). Since $\mathbf{J} = \sigma \mathbf{E}$, in order to have a finite density current $\mathbf{E}$ must be null. From the Faraday's law we also deduce $\mathbf{B} = \mathbf{0}$.
Hence, if there exist a surface charge density ρ_S and a surface current density $\mathbf{J}_S$ on surface S, we must have

$$\mathbf{D}^+ \cdot \mathbf{n}^+ = -\rho_S,$$
$$\mathbf{H}^+ \times \mathbf{n}^+ = \mathbf{J}_S.$$

ii) If both materials have finite conductivities then free surface currents should not exist on the interface, in which case,

$$\mathbf{H}^+ \times \mathbf{n}^+ + \mathbf{H}^- \times \mathbf{n}^- = \mathbf{0}.$$

6.7.1 Surface impedance boundary condition

If one the materials, let say the "–", is a good conductor but not a perfect conductor, then the electromagnetic field penetrates only a thin layer (see also 6.10.4 below). In the harmonic regime, an alternative to computing the field in the good conductor consists in writing a boundary condition on the interface between the two media. This boundary condition has been introduced by *Leontovich* and is called *surface impedance boundary condition*. It reads

$$\mathbf{E} \times \mathbf{n}^+ = -Z\mathbf{n}^+ \times (\mathbf{H} \times \mathbf{n}^+),$$

or, equivalently

$$\mathbf{n}^+ \times (\mathbf{E} \times \mathbf{n}^+) = Z(\mathbf{H} \times \mathbf{n}^+),$$

where $\mathbf{n}^+$ is the unit normal vector pointing inward the good conductor and the complex number Z is the *surface impedance*. In the Leontovich theory it is given by

$$Z = (1+i)\frac{1}{\sigma \delta},$$

with δ being the so-called *skin depth* (see Sect. 6.10.4), namely,

$$\delta = \sqrt{\frac{2}{\omega \mu \sigma}}.$$

6.8 Electromagnetic energy

The objective of this section is to introduce the electromagnetic energy and to write an energy balance. For this purpose we make the scalar product of the local form of the Ampère's law with the electric field **E** and integrate in the whole space. We get

$$\int_{\mathscr{E}} \frac{\partial \mathbf{D}}{\partial t} \cdot \mathbf{E}\, dV - \int_{\mathscr{E}} \mathbf{curl\, H} \cdot \mathbf{E}\, dV + \int_{\mathscr{E}} \mathbf{J} \cdot \mathbf{E}\, dV = 0.$$

Similarly, we make the scalar product of the local form of the Faraday's law with the magnetic field **H** and integrate in $\mathscr{E}$. We obtain

$$\int_{\mathscr{E}} \frac{\partial \mathbf{B}}{\partial t} \cdot \mathbf{H}\, dV + \int_{\mathscr{E}} \mathbf{curl\, E} \cdot \mathbf{H}\, dV = 0.$$

Then we add these two equalities to get

$$\int_{\mathscr{E}} \frac{\partial \mathbf{D}}{\partial t} \cdot \mathbf{E}\, dV + \int_{\mathscr{E}} \frac{\partial \mathbf{B}}{\partial t} \cdot \mathbf{H}\, dV - \int_{\mathscr{E}} \mathbf{curl\, H} \cdot \mathbf{E}\, dV + \int_{\mathscr{E}} \mathbf{curl\, E} \cdot \mathbf{H}\, dV$$
$$+ \int_{\mathscr{E}} \mathbf{J} \cdot \mathbf{E}\, dV = 0 \quad (6.46)$$

and, assuming that **J** has compact support,

$$\int_{\mathscr{E}} \frac{\partial \mathbf{D}}{\partial t} \cdot \mathbf{E}\, dV + \int_{\mathscr{E}} \frac{\partial \mathbf{B}}{\partial t} \cdot \mathbf{H}\, dV + \int_{\mathscr{E}} \mathbf{J} \cdot \mathbf{E}\, dV = 0, \quad (6.47)$$

because

$$\int_{\mathscr{E}} \mathbf{curl\, H} \cdot \mathbf{E}\, dV = \int_{\mathscr{E}} \mathbf{curl\, E} \cdot \mathbf{H}\, dV, \quad (6.48)$$

by using a Green's formula and the decay of fields at infinity. The sum of the first two terms represent the stored electromagnetic power at time t.

For *linear materials*, both electrically and magnetically, we have the constitutive laws

$$\mathbf{D} = \varepsilon \mathbf{E}, \quad \mathbf{B} = \mu \mathbf{H}, \quad (6.49)$$

where ε and μ are functions of x only. Then we can obtain an expression for the electromagnetic energy. Indeed,

$$\int_{\mathscr{E}} \frac{\partial \mathbf{D}}{\partial t} \cdot \mathbf{E}\, dV = \int_{\mathscr{E}} \varepsilon \frac{\partial \mathbf{E}}{\partial t} \cdot \mathbf{E}\, dV = \frac{1}{2} \frac{d}{dt} \int_{\mathscr{E}} \varepsilon |\mathbf{E}|^2\, dV = \frac{1}{2} \frac{d}{dt} \int_{\mathscr{E}} \mathbf{D} \cdot \mathbf{E}\, dV, \quad (6.50)$$

$$\int_{\mathscr{E}} \frac{\partial \mathbf{B}}{\partial t} \cdot \mathbf{H}\, dV = \int_{\mathscr{E}} \mu \frac{\partial \mathbf{H}}{\partial t} \cdot \mathbf{H}\, dV = \frac{1}{2} \frac{d}{dt} \int_{\mathscr{E}} \mu |\mathbf{H}|^2\, dV = \frac{1}{2} \frac{d}{dt} \int_{\mathscr{E}} \mathbf{B} \cdot \mathbf{H}\, dV. \quad (6.51)$$

6.8 Electromagnetic energy

By replacing these expressions in (6.47) and integrating in time from $t = t_1$ to $t = t_2$, we get

$$\frac{1}{2}\int_{\mathcal{E}} \mathbf{D}(x,t_2) \cdot \mathbf{E}(x,t_2)\,dV + \frac{1}{2}\int_{\mathcal{E}} \mathbf{B}(x,t_2) \cdot \mathbf{H}(x,t_2)\,dV$$

$$-\frac{1}{2}\int_{\mathcal{E}} \mathbf{D}(x,t_1) \cdot \mathbf{E}(x,t_1)\,dV - \frac{1}{2}\int_{\mathcal{E}} \mathbf{B}(x,t_1) \cdot \mathbf{H}(x,t_1)\,dV$$

$$+ \int_{t_1}^{t_2} \int_{\mathcal{E}} \mathbf{J} \cdot \mathbf{E}\,dV\,dt = 0. \qquad (6.52)$$

The functions of t

$$\mathfrak{E}_E(t) = \frac{1}{2}\int_{\mathcal{E}} \mathbf{D}(x,t) \cdot \mathbf{E}(x,t)\,dV$$

and

$$\mathfrak{E}_M(t) = \frac{1}{2}\int_{\mathcal{E}} \mathbf{B}(x,t) \cdot \mathbf{H}(x,t)\,dV$$

represent the *electric and magnetic energy* at time t, so equality (6.52) can be rewritten as

$$\mathfrak{E}(t_1) = \mathfrak{E}(t_2) + \int_{t_1}^{t_2} \int_{\mathcal{E}} \mathbf{J} \cdot \mathbf{E}\,dV\,dt, \qquad (6.53)$$

where $\mathfrak{E}(t) = \mathfrak{E}_E(t) + \mathfrak{E}_M(t)$ is the *energy of the electromagnetic field* at time t.

This equality says that the electromagnetic energy is conserved as far as no conductors exist in the space. Otherwise, it says that the electromagnetic energy decreases along time due to the (increasing in time) dissipation term

$$\int_{t_1}^{t_2} \int_{\mathcal{E}} \mathbf{J} \cdot \mathbf{E}\,dV\,dt,$$

that represents the electromagnetic energy transformed into heat by the conductors existing in the space (*Joule effect*). We notice that by using the Ohm's law

$$\int_{\mathcal{E}} \mathbf{J} \cdot \mathbf{E}\,dV = \int_{\mathcal{E}} \sigma|\mathbf{E}|^2\,dV \geq 0. \qquad (6.54)$$

Remark 6.1. In the more general case of nonlinear constitutive laws, in order to be able to give a formula for the electromagnetic energy it is enough to assume that there exist mappings Φ_E and Φ_H giving $\mathbf{E}$ and $\mathbf{H}$ from $\mathbf{D}$ and $\mathbf{B}$, respectively, and such that

$$\Phi_E(\mathbf{d}) = \operatorname{grad}\phi_E(\mathbf{d})$$

and

$$\Phi_H(\mathbf{b}) = \operatorname{grad}\phi_H(\mathbf{b})$$

for some scalar "potential" functions $\phi_E : \mathcal{V} \to \mathbb{R}$ and $\phi_H : \mathcal{V} \to \mathbb{R}$.

Indeed, by using the chain rule we deduce

$$\frac{\partial \phi_E(\mathbf{D})}{\partial t} = \Phi_E(\mathbf{D}) \cdot \frac{\partial \mathbf{D}}{\partial t} = \mathbf{E} \cdot \frac{\partial \mathbf{D}}{\partial t}$$

and, similarly,

$$\frac{\partial \phi_H(\mathbf{B})}{\partial t} = \Phi_H(\mathbf{B}) \cdot \frac{\partial \mathbf{B}}{\partial t} = \mathbf{H} \cdot \frac{\partial \mathbf{B}}{\partial t}.$$

By replacing these equalities in (6.50) and (6.51) we obtain again the conservation principle (6.53) but now the electric and the magnetic energy at time t are defined by

$$\mathfrak{E}_E(t) = \int_{\mathscr{E}} \phi_E(\mathbf{D}(x,t)) \, dV$$

and

$$\mathfrak{E}_M(t) = \int_{\mathscr{E}} \phi_H(\mathbf{B}(x,t)) \, dV,$$

respectively.

6.8.1 Energy balance in a bounded domain

Let us compute the balance of the electromagnetic field in a bounded domain Ω with Lipschitz-continuous boundary Γ.

We can repeat the first part of the computations above but integrating in Ω. Similar to (6.46), we get

$$\int_\Omega \frac{\partial \mathbf{D}}{\partial t} \cdot \mathbf{E} \, dV + \int_\Omega \frac{\partial \mathbf{B}}{\partial t} \cdot \mathbf{H} \, dV - \int_\Omega \operatorname{curl} \mathbf{H} \cdot \mathbf{E} \, dV + \int_\Omega \operatorname{curl} \mathbf{E} \cdot \mathbf{H} \, dV$$

$$+ \int_\Omega \mathbf{J} \cdot \mathbf{E} \, dV = 0, \qquad (6.55)$$

but now the analogous equality to (6.48) is no longer true. Instead, by using the Gauss' theorem we obtain the Green's formula,

$$-\int_\Omega \operatorname{curl} \mathbf{H} \cdot \mathbf{E} \, dV + \int_\Omega \operatorname{curl} \mathbf{E} \cdot \mathbf{H} \, dV = \int_\Omega \operatorname{div}(\mathbf{E} \times \mathbf{H}) \, dV$$

$$= \int_\Gamma \mathbf{E} \times \mathbf{H} \cdot \mathbf{n} \, dA, \qquad (6.56)$$

where $\mathbf{n}$ is an outward unit normal vector to Γ.

Vector field $\mathscr{P} := \mathbf{E} \times \mathbf{H}$ is called the *Poynting vector*. By replacing the previous equality in (6.55) we get

$$\int_\Omega \frac{\partial \mathbf{D}}{\partial t} \cdot \mathbf{E} \, dV + \int_\Omega \frac{\partial \mathbf{B}}{\partial t} \cdot \mathbf{H} \, dV + \int_\Omega \mathbf{J} \cdot \mathbf{E} \, dV = -\int_\Gamma \mathscr{P} \cdot \mathbf{n} \, dA. \qquad (6.57)$$

Let us notice that this equality also holds when the constitutive laws are nonlinear. The right-hand side represents the rate of ingoing electromagnetic energy through boundary Γ. The term involving $\mathbf{J}$ represents the rate of dissipated energy by *Joule effect*. Sometimes it is called the *eddy current losses*.

The first two terms on the left-hand side describe how the electromagnetic energy is stored or dissipated in Ω. Indeed, this energy is partially stored through reversible mechanisms and partially irreversibly transformed into heat because of hysteresis effects. The dissipated energy involved in this term is called *hysteresis losses* (see Chap. 11).

When it is possible to define an energy density function, analogous developments to the previous ones for the whole space can be done and we get the following conservation principle:

$$e^\Omega(t_1) = e^\Omega(t_2) + \int_{t_1}^{t_2} \int_\Omega \mathbf{J} \cdot \mathbf{E} \, dV \, dt + \int_{t_1}^{t_2} \int_\Gamma \mathscr{P} \cdot \mathbf{n} \, dA \, dt,$$

where $e^\Omega(t)$ denotes the electromagnetic energy in domain Ω at time t.

In the general (nonlinear) case we can still integrate Eq. (6.57) in time from t_1 to t_2. We get

$$\int_\Omega \int_{t_1}^{t_2} \frac{\partial \mathbf{D}}{\partial t} \cdot \mathbf{E} \, dt \, dV + \int_\Omega \int_{t_1}^{t_2} \frac{\partial \mathbf{B}}{\partial t} \cdot \mathbf{H} \, dt \, dV + \int_{t_1}^{t_2} \int_\Omega \mathbf{J} \cdot \mathbf{E} \, dV \, dt$$

$$= -\int_{t_1}^{t_2} \int_\Gamma \mathscr{P} \cdot \mathbf{n} \, dA \, dt. \qquad (6.58)$$

For each $x \in \Omega$, when t goes from t_1 to t_2 points $(\mathbf{D}(x,t), \mathbf{E}(x,t))$ describe a curve in the D–E "plane" and, similarly, $(\mathbf{B}(x,t), \mathbf{H}(x,t))$ describes a curve in the B–H "plane". They join "points" $(\mathbf{D}(x,t_1), \mathbf{E}(x,t_1))$ to $(\mathbf{D}(x,t_2), \mathbf{E}(x,t_2))$ and $(\mathbf{B}(x,t_1), \mathbf{H}(x,t_1))$ to $(\mathbf{B}(x,t_2), \mathbf{H}(x,t_2))$, respectively. Sometimes these curvilinear integrals are denoted, respectively, by

$$\oint_{\mathbf{D}(x,t_1)}^{\mathbf{D}(x,t_2)} \mathbf{E} \cdot d\mathbf{D} \quad \text{and} \quad \oint_{\mathbf{B}(x,t_1)}^{\mathbf{B}(x,t_2)} \mathbf{H} \cdot d\mathbf{B} \qquad (6.59)$$

and then (6.58) can be rewritten as

$$\int_\Omega (\oint_{\mathbf{D}(x,t_1)}^{\mathbf{D}(x,t_2)} \mathbf{E} \cdot d\mathbf{D}) \, dV + \int_\Omega (\oint_{\mathbf{B}(x,t_1)}^{\mathbf{B}(x,t_2)} \mathbf{H} \cdot d\mathbf{B}) \, dV + \int_{t_1}^{t_2} \int_\Omega \mathbf{J} \cdot \mathbf{E} \, dV \, dt$$

$$= -\int_{t_1}^{t_2} \int_\Gamma \mathscr{P} \cdot \mathbf{n} \, dA \, dt. \qquad (6.60)$$

The first two terms represent the stored and dissipated electromagnetic energy during the time interval $[t_1, t_2]$.

Let us consider a periodic solution of the Maxwell's system with period $T = t_2 - t_1$. Then the above curves are closed and the values of the integrals in (6.59) are the volumetric densities of the electric (respectively, magnetic) *hysteresis losses* along a period T at point $x \in \Omega$. Moreover, by making an integration by parts and using periodicity,

$$\int_{t_1}^{t_2} \frac{\partial \mathbf{D}}{\partial t}(x,t) \cdot \mathbf{E}(x,t) \, dt$$

$$= \mathbf{D}(x,t_2) \cdot \mathbf{E}(x,t_2) - \mathbf{D}(x,t_1) \cdot \mathbf{E}(x,t_1) - \int_{t_1}^{t_2} \frac{\partial \mathbf{E}}{\partial t}(x,t) \cdot \mathbf{D}(x,t) \, dt$$

$$= -\int_{t_1}^{t_2} \frac{\partial \mathbf{E}}{\partial t}(x,t) \cdot \mathbf{D}(x,t) \, dt$$

and, similarly,

$$\int_{t_1}^{t_2} \frac{\partial \mathbf{B}}{\partial t}(x,t) \cdot \mathbf{H}(x,t)\, dt$$

$$= \mathbf{B}(x,t_2) \cdot \mathbf{H}(x,t_2) - \mathbf{B}(x,t_1) \cdot \mathbf{H}(x,t_1) - \int_{t_1}^{t_2} \frac{\partial \mathbf{H}}{\partial t}(x,t) \cdot \mathbf{B}(x,t)\, dt$$

$$= -\int_{t_1}^{t_2} \frac{\partial \mathbf{H}}{\partial t}(x,t) \cdot \mathbf{B}(x,t)\, dt,$$

so (6.60) yields

$$-\int_\Omega \oint_{\mathbf{E}(x,t_1)}^{\mathbf{E}(x,t_2)} \mathbf{D} \cdot d\mathbf{E}\, dV - \int_\Omega \oint_{\mathbf{H}(x,t_1)}^{\mathbf{H}(x,t_2)} \mathbf{B} \cdot d\mathbf{H}\, dV + \int_{t_1}^{t_2} \int_\Omega \mathbf{J} \cdot \mathbf{E}\, dV\, dt$$

$$= -\int_{t_1}^{t_2} \int_\Gamma \mathscr{P} \cdot \mathbf{n}\, dA\, dt. \tag{6.61}$$

Let us assume isotropic electric (respectively, magnetic) materials. Then **E** and **D** (respectively, **H** and **B**) have the same direction. More specifically, let us suppose that

$$\mathbf{D}(x,t) = D(x,t)\mathbf{e}(x), \quad \mathbf{E}(x,t) = E(x,t)\mathbf{e}(x), \tag{6.62}$$

$$\mathbf{B}(x,t) = B(x,t)\mathbf{d}(x), \quad \mathbf{H}(x,t) = H(x,t)\mathbf{d}(x), \tag{6.63}$$

for some unit vectors $\mathbf{e}(x)$ and $\mathbf{d}(x)$. Then points $(D(x,t), E(x,t))$ (respectively, $(B(x,t), H(x,t))$) describe curves on the D–E plane (respectively, on the B–H plane). Besides

$$\int_{t_1}^{t_2} \frac{\partial \mathbf{D}}{\partial t}(x,t) \cdot \mathbf{E}(x,t)\, dt = \int_{t_1}^{t_2} \frac{\partial D}{\partial t}(x,t) E(x,t)\, dt = \oint_{D(x,t_1)}^{D(x,t_2)} E\, dD, \tag{6.64}$$

$$\int_{t_1}^{t_2} \frac{\partial \mathbf{B}}{\partial t}(x,t) \cdot \mathbf{H}(x,t)\, dt = \int_{t_1}^{t_2} \frac{\partial B}{\partial t}(x,t) H(x,t)\, dt = \oint_{B(x,t_1)}^{B(x,t_2)} H\, dB. \tag{6.65}$$

In the above mentioned periodic case, these curves are closed and included in the so-called electric (respectively, magnetic) *major hysteresis loop* (see Fig. 11.4), respectively. From Proposition B.3 the integrals in (6.64) and (6.65) are equal to the net areas enclosed by these curves. Therefore, according to the above discussion, these net areas are the hysteresis loss densities (electric and magnetic, respectively) along a cycle, at point x.

6.9 Electromagnetic force

The classical expression for the electromagnetic force is given by the Lorentz's law However this expression is not always valid. This is the case, for instance, when they are discontinuities in the magnetic permeability due to the presence of magnetic materials.

6.9 Electromagnetic force

In the opinion of several authors, the most reliable general method for the calculation of electromagnetic forces in *rigid media* is that based on the Maxwell stress tensor which is defined by

$$\Xi = \mathbf{E} \otimes \mathbf{D} + \mathbf{B} \otimes \mathbf{H} - \frac{1}{2}(\mathbf{E} \cdot \mathbf{D} + \mathbf{B} \cdot \mathbf{H})\mathbf{I}, \qquad (6.66)$$

where **I** denotes the identity tensor. According to this method, the electromagnetic force is given by

$$\mathbf{f} = \operatorname{div} \Xi, \qquad (6.67)$$

where the divergence operator must be understood in the sense of distributions, because, in general, some of the fields involved in (6.66) can be discontinuous. This is the case when the media is non-homogeneous and an electromagnetic property like, for instance, the magnetic permeability is discontinuous across the interface between two materials.

Notice that if **D** and **E** are collinear and the same is true for **B** and **H** then the Maxwell stress tensor is symmetric. Under these assumptions, we will compute this force in a subset Ω of the affine space including a volumetric charge density ρ_V, a free current density **J** and an internal material discontinuity across a surface S. Suppose that both the electric permittivity and the magnetic permeability may jump across S. We assume that in the rest of the domain they change smoothly.

Locally, surface S divides Ω into two parts to be called $+$ and $-$. Let us denote by $\mathbf{n}^+$ (respectively $\mathbf{n}^-$) the unit normal vector to S outward to $+$ (respectively, to $-$). Of course, $\mathbf{n}^+ = -\mathbf{n}^-$. Let us introduce the following notations:

$$[a] := a^+ + a^-, \quad \{a\} := a^+ - a^-.$$

Let φ be a vector field in $\overrightarrow{\mathscr{D}}(\Omega)$. By definition we have

$$\langle \operatorname{div} \Xi, \varphi \rangle = -\langle \Xi, \operatorname{\mathbf{grad}} \varphi \rangle$$

and then

$$\langle \mathbf{f}, \varphi \rangle = -\int_{\Omega} \left(\mathbf{E} \otimes \mathbf{D} + \mathbf{B} \otimes \mathbf{H} - \frac{1}{2}(\mathbf{E} \cdot \mathbf{D} + \mathbf{B} \cdot \mathbf{H})\mathbf{I} \right) \cdot \operatorname{\mathbf{grad}} \varphi \, dV. \qquad (6.68)$$

Let $\tilde{\Omega} = \Omega \setminus S$. We have

$$\langle \mathbf{f}, \varphi \rangle = -\int_{\tilde{\Omega}} \left(\mathbf{E} \otimes \mathbf{D} + \mathbf{B} \otimes \mathbf{H} - \frac{1}{2}(\mathbf{E} \cdot \mathbf{D} + \mathbf{B} \cdot \mathbf{H})\mathbf{I} \right) \cdot \operatorname{\mathbf{grad}} \varphi \, dV$$

$$= \int_{\tilde{\Omega}} \operatorname{div} \left(\mathbf{E} \otimes \mathbf{D} + \mathbf{B} \otimes \mathbf{H} - \frac{1}{2}(\mathbf{E} \cdot \mathbf{D} + \mathbf{B} \cdot \mathbf{H})\mathbf{I} \right) \cdot \varphi \, dV$$

$$- \int_{S} \left(\mathbf{D}^+ \cdot \mathbf{n}^+ \mathbf{E}^+ + \mathbf{H}^+ \cdot \mathbf{n}^+ \mathbf{B}^+ - \frac{1}{2}(\mathbf{E}^+ \cdot \mathbf{D}^+ + \mathbf{B}^+ \cdot \mathbf{H}^+)\mathbf{n}^+ \right) \cdot \varphi \, dA$$

$$- \int_{S} \left(\mathbf{D}^- \cdot \mathbf{n}^- \mathbf{E}^- + \mathbf{H}^- \cdot \mathbf{n}^- \mathbf{B}^- - \frac{1}{2}(\mathbf{E}^- \cdot \mathbf{D}^- + \mathbf{B}^- \cdot \mathbf{H}^-)\mathbf{n}^- \right) \cdot \varphi \, dA.$$

102   6 Maxwell's equations in material regions

and then

$$\langle \mathbf{f}, \varphi \rangle = \int_{\tilde{\Omega}} \mathrm{div}\left(\mathbf{E} \otimes \mathbf{D} + \mathbf{B} \otimes \mathbf{H} - \frac{1}{2}(\mathbf{E} \cdot \mathbf{D} + \mathbf{B} \cdot \mathbf{H})\mathbf{I}\right) \cdot \varphi \, dV$$
$$- \int_S \left([\mathbf{D} \cdot \mathbf{n}\,\mathbf{E}] + [\mathbf{H} \cdot \mathbf{n}\,\mathbf{B}] - \frac{1}{2}([(\mathbf{E} \cdot \mathbf{D})\mathbf{n}] + [(\mathbf{H} \cdot \mathbf{B})\mathbf{n}])\right) \cdot \varphi \, dA. \quad (6.69)$$

In what follows we assume that all materials in Ω are electrically and magnetically linear. We recall that surface S has a free charge surface density ρ_S and a free surface current $\mathbf{J}_S$. Hence,

$$[\mathbf{D} \cdot \mathbf{n}] = -\rho_S \quad \text{and} \quad [\mathbf{H} \times \mathbf{n}] = \mathbf{J}_S.$$

From the latter we deduce

$$\{\mathbf{H}_T\} = \mathbf{n}^+ \times \mathbf{H}^+ \times \mathbf{n}^+ - \mathbf{n}^- \times \mathbf{H}^- \times \mathbf{n}^-$$
$$= \mathbf{n}^+[\mathbf{H} \times \mathbf{n}] = \mathbf{n}^+ \times \mathbf{J}_S = -\mathbf{J}_S \times \mathbf{n}^+,$$

where we recall that $\mathbf{H}_T = \mathbf{n} \times \mathbf{H} \times \mathbf{n}$ is the tangential component of $\mathbf{H}$. Then we have

$$[\mathbf{D} \cdot \mathbf{n}\,\mathbf{E}] = \mathbf{D}^+ \cdot \mathbf{n}^+ (\mathbf{E}^+ \cdot \mathbf{n}^+\mathbf{n}^+ + \mathbf{E}_T^+) + \mathbf{D}^- \cdot \mathbf{n}^- (\mathbf{E}^- \cdot \mathbf{n}^-\mathbf{n}^- + \mathbf{E}_T^-)$$
$$= \mathbf{D}^+ \cdot \mathbf{n}^+ \mathbf{E}^+ \cdot \mathbf{n}^+\mathbf{n}^+ + \mathbf{D}^+ \cdot \mathbf{n}^+ \mathbf{E}_T^+ + \mathbf{D}^- \cdot \mathbf{n}^- \mathbf{E}^- \cdot \mathbf{n}^-\mathbf{n}^- + \mathbf{D}^- \cdot \mathbf{n}^- \mathbf{E}_T^-$$
$$= -\rho_S \mathbf{E}_T^+ + \mathbf{D}^+ \cdot \mathbf{n}^+ \frac{1}{\varepsilon^+} \mathbf{D}^+ \cdot \mathbf{n}^+\mathbf{n}^+ + \mathbf{D}^- \cdot \mathbf{n}^- \frac{1}{\varepsilon^-} \mathbf{D}^- \cdot \mathbf{n}^-\mathbf{n}^-$$
$$= -\rho_S \mathbf{E}_T^+ + \frac{1}{\varepsilon^+}|\mathbf{D}^+ \cdot \mathbf{n}^+|^2\mathbf{n}^+ - (\rho_S + \mathbf{D}^+ \cdot \mathbf{n}^+)\frac{1}{\varepsilon^-}(\rho_S + \mathbf{D}^+ \cdot \mathbf{n}^+)\mathbf{n}^+$$
$$= -\rho_S \mathbf{E}_T^+ + \{\frac{1}{\varepsilon}\}|\mathbf{D}^+ \cdot \mathbf{n}^+|^2\mathbf{n}^+ - \rho_S^2 \frac{1}{\varepsilon^-}\mathbf{n}^+ - 2\rho_S \mathbf{D}^+ \cdot \mathbf{n}^+ \frac{1}{\varepsilon^-}\mathbf{n}^+$$
$$= -\rho_S \mathbf{E}_T^+ + \{\frac{1}{\varepsilon}\}|\mathbf{D}^+ \cdot \mathbf{n}^+|^2\mathbf{n}^+ - \rho_S^2 \frac{1}{\varepsilon^-}\mathbf{n}^+ + 2\rho_S(\mathbf{D}^- \cdot \mathbf{n}^- + \rho_S)\frac{1}{\varepsilon^-}\mathbf{n}^+$$
$$= -\rho_S \mathbf{E}_T^+ + \{\frac{1}{\varepsilon}\}|\mathbf{D}^+ \cdot \mathbf{n}^+|^2\mathbf{n}^+ + \rho_S^2 \frac{1}{\varepsilon^-}\mathbf{n}^+ + 2\rho_S \mathbf{E}^- \cdot \mathbf{n}^-\mathbf{n}^+$$

because $\{\mathbf{E}_T\} = 0$ (see (6.42)). Similarly,

$$[\mathbf{H} \cdot \mathbf{n}\,\mathbf{B}] = [\mathbf{B} \cdot \mathbf{n}\,\mathbf{H}] = \mathbf{B}^+ \cdot \mathbf{n}^+ \{\mathbf{H} \cdot \mathbf{n}\mathbf{n} + \mathbf{H}_T\}$$
$$= \mathbf{B}^+ \cdot \mathbf{n}^+ \{\frac{1}{\mu}\mathbf{B} \cdot \mathbf{n}\mathbf{n}\} + \mathbf{B}^+ \cdot \mathbf{n}^+ \{\mathbf{H}_T\}$$
$$= \{\frac{1}{\mu}\}|\mathbf{B}^+ \cdot \mathbf{n}^+|^2\mathbf{n}^+ + \mathbf{B}^+ \cdot \mathbf{n}^+\mathbf{n}^+ \times \mathbf{J}_S,$$

6.9 Electromagnetic force 103

because $[\mathbf{B} \cdot \mathbf{n}] = 0$ (see (6.41)). Let us compute the other terms. We have

$$[(\mathbf{D} \cdot \mathbf{E})\mathbf{n}] = [\mathbf{D} \cdot \mathbf{n}\mathbf{E} \cdot \mathbf{n}\mathbf{n}] + [\mathbf{D}_T \cdot \mathbf{E}_T \mathbf{n}]$$

$$= [\frac{1}{\varepsilon}|\mathbf{D} \cdot \mathbf{n}|^2 \mathbf{n}] + [\varepsilon|\mathbf{E}_T|^2\mathbf{n}]$$

$$= \frac{1}{\varepsilon^+}|\mathbf{D}^+ \cdot \mathbf{n}^+|^2\mathbf{n}^+ + \frac{1}{\varepsilon^-}|\mathbf{D}^- \cdot \mathbf{n}^-|^2\mathbf{n}^- + \{\varepsilon\}|\mathbf{E}_T|^2\mathbf{n}^+$$

$$= \frac{1}{\varepsilon^+}|\mathbf{D}^+ \cdot \mathbf{n}^+|^2\mathbf{n}^+ + \frac{1}{\varepsilon^-}|\mathbf{D}^+ \cdot \mathbf{n}^+ + \rho_S|^2\mathbf{n}^- + \{\varepsilon\}|\mathbf{E}_T|^2\mathbf{n}^+$$

$$= \{\frac{1}{\varepsilon}\}|\mathbf{D}^+ \cdot \mathbf{n}^+|^2\mathbf{n}^+ + \frac{1}{\varepsilon^-}2\rho_S\mathbf{D}^+ \cdot \mathbf{n}^+\mathbf{n}^- + \frac{1}{\varepsilon^-}\rho_S^2\mathbf{n}^- + \{\varepsilon\}|\mathbf{E}_T|^2\mathbf{n}^+$$

$$= \{\frac{1}{\varepsilon}\}|\mathbf{D}^+ \cdot \mathbf{n}^+|^2\mathbf{n}^+ + \frac{1}{\varepsilon^-}2\rho_S(\mathbf{D}^- \cdot \mathbf{n}^- + \rho_S)\mathbf{n}^+ - \frac{1}{\varepsilon^-}\rho_S^2\mathbf{n}^+ + \{\varepsilon\}|\mathbf{E}_T|^2\mathbf{n}^+$$

$$= \{\frac{1}{\varepsilon}\}|\mathbf{D}^+ \cdot \mathbf{n}^+|^2\mathbf{n}^+ + 2\rho_S\mathbf{E}^- \cdot \mathbf{n}^-\mathbf{n}^+ + \frac{1}{\varepsilon^-}\rho_S^2\mathbf{n}^+ + \{\varepsilon\}|\mathbf{E}_T|^2\mathbf{n}^+.$$

Similarly,

$$[(\mathbf{B} \cdot \mathbf{H})\mathbf{n}] = [\mathbf{B} \cdot \mathbf{n}\mathbf{H} \cdot \mathbf{n}\mathbf{n}] + [\mathbf{B}_T \cdot \mathbf{H}_T \mathbf{n}]$$

$$= [\frac{1}{\mu}|\mathbf{B} \cdot \mathbf{n}|^2\mathbf{n}] + [\mu|\mathbf{H}_T|^2\mathbf{n}]$$

$$= \{\frac{1}{\mu}\}|\mathbf{B}^+ \cdot \mathbf{n}^+|^2\mathbf{n}^+ + \mu^+|\mathbf{H}_T^+|^2\mathbf{n}^+ + \mu^-|\mathbf{H}_T^-|^2\mathbf{n}^-.$$

By using the equality $\mathbf{H}_T^- = \mathbf{H}_T^+ - \mathbf{n}^+ \times \mathbf{J}_S$ we obtain

$$[(\mathbf{B} \cdot \mathbf{H})\mathbf{n}] = \{\frac{1}{\mu}\}|\mathbf{B}^+ \cdot \mathbf{n}^+|^2\mathbf{n}^+ + \mu^+|\mathbf{H}_T^+|^2\mathbf{n}^+ + \mu^-|\mathbf{H}_T^+ - \mathbf{n}^+ \times \mathbf{J}_S|^2\mathbf{n}^-$$

$$= \{\frac{1}{\mu}\}|\mathbf{B}^+ \cdot \mathbf{n}^+|^2\mathbf{n}^+ + \{\mu\}|\mathbf{H}_T^+|^2\mathbf{n}^+ + \mu^-\left(2\mathbf{H}_T^+ \cdot (\mathbf{n}^+ \times \mathbf{J}_S) - |\mathbf{n}^+ \times \mathbf{J}_S|^2\right)\mathbf{n}^+$$

$$= \{\frac{1}{\mu}\}|\mathbf{B}^+ \cdot \mathbf{n}^+|^2\mathbf{n}^+ + \{\mu\}|\mathbf{H}_T^+|^2\mathbf{n}^+ + \mu^-\left(2\mathbf{H}_T^+ \cdot (\mathbf{n}^+ \times \mathbf{J}_S) - |\mathbf{J}_S|^2\right)\mathbf{n}^+$$

$$= \{\frac{1}{\mu}\}|\mathbf{B}^+ \cdot \mathbf{n}^+|^2\mathbf{n}^+ + \{\mu\}|\mathbf{H}_T^+|^2\mathbf{n}^+ + \mu^-\left(2\mathbf{J}_S \cdot \mathbf{H}_T^+ \times \mathbf{n}^+ - |\mathbf{J}_S|^2\right)\mathbf{n}^+$$

$$= \{\frac{1}{\mu}\}|\mathbf{B}^+ \cdot \mathbf{n}^+|^2\mathbf{n}^+ + \{\mu\}|\mathbf{H}_T^+|^2\mathbf{n}^+ + \mu^-\left(2\mathbf{J}_S \cdot (-\mathbf{H}_T^- \times \mathbf{n}^- + \mathbf{J}_S) - |\mathbf{J}_S|^2\right)\mathbf{n}^+$$

$$= \{\frac{1}{\mu}\}|\mathbf{B}^+ \cdot \mathbf{n}^+|^2\mathbf{n}^+ + \{\mu\}|\mathbf{H}_T^+|^2\mathbf{n}^+ + \mu^-\left(-2\mathbf{J}_S \cdot \mathbf{H}_T^- \times \mathbf{n}^- + |\mathbf{J}_S|^2\right)\mathbf{n}^+$$

$$= \{\frac{1}{\mu}\}|\mathbf{B}^+ \cdot \mathbf{n}^+|^2\mathbf{n}^+ + \{\mu\}|\mathbf{H}_T^+|^2\mathbf{n}^+ + \left(2\mathbf{B}_T^- \cdot \mathbf{n}^+ \times \mathbf{J}_S + \mu^-|\mathbf{J}_S|^2\right)\mathbf{n}^+.$$

6 Maxwell's equations in material regions

By replacing the expressions above in (6.69) we get

$$\langle f, \varphi \rangle = \int_{\tilde{\Omega}} \operatorname{div}\left(\mathbf{E} \otimes \mathbf{D} + \mathbf{B} \otimes \mathbf{H} - \frac{1}{2}(\mathbf{E} \cdot \mathbf{D} + \mathbf{B} \cdot \mathbf{H})\mathbf{I}\right) \cdot \varphi \, dV$$

$$- \frac{1}{2} \int_S \left(\{\frac{1}{\varepsilon}\}|\mathbf{D}^+ \cdot \mathbf{n}^+|^2 + \{\frac{1}{\mu}\}|\mathbf{B}^+ \cdot \mathbf{n}^+|^2 - \{\varepsilon\}|\mathbf{E}_T|^2 - \{\mu\}|\mathbf{H}_T|^2\right)\mathbf{n}^+ \cdot \varphi \, dA$$

$$- \frac{1}{2} \int_S \left(-2\rho_S \mathbf{E}_T^+ + \frac{1}{\varepsilon^-}\rho_S^2 \mathbf{n}^+ + 2\rho_S \mathbf{E}^- \cdot \mathbf{n}^- \mathbf{n}^+ + 2\mathbf{B}^+ \cdot \mathbf{n}^+ \mathbf{n}^+ \times \mathbf{J}_S \right.$$

$$\left. - 2\mathbf{B}_T^- \cdot \mathbf{n}^+ \times \mathbf{J}_S \mathbf{n}^+ - \mu^- |\mathbf{J}_S|^2 \mathbf{n}^+\right) \cdot \varphi \, dA.$$

By using Maxwell's equations, we compute the field

$$\operatorname{div}\left(\mathbf{E} \otimes \mathbf{D} + \mathbf{B} \otimes \mathbf{H} - \frac{1}{2}(\mathbf{E} \cdot \mathbf{D} + \mathbf{B} \cdot \mathbf{H})\mathbf{I}\right)$$

in the open set $\tilde{\Omega}$ where all of the fields are supposed to be smooth. Firstly, we have

$$\operatorname{div}(\mathbf{E} \otimes \mathbf{D}) = \mathbf{E} \operatorname{div} \mathbf{D} + \operatorname{\mathbf{grad}} \mathbf{E} \mathbf{D}. \tag{6.70}$$

Moreover, by definition of the **curl** operator:

$$\operatorname{\mathbf{curl}} \mathbf{E} \times \mathbf{E} = (\operatorname{\mathbf{grad}} \mathbf{E} - \operatorname{\mathbf{grad}} \mathbf{E}^t)\mathbf{E} = \operatorname{\mathbf{grad}} \mathbf{E} \mathbf{E} - \frac{1}{2} \operatorname{\mathbf{grad}} |\mathbf{E}|^2.$$

By using this equality and the Gauss' law for the electric charge in (6.70) we deduce

$$\operatorname{div}(\mathbf{E} \otimes \mathbf{D}) = \rho_V \mathbf{E} + \operatorname{\mathbf{curl}} \mathbf{E} \times \mathbf{D} + \frac{\varepsilon}{2} \operatorname{\mathbf{grad}} |\mathbf{E}|^2, \tag{6.71}$$

from which the following equality holds

$$\operatorname{div}(\mathbf{E} \otimes \mathbf{D} - \frac{1}{2}\mathbf{D} \cdot \mathbf{E} \mathbf{I}) = \rho_V \mathbf{E} + \operatorname{\mathbf{curl}} \mathbf{E} \times \mathbf{D} - \frac{1}{2}|\mathbf{E}|^2 \operatorname{\mathbf{grad}} \varepsilon. \tag{6.72}$$

Finally, by using Faraday's law we get

$$\operatorname{div}(\mathbf{E} \otimes \mathbf{D} - \frac{1}{2}\mathbf{D} \cdot \mathbf{E} \mathbf{I}) = \rho_V \mathbf{E} + \mathbf{D} \times \frac{\partial \mathbf{B}}{\partial t} - \frac{1}{2}|\mathbf{E}|^2 \operatorname{\mathbf{grad}} \varepsilon. \tag{6.73}$$

Similarly, we transform the terms involving the magnetic field. Since $\mathbf{B} = \mu \mathbf{H}$, we have

$$\operatorname{div}(\mathbf{B} \otimes \mathbf{H}) = \operatorname{div}(\mathbf{H} \otimes \mathbf{B}) = \mathbf{H} \operatorname{div} \mathbf{B} + \operatorname{\mathbf{grad}} \mathbf{H} \mathbf{B} = \operatorname{\mathbf{grad}} \mathbf{H} \mathbf{B}. \tag{6.74}$$

By definition of the **curl** operator:

$$\operatorname{\mathbf{curl}} \mathbf{H} \times \mathbf{B} = (\operatorname{\mathbf{grad}} \mathbf{H} - \operatorname{\mathbf{grad}} \mathbf{H}^t)\mathbf{B} = \operatorname{\mathbf{grad}} \mathbf{H} \mathbf{B} - \frac{\mu}{2} \operatorname{\mathbf{grad}} |\mathbf{B}|^2.$$

By using this equality in (6.74) we obtain

$$\text{div}(\mathbf{B} \otimes \mathbf{H}) = \mathbf{curl}\,\mathbf{H} \times \mathbf{B} + \frac{\mu}{2}\,\text{grad}\,|\mathbf{H}|^2, \tag{6.75}$$

and using the Ampère's law

$$\text{div}(\mathbf{B} \otimes \mathbf{H}) = \frac{\partial \mathbf{D}}{\partial t} \times \mathbf{B} + \mathbf{J} \times \mathbf{B} + \frac{\mu}{2}\,\text{grad}\,|\mathbf{H}|^2. \tag{6.76}$$

Therefore,

$$\text{div}\left(\mathbf{B} \otimes \mathbf{H} - \frac{1}{2}\mathbf{B}\cdot\mathbf{H}\mathbf{I}\right) = \frac{\partial \mathbf{D}}{\partial t} \times \mathbf{B} + \mathbf{J} \times \mathbf{B} + \frac{\mu}{2}\,\text{grad}\,|\mathbf{H}|^2 - \frac{1}{2}\,\text{grad}(\mu|\mathbf{H}|^2)$$

$$= \frac{\partial \mathbf{D}}{\partial t} \times \mathbf{B} + \mathbf{J} \times \mathbf{B} - \frac{1}{2}|\mathbf{H}|^2\,\text{grad}\,\mu. \tag{6.77}$$

By adding (6.73) and (6.77) we obtain

$$\text{div}\left(\mathbf{E} \otimes \mathbf{D} + \mathbf{B} \otimes \mathbf{H} - \frac{1}{2}(\mathbf{E}\cdot\mathbf{D} + \mathbf{B}\cdot\mathbf{H})\mathbf{I}\right) = \rho_V \mathbf{E} + \mathbf{J} \times \mathbf{B} + \frac{\partial}{\partial t}(\mathbf{D} \times \mathbf{B})$$

$$- \frac{1}{2}|\mathbf{E}|^2\,\text{grad}\,\varepsilon - \frac{1}{2}|\mathbf{H}|^2\,\text{grad}\,\mu.$$

Finally, the electromagnetic density force in domain Ω is the vector distribution,

$$\mathbf{f} = \rho_V \mathbf{E} + \mathbf{J} \times \mathbf{B} + \frac{\partial}{\partial t}(\mathbf{D} \times \mathbf{B}) - \frac{1}{2}|\mathbf{E}|^2\,\text{grad}\,\varepsilon - \frac{1}{2}|\mathbf{H}|^2\,\text{grad}\,\mu$$

$$- \frac{1}{2}\left(\{\frac{1}{\varepsilon}\}|\mathbf{D}^+\cdot\mathbf{n}^+|^2 + \{\frac{1}{\mu}\}|\mathbf{B}^+\cdot\mathbf{n}^+|^2 - \{\varepsilon\}|\mathbf{E}_T|^2 - \{\mu\}|\mathbf{H}_T|^2\right)\mathbf{n}^+\delta_S$$

$$- \frac{1}{2}\left(-2\rho_S \mathbf{E}_T^+ + \frac{1}{\varepsilon^-}\rho_S{}^2\mathbf{n}^+ + 2\rho_S \mathbf{E}^-\cdot\mathbf{n}^-\mathbf{n}^+ + 2\mathbf{B}^+\cdot\mathbf{n}^+\mathbf{n}^+ \times \mathbf{J}_S\right.$$

$$\left. - 2\mathbf{B}_T^-\cdot\mathbf{n}^+ \times \mathbf{J}_S \mathbf{n}^+ - \mu^-|\mathbf{J}_S|^2\mathbf{n}^+\right)\delta_S,$$

where δ_S denotes de Dirac distribution on surface S. More precisely, given a vector field $\mathbf{g}$ defined on S,

$$\langle \mathbf{g}\delta_S, \varphi \rangle = \int_S \mathbf{g}(x) \cdot \varphi(x)\,dA.$$

6.10 Plane waves in an unbounded conductive region

In Sect. 4.9 we have computed harmonic plane wave solutions of Maxwell's equations in the empty space. Now we make similar computations for material regions. The presence of conductors leads to dissipation of energy and damping phenomena.

6.10.1 Harmonic regime

Let us recall that, in the harmonic regime, we suppose that the sources are of the form

$$\rho_V(x,t) = \mathrm{Re}(e^{i\omega t}\widehat{\rho}_V(x)),$$
$$\mathbf{J}(x,t) = \mathrm{Re}(e^{i\omega t}\widehat{\mathbf{J}}(x)),$$

and look for solutions with the same time dependency. By replacing in Maxwell's equations we get

$$\mathrm{div}(\varepsilon\widehat{\mathbf{E}}) = \widehat{\rho}_V,$$
$$\mathrm{div}\,\widehat{\mathbf{B}} = 0,$$
$$\mathrm{curl}\,\widehat{\mathbf{E}} = -i\omega\mu\widehat{\mathbf{H}},$$
$$\mathrm{curl}\,\widehat{\mathbf{H}} = \widehat{\mathbf{J}} + i\omega\varepsilon\widehat{\mathbf{E}} = \sigma\widehat{\mathbf{E}} + i\omega\varepsilon\widehat{\mathbf{E}} = i\omega\left(\varepsilon - \frac{\sigma}{\omega}i\right)\widehat{\mathbf{E}}.$$

These equations are analogous to the harmonic Maxwell's equations for free space. In particular, the above Faraday's and Ampère's equations can be obtained from the respective ones for the empty space by replacing,

$$\mu_0 \text{ with } \mu \text{ and } \varepsilon_0 \text{ with } \varepsilon - \frac{\sigma}{\omega}i.$$

Hence, by eliminating successively $\widehat{\mathbf{H}}$ and $\widehat{\mathbf{E}}$, we get

$$\Delta\widehat{\mathbf{E}} + \omega^2\mu\left(\varepsilon - \frac{\sigma}{\omega}i\right)\widehat{\mathbf{E}} = 0,$$
$$\Delta\widehat{\mathbf{H}} + \omega^2\mu\left(\varepsilon - \frac{\sigma}{\omega}i\right)\widehat{\mathbf{H}} = 0.$$

As in free-space, we look for plane wave solutions. By making the previous replacements in the computations of Sect. 4.9, Eq. (4.45) becomes,

$$\widehat{\mathbf{E}}(x) = \left[\widehat{E}_m^+ e^{-i\omega\sqrt{\mu(\varepsilon-\frac{\sigma}{\omega}i)}x_3} + \widehat{E}_m^- e^{i\omega\sqrt{\mu(\varepsilon-\frac{\sigma}{\omega}i)}x_3}\right]\mathbf{e}_1.$$

6.10.2 Propagation constant. Attenuation constant. Phase constant

The complex number $\omega\sqrt{\mu\left(\varepsilon - \frac{\sigma}{\omega}i\right)}$ replaces parameter $\beta_0 = \omega\sqrt{\mu_0\varepsilon_0}$ of the free-space. The number,

$$\gamma = i\omega\sqrt{\mu\left(\varepsilon - \frac{\sigma}{\omega}i\right)},$$

is called the *propagation constant*. It can be separated into its real and imaginary parts,

$$\gamma = \alpha + i\beta \quad (\mathrm{m}^{-1}).$$

The real part α is called the *attenuation constant* while β is termed the *phase constant* of the plane wave. The following positive values of α and β are obtained:

$$\alpha = \frac{\omega\sqrt{\mu\varepsilon}}{\sqrt{2}} \left[\sqrt{1+\left(\frac{\sigma}{\omega\varepsilon}\right)^2} - 1\right]^{\frac{1}{2}} \quad (\text{m}^{-1}),$$

$$\beta = \frac{\omega\sqrt{\mu\varepsilon}}{\sqrt{2}} \left[\sqrt{1+\left(\frac{\sigma}{\omega\varepsilon}\right)^2} + 1\right]^{\frac{1}{2}} \quad (\text{m}^{-1}).$$
(6.78)

We notice that, if $\sigma = 0$, $\varepsilon = \varepsilon_0$ and $\mu = \mu_0$, then α is zero and $\beta = \beta_0$.

The corresponding solution of Maxwell's equations is

$$\mathbf{E}(x,t) = \left[E_m^+ e^{-\alpha x_3}\cos(\omega t - \beta x_3 + \phi^+) + E_m^- e^{\alpha x_3}\cos(\omega t + \beta x_3 + \phi^-)\right]\mathbf{e}_1.$$

A comparison of the conductive region wave solution with the one in empty-space reveals the presence of two real factors $e^{-\alpha x_3}$ and $e^{\alpha x_3}$ accounting for the amplitude decay as the positive x_3 and the negative x_3 travelling waves proceed in their corresponding direction of propagation with increasing time.

6.10.3 Wavelength. Propagation velocity

The *wavelength* is given by

$$\lambda = \frac{2\pi}{\beta},$$

and the *propagation velocity* (or *phase velocity*) by

$$v = \frac{\omega}{\beta},$$

with

$$\beta = \frac{\omega\sqrt{\mu\varepsilon}}{\sqrt{2}} \left[\sqrt{1+\left(\frac{\sigma}{\omega\varepsilon}\right)^2} + 1\right]^{1/2}.$$

We notice that the phase velocity depends on ω if $\sigma \neq 0$. This fact does not occur in the empty space and means that, in conductive media, waves are *dispersive*.

6.10.4 Skin depth

Looking at the expression for α we see that the wave attenuation in conductive regions is governed by the size of the non-dimensional term $\sigma/(\omega\varepsilon)$ relative to unity.

Moreover, we notice that the wave penetrates the conductive region in such a way that for $x_3 = 1/\alpha$ its amplitude decays to e^{-1} of its value at the reference surface $x_3 = 0$. This motivates the definition of *depth of penetration* or *skin depth* by,

$$\delta = \frac{1}{\alpha} \quad (\text{m}^{-1}).$$

In the limit case of a perfect conductor ($\sigma \to \infty$), the limit of α is also ∞ and then the skin depth is 0. This means that current concentrates on the surface ($x_3 = 0$).

If $\frac{\sigma}{\omega\varepsilon} \gg 1$, then α can be approximated by

$$\alpha = \frac{\omega\sqrt{\mu\varepsilon}}{\sqrt{2}}\sqrt{\frac{\sigma}{\omega\varepsilon}} = \sqrt{\frac{\omega^2\mu\varepsilon\sigma}{2\omega\varepsilon}} = \sqrt{\frac{\omega\mu\sigma}{2}}.$$

In its turn, the skin depth can be approximated by

$$\delta \simeq \sqrt{\frac{2}{\omega\mu\sigma}} \quad \text{(m)}. \tag{6.79}$$

In fact, this is the usual definition of skin depth.

6.10.5 Intrinsic wave impedance

Let us recall that for waves in the empty space we have defined the intrinsic wave impedance as,

$$\eta_0 = \sqrt{\frac{\mu_0}{\varepsilon_0}}.$$

For materials, μ_0 has to be replaced with μ and ε_0 with $\varepsilon - \frac{\sigma}{\omega}i$. Thus the *intrinsic wave impedance* is defined by,

$$\widehat{\eta} = \sqrt{\frac{\mu}{\varepsilon - \frac{\sigma}{\omega}i}}. \tag{6.80}$$

In polar form, this complex number becomes

$$\widehat{\eta} = \eta e^{i\theta},$$

where

$$\eta = \frac{\sqrt{\frac{\mu}{\varepsilon}}}{\left[1 + \left(\frac{\sigma}{\omega\varepsilon}\right)^2\right]^{1/4}}, \quad \theta = \frac{1}{2}\arctan\frac{\sigma}{\omega\varepsilon}.$$

We notice that in nonconductive (dielectric) media, $\sigma = 0$, so $\widehat{\eta}$ reduces to the real number $\sqrt{\mu/\varepsilon}$.

6.11 Electric classification of materials

6.11.1 Conductive media

According to the discussion in Sect. 6.10, conductive materials can be classified with reference to the magnitude of the conduction current density term, $\mathbf{J} = \sigma\mathbf{E}$, relative to the displacement current density term $\partial\mathbf{D}/\partial t$ ($i\omega\widehat{\mathbf{D}}$ in the frequency domain) appearing in the Ampère's law:

$$\text{curl}\,\widehat{\mathbf{H}} = \sigma\widehat{\mathbf{E}} + i\omega\varepsilon\widehat{\mathbf{E}} = i\omega(\varepsilon - \frac{\sigma}{\omega}i)\widehat{\mathbf{E}}.$$

Let us introduce the *complex permittivity*, $\widehat{\varepsilon}$, by

$$\widehat{\varepsilon} = \varepsilon - \frac{\sigma}{\omega}i \quad (\text{F/m}).$$

This complex number has an argument δ_d such that,

$$\tan \delta_d = -\frac{\sigma}{\omega\varepsilon}.$$

We notice that the number $\sigma/(\omega\varepsilon)$ appears in the expressions for α and β given in (6.78), as well as in the intrinsic wave impedance of plane waves propagating in a conductive material, given in (6.80).

For a good conductor $\frac{\sigma}{\omega\varepsilon} \gg 1$, and then α and β can be approximated by

$$\alpha \simeq \sqrt{\frac{\omega\mu\sigma}{2}},$$

$$\beta \simeq \sqrt{\frac{\omega\mu\sigma}{2}}.$$

and $\widehat{\eta}$ by

$$\widehat{\eta} \simeq (1+i)\sqrt{\frac{\mu\omega}{2\sigma}} \quad (\text{m}).$$

We recall that in the above expression for α, the skin depth in a good conductor can be approximated by Eq. (6.79).

Example 6.1. Copper has a conductivity of $5.8 \; 10^7$ S/m and a relative permeability ($\mu_r = 1$). Thus the skin depth at 1000 Hz is about $2 \; 10^{-3}$ m.

6.11.2 Lossy dielectrics

We recall that when a dielectric material is subjected to an applied static electric field, the centroids of the positive and negative charges at atomic level are displaced relative to each other forming electric dipoles. When the material is examined macroscopically, the presence of all the dielectric dipoles is accounted for by introducing an electric polarization vector **P**. In fact, the electrical permittivity ε is introduced to account for the presence of **P**.

When the applied electric field changes with time (for instance it is time-harmonic), the polarization field **P** is affected becoming a function of the frequency of the applied electric field. By this action of the alternating fields, damping (loss) effects appear similar to those produced by conduction currents. They are responsible, for instance, for the heating of materials using microwaves (see Sect. 6.12 below).

The classical model for this phenomena, inspired by experimental measurements on dielectric materials, assumes an oscillating system of interacting atomic or molecular particles in which the response of the material to the applied electric fields involves damping mechanisms plus resonances about certain frequencies. The damping mechanism is taken as proportional to the velocity of particles, so the model is similar to the forced damped harmonic oscillator. In other words, the response of the dielectric to the applied field resembles that of a three-dimensional system of masses interconnected through springs and dashpots and subjected to applied distributed vibrational forces or, analogously, a network of reactive and resistive circuit elements excited by sinusoidal voltages with maximum losses occurring at resonant frequencies. For typical dielectric materials, the lowest resonance is usually in or above the microwave range.

In order to model the macroscopic effect of these interaction phenomena, we make the permittivity of the dielectric to become complex:

$$\widehat{\varepsilon} = \varepsilon' - \varepsilon'' i.$$

We notice that a complex permittivity has already been defined by $\widehat{\varepsilon} = \varepsilon - \frac{\sigma}{\omega} i$, in connection with the loss mechanism in a conductive region. Comparing both expressions we observe that the above described damping mechanism is equivalent to a conduction loss where $\omega \varepsilon''$ assumes the role of conductivity σ in a lossy dielectric. Moreover ε' is identical with the real electric permittivity ε. Summarizing,

$$\varepsilon = \varepsilon',$$
$$\sigma = \omega \varepsilon''.$$

Accordingly, the loss tangent in a lossy dielectric is given by,

$$|\tan \delta_d| = \frac{\varepsilon''}{\varepsilon},$$

and the attenuation constant by,

$$\alpha = \frac{\omega \sqrt{\mu \varepsilon'}}{\sqrt{2}} \left[\sqrt{1 + \left(\frac{\varepsilon''}{\varepsilon'}\right)^2} - 1 \right]^{1/2}.$$

Similarly the phase constant β and the wave impedance $\widehat{\eta}$ are,

$$\beta = \frac{\omega \sqrt{\mu \varepsilon'}}{\sqrt{2}} \left[\sqrt{1 + \left(\frac{\varepsilon''}{\varepsilon'}\right)^2} + 1 \right]^{1/2},$$

$$\widehat{\eta} = \frac{\sqrt{\mu/\varepsilon'}}{\left[\sqrt{1 + \left(\frac{\varepsilon''}{\varepsilon'}\right)^2} \right]^{1/4}} e^{i(1/2)\arctan(\varepsilon''/\varepsilon')}.$$

For *lossy dielectrics* $\varepsilon''/\varepsilon' \ll 1$ and the above expressions can be approximated by

$$\alpha = \frac{\omega\sqrt{\mu\varepsilon'}}{2}\left(\frac{\varepsilon''}{\varepsilon'}\right),$$

$$\beta = \omega\sqrt{\mu\varepsilon'}\left[1 + \frac{1}{8}\left(\frac{\varepsilon''}{\varepsilon'}\right)^2\right],$$

$$\widehat{\eta} = \sqrt{\frac{\mu}{\varepsilon'}}\left[1 - \frac{3}{8}\left(\frac{\varepsilon''}{\varepsilon'}\right)^2 + \frac{1}{2}\frac{\varepsilon''}{\varepsilon'}i\right].$$

We notice that in the limiting case of *lossless dielectrics* ($\varepsilon'' = 0$, $\varepsilon' = \varepsilon$), they reduce to

$$\alpha = 0, \ \beta = \omega\sqrt{\mu\varepsilon}, \ \widehat{\eta} = \sqrt{\mu/\varepsilon},$$

as expected.

6.12 Microwave heating

In this section we introduce a model to simulate microwave heating. As previously mentioned, microwave heating takes place due to the polarization effect of electromagnetic radiation at frequencies between 300 MHz and 300 GHz. Started as a by-product of the radar technology developed during World War II, microwave heating is now used in many homes but it has also found important industrial applications.

Microwave heating is a complicated physical process which depends upon:

- the propagation of microwaves governed by Maxwell's equations for electromagnetic waves;
- on the interactions between microwaves and materials determined by their dielectric properties;
- heat dissipation governed by the heat transfer equation.

Thus, mathematical modelling of microwave heating requires to compute first the propagation of the electromagnetic field and then to determine the temperature in the heated workpiece. In practice, both the electromagnetic and thermal models are coupled because sometimes dielectric properties can be strongly dependent on temperature. In what follows we write the electromagnetic model in the harmonic case.

6.12.1 Electromagnetic model

We suppose that sources are harmonic so we are led to solve the harmonic Maxwell's equations. In terms of the electric field they read (see Sect. 6.10.1):

$$-\omega^2\mu\left(\varepsilon - \frac{\sigma}{\omega}i\right)\widehat{\mathbf{E}} + \mathbf{curl}\,\mathbf{curl}\,\widehat{\mathbf{E}} = 0.$$

Since we have to deal with lossy dielectrics, the electric permittivity is a complex number so we replace ε by $\varepsilon' - i\varepsilon''$ to get

$$-\omega^2 \mu \varepsilon' \widehat{\mathbf{E}} + i\omega\mu(\sigma + \omega\varepsilon'')\widehat{\mathbf{E}} + \mathbf{curl}\,\mathbf{curl}\,\widehat{\mathbf{E}} = \mathbf{0}.$$

6.12.1.1 Boundary conditions

In order to solve the above partial differential equations we need to introduce boundary conditions. We have several choices:

- Let us suppose that the internal surface of the oven is made with a perfect conductor. Then the boundary condition on this surface is (*perfect conductor condition*),

$$\widehat{\mathbf{E}} \times \mathbf{n} = \mathbf{0},$$

where $\mathbf{n}$ is the outward unit normal vector.
More accurate is the impedance boundary condition (see Sect. 6.7.1), namely,

$$\mathbf{n} \times (\widehat{\mathbf{E}} \times \mathbf{n}) + i\frac{Z}{\omega\mu}(\mathbf{curl}\,\widehat{\mathbf{E}} \times \mathbf{n}) = \mathbf{0},$$

where Z is the *surface impedance*.
- On symmetry boundaries we have,

$$\widehat{\mathbf{H}} \times \mathbf{n} = \mathbf{0}.$$

- The source wave is imposed on a section of the waveguide connecting the device producing waves (the so-called *magnetron*) to the cavity. By properly choosing this waveguide (actually by choosing its section), and depending on the excitation frequency, it is possible to get a waveguide where only the so-called transversal electric mode TE_{10} propagates. Neglecting reflections we can prescribe the analytical expression of the tangential component of this mode, $\widehat{\mathbf{E}}_{10} \times \mathbf{n}$, as a Dirichlet boundary condition.

6.12.2 A simplified solution: Lambert's law

Sometimes, instead of solving the above electromagnetic model an approximation of the electric field is used. It is nothing but the plane wave solution obtained in Chap. 4, namely,

$$\mathbf{E}(x,t) = E_m^+ e^{-\alpha x_3} \cos(\omega t - \beta x_3 + \phi^+)\mathbf{e}_1$$

In fact, in the so-called *Lambert's law* this expression is further simplified to

$$\mathbf{E}(x,t) = E_m^+ e^{-\alpha x_3} \mathbf{e}_1,$$

with

$$\alpha = \frac{\omega\sqrt{\mu\varepsilon'}}{\sqrt{2}} \left[\sqrt{1 + \left(\frac{\varepsilon''}{\varepsilon'}\right)^2} - 1\right]^{1/2}.$$

6.12.3 Heating power

The power density released by the electromagnetic field in the material is given by

$$P = \frac{1}{2}(\sigma + \varepsilon''\omega)|\widehat{\mathbf{E}}|^2.$$

This power is the source term in the heat transfer equation to be solved in order to determine the temperature in the workpiece.

6.13 Properties of materials: homogeneity, isotropy, linearity

A material having parameters μ, ε and σ independent of the position is termed *homogeneous*.

Conversely, if one or more of these parameters is space-dependent then the material is called *inhomogeneous*.

In some physical materials such as crystalline substances possessing a well-ordered atomic or molecular lattice, the polarization fields **P** or **M** resulting from the application of electric or magnetics fields may not necessarily have the same directions as the applied fields. Such material are called *anisotropic*.

This fact can be modelled by taking tensors rather than scalars for the characteristic parameters ε, μ and σ. As an example, the constitutive law $\mathbf{D} = \varepsilon\mathbf{E}$ should be replaced by (in coordinates),

$$\begin{pmatrix} D_1 \\ D_2 \\ D_3 \end{pmatrix} = \begin{pmatrix} \varepsilon_{11} & \varepsilon_{12} & \varepsilon_{13} \\ \varepsilon_{21} & \varepsilon_{22} & \varepsilon_{23} \\ \varepsilon_{31} & \varepsilon_{32} & \varepsilon_{33} \end{pmatrix} \begin{pmatrix} E_1 \\ E_2 \\ E_3 \end{pmatrix}.$$

Finally, a material is termed nonlinear if one or more of the parameters ε, μ or σ are dependent on the level of the applied fields. In that case, the constitutive laws would become,

$$\mathbf{D} = \varepsilon(|\mathbf{E}|)\mathbf{E},$$
$$\mathbf{B} = \mu(|\mathbf{H}|)\mathbf{H},$$
$$\mathbf{J} = \sigma(|\mathbf{E}|)\mathbf{E}.$$

7
Electrostatics

In many practical applications one does not have to deal with the full system of Maxwell's equations because there are assumptions allowing for substantial simplifications. For example, in electrostatics, charges do not move so there are no currents and then the magnetic field is null. In this chapter, we will study this model in terms of the electrostatic potential and introduce the concept of capacitance.

7.1 The electrostatics model. Electrostatic potential

If we assume that there are only charges at rest, Maxwell's equations reduce to the electrostatics model which consists in finding the electric field satisfying:

$$\text{div}\,\mathbf{D} = \rho_V, \tag{7.1}$$

$$\mathbf{curl}\,\mathbf{E} = 0, \tag{7.2}$$

with $\mathbf{D} = \varepsilon \mathbf{E}$.

As in the empty space, the electric field is curl-free which implies the existence of a potential field V such that,

$$\mathbf{E} = -\,\text{grad}\,V.$$

By replacing in the Gauss' law for electric charge we get

$$-\,\text{div}(\varepsilon\,\text{grad}\,V) = \rho_V. \tag{7.3}$$

This is an elliptic partial differential equation of Poisson's type, which, in general, is easier to solve than Eqs. (7.1)–(7.2).

Remark 7.1. As explained in Chap. 4, charges are represented by scalar density fields. These densities can be volume, surface or line densities. We notice that if the charge is not volumetric, then the charge density has to be represented by a distribution and Eq. (7.3) has to be understood in the distributional sense. In 3D, we will also consider point charges represented by Dirac measures.

The electric permittivity needs not to be constant in the whole space but, if this were the case, Eq. (7.3) would be equivalent to

$$-\Delta V = \frac{\rho_V}{\varepsilon},$$

which is a Poisson's equation similar to that obtained for empty space. Let us recall that the solution of this problem can be obtained by convolution of ρ_V/ε with the *fundamental solution*, i.e., the solution of the problem

$$-\Delta V = \delta_o,$$

$$V(x) \to 0 \text{ as } r(x) \to \infty.$$

Actually, we have (see [33])

$$V(x) = \frac{1}{4\pi\varepsilon} \int_{\mathscr{E}} \frac{\rho_V(y)}{|x-y|} dV(y).$$

Then the electric field $\mathbf{E} = -\text{grad}\,V$ is given by

$$\mathbf{E}(x) = \frac{1}{4\pi\varepsilon} \int_{\mathscr{E}} \rho_V(y) \frac{x-y}{|x-y|^3} dV(y).$$

Indeed,

$$\mathbf{E} = -\text{grad}_x \int_{\mathscr{E}} \frac{1}{4\pi\varepsilon} \frac{\rho_V(y)}{|x-y|} dV(y) = -\frac{1}{4\pi\varepsilon} \int_{\mathscr{E}} \rho_V(y) \text{grad}_x \left(\frac{1}{|x-y|} \right) dV(y).$$

But we have shown that (see Sect. C.8)

$$\text{grad}_x \frac{1}{|x-y|} = -\frac{x-y}{|x-y|^3}.$$

Then,

$$\mathbf{E}(x) = \frac{1}{4\pi\varepsilon} \int_{\mathscr{E}} \rho_V(y) \frac{x-y}{|x-y|^3} dV.$$

Remark 7.2. In a similar way as for volumetric charge distributions, we can obtain the electric potential for line and surface charge distributions by

$$V(x) = \int_l \frac{1}{4\pi\varepsilon} \frac{\rho_l(y)}{|x-y|} dl(y), \tag{7.4}$$

$$V(x) = \int_S \frac{1}{4\pi\varepsilon} \frac{\rho_S(y)}{|x-y|} dA(y), \tag{7.5}$$

and then, by differentiation, the respective electric fields are given by

$$\mathbf{E}(x) = \int_l \frac{\rho_l(y)}{4\pi\varepsilon} \frac{x-y}{|x-y|^3} dl(y), \tag{7.6}$$

$$\mathbf{E}(x) = \int_S \frac{\rho_S}{4\pi\varepsilon} \frac{x-y}{|x-y|^3} dA(y). \tag{7.7}$$

7.2 Energy of the electrostatic field

In Sect. 6.8 we have studied the electromagnetic energy. In particular, we have introduced expressions for the electrical and magnetic energies. The goal of this section is to interpret and justify the adopted definition for the electrical energy, by considering first a finite system of point charges.

In electrostatics, the charge positions determine the potential energy of the system. Let us start by a system consisting of M point charges located at points $x^1,\ldots,x^M$. In order to establish this system, mechanical work must be done by some external agent in bringing the charges to their final positions from infinity. Indeed, whenever two charges are brought within a distance of each other, work is done against the Coulomb's force. Once the charges are in place, the persistence of the Coulomb's force makes the stored energy potentially available.

In order to determine this stored energy we assume initially that all charges, $Q_1,\ldots,Q_M$, are located at infinity, in their zero potential state. On bringing the first charge to position x^1 no work is done because Q_1 only is present. At least, two charges are needed to produce Coulomb's forces.

Next we bring Q_2 from infinity to x^2. The work done on the field is the negative of the work done by the field, i.e.,

$$\mathscr{U}_2 = -\int_{l(\infty,x^1)} Q_2 \mathbf{E}_1 \cdot d\mathbf{l}$$

where $\mathbf{E}_1$, the electric field created by charge Q_1, is given by

$$\mathbf{E}_1(y) = -\operatorname{grad} V_1(y),$$

with

$$V_1(y) = \frac{1}{4\pi\varepsilon_0} \frac{Q_1}{|y - x^1|}.$$

Therefore,

$$\mathscr{U}_2 = \frac{1}{4\pi\varepsilon_0} \frac{Q_1 Q_2}{|x^2 - x^1|} = \frac{1}{2}\frac{1}{4\pi\varepsilon_0} \sum_{\substack{i,j=1 \\ i \neq j}}^{2} \frac{Q_i Q_j}{|x^i - x^j|}.$$

It is straightforward to see that for the M charges the energy is given by

$$\mathscr{U} = \frac{1}{2}\sum_{\substack{i,j=1 \\ i \neq j}}^{M} \frac{1}{4\pi\varepsilon_0} \frac{Q_i Q_j}{|x^i - x^j|} = \frac{1}{2}\sum_{i=1}^{M} Q_i \left(\frac{1}{4\pi\varepsilon_0} \sum_{\substack{j=1 \\ j \neq i}}^{M} \frac{Q_j}{|x^i - x^j|} \right)$$

$$= \frac{1}{2}\sum_{i=1}^{M} Q_i V_i(x^i) \quad \text{(J)},$$

where $V_i(x)$ denotes the potential due to all the charges except the i-th, at point x^i.

Similarly, if we have a charge distribution ρ_V, ρ_S, or ρ_l, then the electrostatic energy is given, respectively, by

$$\mathscr{U} = \frac{1}{2}\int_V \rho_V V dV \quad (J),$$

$$\mathscr{U} = \frac{1}{2}\int_S \rho_S V dA \quad (J),$$

$$\mathscr{U} = \frac{1}{2}\int_l \rho_l V dl \quad (J).$$

According to the Gauss' law for electric charge, the distribution defined by the charge density is equal to div **D**. Then the electrostatic energy can also be written as

$$\mathscr{U} = \frac{1}{2}\int_{\mathscr{E}} \text{div}\,\mathbf{D} V dV = -\frac{1}{2}\int_{\mathscr{E}} \mathbf{D} \cdot \text{grad}\, V dV = \frac{1}{2}\int_{\mathscr{E}} \mathbf{D} \cdot \mathbf{E} dV,$$

and, since $\mathbf{D} = \varepsilon \mathbf{E}$,

$$\mathscr{U} = \frac{1}{2}\int_{\mathscr{E}} \varepsilon |\mathbf{E}|^2 dV.$$

7.3 The electrostatics model in bounded domains

Up to now, the electrostatics problems we have solved assume they are stated in the whole space. In previous sections, the solutions of electrostatic field problems have been obtained by either:

- the use of Gauss' law in integral form;
- the use of the electric potential obtained from the charge distribution in the whole space by convolution with the fundamental solution.

Both of these methods require a specification of the charge in the whole space. However, in many practical applications, we can write and solve a model in a bounded domain. This requires, of course, to prescribe boundary conditions on its boundary. These conditions can be of the following two types:

- *Dirichlet*: the value V_d of the potential on specified parts of the boundary is given;
- *Neumann*: the normal derivative of the potential is given on specified parts of the boundary:

$$-\varepsilon \frac{\partial V}{\partial \mathbf{n}} = \varepsilon \mathbf{E} \cdot \mathbf{n} = -g.$$

The corresponding boundary-value problems for the Poisson's equation are solved, in general, by using numerical techniques. Only if the domain has special geometry (disk, rectangle) the problem can be solved analytically.

7.4 Weak formulation of electrostatics problems

In the following sections we will derive the weak formulation of electrostatics problems in two and three dimensions and the finite element space to be used in the numerical approximation.

7.4.1 Weak formulation of the 3D electrostatics model

Let Ω be a bounded three dimensional domain with boundary Γ. Let us assume there is a volume charge density ρ_V in Ω, a surface charge density ρ_S on a surface $S \subset \Omega$, a linear charge density ρ_l along a line $l \subset \Omega$ and a point charge Q_p at a point $p \in \Omega$. For the sake of simplicity, let us consider homogeneous Dirichlet boundary condition, namely,

$$V = 0 \text{ on } \Gamma.$$

This is a suitable boundary condition as far as domain Ω is taken large enough.

Then, the problem to be solved is the following:

Find $V : \Omega \to \mathbb{R}$ *such that*

$$-\text{div}(\varepsilon \, \text{grad} \, V) = \rho_V + T_{\rho_S} + T_{\rho_l} + T_{Q_p} \quad \text{in } \Omega, \tag{7.8}$$

$$V = 0 \quad \text{on } \Gamma, \tag{7.9}$$

where T_{ρ_S}, T_{ρ_l} *and* T_{Q_p} *are distributions supported on* S, l *and* p, *respectively, defined by*

$$\langle T_{\rho_S}, \psi \rangle = \int_S \rho_S \, \psi \, dS, \tag{7.10}$$

$$\langle T_{\rho_l}, \psi \rangle = \int_l \rho_l \, \psi \, dl, \tag{7.11}$$

$$\langle T_{Q_p}, \psi \rangle = Q_p \, \psi(p), \tag{7.12}$$

for $\psi \in \mathscr{D}(\Omega)$.

Since the source is singular, we cannot expect a solution of problem (7.8)–(7.9) to be in the Sobolev space $H^1(\Omega)$. Then, building and analyzing a weak formulation of this problem is not so standard and should be made by using *transposition methods*. Let us introduce the space $C_0(\bar{\Omega})$ of continuous functions from $\bar{\Omega}$ in $\mathbb{R}$, null on Γ. Endowed with the maximum norm it is a Banach space. Moreover, for spatial dimension $n \leq 3$ we have the inclusion

$$H^2(\Omega) \cap H_0^1(\Omega) \subset C_0(\bar{\Omega}).$$

Let us assume that the electric permittivity, ε, is Lipschitz-continuous in $\bar{\Omega}$ and that the charge densities are in L^1-spaces, more specifically, $\rho_V \in L^1(\Omega)$, $\rho_S \in L^1(S)$ and

$\rho_l \in L^1(l)$, where S and l are closed sets included in Ω. Then their respective distributions can be extended to the space $C(\bar{\Omega})$, in other words, they are also *regular Borel measures*.

By making the duality product of (7.8) by a test function W and then using a Green's formula twice we are led to the following *weak formulation*:

We say that $V \in L^2(\Omega)$ *is a weak solution of* (7.8)–(7.9) *if*

$$-\int_\Omega V \operatorname{div}(\varepsilon \operatorname{grad} w)\, dV = \int_\Omega \rho_V w\, dV + \int_S \rho_S w\, dA + \int_l \rho_l w\, dl$$
$$+ Q_p w(p)$$

for all $w \in H^2(\Omega) \cap H_0^1(\Omega)$.

We notice that this formulation makes sense. Indeed, the term

$$\operatorname{div}(\varepsilon \operatorname{grad} w)$$

belongs to $L^2(\Omega)$ because ε is Lipschitz-continuous.

One can prove that this weak formulation has a unique solution. Moreover, for $1 < p < \frac{n}{n-1}$, it belongs to the Sobolev space $W_0^{1,p}(\Omega)$ of real-valued functions null on Γ which, together with all their first order partial distributional derivatives are in $L^p(\Omega)$.

Remark 7.3. The numerical discretization of this problem can be done by using continuous piecewise linear elements on a tetrahedral mesh of the domain. Numerical analysis including error estimates can be seen, for instance, in [26].

7.4.2 Weak formulation of the in-plane 2D electrostatics model

Taking into account the geometry of the domain and the symmetries of the problem, sometimes the solution of a (in principle 3D) electrostatics problem can be approximated by an in-plane 2D model; in particular, this is the case where the charges and the physical properties of the domain are invariant under translation in one privileged direction, let us say, the x_3-direction. In this situation, the problem can be written in an orthogonal section to the x_3-axis which will be denoted by $\widehat{\Omega}$. For symmetry reasons, the electric field in $\widehat{\Omega}$ has to be of the form

$$\mathbf{E} = E_1(x_1, x_2)\mathbf{e}_1 + E_2(x_1, x_2)\mathbf{e}_2,$$

where $\mathbf{e}_1$ and $\mathbf{e}_2$ denote the cartesian basis vectors. As a consequence, the electric scalar field V only depends on x_1 and x_2.

Let us notice that volumetric, surface and line charges in 3D become, respectively, surface, line and point charges in 2D. Thus, we obtain a 2D Poisson's problem similar to (7.8) but defined in a two dimensional domain. Let us denote by $\widehat{\Gamma}$ the boundary of $\widehat{\Omega}$. As in the 3D case, we consider homogeneous Dirichlet boundary condition.

Then, the problem to be solved consists in

Finding $V: \widehat{\Omega} \to \mathbb{R}$ *such that*

$$-\operatorname{div}(\varepsilon \operatorname{grad} V) = \rho_V + \widehat{T}_{\rho_S} + \widehat{T}_{\rho_l} \text{ in } \widehat{\Omega}, \tag{7.13}$$

$$V = 0 \text{ on } \widehat{\Gamma}, \tag{7.14}$$

where $\widehat{T}_{\rho_S}$ and $\widehat{T}_{\rho_l}$ are distributions supported on line $\widehat{l} = S \cap \widehat{\Omega}$ and at point $\widehat{p} = l \cap \widehat{\Omega}$, respectively, defined as follows:

$$\langle \widehat{T}_{\rho_l}, \psi \rangle = \int_{\widehat{l}} \rho_S \psi \, dl, \tag{7.15}$$

$$\langle \widehat{T}_{Q_p}, \psi \rangle = \rho_l \psi(\widehat{p}), \tag{7.16}$$

for $\psi \in \mathscr{D}(\widehat{\Omega})$.

A weak formulation of the above boundary-value problem, similar to the one of the 3D case, can be obtained and then discretized by continuous piecewise linear finite elements on a triangular mesh of $\widehat{\Omega}$.

7.4.3 The axisymmetric case

Let us assume that Ω is a 3D domain generated by the rotation about the x_3-axis of a set $\widehat{\Omega}$. If the electrostatic problem has cylindrical symmetry, i.e, if all the fields and physical parameters are independent of the θ-component, then it can be solved on a radial section $\widehat{\Omega}$. Thus, by applying the Laplace operator in cylindrical coordinates (see Sect. B.2.8.2), Eq. (7.8) writes as follows

$$-\frac{1}{\rho}\frac{\partial}{\partial \rho}\left(\varepsilon \rho \frac{\partial V}{\partial \rho}\right) - \frac{\partial}{\partial z}\left(\varepsilon \frac{\partial V}{\partial z}\right) = \rho_V, \tag{7.17}$$

where $V = V(\rho, z)$. Multiplying Eq. (7.17) by ρ, we obtain

$$-\frac{\partial}{\partial \rho}\left(\varepsilon \rho \frac{\partial V}{\partial \rho}\right) - \frac{\partial}{\partial z}\left(\varepsilon \rho \frac{\partial V}{\partial z}\right) = \rho_V \rho \tag{7.18}$$

Hence, by setting $\tilde{\varepsilon} = \varepsilon \rho$ and $\tilde{\rho} = \rho_V \rho$ the problem becomes similar to the above 2D problem. Thus, the boundary conditions, weak formulation and numerical approximation are analogous to those for problem (7.13)–(7.14).

7.5 Electric field created by a set of charged conductors. Capacitance matrix

This is a case where we do not need to know the charge distribution in order to determine the electric field.

7 Electrostatics

Let us consider M bounded conductors $\Omega_1,\ldots,\Omega_M$ having total charges $Q_1,\ldots,Q_M$, respectively. Let us assume that the complementary set $\Omega = \mathscr{E}\setminus(\Omega_1\cup\ldots\cup\Omega_M)$ is filled with a free of charge dielectric material, with electric permittivity, $\varepsilon(x)$, which may depend on the position (non-homogenous media).

According to the local form of the Gauss' law, we have

$$-\operatorname{div}(\varepsilon\operatorname{grad}V) = 0 \text{ in } \Omega, \tag{7.19}$$

because the dielectric is assumed to be charge-free.

In order to get a well-posed problem we need to write boundary conditions on the boundaries $\Gamma_1,\ldots,\Gamma_M$ of domains $\Omega_1,\ldots,\Omega_M$, respectively.

For this we observe that, in electrostatics, the electric field in conductors must be null because otherwise we would have an electric current different from zero (i.e., charges in motion) by Ohm's law.

Furthermore, since $\mathbf{E} = 0$ in Ω_i, $i = 1,\ldots,M$ then the potential has to be constant in each conductor. Let $V(x) = V_i\ \forall x \in \Omega_i$, for $i = 1,\ldots,M$.

If these potentials were known, then they could be Dirichlet data for the Poisson's equation (7.2) and we were led to solve the boundary-value problem,

$$-\operatorname{div}(\varepsilon\operatorname{grad}V) = 0 \text{ in } \Omega, \tag{7.20}$$

$$V = V_i \text{ on } \Gamma_i, \tag{7.21}$$

$$V \to 0 \text{ as } r \to \infty. \tag{7.22}$$

Let us introduce the weighted Sobolev space,

$$\mathrm{W}^1(\Omega) = \left\{w : \int_\Omega \frac{|w(x)|^2}{1+r^2(x)}\,dV < \infty,\ \operatorname{grad}w \in \mathbf{L}^2(\Omega)\right\},$$

and its closed subspace consisting of functions with constant trace on each Γ_i, namely,

$$\mathscr{Y} = \{w \in \mathrm{W}^1(\Omega) :\ w_{|\Gamma_i} = c_i,\ i = 1,\ldots,M \text{ with } (c_1,\ldots,c_M) \in \mathbb{R}^M\}.$$

Let us denote by $\mathscr{Y}_0$ the closed subspace of $\mathscr{Y}$ defined by

$$\mathscr{Y} = \{w \in \mathrm{W}^1(\Omega) :\ w_{|\Gamma_i} = 0,\ i = 1,\ldots,M\}.$$

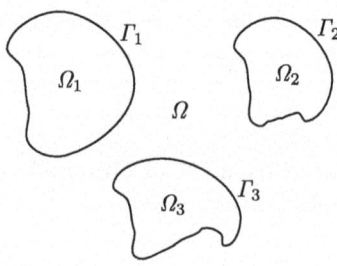

Fig. 7.1. Example of domain composed by a set of charged conductors

7.5 Electric field created by a set of charged conductors. Capacitance matrix

Multiplying Eq. (7.20) by a test function $w \in \mathcal{Y}_0$, integrating in Ω and using a Green's formula we get the following weak formulation:

Find $V \in \mathcal{Y}$ with $V_{|\Gamma_l} = V_l$, $l = 1, \ldots, M$ and such that

$$\int_\Omega \varepsilon(x) \operatorname{grad} V(x) \cdot \operatorname{grad} w(x) \, dV = 0 \quad \forall w \in \mathcal{Y}_0. \tag{7.23}$$

If Ω is connected, the continuous bilinear form

$$a(v,w) = \int_\Omega \varepsilon(x) \operatorname{grad} v(x) \cdot \operatorname{grad} w(x) \, dV$$

is coercive in the space $W^1(\Omega)$ (and then in $\mathcal{Y}$), that is,

$$a(v,v) \geq \|v\|^2_{W^1(\Omega)} \quad \forall v \in W^1(\Omega)$$

(see [57]). Hence, the Lax-Milgran Lemma shows the existence and uniqueness of a solution to (7.23).

Let us recall that we do not know the vector of potentials in conductors, $(V_1, \ldots, V_M)$, so we cannot use (7.23). Then we proceed as follows. Firstly, since $\mathbf{E} = 0$ in Ω_i, then $\operatorname{div}(\varepsilon \mathbf{E}) = 0$ in Ω_i so the charge density must be null in the interior of each conductor. This implies that the charge must be concentrated on the surface Γ_i where the surface density, ρ_S^i, may depend on each particular point. Furthermore (see (6.45) in Sect. 6.7),

$$\varepsilon^+ \mathbf{E}^+ \cdot \mathbf{n}^+ + \varepsilon^- \mathbf{E}^- \cdot \mathbf{n}^- = -\rho_S^i.$$

Let the "−" region be the interior of Ω_i and the "+" region its complementary set (see Fig. 7.2). Since $\mathbf{E}^- = 0$, then the previous equation yields

$$\varepsilon^+ \mathbf{E}^+ \cdot \mathbf{n}^+ = -\rho_S^i,$$

or

$$\varepsilon^+ \frac{\partial V^+}{\partial \mathbf{n}^+} = \rho_S^i.$$

This is a Neumann boundary condition for Laplace's equation (7.19).

Again, the difficulty is that surface densities ρ_S^i are not known (recall that the data are the total charges $Q_1, \ldots, Q_M$ which are the integrals of ρ_S^i on Γ_i). However, since the solution must be constant along each boundary we can solve the problem. Indeed,

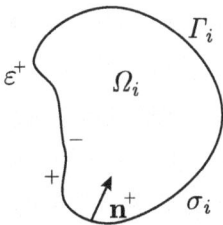

Fig. 7.2. Conductor Ω_i

multiplying Eq. (7.20) by a test function $w \in \mathscr{Y}$, integrating in Ω and using a Green's formula we get

$$\int_\Omega \varepsilon(x) \operatorname{grad} V(x) \cdot \operatorname{grad} w \, dV = \sum_{l=1}^M \int_{\Gamma_l} \rho_S^l w \, dA = \sum_{l=1}^M Q_l w_l, \qquad (7.24)$$

where w_l denotes the constant value of w on Γ_l. Thus, we have built another *weak formulation* of the problem in terms of the true data $(Q_1, \ldots, Q_M)$:

Find $V \in \mathscr{Y}$ such that

$$\int_\Omega \varepsilon(x) \operatorname{grad} V(x) \cdot \operatorname{grad} w \, dV = \sum_{l=1}^M Q_l w_l \quad \forall w \in \mathscr{Y}. \qquad (7.25)$$

Notice that $\forall w \in \mathscr{Y}_0$, (7.25) yields

$$\int_\Omega \varepsilon(x) \operatorname{grad} V(x) \cdot \operatorname{grad} w \, dV = 0.$$

Hence, taking in problem (7.23) the Dirichlet data V_l, $l = 1, \ldots, M$, as the (constant) traces of the solution of (7.25), we deduce that both problems (7.23) and (7.25) have the same solution.

Now, for each $j \in \{1, \ldots, M\}$ let $v^j \in \mathscr{Y}$ be the unique solution to problem (7.23) corresponding to the potentials $V_i = \delta_{il}$, $l = 1, \ldots, M$. Then the solution of (7.23) for any given potentials $V_1, \ldots, V_M$ is the linear combination

$$V(x) = \sum_{j=1}^M V_j v^j(x).$$

By replacing this expression in (7.25) and taking $w = v^i$ as test function we get

$$\sum_{j=1}^M V_j \int_\Omega \varepsilon(x) \operatorname{grad} v^j(x) \cdot \operatorname{grad} v^i(x) \, dV = \sum_{l=1}^M Q_l v^i|_{\Gamma_l} = \sum_{l=1}^M Q_l \delta_{il} = Q_i.$$

These equalities show that there exists a linear relation between the potentials of the conductors and their respective total charges. More precisely, let $\mathscr{C}$ be the matrix whose elements c_{ij} are defined by

$$c_{ij} = \int_\Omega \varepsilon(x) \operatorname{grad} v^j(x) \cdot \operatorname{grad} v^i(x) \, dV. \qquad (7.26)$$

Then

$$\vec{Q} = \mathscr{C} \vec{V},$$

where

$$\vec{V} = \begin{pmatrix} V_1 \\ \vdots \\ V_M \end{pmatrix}$$

7.5 Electric field created by a set of charged conductors. Capacitance matrix

and

$$\vec{Q} = \begin{pmatrix} Q_1 \\ \vdots \\ Q_M \end{pmatrix}.$$

Since the bilinear form a is symmetric and coercive, then matrix $\mathscr{C}$ is symmetric and positive definite. It is called the *capacitance matrix* of the set of conductors and its entries have units coulomb per volt or farad (F). We notice that it only depends on the electric permittivity of the dielectric filling Ω and on the geometry of conductors Ω_i, $i = 1, \ldots, M$.

7.5.1 An important case: the capacitor

This corresponds to $M = 2$ and $Q_1 = -Q_2 = Q$. Under these circumstances, in addition to the 2×2 capacitance matrix we can define a *reduced capacitance* which is the single complex number C such that

$$Q = C(V_1 - V_2).$$

In order to determine this reduced capacitance let us recall that the capacitance matrix $\mathscr{C}$ is symmetric and we have

$$\begin{pmatrix} c_{11} & c_{12} \\ c_{21} & c_{22} \end{pmatrix} \begin{pmatrix} V_1 \\ V_2 \end{pmatrix} = \begin{pmatrix} Q \\ -Q \end{pmatrix}.$$

By taking the inverse of $\mathscr{C}$,

$$\mathscr{C}^{-1} = \frac{1}{c_{11}c_{22} - c_{12}^2} \begin{pmatrix} c_{22} & -c_{12} \\ c_{12} & c_{11} \end{pmatrix},$$

we deduce

$$\begin{pmatrix} V_1 \\ V_2 \end{pmatrix} = \frac{Q}{c_{11}c_{22} - c_{12}^2} \begin{pmatrix} c_{22} + c_{12} \\ -c_{12} - c_{11} \end{pmatrix}$$

from which it follows that

$$V_1 - V_2 = \frac{Q}{c_{11}c_{22} - c_{12}^2} (c_{22} + 2c_{12} + c_{11}).$$

Thus the *reduced capacitance* is given by

$$C = \frac{c_{11}c_{22} - c_{12}^2}{c_{22} + 2c_{12} + c_{11}}.$$

Sometimes one can show that $c_{11} = c_{22}$. This is the case, for instance, if conductors Ω_1 and Ω_2 are symmetric with respect to a plane. Then,

$$C = \frac{c_{11}^2 - c_{12}^2}{2(c_{11} + c_{12})} = \frac{c_{11} - c_{12}}{2}.$$

For simple systems of conductors, matrix $\mathscr{C}$ can be easily obtained by hand. However, in general, it needs to be computed by solving numerically Poisson's equations.

Example 7.1. Show that the capacitance of a parallel plate capacitor, consisting of two concentric spheres of radius a and b with $a < b$ (see Fig. 7.3) is given by

$$C = \frac{4\pi\varepsilon}{\frac{1}{a} - \frac{1}{b}} \text{ (F)}.$$

Indeed, let us call Γ_1 the interior sphere and Γ_2 the exterior one. In order to compute the capacitance matrix we have to solve the two following boundary-value problems:

$$\begin{cases} -\operatorname{div}(\varepsilon \operatorname{grad} v^1) = 0, \\ v^1 = 1 \quad \text{on } \Gamma_1, \\ v^1 = 0 \quad \text{on } \Gamma_2, \end{cases}$$

$$\begin{cases} -\operatorname{div}(\varepsilon \operatorname{grad} v^2) = 0, \\ v^2 = 0 \quad \text{on } \Gamma_1, \\ v^2 = 1 \quad \text{on } \Gamma_2. \end{cases}$$

From the spherical symmetry of the problems, the potentials only depends on the radial coordinate, that is, $v_i(x) = \hat{v}_i(r(x))$, $i = 1, 2$, and hence the expression of the Laplacian operator reduces to (see B.2.8.3),

$$\Delta \varphi = \frac{1}{r^2} \frac{d}{dr} \left(r^2 \frac{d\hat{\varphi}}{dr} \right)$$

and the above problems become: find v^1, v^2 such that

$$\begin{cases} \frac{d}{dr} \left(r^2 \frac{d\hat{v}^1(r)}{dr} \right) = 0 \text{ in } (-\infty, a) \cup (a, b) \cup (b, \infty), \\ \hat{v}^1(a) = 1, \\ \hat{v}^1(b) = 0, \end{cases}$$

$$\begin{cases} \frac{d}{dr} \left(r^2 \frac{d\hat{v}^2(r)}{dr} \right) = 0 \text{ in } (-\infty, a) \cup (a, b) \cup (b, \infty), \\ \hat{v}^2(a) = 0, \\ \hat{v}^2(b) = 1. \end{cases}$$

The general solution of problem

$$\frac{1}{r^2} \frac{d}{dr} \left(r^2 \frac{d\varphi}{dr} \right) = 0$$

is

$$\varphi(r) = \frac{C_1}{r} + C_2.$$

Then, taking into account the boundary conditions and that $\hat{v}^1(0) < \infty$, we get

$$\hat{v}^1(r) = \begin{cases} 1 & r \in (0, a), \\ -\frac{a}{b-a} + \frac{ab}{(b-a)r} & r \in (a, b), \\ 0 & r \in (b, \infty). \end{cases}$$

7.5 Electric field created by a set of charged conductors. Capacitance matrix

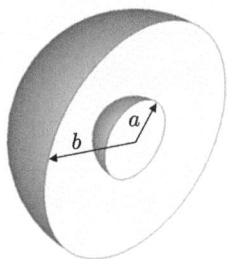

Fig. 7.3. Sketch of a parallel plate spherical capacitor

Similarly, we obtain the following expression for $\hat{v}^2$,

$$\hat{v}^2(r) = \begin{cases} 0 & r \in (0,a), \\ \frac{b}{b-a} - \frac{ab}{(b-a)r} & r \in (a,b), \\ \frac{b}{r} & r \in (b,\infty), \end{cases}$$

because the space $W^1(\mathscr{E})$ does not contain the constant functions.

From the above expressions for $\hat{v}^1$ and $\hat{v}^2$ we get

$$\frac{d\hat{v}^1}{dr}(r) = \begin{cases} -\frac{ab}{(b-a)r^2} & r \in (a,b), \\ 0 & r \in (0,a) \cup (b,\infty), \end{cases}$$

$$\frac{d\hat{v}^2}{dr}(r) = \begin{cases} 0 & r \in (0,a), \\ \frac{ab}{(b-a)r^2} & r \in (a,b), \\ \frac{-b}{r^2} & r \in (b,\infty). \end{cases}$$

Hence,

$$-c_{12} = c_{11} = \int_a^b \int_0^\pi \int_0^{2\pi} (r^2 \sin\phi)\varepsilon \left(\frac{ab}{(b-a)r^2}\right)^2 dr d\phi d\theta = \frac{4\pi\varepsilon}{\frac{1}{a} - \frac{1}{b}}$$

$$c_{22} = \int_a^b \int_0^\pi \int_0^{2\pi} (r^2 \sin\phi)\varepsilon \left(\frac{ab}{(b-a)r^2}\right)^2 dr d\phi d\theta$$

$$+ \int_c^\infty \int_0^\pi \int_0^{2\pi} (r^2 \sin\phi)\varepsilon \left(\frac{-c}{r^2}\right)^2 dr d\phi d\theta$$

$$= \frac{4\pi\varepsilon}{\frac{1}{a} - \frac{1}{b}} + 4\pi\varepsilon \int_c^\infty \frac{c^2}{r^2} dr = \frac{4\pi\varepsilon}{\frac{1}{a} - \frac{1}{b}} + 4\pi\varepsilon c$$

Finally, the capacitance matrix is

$$\mathscr{C} = \begin{pmatrix} \dfrac{4\pi\varepsilon}{\frac{1}{a}-\frac{1}{b}} & -\dfrac{4\pi\varepsilon}{\frac{1}{a}-\frac{1}{b}} \\ -\dfrac{4\pi\varepsilon}{\frac{1}{a}-\frac{1}{b}} & \dfrac{4\pi\varepsilon}{\frac{1}{a}-\frac{1}{b}} + 4\pi\varepsilon c \end{pmatrix}$$

and the *reduced capacitance* is

$$C = \frac{c_{11}c_{22} - c_{12}^2}{c_{22} + 2c_{12} + c_{11}} = \frac{4\pi\varepsilon}{\frac{1}{a} - \frac{1}{b}}.$$

Example 7.2. Let us consider the two coaxial conductors depicted in Fig. 7.4. Let us suppose they are infinite in the orthogonal direction to the figure. In fact, this case is not covered by the previous theory because the conductors themselves are not bounded, and the total charge is not finite so we cannot affirm that the potential belongs to the 3D function space $\mathbf{W}^1(\Omega)$. However, since there is translational symmetry in the x_3-direction, it is possible to work in a 2D orthogonal section to that direction (see Fig. 7.4). But in 2D we have to allow for functions having logarithmic growth at infinity, so we cannot use the 2D analogous space to $\mathbf{W}^1(\Omega)$ but the following one (see [57]):

$$\mathbf{W}^{1,-1}(\Omega) := \left\{ w : \int_\Omega \frac{|w(x)|^2}{(1+\rho(x)^2)(\log(2+\rho(x)))^2} dV < \infty, \; \text{grad}\, w \in \mathbf{L}^2(\Omega) \right\}.$$

Let us notice that constant functions belong to this space.

It turns out that the bilinear form a is not coercive in this space (indeed, constant functions belongs to its kernel). This is why one can only show that the capacitance matrix is positive semi-definite. Actually, it can be singular as the present example shows.

Let function v^1 be the solution to the problem

$$\begin{cases} -\text{div}(\varepsilon \,\text{grad}\, v^1) = 0 \text{ in } \Omega, \\ v^1 = 1 \text{ on } \Gamma_1, \\ v^1 = 0 \text{ on } \Gamma_2^1 \cup \Gamma_2^2. \end{cases}$$

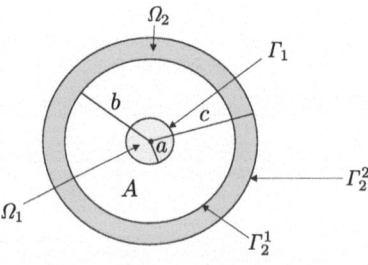

Fig. 7.4. Two coaxial conductors

7.5 Electric field created by a set of charged conductors. Capacitance matrix

If ε is constant this equation can be analytically solved by using polar coordinates. Indeed, firstly for symmetry reasons the solution only depends on the radial coordinate, i. e., $v^1(x) = \hat{v}^1(\rho(x))$, and then

$$\Delta \hat{v}^1 = \frac{1}{\rho} \frac{d}{d\rho} \left(\rho \frac{d\hat{v}^1}{d\rho} \right) = 0, \ \rho \in (a,b) \cup (c, \infty), \tag{7.27}$$

with

$$\hat{v}^1(a) = 1, \ \hat{v}^1(b) = \hat{v}^1(c) = 0.$$

By integrating (7.27) we get

$$\hat{v}^1(\rho) = R_1 \ln \rho + S_1,$$

and using the boundary conditions we obtain

$$\hat{v}^1(\rho) = \frac{1}{\ln\left(\frac{a}{b}\right)} \ln\left(\frac{\rho}{b}\right), \ a < \rho < b,$$

$$\hat{v}^1(\rho) = 0, \ \rho > c.$$

Then

$$\frac{d\hat{v}^1}{d\rho}(\rho) = \frac{1}{\rho} \frac{1}{\ln\left(\frac{a}{b}\right)}, \ a < \rho < b,$$

$$\frac{d\hat{v}^1}{d\rho}(\rho) = 0, \ \rho > c.$$

Similarly, one can obtain $\hat{v}^2$ which is given by

$$\hat{v}^2(\rho) = \frac{1}{\ln\left(\frac{b}{a}\right)} \ln\left(\frac{\rho}{a}\right), \ a < \rho < b,$$

$$\hat{v}^2(\rho) = 1, \ \rho > c.$$

Then

$$\frac{d\hat{v}^2}{d\rho}(\rho) = \frac{1}{\rho} \frac{1}{\ln\left(\frac{b}{a}\right)}, \ a < \rho < b,$$

$$\frac{d\hat{v}^2}{d\rho}(\rho) = 0, \ \rho > c,$$

7 Electrostatics

and hence the capacitance matrix is

$$c_{11} = 2\pi \int_a^b \varepsilon\rho \frac{1}{\rho^2} \frac{1}{[\ln(\frac{a}{b})]^2} d\rho = \frac{2\pi\varepsilon}{[\ln(\frac{a}{b})]^2} \ln\left(\frac{b}{a}\right) = \frac{2\pi\varepsilon}{\ln(\frac{b}{a})},$$

$$c_{22} = 2\pi \int_a^b \varepsilon\rho \frac{1}{\rho^2} \frac{1}{[\ln(\frac{b}{a})]^2} d\rho = \frac{2\pi\varepsilon}{[\ln(\frac{b}{a})]^2} \ln\left(\frac{b}{a}\right) = \frac{2\pi\varepsilon}{\ln(\frac{b}{a})},$$

$$c_{12} = c_{21} = 2\pi \int_a^b \varepsilon\rho \frac{1}{\rho^2} \left(-\frac{1}{[\ln(\frac{b}{a})]^2}\right) d\rho = -\frac{2\pi\varepsilon}{[\ln(\frac{b}{a})]^2} \ln\left(\frac{b}{a}\right)$$

$$= -\frac{2\pi\varepsilon}{\ln(\frac{b}{a})}.$$

Finally, since $c_{11} = c_{22}$ we have

$$C = \frac{c_{11} - c_{12}}{2} = \frac{2\pi\varepsilon}{\ln(\frac{b}{a})} \quad (\text{F/m}).$$

Of course, in this particular example, the reduced capacitance can also be obtained by a much simpler procedure based on the Gauss' law in integral form rather than in differential form. Thus, we determine first the electric field and then the potential. Indeed, due to symmetry, field **D** (and then **E**) must be radial and independent of θ and z in cylindrical coordinates; namely, it is of the form

$$\mathbf{D}_\rho(x) = D_\rho(\rho(x))\mathbf{e}_\rho(x).$$

We take a coaxial cylinder of radius ρ and unit length enclosing the inner conductor and call S_0 its lateral surface. We have

$$\int_{S_0} D_\rho \mathbf{e}_\rho \cdot \mathbf{e}_\rho dA = Q,$$

where Q is the charge in the inner conductor per unit length. From this we deduce

$$D_\rho(\rho) = \frac{Q}{2\pi\rho},$$

and then

$$\mathbf{E} = E_\rho \mathbf{e}_\rho \quad \text{with} \quad E_\rho(\rho) = \frac{Q}{2\pi\rho\varepsilon},$$

for $a < \rho < b$.

The difference of potential between a reference point p_2 in the outer conductor and another point p_1 on the inner conductor (we take p_1 and p_2 along the same radius ad in Fig. 7.5) is given by

$$V_1 - V_2 = -\int_b^a \frac{Q}{2\pi\varepsilon\rho} \mathbf{e}_\rho \cdot \mathbf{e}_\rho d\rho = \frac{Q}{2\pi\varepsilon} \ln\left(\frac{b}{a}\right).$$

7.5 Electric field created by a set of charged conductors. Capacitance matrix

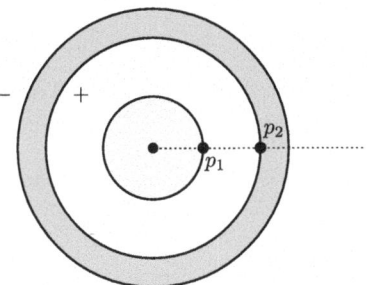

Fig. 7.5. Section of a coaxial conductor

Then the reduced capacitance per unit length is given by

$$C = \frac{Q}{V_1 - V_2} = \frac{Q}{\frac{Q}{2\pi\varepsilon}\ln\left(\frac{b}{a}\right)} = \frac{2\pi\varepsilon}{\ln\left(\frac{b}{a}\right)} \text{ (F/m)}.$$

We observe that it only depends on the geometrical parameters and the electric permittivity of the dielectric inside. We also notice that the above computations assume an infinite cylinder. Otherwise there are "end effects" and the potential would be determined by numerical solution of the Poisson's equation.

Example 7.3. Let us consider two infinite parallel plates separated a distance d which is filled with a dielectric of permittivity ε. Again, this case does not fit in the 3D theory developed above, because the load is not compactly supported. We also have translational symmetry but now in two directions. Then one can work in the orthogonal direction so the potential problem is the 1D boundary-value problem defined below.

In 1D we have to allow for functions having linear growth at infinity. Since this space include the constant functions, the bilinear form a is again not coercive (indeed, constant functions also belongs to its kernel). Thus, the capacitance matrix is only positive semi-definite. In this particular example it is singular. From the boundary-value problem

$$\varepsilon\frac{d^2V}{dx^2} = 0 \text{ in } (-\infty,0) \cup (0,d) \cup (d,\infty),$$

$$V(0) = V_1,$$

$$V(d) = V_2.$$

functions v^1 and v^2 are easily computed:

$$v^1(x) = 1, \ x \in (-\infty,0), \ v^1(x) = 1 - \frac{x}{d}, \ x \in (0,d), \ v^1(x) = 0, \ x \in (d,\infty),$$

$$v^2(x) = 0, \ x \in (-\infty,0), \ v^2(x) = \frac{x}{d}, \ x \in (0,d), \ v^2(x) = 1, \ x \in (d,\infty),$$

and hence,

$$\frac{dv^1}{dx} = -\frac{1}{d}, \quad x \in (0,d) \quad \frac{dv^1}{dx} = 0 \text{ outside},$$

$$\frac{dv^2}{dx} = \frac{1}{d}, \quad x \in (0,d) \quad \frac{dv^2}{dx} = 0 \text{ outside}.$$

Then,

$$c_{11} = \varepsilon \int_0^d \left(-\frac{1}{d}\right)^2 dx = \frac{\varepsilon}{d},$$

$$c_{22} = \varepsilon \int_0^d \left(\frac{1}{d}\right)^2 dx = \frac{\varepsilon}{d},$$

$$c_{12} = c_{21} = \varepsilon \int_0^d \left(-\frac{1}{d}\right)\left(\frac{1}{d}\right) dx = -\frac{\varepsilon}{d}.$$

Since $c_{11} = c_{22}$, the reduced capacitance per unit surface is

$$C = \frac{c_{11} - c_{12}}{2} = \frac{\varepsilon}{d} b \quad (F/m^2).$$

7.5.2 Energy

Let us take $w = V$ in (7.25). We get

$$\int_\Omega \varepsilon(x) |\operatorname{grad} V(x)|^2 dV = \sum_{l=1}^M Q_l V_{|\Gamma_l} = \sum_{l=1}^M Q_l V_l.$$

Then the *electrostatic energy* of the system is given by

$$\mathscr{U} = \frac{1}{2} \int_\Omega \mathbf{D} \cdot \mathbf{E} \, dV = \frac{1}{2} \int_\Omega \varepsilon |\operatorname{grad} V|^2 dV = \frac{1}{2} \sum_{l=1}^M Q_l V_l$$

$$= \sum_{l=1}^M \sum_{j=1}^M c_{lj} V_j V_l = \frac{1}{2} \vec{V}^t \mathscr{C} \vec{V}.$$

Problems

7.1. Use Gauss' law to determine the electric field intensity created by a charged spherical conducting shell of radius r_0 with a surface charge density ρ_S. What can you conclude? Compute the electric field inside any closed hollow conductor assuming that the region enclosed by the conductor contains no charges.

7.5 Electric field created by a set of charged conductors. Capacitance matrix

Solution: You must conclude that the electric field is the same as that generated when all the charge is concentrated at the center of the shell. Moreover, the electric field inside the shell is zero. Same reasoning applies to any closed hollow conductor assuming it contains no charges inside.

7.2. An insulating sphere of radius r_0 carries a volume charge density that is proportional to the distance from the center $\rho_V = \rho_0 r$ where ρ_0 is a constant which is null for $r > r_0$. Use the Gauss' law to find the electric field and the electric potential inside and outside the sphere.

Solution: See Sect. 13.1.4.

7.3. Compute the electric field and the electric potential at point p created by a line charge of finite length $2L$ centered on the x_3-axis. **Hint:** Employ expressions (7.6) and (7.4) from Sect. 7.1.

Solution: See Sect. 13.1.6.

7.4. Consider two infinitely long concentric cylindrical shells. The inner shell, with radius r_1 carries an uniform surface charge density ρ_{S_1} and the outer shell, with radius r_2, carries an uniform surface charge density ρ_{S_2}.

(a) Use Gauss' law to find the electric field inside the first cylinder, between and outside the cylinders.
(b) What would the electric field between the shells be if the electric field is zero outside the cylinders?

Solution: $E_\rho(x) = 0$ for $\rho(x) < r_1$, $E_\rho(x) = \frac{\rho_{S_1} r_1}{\varepsilon_0 \rho(x)}$ for $r_1 < \rho(x) < r_2$ and $E_\rho(x) = \frac{\rho_{S_1} r_1 + \rho_{S_2} r_2}{\varepsilon_0 \rho(x)}$ for $\rho(x) > r_2$ (b) $\frac{\rho_{S_1} r_1}{\varepsilon_0 \rho(x)}$ or $-\frac{\rho_{S_2} r_2}{\varepsilon_0 \rho(x)}$.

7.5. A circular ring of radius a carries a uniform charge density per unit length ρ_l and it is placed on the $x_1 x_2$-plane with axis the x_3-axis.

(a) Compute the electric field at a point $p = (0,0,h)$.
(b) What values of h gives the maximum value of **E**?
(c) If the total charge on the ring is Q, find **E** as $a \to 0$. What can you conclude?

Solution: (a) $\mathbf{E}(0,0,h) = \frac{\rho_l a h}{2\varepsilon_0 (h^3 + a^2)^{3/2}} \mathbf{e}_z$ (b) $h = \pm a/\sqrt{(2)}$ (c) $\mathbf{E} = \frac{Q}{4\pi \varepsilon_0 h^2} \mathbf{e}_z$. It is the same as that of a point charge.

7.6. A circular wire of radius r has a uniform linear charge density $\rho_l = \rho_0 \cos^2 \theta$. Compute the total charge on the wire.

Solution: $\pi \rho_0 r$.

7.7. A circular disk of radius a is uniformly charged with ρ_S C/m². If the disk lies on the $x_3 = 0$ plane with its axis along the x_3-axis,

(a) Compute the electric field **E** at point $p = (0,0,h)$.
(b) From this expression, derive the electric field due to an infinite sheet of charge on the $x_3 = 0$ plane.

(c) If $a \ll h$, show that **E** is similar to the field due to a point charge.

Solution: See Sect. 13.1.10.

7.8. The finite sheet $0 < x_1 < 1, 0 < x_2 < 1$ on the $x_3 = 0$ plane has a charge density $\rho_S = x_1 x_2 (x_1^2 + x_2^2 + 25)^{3/2}$ C/m². Find

(a) The total charge Q on the sheet.
(b) The electric field at point $p = (0,0,5)$.
(c) The magnitude of the force experienced by a 1 μC charge located at point $(0,0,5)$.

Solution: (a) $Q = 3.315\text{e-}8$ C (b) $\mathbf{E} = -1.5\mathbf{e}_1 - 1.5\mathbf{e}_2 + 11.25\mathbf{e}_3$ N/C (c) 0.01145 N.

7.9. A thin rod of length $2l$ is placed along the x_2-axis in the $x_1 x_2$-plane carrying a uniform charge density ρ_l. A point p_1 is located at $(0, 2l)$ and p_2 at $(d, 0)$.

(a) Find d if the potentials at p_1 and p_2 are equal.
(b) Find the corresponding potential.

Solution: (a) $d = \sqrt{3}a$ (b) 9.89×10^9 V.

7.10. Consider two charges, one positive and the other negative, of the same magnitude, separated by a distance d (this configuration is also called electric dipole in many physical books). Find the electric potential for this configuration assuming a magnitude charge 10^{-10} C located at points $p = (2,0,0)$ and $q = (-2,0,0)$.

Solution: See Sect. 13.1.2.

7.11. A linear quadrupole consists of charge +2Q at the origin and two charges -Q at $(-d, 0)$ and $(+d, 0)$. Compute the magnitude of the electric field at a point on the x_1-axis where $x_1 > d$.

Solution: $E = -\dfrac{2Qd^2(3x_1^2 - d^2)}{4\pi\varepsilon_0 x_1^2 (x_1^2 - d^2)^2}$.

7.12. Suppose we have a number of identical charges Q located at the vertex of a regular n-sided polygon. Determine the electric field at the center of the polygon depending on n is even or odd.

Solution: Zero in both cases. If the number of charges is even, the fact is clear. The field produced by each charge is canceled with the field due to the charge on the opposite vertex and the result is a null field. The result is also true if n is odd. In that case, the resultant field can be obtained by vectorially summing the individual fields produced by contiguous charges. The result is another regular polygon in which the end of the last vector coincides with the origin of the first one, and then the resultant is null.

7.13. Two-infinite and uniformly charged planes which lie parallel to each one and to the $x_2 x_3$-plane are located at distance d. The first one is at $x_1 = -a$ (m) and has a surface charge density ρ_{S_1}. The other is at $x_1 = a$ (m) and has a surface charge density ρ_{S_2}. Find the electric fields for $x_1 < -a$, $-a < x_1 < a$, $x_1 > a$ in the following cases

7.5 Electric field created by a set of charged conductors. Capacitance matrix

(a) $\rho_{S_1} = \rho_S$ a positive constant and $\rho_{S_2} = \rho_{S_1}$.
(b) $\rho_{S_1} = \rho_S$ a positive constant and $\rho_{S_2} = -\rho_{S_1}$.

Solution: (a) $\mathbf{E} = -\rho_S/\varepsilon_0 \mathbf{e}_1$ for $x_1 < -a$, zero between the planes and $\mathbf{E} = \rho_S/\varepsilon_0$ for $x > a$ (b) $E_x = \rho_S/\varepsilon_0$ between the planes and zero outside.

7.14. Consider a uniform electric field $\mathbf{E} = 4 \times 10^3 \mathbf{e}_1$ N/C. Compute

(a) The flux of this field through a square of side 10 cm in a plane parallel to the $x_2 x_3$-plane.
(b) The flux through the same square if the normal to its plane makes a 45° angle with the x_1-axis.

Solution: (a) 40 N/C m² (b) 28.3 N/C m².

7.15. An insulating spherical shell of inner radius r_1 and outer radius r_2 is charged so that its volume charge density is given by $\rho_V = a/r$, where a is a non positive constant for $r_1 < r < r_2$ and zero otherwise. Find the electric field due to the shell throughout all space.

Solution: $E_r = 0$, for $0 \le r < r_1$, $E_r = \frac{a}{2\varepsilon_0 r^2}(r^2 - r_1^2)$ for $r_1 \le r \le r_2$ and $E_r = \frac{a}{2\varepsilon_0 r^2}(r_2^2 - r_1^2)$ for $r > r_2$.

7.16. Inside a sphere of radius r_0 and uniformly charged with a volume charge density ρ_V, there is a neutral spherical cavity of radius r_1 with its center at distance a from the center of the charged sphere. If $(r_1 + a) < r_0$, find the magnitude of the electric field inside the cavity.

Solution: $\rho_V a/(3\varepsilon_0)$.

7.17. In the ESTA® patented separation process, the salt minerals in the crude salt are separated using an electrical voltage field as shown in Fig. 7.6. This method utilizes the different electrical properties of the surfaces of individual salts. When passing through a high-voltage field, the positively charged particles (the mixture of substances) are diverted in one direction and the negatively charged particles (the residue) in another. Assuming zero initial velocity and displacement, determine the separation between the particles after falling 80 cm. Take the electric field intensity

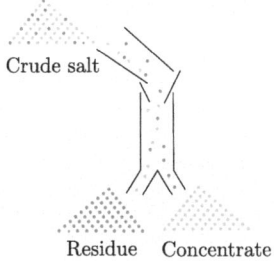

Fig. 7.6. Salt separation process in Problem 7.17

$E = 500$ kV/m and $Q/m = 9\mu$C/kg for both positively and negatively charged particles.

Solution: 0.7347 m.

7.18. When painting is done with an ordinary spray gun, part of the paint escapes deposition. The fraction lost depends on the shape of the surface, on drafts, etc., and can be as high as 80%. The use of the ordinary spray gun in large-scale industrial processes would therefore result in intolerable waste and pollution. The efficiency of spray painting can be increased to nearly 100%, and the pollution reduced by a large factor, by charging electrically the droplets of paint, and applying a voltage difference between the gun and the object to be coated. It is found that, in such devices, the droplets carry a specific charge of roughly one coulomb per kilogram. Assuming that the electric field intensity in the region between the gun and the part is at least 10 kilovolts per meter, what is the minimum ratio of the electric force to the gravitational force?

Solution: $QE/mg = (Q/m)(E/g) = E/g >= 10^4/9.8 \approx 10^3$.

7.19. A 1 m length RG58 coaxial cable consists of a center conductor of about 0.4 mm radius and a polyethylene dielectric of 1.5 mm radius surrounded by an outer copper braid. Assuming that the permittivity of the dielectric is $\varepsilon = 2.26\varepsilon_0 = 20$ pF/m, compute the capacitance of the cable.

Solution: 95 pF/m.

7.20. Two dielectrics with dielectric constants ε_1 and ε_2 each fill half the space between the plates of a parallel plate capacitor as shown in Fig. 7.7. Each plate has an area A and the plates are separated by a distance d. Compute the capacitance of the system. *Hint:* Treat the system as being composed of two capacitors with charges $\pm Q_1$ and $\pm Q_2$ on each half.

Solution: $C = A(\varepsilon_1 + \varepsilon_2)/(2d)$.

7.21. Consider two conducting concentric spherical shells with inner radius a and outer radius c. Both shells have the same charge Q uniformly distributed over its surface but opposite in sign. Let us assume that the space between two surfaces is filled with two different dielectric materials so that the dielectric relative constant is ε_{r_1} between a and b, and ε_{r_2} between b and c, with $a < b < c$.
(a) Determine the capacitance of this system.
(b) Check the limit when $\varepsilon_{r_1}, \varepsilon_{r_2} \to 1$. What can you conclude?

Fig. 7.7. Parallel capacitor with two dielectrics

7.5 Electric field created by a set of charged conductors. Capacitance matrix

Hint: The system can be treated as two capacitors connected in series, since the total potential difference across the capacitors is the sum of potential differences across individual capacitors, $\Delta V = \Delta V_1 + \Delta V_2$.

Solution: (a) $C = \dfrac{Q}{\Delta V_1 + \Delta V_2} = \dfrac{4\pi\varepsilon_0 \varepsilon_{r_1} \varepsilon_{r_2} abc}{\varepsilon_{r_1} c(b-a) + \varepsilon_{r_2} a(c-b)}$ (b) The expression agrees with that for a spherical capacitor of inner radius a and outer radius c.

8
Direct current

In this chapter we deal with the problem of determining the electric fields in conducting domains under the assumption that sources, i.e., charges and currents do not depend on time but, unlike electrostatics, currents are not null. In this case, the electric and magnetic fields are not coupled. Thus, we can compute first the current density which can be used later in the magnetostatic problem (Chap. 9) to obtain the magnetic fields. We will state the direct current problem in terms of the scalar electric potential and introduce the resistance and conductance matrices. At the end of the chapter we give a brief description of lumped direct current circuits.

8.1 Maxwell's equations for direct current

We assume that charges and currents do not depend on time and then all fields must be independent on time. In this case, Maxwell's equations become

$$\operatorname{div} \mathbf{D} = \rho_V,$$
$$\operatorname{div} \mathbf{B} = 0,$$
$$\operatorname{\mathbf{curl}} \mathbf{E} = \mathbf{0},$$
$$\operatorname{\mathbf{curl}} \mathbf{H} = \mathbf{J},$$

with $\mathbf{D} = \varepsilon \mathbf{E}$, $\mathbf{B} = \mu \mathbf{H}$ and $\mathbf{J} = \sigma \mathbf{E}$.

The interesting feature of this case is that we can compute first the electric field and the current density, and then the magnetic field, because electric and magnetic field equations are uncoupled. The direct current problem deals with the computation of the former.

First, we notice that, by taking the divergence, Ampère's law yields

$$\operatorname{div} \mathbf{J} = 0,$$

A. Bermúdez, D. Gómez, P. Salgado: *Mathematical Models and Numerical Simulation in Electromagnetism.* UNITEXT – La Matematica per il 3+2 74
DOI 10.1007/978-3-319-02949-8_8, © Springer International Publishing Switzerland 2014

and from Ohm's law,
$$\operatorname{div}(\sigma \mathbf{E}) = 0 \tag{8.1}$$
in conductors.

Moreover $\mathbf{curl\,E} = 0$ implies the existence of an electric potential V such that $\mathbf{E} = -\operatorname{grad} V$. By replacing in (8.1) we get
$$-\operatorname{div}(\sigma \operatorname{grad} V) = 0, \tag{8.2}$$
which is a Laplace-like partial differential equation similar to the one of electrostatics.

In order to solve this equation, which is only valid in conductors (otherwise $\sigma = 0$), we consider boundary conditions. They can be either *Dirichlet* or *Neumann* conditions:

- Dirichlet: $V = V_d$ on Γ_d.
- Neumann: $-\sigma \frac{\partial V}{\partial \mathbf{n}} = \sigma \mathbf{E} \cdot \mathbf{n} = \mathbf{J} \cdot \mathbf{n} = g$ on Γ_n.

8.1.1 Weak formulation of the 3D direct current model

The weak formulation of the above boundary value problem can be obtained by standard procedures. It reads as follows:

Find $V \in H^1(\Omega)$ with $V = V_d$ on Γ_d such that
$$\int_\Omega \sigma \operatorname{grad} V \cdot \operatorname{grad} w \, dV = -\int_{\Gamma_n} g w \, dA \quad \forall w \in H^1(\Omega) \quad \text{with} \quad w|_{\Gamma_d} = 0.$$

Remark 8.1. The numerical discretization of this problem can be done by using continuous piecewise linear elements on a tetrahedral mesh of the domain (see Appendix E).

8.1.2 Weak formulation of the in-plane 2D direct current model

Taking into account the geometry of the domain and the symmetries of the problem, sometimes the solution of a (in principle 3D) direct current problems can be approximated by an in-plane 2D model. In particular, this is the case when the electrical conductivity of the domain and the boundary conditions are invariant under translation in one privileged direction, let us say, in the x_3-direction, and the current density does not have x_3-component. In this situation, the problem can be written in an orthogonal section to the x_3-axis that will be denoted by $\widehat{\Omega}$. Let us suppose the electric field (and then the current density) has the form
$$\mathbf{E} = E_1(x_1, x_2) \mathbf{e}_1 + E_2(x_1, x_2) \mathbf{e}_2,$$
where $\mathbf{e}_1$ and $\mathbf{e}_2$ denote the cartesian basis vectors. As a consequence, the electric scalar field V only depends on x_1 and x_2.

Let us denote by $\widehat{\Gamma}_d$ and $\widehat{\Gamma}_n$ the two disjoint sets that form the boundary of $\widehat{\Omega}$. As in the 3D case, the boundary conditions can be either Dirichlet or Neumann.

Then, the problem to be solved is the following:

Find $V: \widehat{\Omega} \to \mathbb{R}$ such that

$$-\text{div}(\sigma \, \text{grad} \, V) = 0 \text{ in } \widehat{\Omega}, \tag{8.3}$$

$$V = V_d \text{ on } \widehat{\Gamma}_d, \tag{8.4}$$

$$-\sigma \frac{\partial V}{\partial \mathbf{n}} = g \text{ on } \widehat{\Gamma}_n, \tag{8.5}$$

where $g = \mathbf{J} \cdot \mathbf{n}$ on Γ_n. The weak formulation of the above boundary-value problem states as follows:

Find a scalar field $V \in H^1(\widehat{\Omega})$ with $V = V_d$ on $\widehat{\Gamma}_d$ such that

$$\int_{\widehat{\Omega}} \sigma \, \text{grad} \, V \cdot \text{grad} \, w \, dV = -\int_{\widehat{\Gamma}_n} g w \, dA,$$

$\forall w \in H^1(\widehat{\Omega})$ with $w|_{\widehat{\Gamma}_d} = 0$.

Remark 8.2. The numerical discretization of this problem can be done by using continuous piecewise linear elements on a triangular mesh of the 2D domain $\widehat{\Omega}$ (see Appendix E).

8.1.3 The axisymmetric case

Let us assume that Ω is a 3D domain generated by the rotation about the x_3-axis of a 2D set $\widehat{\Omega}$. Let us assume that the direct current problem has cylindrical symmetry in the sense that the current density does not depend on the azimuthal variable and does not have tangential component. We also assume that the electrical conductivity and the boundary conditions are independent of the azimuthal variable. Under these circumstances, the problem can be solved in $\widehat{\Omega}$. By using the expression of the Laplace operator in cylindrical coordinates (see Sect. B.2.8.2), Eq. (8.2) writes as follows

$$-\frac{1}{\rho} \frac{\partial}{\partial \rho}\left(\sigma \rho \frac{\partial V}{\partial \rho}\right) - \frac{\partial}{\partial z}\left(\sigma \frac{\partial V}{\partial z}\right) = 0,$$

where $V = V(\rho, z)$. Multiplying the previous equation by ρ, we obtain

$$-\frac{\partial}{\partial \rho}\left(\sigma \rho \frac{\partial V}{\partial \rho}\right) - \frac{\partial}{\partial z}\left(\sigma \rho \frac{\partial V}{\partial z}\right) = 0.$$

Hence, by setting $\tilde{\sigma} = \sigma \rho$ this equation is similar to the 2D problem introduced in Sect. 8.1.2.

Remark 8.3. The numerical discretization of this problem can be done by using continuous piecewise linear elements on a triangular mesh of the 2D meridional section $\widehat{\Omega}$ (see Appendix E).

8.2 Conductance matrix. Resistance matrix

Let us consider a conducting domain Ω with boundary Γ. Let $\Gamma_0,\ldots,\Gamma_P$ be the ports of Ω, i.e., the connected subsets of Γ through which direct currents enter or leave Ω (see Fig. 8.1).

We assume that no current exists across the rest of the boundary $\Gamma_n = \Gamma \setminus (\Gamma_0 \cup \cdots \cup \Gamma_P)$. We also suppose that current enters or leaves the domain Ω through the ports perpendicularly to them (i.e., $\mathbf{J} \times \mathbf{n} = 0$).

These conditions mean,

$$\sigma \frac{\partial V}{\partial \mathbf{n}} = -\mathbf{J} \cdot \mathbf{n} = 0 \text{ on } \Gamma_n,$$

$$\mathbf{E} \times \mathbf{n} = 0 \text{ on } \Gamma_0 \cup \ldots \cup \Gamma_P.$$

Since $\mathbf{E} = -\operatorname{grad} V$, the latter condition yields

$$V = V_i \text{ (constant) on } \Gamma_i,\ i = 0,\ldots,P,$$

because each Γ_i is supposed to be connected.

Knowing the values V_i, $i = 0,\ldots,P$, we can state the well-posed boundary-value problem

$$-\operatorname{div}(\sigma \operatorname{grad} V) = 0 \text{ in } \Omega, \tag{8.6}$$

$$\sigma \frac{\partial V}{\partial \mathbf{n}} = 0 \text{ on } \Gamma_n, \tag{8.7}$$

$$V = V_i \text{ on } \Gamma_i,\ i = 0,\ldots,P, \tag{8.8}$$

and then solve it by numerical methods.

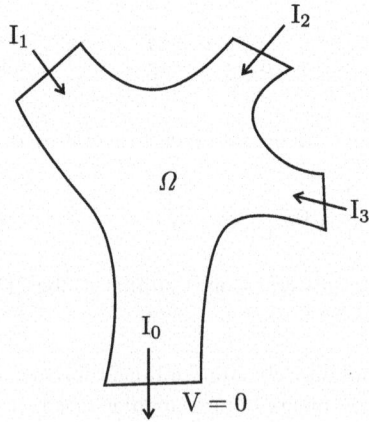

Fig. 8.1. Conducting domain with several ports

8.2 Conductance matrix. Resistance matrix

In order to write a weak formulation of this problem let us introduce the functional spaces

$$\mathscr{L} = \{w \in H^1(\Omega) : w_{|\Gamma_i} = c_i,\ i = 0,\ldots,P \text{ with } (c_0,\ldots,c_P) \in \mathbb{R}^{P+1}\}$$
$$\mathscr{L}_0 = \{w \in H^1(\Omega) : w_{|\Gamma_i} = 0,\ i = 0,\ldots,P\},$$

which are both closed subspaces of $H^1(\Omega)$ and $\mathscr{L}_0 \subset \mathscr{L}$. By multiplying (8.6) by a test function $w \in \mathscr{L}_0$, integrating in Ω and using a Green's formula we get the following weak formulation:

Find $V \in \mathscr{L}$ *such that* $V_{|\Gamma_i} = V_i,\ i = 0,\ldots,P$ *and*

$$\int_\Omega \sigma \operatorname{grad} V \cdot \operatorname{grad} w \, dV = 0 \quad \forall w \in \mathscr{L}_0. \tag{8.9}$$

However, very often we know the intensities through each Γ_i instead of the potentials. Firstly, let us recall that the intensities entering the domain are given by

$$I_i = -\int_{\Gamma_i} \mathbf{J} \cdot \mathbf{n}\, dA = \int_{\Gamma_i} \sigma \frac{\partial V}{\partial \mathbf{n}}\, dA,\ i = 1,\ldots,P.$$

Hence, by using Gauss' theorem we get

$$\sum_{i=0}^P -\int_{\Gamma_i} \mathbf{J} \cdot \mathbf{n}\, dA = -\int_\Gamma \mathbf{J} \cdot \mathbf{n}\, dA = -\int_\Omega \operatorname{div} \mathbf{J}\, dV = 0. \tag{8.10}$$

Moreover, since V is only defined up to a constant, we can choose, for instance, $V_0 = 0$, i.e., $V(x) = 0$ on Γ_0.

In order to write a weak formulation of the problem when the intensities are prescribed, we introduce the functional space

$$\mathscr{W} = \{w \in H^1(\Omega) : w_{|\Gamma_0} = 0,\ w_{|\Gamma_i} = c_i,\ i = 1,\ldots,P \text{ with } (c_1,\ldots,c_P) \in \mathbb{R}^P\}.$$

Multiplying (8.6) by a test function $w \in \mathscr{W}$, integrating in Ω and using a Green's formula we get

$$\int_\Omega \sigma \operatorname{grad} V \cdot \operatorname{grad} w \, dV = -\sum_{i=1}^P \int_\Gamma \mathbf{J} \cdot \mathbf{n} w\, dA = -\sum_{i=1}^P w_{|\Gamma_j} \int_{\Gamma_j} \mathbf{J} \cdot \mathbf{n}\, dA = \sum_{i=1}^P I_j w_{|\Gamma_j}.$$

Hence, the weak formulation of the problem is the following:

Find $V \in \mathscr{W}$ *such that*

$$\int_\Omega \sigma \operatorname{grad} V \cdot \operatorname{grad} w \, dV = \sum_{i=1}^P I_j w_{|\Gamma_j} \quad \forall w \in \mathscr{W}. \tag{8.11}$$

We notice that, since $\mathscr{L}_0 \subset \mathscr{W}$, from (8.11) we deduce

$$\int_\Omega \sigma \operatorname{grad} V \cdot \operatorname{grad} w \, dV = 0 \quad \forall w \in \mathscr{L}_0.$$

Hence, taking in problem (8.9) $V_1,\ldots,V_P$ as the (constant) traces of the solution of (8.11) we deduce that both problems (8.9) and (8.11) have the same solution. Now, for each $j \in \{1,\ldots,P\}$ let v^j be the unique solution to problem (8.9) corresponding to the potentials at the ports given by $V_i = \delta_{ij}$, $j = 1,\ldots,P$, and $V_0 = 0$. Then, for any given vector of potentials $\vec{V}$ the solution of (8.9) is the linear combination

$$V(x) = \sum_{j=1}^{P} V_j v^j(x).$$

By replacing this expression in (8.11) and taking $w = v^i$ as test function we get

$$\sum_{j=1}^{P} V_j \int_{\Omega} \sigma(x) \operatorname{grad} v^j(x) \cdot \operatorname{grad} v^i(x) \, dV = \sum_{l=1}^{P} I_l v^i_{|\Gamma_l} = \sum_{l=1}^{P} I_l \delta_{il} = I_i.$$

These equalities show that there exists a linear relation between potentials and current intensities at ports $\Gamma_1,\ldots,\Gamma_P$. More precisely, let $\mathcal{G}$ be the matrix whose elements g_{ij} are defined by

$$g_{ij} = \int_{\Omega} \sigma \operatorname{grad} v^j \cdot \operatorname{grad} v^i \, dV. \tag{8.12}$$

Then

$$\vec{I} = \mathcal{G}\vec{V},$$

where

$$\vec{I} = \begin{pmatrix} I_1 \\ \vdots \\ I_P \end{pmatrix}$$

and

$$\vec{V} = \begin{pmatrix} V_1 \\ \vdots \\ V_P \end{pmatrix}.$$

Since the bilinear form

$$a(v,w) = \int_{\Omega} \sigma(x) \operatorname{grad} v(x) \cdot \operatorname{grad} w(x) \, dV$$

is symmetric and coercive in $\mathcal{W}$, then matrix $\mathcal{G}$ is symmetric and positive definite. It is called the *conductance matrix* of the domain Ω and its entries have units ampere per volt (A/V). We notice that it only depends on the geometry, Ω, and on the electrical conductivity.

The inverse of $\mathcal{G}$, denoted by $\mathcal{R}$, is called *resistance matrix*. Since we have

$$\vec{V} = \mathcal{G}^{-1}\vec{I} = \mathcal{R}\vec{I},$$

it allows us to obtain potentials from intensities. The previous equality is called *Ohm's law* for circuits. The unit for the elements of $\mathcal{R}$ is ohm (Ω).

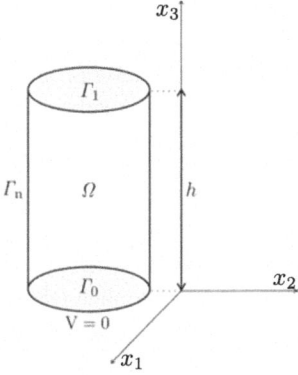

Fig. 8.2. Conducting cylinder with two ports Γ_0 and Γ_1

Example 8.1. Let us consider a conducting cylinder with height h, radius R and constant electric conductivity σ (see Fig. 8.2).

Since $P = 1$, the resistance matrix is just a number. In order to determine it let us consider the boundary-value problem

$$-\text{div}(\sigma \,\text{grad}\, V) = 0 \text{ in } \Omega$$
$$V = 1 \text{ on } \Gamma_1,$$
$$V = 0 \text{ on } \Gamma_0,$$
$$\sigma \frac{\partial V}{\partial \mathbf{n}} = 0 \text{ on } \Gamma_n.$$

It is easy to see that the solution to this problem is

$$V(x) = \frac{x_3}{h},$$

and the corresponding intensity is

$$I = I_1 = \int_{\Gamma_1} \sigma \frac{\partial V}{\partial \mathbf{n}} = \frac{\sigma}{h} \pi R^2.$$

Thus, $\mathscr{G}$ is the 1x1 matrix

$$\mathscr{G} = \frac{\sigma \pi R^2}{h} \text{ (A/V)},$$

and hence

$$\mathscr{R} = \frac{h}{\sigma \pi R^2} \text{ (}\Omega\text{)}.$$

This resistance coincides with the approximate resistance R_{ap} given below for thin electric conductors since the current density in this case is constant along the conductor.

8.3 Approximate resistance

We have previously seen that the computation of the resistance matrix requires to solve PDE problems which must be done, in general, by using numerical methods. However, if the conductors are thin enough (i.e., with a small diameter compared to its length) the resistance may be approximated in a simple way.

Let us consider a thin conductor carrying a direct current $\mathbf{J}$ as sketched in Fig. 8.3. In the conductor we know that $\mathbf{J} = \sigma \mathbf{E}$ and the current intensity through any cross section Σ is given by $I = \int_\Sigma \mathbf{J} \cdot \mathbf{n} \, dA$. If we denote by $\mathbf{E}$ and $\mathbf{E}_g$ the electric field in the conductor and in the generator respectively, then taking into account that for direct currents the electric field is the gradient of a potential, we have

$$\int_l \mathbf{E} \cdot d\mathbf{l} = \int_{p_1}^{p_2} \mathbf{E} \cdot d\mathbf{l} + \int_{p_2}^{p_1} \mathbf{E}_g \cdot d\mathbf{l} = 0$$

along any closed path l of the circuit of Fig. 8.3. If we assume that the battery does not have an internal resistance, the second integral is the negative source voltage of the generator, i. e., $-\mathrm{E}$. Thus,

$$\int_{p_1}^{p_2} \frac{\mathbf{J}}{\sigma} \cdot d\mathbf{l} = \mathrm{E}. \tag{8.13}$$

Since the conductor is assumed to be very thin, $\mathbf{J}$ can be approximated by an average value $\mathbf{J}_{ap}$ tangent to the centerline l as shown in Fig. 8.3; namely, let us define,

$$\mathbf{J}_{ap}(l) := \left(\frac{1}{\mathrm{meas}(\Sigma(l))} \int_{\Sigma(l)} \mathbf{J} \cdot \mathbf{n} \, dA \right) \mathbf{e}_l(l) = \frac{I}{\mathrm{meas}(\Sigma(l))} \mathbf{e}_l(l),$$

$\mathbf{e}_l$ being a unit tangent vector to the centerline l. Notice that notation $\Sigma(l)$ means that the cross section of the conductor may be non-uniform along the line l.

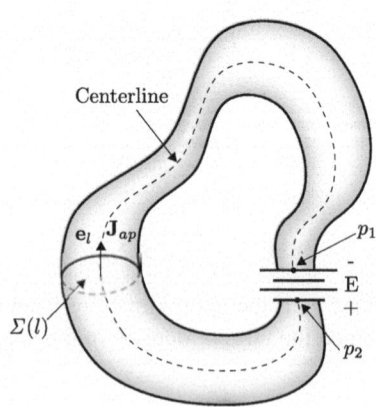

Fig. 8.3. A thin direct current electric circuit.

Thus, by approximating $\mathbf{J}$ by $\mathbf{J}_{ap}$ in Eq. (8.13) we obtain

$$\int_{p_1}^{p_2} \frac{\mathbf{J}_{ap}}{\sigma} \cdot d\mathbf{l} = \int_{p_1}^{p_2} \frac{I \, dl}{\sigma \operatorname{meas}(\Sigma(l))} \approx E,$$

which yields

$$I \approx \frac{E}{\int_{p_1}^{p_2} \frac{dl}{\sigma \operatorname{meas}(\Sigma(l))}}.$$

As a consequence, the resistance of a thin conductor can be approximated by

$$R_{ap} = \int_{p_1}^{p_2} \frac{dl}{\sigma \operatorname{meas}(\Sigma(l))}.$$

In particular, for a conductor of constant cross section, the approximate resistance is given by

$$R_{ap} = \frac{\text{mean length of the conductor}}{\sigma \operatorname{meas}(\Sigma)}.$$

Notice that for a thin conducting cylinder this approximate value coincides with the one computed in Example 8.1.

Remark 8.4. A real battery usually carries an internal resistance r and in that case, the potential difference between the terminals of the battery is equal to $E - Ir$.

8.4 Lumped direct current circuits

A direct current circuit is a particular case of the linear electric circuits studied in Chap. 3 by using graph theory. In particular, direct current circuits only have resistors and generators. Thus, by using the notation of Chap. 3, for an edge j with nodes m_{1j} and m_{2j} where there is a source voltage E_j with internal resistance r_j, we have

$$V_{m_{1j}} - V_{m_{2j}} = r_j I_j - E_j.$$

Similarly, along an edge j where there is only a resistor, we have

$$V_{m_{1j}} - V_{m_{2j}} = R_j I_j.$$

In the following examples, we will introduce the concept of equivalent resistance in direct current circuits connected *in series* or *in parallel*.

Example 8.2. Let us consider an electrical circuit with two resistors with resistances R_1 and R_2 connected end to end to form a single path for flowing electric current (see, for instance, Fig. 8.4); in this case, we say that the resistors are connected *in series* to a source voltage E. By the conservation of charge, in a series direct current circuit the same current I flows through all parts of the circuit. On the other hand, the voltage drop from a to c is the sum of the voltage drops across the individual resistors. Thus, if we assume that there is no internal resistance in the battery, we have

$$E = R_1 I + R_2 I = (R_1 + R_2) I.$$

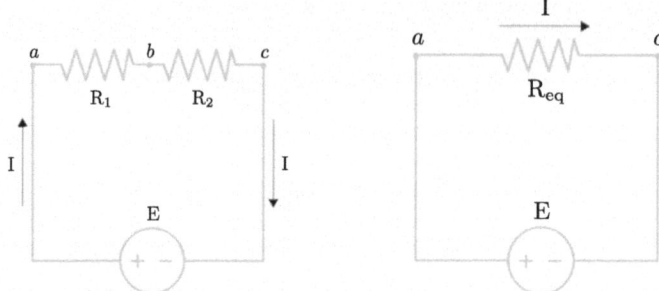

Fig. 8.4. A series direct current circuit (left). Equivalent circuit (right)

As a consequence, the two resistors in series could be replaced by an equivalent resistor with resistance R_{eq} given by

$$R_{eq} = R_1 + R_2,$$

and the voltage drop is identical, that is, $E = R_{eq}I$.

This argument can be extended to n resistors connected in series, where we can talk about an equivalent resistance which is the sum of the original resistances:

$$R_{eq} = \sum_{i=1}^{n} R_i.$$

Example 8.3. Let us consider an electrical circuit with two resistors with resistances R_1 and R_2 which are connected *in parallel* to a source voltage E (see Fig. 8.5). By charge conservation, current I through the source must divide into a current I_1 and a current I_2 which go, respectively, through resistors R_1 and R_2. Each resistor satisfies the Ohm's law and, as consequence, the voltage drop across any resistive component is equal to the product of its electrical resistance and the current intensity through it. On the other hand, the voltage drop along every component connected in parallel is

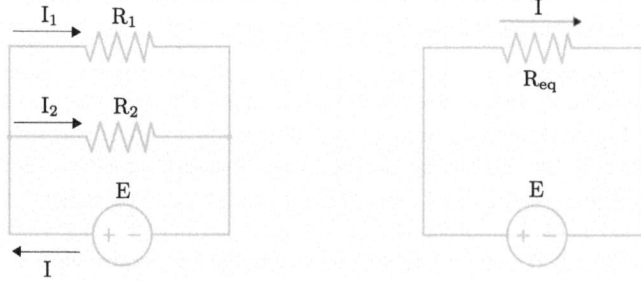

Fig. 8.5. Two resistors in parallel (left). Equivalent circuit (right)

the same and equal to E, thus

$$I = I_1 + I_2 = \frac{E}{R_1} + \frac{E}{R_2} = E(\frac{1}{R_1} + \frac{1}{R_2}).$$

Therefore, the two resistors in parallel can be replaced by an equivalent resistor with resistance given by R_{eq} (see Fig. 8.5),

$$\frac{1}{R_{eq}} = \frac{1}{R_1} + \frac{1}{R_2}.$$

This argument can be extended to n resistors connected in parallel:

$$\frac{1}{R_{eq}} = \sum_{i=1}^{n} \frac{1}{R_i}.$$

Problems

8.1. A resistor is made from a truncated cone of a material with uniform electrical conductivity σ. The minor radius of the cone is a, the major radius b and the height h. Let us assume that each cross section of the cone is an equipotential surface, that is, **E** is parallel to the central axis of the cone. Compute the approximate resistance of the cone.

Solution: $R_{ap} = \dfrac{h}{\sigma \pi ab} \Omega$.

8.2. Consider a hollow conducting cylinder of inner radius a, outer radius b, length h and electrical conductivity σ. A potential difference is applied between the ends of the cylinder that produces a current flowing parallel to its axis. Compute the resistance R_{ap}.

Solution: $R_{ap} = \dfrac{h}{\sigma \pi (b^2 - a^2)} \Omega$.

8.3. Compute the resistance of the hollow cylinder in the previous exercise if the potential difference is applied between the inner and outer surfaces so that current flows radially outwards.

Solution: $R_{ap} = \dfrac{\ln(\frac{b}{a})}{2\pi h \sigma} \Omega$.

8.4. Two concentric metallic spherical shells of radius a and b, respectively, are separated by a conducting material of conductivity σ. If they are maintained at a potential difference V, compute the resistance between the shells.

Solution: $R_{ap} = \dfrac{(b-a)}{4\pi \sigma ab} \Omega$.

9

Magnetostatics

In this chapter we deal with the problem of determining the magnetic field if we know a time independent current density **J**. This problem is known as *magnetostatics problem* and it will be studied in a bounded domain in terms of different scalar and vector fields. In the last part of the chapter we will introduce distributed and lumped magnetic circuits.

9.1 Maxwell's equations for magnetostatics

If we know the static, i.e., time independent current density **J**, the magnetostatic problem consists in finding the magnetic field **H** and the magnetic induction **B** satisfying

$$\mathbf{curl\,H} = \mathbf{J}, \tag{9.1}$$

$$\mathrm{div}\,\mathbf{B} = 0, \tag{9.2}$$

$$\mathbf{B} = \mu \mathbf{H}. \tag{9.3}$$

Firstly, let us introduce an example which can be solved by hand.

Example 9.1. Let us consider a n-turn, closely wound toroidal coil with rectangular cross-section filled with a magnetic material of constant permeability μ from a to b as in Fig. 9.1. The coil carries a static current I and we will compute the magnetostatic field inside and outside the torus.

For any point $\bar{x} = (\bar{x}_1, \bar{x}_2, \bar{x}_3)$ inside the torus, let us consider the circumference through $\bar{x}$ on the plane $x_3 = \bar{x}_3$ and centered at point $(0, 0, \bar{x}_3)$.

Let $\rho = \sqrt{(\bar{x}_1)^2 + (\bar{x}_2)^2}$ be the radius of this circumference. Notice that $a < \rho < c$. By symmetry reasons, the magnetic field **H** on this circumference must be of the form

$$\mathbf{H} = H_\theta(\rho)\mathbf{e}_\theta.$$

A. Bermúdez, D. Gómez, P. Salgado: *Mathematical Models and Numerical Simulation in Electromagnetism.* UNITEXT – La Matematica per il 3+2 74
DOI 10.1007/978-3-319-02949-8_9, © Springer International Publishing Switzerland 2014

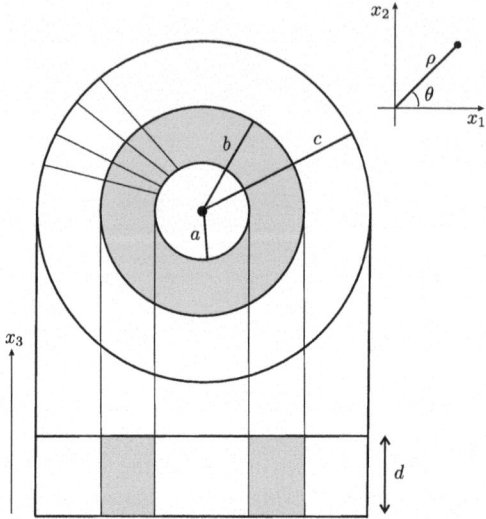

Fig. 9.1. Toroidal coil with a rectangular cross-section

Moreover, from the Ampère's law in integral form we have,

$$\int_l H_\theta(\rho) \mathbf{e}_\theta \cdot \mathbf{e}_\theta dl = nI,$$

which yields

$$H_\theta(\rho) = \frac{nI}{2\pi\rho}.$$

A similar reasoning shows that the magnetic field is null outside the torus.
Hence,

$$\mathbf{B}(\rho) = \begin{cases} \dfrac{\mu nI}{2\pi\rho}\mathbf{e}_\theta & \text{if } a < \rho < b, \\ \dfrac{\mu_0 nI}{2\pi\rho}\mathbf{e}_\theta & \text{if } b < \rho < c, \end{cases}$$

inside the torus, and $\mathbf{B} = 0$ outside the torus.

9.2 Magnetic vector potential

In this section we introduce a vector field, called *magnetic vector potential*, which will be very useful to solve magnetostatic problems. Existence of this vector field relies upon the Gauss' law for magnetic field.

Indeed, since $\text{div}\,\mathbf{B} = 0$ there exists a vector field $\mathbf{A}$ such that $\mathbf{B} = \text{curl}\,\mathbf{A}$. In fact, there exist many of such vector fields: indeed, if $\text{curl}\,\mathbf{A} = \mathbf{B}$, then $\text{curl}(\mathbf{A} + \text{grad}\,\varphi) = \mathbf{B}$ for all scalar field φ. In order to uniquely determine $\mathbf{A}$ a so-called *gauge condition* should be added. An example is the *Coulomb's gauge*:

$$\text{div}\,\mathbf{A} = 0.$$

Moreover, by using the constitutive law $\mathbf{B} = \mu \mathbf{H}$, Ampère's law yields

$$\mathbf{curl}\left(\frac{1}{\mu}\mathbf{curl}\,\mathbf{A}\right) = \mathbf{J}.$$

Now we have to solve this equation together with the gauge condition. Firstly, we notice that things becomes much simpler if μ is constant in the whole space, let say, $\mu = \mu_0$. Indeed, in this case we have,

$$\mathbf{curl}(\mathbf{curl}\,\mathbf{A}) = \mu_0 \mathbf{J}. \tag{9.4}$$

By subtracting the term $\text{grad}(\text{div}\,\mathbf{A})$ (which is null by the Coulomb's gauge) to the left-hand side, and using the vector equality,

$$-\Delta \mathbf{A} = \mathbf{curl}\,\mathbf{curl}\,\mathbf{A} - \text{grad}(\text{div}\,\mathbf{A}),$$

Eq. (9.4) yields

$$-\Delta \mathbf{A} = \mu_0 \mathbf{J}.$$

In a fixed cartesian system of coordinates, this equation is equivalent to

$$-\Delta A_i = \mu_0 J_i, \ i = 1, 2, 3. \tag{9.5}$$

Then we can use the fundamental solution of the Poisson's equation to write the solutions of (9.5) by convolution with their right-hand sides. We have

$$A_i(x) = \int_\Omega \frac{\mu_0 J_i(y)}{4\pi |x-y|} dV(y), \ i = 1, 2, 3,$$

and hence,

$$\mathbf{A}(x) = \int_\Omega \frac{\mu_0 \mathbf{J}(y)}{4\pi |x-y|} dV(y) \ (\text{Wb/m}). \tag{9.6}$$

In these integrals, Ω denotes any bounded domain in the affine space containing the support of $\mathbf{J}$. Of course, $\mathbf{J}$ may also be a distribution supported on a surface S or on a line l in which case Ω should be replaced in the above formula by S or l, respectively.

In general, by taking the **curl** in (9.6), we get

$$\mathbf{B}(x) = \mathbf{curl}_x\,\mathbf{A}(x) = \mathbf{curl}_x \int_\Omega \frac{\mu_0 \mathbf{J}(y)}{4\pi|x-y|} dV(y) = \int_\Omega \frac{\mu_0}{4\pi} \mathbf{curl}_x\left(\frac{\mathbf{J}(y)}{|x-y|}\right) dV(y).$$

But $\mathbf{curl}(\phi \mathbf{u}) = \text{grad}\,\phi \times \mathbf{u} + \phi\,\mathbf{curl}\,\mathbf{u}$. Besides, in our case $\mathbf{J}(y)$ does not depend on x, so

$$\mathbf{curl}_x\left(\frac{\mathbf{J}(y)}{|x-y|}\right) = \text{grad}_x\left(\frac{1}{|x-y|}\right) \times \mathbf{J}(y) = -\frac{x-y}{|x-y|^3} \times \mathbf{J}(y).$$

Hence,
$$\mathbf{B}(x) = \int_\Omega \frac{\mu_0}{4\pi} \frac{\mathbf{J}(y) \times (x-y)}{|x-y|^3} dV(y). \tag{9.7}$$

This integral for **B**, expressed directly in terms of the static current distribution **J** in free space, is known as the *Biot-Savart* law.

In the general case, when μ is not constant in Ω, introducing the gauge condition and solving the problem is more difficult. A way to do it is described below.

For the sake of simplicity, let us suppose that Ω is a connected and simply connected bounded domain with a connected Lipschitz boundary Γ

Let us consider the boundary condition,

$$\mathbf{B} \cdot \mathbf{n} = g \text{ on } \Gamma, \tag{9.8}$$

where $g \in \mathbf{H}^{-1/2}(\Gamma)$. The latter is needed because of the divergence-free condition for **B**. Moreover, we notice that div **B** = 0 implies

$$\int_\Gamma \mathbf{B} \cdot \mathbf{n} \, dA = 0$$

by using Gauss' theorem. Hence, condition $\int_\Omega g \, dA = 0$ is needed for g.

In what follows we assume that $\mu \in L^\infty(\Omega)$ and there exist a positive constant $\bar{\mu}$ such that

$$\mu(x) \geq \bar{\mu}, \text{ a.e. in } \Omega.$$

Let us introduce the scalar potential problem

$$-\text{div}(\mu \text{ grad } \varphi) = 0 \text{ in } \Omega, \tag{9.9}$$

$$\mu \frac{\partial \varphi}{\partial \mathbf{n}} = g \text{ on } \Gamma. \tag{9.10}$$

This problem has a solution in $H^1(\Omega)$ which is unique up to a constant. It is obvious that **H** is a solution of the above magnetostatics problem if and only if $\hat{\mathbf{H}} := \mathbf{H} - \text{grad } \varphi$ is a solution of an analogous one but with $g = 0$. Thus, for the sake of simplicity, in what follows we will assume that $g = 0$ without loss of generality.

At this point, we summarize the magnetostatic problem defined in a bounded domain Ω to be considered in the next sections. Given **J** and g, find **H** and **J** satisfying

$$\mathbf{curl\,H} = \mathbf{J} \quad \text{in } \Omega, \tag{9.11}$$

$$\text{div } \mathbf{B} = 0 \quad \text{in } \Omega, \tag{9.12}$$

$$\mathbf{B} = \mu \mathbf{H} \quad \text{in } \Omega, \tag{9.13}$$

$$\mathbf{B} \cdot \mathbf{n} = g \quad \text{on } \Gamma. \tag{9.14}$$

9.3 A formulation in magnetic vector potential

Here, we will introduce a weak formulation of the problem (9.11)–(9.14) in terms of the magnetic vector potential **A**.

If $\mathbf{B} \in \mathbf{L}^2(\Omega)$ with $\operatorname{div} \mathbf{B} = 0$ and $\mathbf{B} \cdot \mathbf{n} = 0$ we deduce from [37, Theorem 1.3.6] the existence of a unique vector field $\mathbf{A} \in \mathbf{H}(\mathbf{curl}, \Omega) \cap \mathbf{H}(\operatorname{div}, \Omega)$ such that

$$\mathbf{curl}\,\mathbf{A} = \mathbf{B} \text{ in } \Omega, \tag{9.15}$$

$$\operatorname{div} \mathbf{A} = 0 \text{ in } \Omega, \tag{9.16}$$

$$\mathbf{A} \times \mathbf{n} = 0 \text{ on } \Gamma. \tag{9.17}$$

Furthermore, if Ω is smooth (in particular of class $\mathscr{C}^{1,1}$) or convex then $\mathbf{A} \in \mathbf{H}^1(\Omega)$.

By using (9.15) in Ampére's law we have

$$\mathbf{curl}(\frac{1}{\mu}\mathbf{curl}\,\mathbf{A}) = \mathbf{J}. \tag{9.18}$$

In order to make a weak formulation of the above problem, we first introduce the function space

$$\mathscr{V} = \{\mathbf{A} \in \mathbf{H}(\mathbf{curl}, \Omega) \cap \mathbf{H}(\operatorname{div}, \Omega), \operatorname{div} \mathbf{A} = 0, \mathbf{A} \times \mathbf{n} = 0 \text{ on } \Gamma\}.$$

Let us multiply (9.18) by a test function $\phi \in \mathscr{V}$, integrate in Ω and use a Green's formula. We get the weak formulation

Find $\mathbf{A} \in \mathscr{V}$ such that

$$\int_\Omega \frac{1}{\mu} \mathbf{curl}\,\mathbf{A} \cdot \mathbf{curl}\,\phi\, dV = \int_\Omega \mathbf{J} \cdot \phi\, dV \quad \forall \phi \in \mathscr{V}. \tag{9.19}$$

Since the bilinear form

$$a(\mathbf{A}, \phi) = \int_\Omega \frac{1}{\mu} \mathbf{curl}\,\mathbf{A} \cdot \mathbf{curl}\,\phi\, dV$$

is continuous and coercive in $\mathscr{V}$ (see, for instance, [5, Cor. 3.19]), the above problem has a unique solution **A**.

For numerical solution, the divergence-free condition in $\mathscr{V}$ is difficult to handle. Fortunately, we can prove that the following problem is equivalent to (9.19):

Find $\mathbf{A} \in \mathscr{Z}$ such that

$$\int_\Omega \frac{1}{\mu} \mathbf{curl}\,\mathbf{A} \cdot \mathbf{curl}\,\phi\, dV + \int_\Omega \operatorname{div} \mathbf{A}\, \operatorname{div} \phi\, dV = \int_\Omega \mathbf{J} \cdot \phi\, dV \quad \forall \phi \in \mathscr{Z}, \tag{9.20}$$

where

$$\mathscr{Z} = \{\mathbf{A} \in \mathbf{H}(\mathbf{curl}, \Omega) \cap \mathbf{H}(\operatorname{div}, \Omega), \mathbf{A} \times \mathbf{n} = 0 \text{ on } \Gamma\}.$$

First of all, let us notice that this problem also has a unique solution because the bilinear form in the left-hand side is coercive in $\mathscr{X}$. Next, we prove that this solution is divergence-free.

Indeed, for any $\varphi \in L^2(\Omega)$, the Dirichlet boundary-value problem

$$\begin{cases} -\Delta \psi = \varphi & \text{in } \Omega, \\ \psi = 0 & \text{on } \Gamma, \end{cases}$$

has a unique solution.

Let us take $\phi = \text{grad}\,\psi$. Then $\phi \in \mathscr{X}$ because $\text{curl}\,\text{grad}\,\psi = 0$, $\text{div}\,\phi = \Delta \psi = \varphi \in L^2(\Omega)$ and $\phi \times \mathbf{n} = \mathbf{0}$ on Γ. By replacing in (9.20), we get

$$\int_\Omega \text{div}\,\mathbf{A} \Delta \psi \, dV = \int_\Omega \mathbf{J} \cdot \text{grad}\,\psi \, dV. \tag{9.21}$$

But,

$$\int_\Omega \mathbf{J} \cdot \text{grad}\,\psi \, dV = \int_\Gamma \mathbf{J} \cdot \mathbf{n} \psi \, dA - \int_\Omega \text{div}\,\mathbf{J} \psi \, dV = 0,$$

because $\text{div}\,\mathbf{J} = 0$ and $\psi = 0$ on Γ. Then (9.21) implies that

$$\int_\Omega \text{div}\,\mathbf{A} \varphi \, dV = 0 \; \forall \varphi \in L^2(\Omega),$$

from which it follows that $\text{div}\,\mathbf{A} = 0$ in Ω.

Remark 9.1. If Ω is convex or has a "smooth boundary", then $\mathscr{X} \subset \mathbf{H}^1(\Omega)$ and one can use continuous piecewise linear finite elements to discretize each component of $\mathbf{A}$.

Otherwise, if the boundary of Ω is only Lipschitz-continuous, then $\mathbf{A}$ can be singular at reentrant corners and this standard nodal finite element method does not converge, in general.

An alternative approach consists of using edge Nédélec finite elements for the mixed formulation:

Find $\mathbf{A} \in \mathscr{W}$ and $\psi \in \mathscr{U}$ such that

$$\int_\Omega \frac{1}{\mu} \text{curl}\,\mathbf{A} \cdot \text{curl}\,\phi \, dV + \int_\Omega \text{grad}\,\psi \cdot \phi \, dV = \int_\Omega \mathbf{J} \cdot \phi \, dV \; \forall \phi \in \mathscr{W}, \tag{9.22}$$

$$\int_\Omega \mathbf{A} \cdot \text{grad}\,\varphi \, dV = 0 \; \forall \varphi \in \mathscr{U}, \tag{9.23}$$

where
$$\mathscr{W} = \mathbf{H}_0(\text{curl}, \Omega) := \{\phi \in \mathbf{H}(\text{curl}, \Omega) : \phi \times \mathbf{n} = \mathbf{0} \text{ on } \Gamma\}$$
and $\mathscr{U} = H_0^1(\Omega)$.

9.3 A formulation in magnetic vector potential

Remark 9.2. By taking $\phi = \operatorname{grad} \psi$ in the first equation and using a Green's formula in the right-hand side, it is easy to see that $\psi \equiv 0$ in Ω, because $\operatorname{div} \mathbf{J} = 0$ in Ω.

However, the weak problem

$$\int_\Omega \frac{1}{\mu} \operatorname{curl} \mathbf{A} \cdot \operatorname{curl} \phi \, dV = \int_\Omega \mathbf{J} \cdot \phi \, dV \ \forall \phi \in \mathscr{W}, \tag{9.24}$$

is not well-posed because the bilinear form on the right-hand side is null for any $\mathbf{A} = \operatorname{grad} \psi$ with $\psi \in H_0^1(\Omega)$. In particular, it is not coercive in $\mathscr{W}$.

Remark 9.3. For numerical solution, the mixed formulation (9.22), (9.23) can be discretized by using *edge* Nédélec finite elements to approximate $\mathbf{A}$ and *nodal* piecewise linear continuous finite elements to approximate ψ (see Appendix E).

In this case, it is not difficult to see that the solution of the discrete problem also has the ψ component identically zero. This is a consequence of the fact that the gradient of any continuous piecewise linear function belongs to the Nédélec finite element space. In practice one can solve the singular problem:

Find $\mathbf{A} \in \mathscr{W}$ *such that*

$$\int_\Omega \frac{1}{\mu} \operatorname{curl} \mathbf{A} \cdot \operatorname{curl} \phi \, dV = \int_\Omega \mathbf{J} \cdot \phi \, dV \ \forall \phi \in \mathscr{W}, \tag{9.25}$$

discretized by the first order Nédélec finite elements, by using iterative conjugate gradient-like methods.

We notice that the kernel of the associated matrix is the set of curl-free Nédélec finite elements which, in its turn, coincides with the set of gradients of the nodal finite element space. Moreover, since $\mathbf{J}$ is divergence-free, the right hand side of the discrete linear system is orthogonal to this kernel so it belongs to the image set of its matrix. This is a sufficient condition for conjugate gradient-like iterative algorithms to converge. Moreover, we remark that any solution to problem (9.25) yields a solution of the magnetostatics problem. Indeed, fields $\mathbf{B} := \operatorname{curl} \mathbf{A}$ and $\mathbf{H} := \frac{1}{\mu}\mathbf{B}$ satisfy Eqs. (9.11) to (9.14).

We should also emphasize that these results are no longer true if nodal piecewise linear finite elements are used to discretize each component of $\mathbf{A}$ instead of edge Nédélec ones.

9.4 A formulation in magnetic field

In this section we introduce another formulation of the magnetostatics problem (9.11)–(9.14) in terms of a more interesting unknown from the physical point of view, namely, the magnetic field intensity **H**. The weak formulation is the following:

Find $\mathbf{H} \in \mathbf{H}(\mathbf{curl},\Omega)$ and $\psi \in H^1(\Omega)/\mathbb{R}$ such that

$$\int_\Omega \mathbf{curl}\,\mathbf{H} \cdot \mathbf{curl}\,\phi\,dV + \int_\Omega \mu\,\mathrm{grad}\,\psi \cdot \phi\,dV = \int_\Omega \mathbf{J} \cdot \mathbf{curl}\,\phi\,dV$$
$$\forall \phi \in \mathbf{H}(\mathbf{curl},\Omega), \qquad (9.26)$$

$$\int_\Omega \mu \mathbf{H} \cdot \mathrm{grad}\,\varphi\,dV = 0 \;\; \forall \varphi \in H^1(\Omega)/\mathbb{R}. \qquad (9.27)$$

Again, by taking $\phi = \mathrm{grad}\,\psi$ as a test function in the first equation we get

$$\int_\Omega \mu |\mathrm{grad}\,\psi|^2\,dV = 0,$$

so $\psi \equiv 0$.

Remark 9.4. As for the **A**-formulation, for numerical solution we can approximate **H** by edge Nédélec finite elements and ψ by nodal continuous piecewise linear finite elements (see [44]).

Thus, a first alternative consists in solving the linear system associated with the above mixed discrete problem.

However, as for the continuous problem, the discrete ψ is also null and the right-hand side of the first equation is orthogonal to the discrete kernel of the **curl** operator. Hence, for the Nédélec discretization of the problem:

Find $\mathbf{H} \in \mathbf{H}(\mathbf{curl},\Omega)$ such that

$$\int_\Omega \mathbf{curl}\,\mathbf{H} \cdot \mathbf{curl}\,\phi\,dV = \int_\Omega \mathbf{J} \cdot \mathbf{curl}\,\phi\,dV \;\; \forall \phi \in \mathbf{H}(\mathbf{curl},\Omega), \qquad (9.28)$$

conjugate gradient-like iterative methods converges to one solution. Of course, there is no reason for this solution satisfies (9.27). Thus, in a second step we have to solve the following scalar Poisson's-like problem:

Find $\psi \in H^1(\Omega)$ such that,

$$\int_\Omega \mu\,\mathrm{grad}\,\psi \cdot \mathrm{grad}\,\varphi\,dV = \int_\Omega \mu \mathbf{H} \cdot \mathrm{grad}\,\varphi\,dV \;\; \forall \varphi \in H^1_0(\Omega).$$

After this, the magnetic field we are looking for is $\mathbf{H} - \mathrm{grad}\,\psi$. Thus, similar to the reduced scalar magnetic potential method to be introduced below, the present one may suffer from the cancelation error. Anyway, solving (9.28) by edge finite elements is a much cheaper alternative than the Biot-Savart formula to compute the field **T** below.

9.5 A formulation in terms of the reduced scalar magnetic potential

Another technique to solve the magnetostatics problem (9.11)–(9.14) consists in introducing a scalar potential.

Let **T** be a vector field such that **curl T** = **J**. Notice that **T** can be obtained, for instance, by using the Biot-Savart law (see (9.7), or by solving problem (9.25) with edge finite elements and conjugate gradient iterative methods.

We notice that, in general, $\mathbf{H}(x) \neq \mathbf{T}(x)$ because $\text{div}(\mu \mathbf{T})$ needs not to be null (recall that μ is not supposed to be spatially constant). However, **curl H** = **curl T** so there exists a scalar field φ^R such that

$$\mathbf{H} = \mathbf{T} - \text{grad}\,\varphi^R.$$

Field φ^R is called *reduced* scalar magnetic potential.

Then we seek for φ^R such that,

$$\text{div}(\mu(\mathbf{T} - \text{grad}\,\varphi^R)) = 0,$$

or

$$-\text{div}(\mu\,\text{grad}\,\varphi^R) = -\text{div}(\mu \mathbf{T}) \text{ in } \Omega,$$

satisfying the boundary condition

$$\mu \frac{\partial \varphi^R}{\partial \mathbf{n}} = \mu \frac{\partial \mathbf{T}}{\partial \mathbf{n}} \text{ on } \Gamma.$$

The weak formulation of this problem is the following

Find $\varphi^R \in H^1(\Omega)$ such that

$$\int_\Omega \mu\,\text{grad}\,\varphi^R \cdot \text{grad}\,\varphi\,dV = \int_\Omega \mu\,\text{grad}\,\mathbf{T} \cdot \text{grad}\,\varphi\,dV \quad \forall \varphi \in H^1(\Omega).$$

We notice that this method is similar to the one in the previous section. The difference is that now the magnetic field **T** is obtained from the Biot-Savart law instead of (9.28).

As already mentioned, the inconvenient of this method is that it suffers from the so-called *cancelation error* because the two terms **T** and $-\text{grad}\,\varphi^R$ are of the same order of magnitude and opposite direction in magnetic materials.

9.6 A formulation with two scalar potentials

Now we present an alternative approach. Firstly, we split the set Ω into three subdomains: the support of the current density **J**, Ω_J, the set occupied by air and other non-magnetic materials, more precisely, those for which $\mu = \mu_0$ (the permeability of the empty space) to be called Ω_A, and the domain filled with magnetic materials to be denoted by Ω_M, where we assume **J** = **0**. Let $\Gamma_I := \bar{\Omega}_R \cap \bar{\Omega}_M$

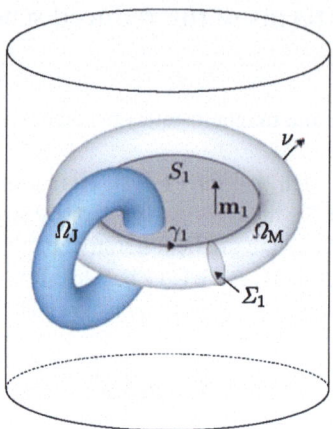

Fig. 9.2. Cut surface and geometry description

Domain Ω_M is supposed to be connected but it may be multiply connected. We assume that there exists a finite number of open connected surfaces (so-called "cuts") Σ_j, $j = 1, \ldots, J$, such that:

- $\Sigma_j \subset \Omega_M$ and $\partial \Sigma_j \subset \partial \Omega_M$.
- $\bar{\Sigma}_j \cap \bar{\Sigma}_k = \emptyset$, for $j \neq k$.
- The open set $\widetilde{\Omega}_M := \Omega_M \setminus \bigcup_{j=1}^{j=J} \Sigma_j$ is simply connected.

Figure 9.2 shows a particular case for $J = 1$. We also assume that for each cut Σ_j there exists a simple closed curve $\gamma_j \subset \partial \Omega_M$, crossing $\bar{\Sigma}_j$ once and only once, and such that γ_j is the boundary of an open surface $S_j \subset \Omega_R$ (again, see Fig. 9.2).

Let $\widetilde{\Omega}_M^j := \Omega_M \setminus \bar{\Sigma}_j$, $j = 1, \ldots, J$. We fix a unit normal vector $\mathbf{n}_j$ to Σ_j and denote its two faces by Σ_j^- and Σ_j^+, with $\mathbf{n}_j$ being outward $\widetilde{\Omega}_M^j$ along Σ_j^+. We choose an orientation for each γ_j by taking its initial and end points on $\bar{\Sigma}_j^-$ and $\bar{\Sigma}_j^+$, respectively. We denote by $\mathbf{t}_j$ the unit vector tangent to γ_j and by $\mathbf{m}_j$ a unit normal vector to surface S_j oriented with respect to $\mathbf{t}_j$ as in the Stokes' theorem. On the other hand, let $\partial \Omega_M$ be the boundary of Ω_M and $\boldsymbol{\nu}$ be the unit normal vector to $\partial \Omega_M$ pointing outwards Ω_M.

For any function $\widetilde{\psi} \in H^1(\widetilde{\Omega}_M)$, we denote by

$$[\![\widetilde{\psi}]\!]_{\Sigma_j} := \widetilde{\psi}|_{\Sigma_j^-} - \widetilde{\psi}|_{\Sigma_j^+}$$

the jump of $\widetilde{\psi}$ through Σ_j along $\mathbf{n}_j$. The gradient of $\widetilde{\psi}$ in $\mathscr{D}'(\widetilde{\Omega}_M)$ can be extended to $L^2(\Omega_M)^3$ and will be denoted by $\widetilde{\operatorname{grad}\widetilde{\psi}}$.

Let Θ be the subspace of $H^1(\widetilde{\Omega}_M)$ defined by

$$\Theta = \left\{ \widetilde{\psi} \in H^1(\widetilde{\Omega}_M) : [\![\widetilde{\psi}]\!]_{\Sigma_j} = \text{constant}, \ j = 1, \ldots, J \right\}. \tag{9.29}$$

9.6 A formulation with two scalar potentials

For all $\widetilde{\psi} \in H^1(\widetilde{\Omega}_M)$, one can show that $\widetilde{\psi} \in \Theta$ if and only if $\mathbf{curl}\left(\widetilde{\mathrm{grad}\,\widetilde{\psi}}\right) = \mathbf{0}$ in Ω_M (see, for instance, [5]).

Now we are in a position to introduce the additional scalar potential that will be used to solve the magnetostatic problem. Indeed, since $\mathbf{J}$ vanishes in Ω_M then $\mathbf{H}$ is curl-free in Ω_M and hence

$$\mathbf{H}|_{\Omega_M} = -\widetilde{\mathrm{grad}\,\widetilde{\varphi}} \quad \text{for some } \widetilde{\varphi} \in \Theta.$$

The scalar multivalued function $\widetilde{\varphi}$ is known as the *total scalar potential*.

Remark 9.5. The jumps of this potential have a precise physical meaning. Indeed, provided $\mathbf{H}$ is sufficiently smooth, by using the Stokes' theorem, Ampère's law and the fact that $\gamma_j \subset \Gamma_{\!I}$, we have

$$[\![\widetilde{\varphi}]\!]_{\Sigma_j} = -\int_{\gamma_j} \widetilde{\mathrm{grad}\,\widetilde{\varphi}} \cdot \mathbf{t}_j = \int_{\gamma_j} \mathbf{H} \cdot \mathbf{t}_j = \int_{S_j} \mathbf{curl}\,\mathbf{H} \cdot \mathbf{m}_j = \int_{S_j} \mathbf{J} \cdot \mathbf{m}_j =: \mathrm{I}_j,$$

Thus, the jump of the total scalar potential $\widetilde{\varphi}$ across each cut surface S_j is completely determined by the net current intensity I_j flowing through surface S_j (as defined in the equation above). Notice that, in particular, if $S_j \cap \Omega_J = \emptyset$, then $[\![\widetilde{\varphi}]\!]_{\Sigma_j} = 0$.

Our next step is to write the magnetostatics problem in terms of the reduced and the total scalar potentials, $\widetilde{\varphi} \in \Theta$ and $\varphi^R \in H^1(\Omega_R)$, so that

$$\mathbf{H} = \begin{cases} -\widetilde{\mathrm{grad}\,\widetilde{\varphi}} & \text{in } \Omega_M, \\ \mathbf{T} - \mathrm{grad}\,\varphi^R & \text{in } \Omega_R. \end{cases} \qquad (9.30)$$

Boundary and interface conditions must be written in terms of these potentials, as well. With this purpose, we notice that for $\mathbf{J} \in L^2(\Omega_I)^3$ we have that $\mathbf{H} \in \mathbf{H}(\mathbf{curl},\Omega)$ and hence $\mathbf{H} \times \mathbf{v}$ does not jump across $\Gamma_{\!I}$. Analogously, $\mu \mathbf{H} \cdot \mathbf{v}$ does not have jumps across $\Gamma_{\!I}$ either, because $\mathbf{B} \in \mathbf{H}(\mathrm{div},\Omega)$. Consequently, we have to search for $\widetilde{\varphi} \in \Theta$ and $\varphi^R \in H^1(\Omega_R)$, satisfying the following equations:

$$-\mathrm{div}\left(\mu_0 \,\mathrm{grad}\,\varphi^R\right) = 0 \quad \text{in } \Omega_R, \qquad (9.31)$$

$$-\mathrm{div}\left(\mu\,\widetilde{\mathrm{grad}\,\widetilde{\varphi}}\right) = 0 \quad \text{in } \Omega_M, \qquad (9.32)$$

$$\mu \,\mathrm{grad}\,\varphi^R \cdot \mathbf{n} = \mu_0 \mathbf{T} \cdot \mathbf{n} \quad \text{on } \Gamma, \qquad (9.33)$$

$$\mu_0 \,\mathrm{grad}\,\varphi^R \cdot \mathbf{v} - \mu\,\widetilde{\mathrm{grad}\,\widetilde{\varphi}} \cdot \mathbf{v} = \mu_0 \mathbf{T} \cdot \mathbf{v} \quad \text{on } \Gamma_{\!I}, \qquad (9.34)$$

$$\mathrm{grad}\,\varphi^R \times \mathbf{v} - \widetilde{\mathrm{grad}\,\widetilde{\varphi}} \times \mathbf{v} = \mathbf{T} \times \mathbf{v} \quad \text{on } \Gamma_{\!I}. \qquad (9.35)$$

To obtain the weak formulation, we multiply Eqs. (9.31) and (9.32) by a sufficiently smooth test function ψ, integrate by parts and use (9.33) and (9.34). Thus, we

9 Magnetostatics

obtain

$$\int_{\Omega_M} \mu \, \widetilde{\operatorname{grad} \widetilde{\varphi}} \cdot \nabla \psi + \int_{\Omega_R} \mu_0 \operatorname{grad} \varphi^R \cdot \operatorname{grad} \psi = \int_\Gamma \mu_0 \mathbf{T} \cdot \mathbf{n} \, \psi - \int_{\Gamma_I} \mu_0 \mathbf{T} \cdot \mathbf{v} \, \psi.$$

For the subsequent analysis, Eq. (9.35) will be imposed as an essential condition. We introduce the space

$$\mathscr{X} := \Theta/\mathbb{R} \times H^1(\Omega_R)/\mathbb{R},$$

the closed linear manifold

$$\mathscr{V}(\mathbf{T}) := \left\{ (\widetilde{\psi}, \psi^R) \in \mathscr{X} : \operatorname{grad} \psi^R \times \mathbf{v} - \widetilde{\operatorname{grad} \widetilde{\psi}} \times \mathbf{v} = \mathbf{T} \times \mathbf{v} \text{ on } \Gamma_I \right\}$$

and its corresponding closed subspace

$$\mathscr{V}(\mathbf{0}) := \left\{ (\widetilde{\psi}, \psi^R) \in \mathscr{X} : \operatorname{grad} \psi^R \times \mathbf{v} - \widetilde{\operatorname{grad} \widetilde{\psi}} \times \mathbf{v} = \mathbf{0} \text{ on } \Gamma_I \right\}.$$

Then we can state the weak formulation of the magnetostatics problem (9.31)-(9.35):

Find $(\widetilde{\varphi}, \varphi^R) \in \mathscr{V}(\mathbf{T})$, such that

$$\int_{\Omega_M} \mu \, \widetilde{\operatorname{grad} \widetilde{\varphi}} \cdot \widetilde{\operatorname{grad} \widetilde{\psi}} + \int_{\Omega_R} \mu_0 \operatorname{grad} \varphi^R \cdot \operatorname{grad} \psi^R \tag{9.36}$$

$$= \int_\Gamma \mu_0 \mathbf{T} \cdot \mathbf{n} \, \psi^R - \int_{\Gamma_I} \mu_0 \mathbf{T} \cdot \mathbf{v} \, \psi^R \quad \forall (\widetilde{\psi}, \psi^R) \in \mathscr{V}(\mathbf{0}).$$

Handling directly the constraint appearing in the definition of $\mathscr{V}(\mathbf{T})$ is not easy. This is why we introduce the following weak formulation involving a Lagrange multiplier:

Find $(\widetilde{\varphi}, \varphi^R) \in \mathscr{X}$ and $\boldsymbol{\lambda} \in \mathbf{H}^{1/2}(\Gamma_I)$ such that

$$\int_{\Omega_M} \mu \, \widetilde{\operatorname{grad} \widetilde{\varphi}} \cdot \widetilde{\operatorname{grad} \widetilde{\psi}} + \int_{\Omega_R} \mu_0 \operatorname{grad} \varphi^R \cdot \operatorname{grad} \psi^R + \langle \boldsymbol{\lambda}, \operatorname{grad} \widetilde{\psi} \times \mathbf{n} - \operatorname{grad} \psi^R \times \mathbf{n} \rangle_\Gamma$$

$$= \int_\Gamma \mu_0 \mathbf{T} \cdot \mathbf{n} \, \psi^R - \int_{\Gamma_I} \mu_0 \mathbf{T} \cdot \mathbf{v} \, \psi^R \quad \forall (\widetilde{\psi}, \psi^R) \in \mathscr{X}$$

$$\langle -\operatorname{grad} \varphi^R \times \mathbf{v} + \mathbf{T} \times \mathbf{v} + \operatorname{grad} \widetilde{\varphi} \times \mathbf{v}, \boldsymbol{\beta} \rangle_\Gamma = 0 \quad \forall \boldsymbol{\beta} \in \mathbf{H}^{1/2}(\Gamma_I).$$

Remark 9.6. The above formulation can be discretized by using 3D continuous piecewise linear finite elements on a tetrahedral mesh of Ω for $\widetilde{\varphi}$ and φ^R and 2D continuous piecewise linear finite elements on boundary Γ_I for the Lagrange multiplier $\boldsymbol{\lambda}$ (see Appendix E). Further details can be found in [19].

9.7 Distributed magnetic circuits: reluctance matrix

For magnetic circuits, *reluctance* is similar to resistance for electrical circuits. Recall that, for direct currents, the electric field is curl-free in the whole space so it is equal to the gradient of a scalar electric potential. However, the magnetic field is only curl-free in the regions of the space where $\mathbf{J} = 0$, in general in dielectrics. As we have seen in the previous section, this region needs not to be simply connected so the scalar magnetic potential is likely to be discontinuous through some internal surfaces. This makes the study of magnetic circuits and reluctance more difficult than the one of electric circuits and resistance.

We follow the notations of the previous section. Furthermore, let us suppose $\partial \Omega_M$ is decomposed as follows:

$$\partial \Omega_M = \Gamma_0 \cup \cdots \cup \Gamma_L.$$

We will assume that Γ_0 is impermeable to the magnetic flux but some given magnetic fluxes enter domain Ω_M through the other parts of the boundary, Γ_l, $l = 1, \ldots, L$. Figure 9.3 shows a particular case for $J = 3$ and $L = 0$.

Let us consider the following magnetostatics problem in Ω_M:

For given real numbers Φ_i, $i = 1, \ldots, L$ and I_j, $j = 1, \ldots, J$ find $\mathbf{H} \in \mathrm{H}(\mathbf{curl}, \Omega_M)$ such that,

$$\mathbf{curl}\,\mathbf{H} = 0 \text{ in } \Omega_M, \tag{9.37}$$

$$\mathrm{div}(\mu \mathbf{H}) = 0 \text{ in } \Omega_M, \tag{9.38}$$

$$\mathbf{B} \cdot \mathbf{v} = 0 \text{ on } \Gamma_0, \tag{9.39}$$

$$\mathbf{H} \times \mathbf{v} = \mathbf{0} \text{ on } \Gamma_1 \cup \cdots \cup \Gamma_L, \tag{9.40}$$

$$-\int_{\Gamma_i} \mathbf{B} \cdot \mathbf{v}\, dA = \Phi_i, \ i = 1, \ldots, L \tag{9.41}$$

$$\int_{\gamma_j} \mathbf{H} \cdot \mathbf{t}_j\, dl = I_j, \ j = 1, \ldots, J. \tag{9.42}$$

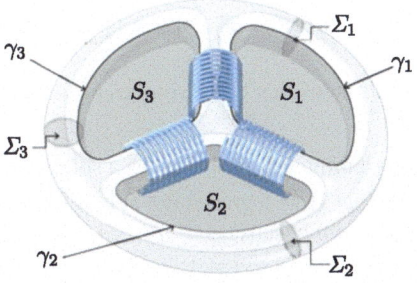

Fig. 9.3. Example of domain corresponding to a distributed magnetic circuit

Equation (9.39) means that Γ_0 is a *magnetic wall*, i.e., the magnetic flux is null through this boundary. Equation (9.40) means that the magnetic flux enters the domain perpendicularly to the *magnetic ports* $\Gamma_1,\ldots,\Gamma_L$. Equation (9.41) prescribes the input magnetic flux through these ports. Finally, in order to see the physical meaning of (9.42) we notice that, by using the Stokes' theorem and the fact that $\gamma_j \subset \partial \Omega_M$, we have (see Fig. 9.2),

$$\int_{\gamma_j} \mathbf{H}\cdot \mathbf{t}_j = \int_{S_j} \mathbf{curl\,H}\cdot \mathbf{m}_j = \int_{S_j} \mathbf{J}\cdot \mathbf{m}_j := I_j, \tag{9.43}$$

so (9.42) imposes the current intensity across surface S_j.

By using the results concerning the curl-free fields in Ω_M from the previous section, we are going to build a solution to this problem. For this purpose we first introduce the function space:

$$\mathcal{T} = \{\widetilde{\psi} \in \Theta : \widetilde{\psi}_{|\Gamma_i} = constant,\ i=1,\ldots,L\},$$

where Θ is defined in (9.29). Let us consider the following problem:

Find $\widetilde{\varphi} \in \mathcal{T}$ such that,

$$\mathrm{div}(\mu\,\widetilde{\mathrm{grad}\,\varphi}) = 0 \text{ in } \widetilde{\Omega}_M, \tag{9.44}$$

$$\mu\frac{\partial \widetilde{\varphi}}{\partial \nu} = 0 \text{ on } \Gamma_0, \tag{9.45}$$

$$[\![\widetilde{\varphi}]\!]_{\Sigma_j} = I_j,\ j=1,\ldots,J, \tag{9.46}$$

$$[\![\mu \frac{\partial \widetilde{\varphi}}{\partial \mathbf{n}_j}]\!]_{\Sigma_j} = 0,\ j=1,\ldots,J, \tag{9.47}$$

$$\int_{\Gamma_i} \mu \frac{\partial \widetilde{\varphi}}{\partial \nu}\,dA = \Phi_i,\ i=1,\ldots,L. \tag{9.48}$$

We have,

$$[\![\widetilde{\varphi}]\!]_{\Sigma_j} = -\int_{\gamma_j} \widetilde{\mathrm{grad}\,\varphi}\cdot \mathbf{t}_j = \int_{\gamma_j} \mathbf{H}\cdot \mathbf{t}_j.$$

Thus, from (9.43) we deduce that the jump of the scalar potential $\widetilde{\varphi}$ across each cut surface Σ_j is the net current intensity I_j flowing through surface S_j (as defined in the equation above) in the direction $\mathbf{m}_j$. Notice that, in particular, if $S_j \cap \Omega_J = \emptyset$, then $[\![\widetilde{\varphi}]\!]_{\Sigma_j} = 0$.

Moreover, since current intensities I_j are sources of magnetic field, they are usually called *magnetomotive forces*.

From the Lax-Milgram lemma this problem has a solution, $\widetilde{\varphi}$, which is unique up to a constant. Let us show that $\mathbf{H} = -\,\widetilde{\mathrm{grad}\,\varphi}$ is a solution to problem (9.37)–(9.42).

Indeed, first of all $\mathbf{curl\,H} = 0$ because $\widetilde{\varphi} \in \mathcal{T} \subset \Theta$. Moreover, Eq. (9.40) follows from the fact that $\widetilde{\varphi}$ is constant on $\Gamma_j,\ j=1,\ldots,L$. Equation (9.47) guarantees that $\mathbf{B}\cdot \mathbf{n}_j$ is continuous through $\Sigma_j,\ j=1,\ldots,J$ which is needed in order for $\mathbf{B}$ to be divergence-free. The rest of the equations easily follows from (9.44)–(9.47).

9.7 Distributed magnetic circuits: reluctance matrix

The weak formulation of problem (9.44)–(9.47) is the following:

Find $\widetilde{\varphi} \in \mathscr{T}$ with $[\![\widetilde{\varphi}]\!]_{\Sigma_j} = I_j$, $j = 1,\ldots,J$, such that,

$$\int_{\widetilde{\Omega}_M} \mu \, \widetilde{\mathrm{grad}\,\varphi} \cdot \mathrm{grad}\, \widetilde{\psi} = \sum_{j=1}^{L} \Phi_j \psi_{|\Gamma_j} \quad \forall \widetilde{\psi} \in \Psi, \tag{9.49}$$

where

$$\Psi = \{\psi \in H^1(\Omega_M) : \psi_{|\Gamma_j} = constant,\ j = 1,\ldots,L\}. \tag{9.50}$$

Associated with this problem we are going to define the so-called *reluctance matrix*. For this purpose we first consider a slightly different problem from (9.44)-(9.47). Given constants φ_i, $i = 1,\ldots,L$ and I_i, $i = 1,\ldots,J$, it consists in solving,

$$\mathrm{div}(\mu \, \widetilde{\mathrm{grad}\, \varphi}) = 0 \text{ in } \widetilde{\Omega}_M, \tag{9.51}$$

$$\mu \frac{\partial \widetilde{\varphi}}{\partial \nu} = 0 \text{ on } \Gamma_0, \tag{9.52}$$

$$[\![\widetilde{\varphi}]\!]_{\Sigma_j} = I_j,\ j = 1,\ldots,J, \tag{9.53}$$

$$[\![\mu \frac{\partial \widetilde{\varphi}}{\partial \mathbf{n}_j}]\!]_{\Sigma_j} = 0,\ j = 1,\ldots,J, \tag{9.54}$$

$$\widetilde{\varphi}_{|\Gamma_i} = \varphi_i,\ i = 1,\ldots,L. \tag{9.55}$$

From this problem one can define a linear mapping, $\widetilde{\mathscr{Q}} : \mathbb{R}^{L+J} \longrightarrow \mathbb{R}^{L+J}$ by

$$\widetilde{\mathscr{Q}}(\varphi_1,\ldots,\varphi_L, I_1,\ldots,I_J) = (\Phi_1,\ldots,\Phi_L,\Phi_{L+1},\ldots,\Phi_{L+J}), \tag{9.56}$$

with

$$\Phi_i = \int_{\Gamma_i} \mu \frac{\partial \widetilde{\varphi}}{\partial \nu} dA,\ i = 1,\ldots,L,$$

$$\Phi_i = \int_{\Sigma_i} \mu \frac{\partial \widetilde{\varphi}}{\partial \mathbf{n}_i} dA,\ i = L+1,\ldots,L+J.$$

We notice that $\widetilde{\mathscr{Q}}$ depends on the geometry and the materials (through μ) but not on the excitations of the system.

Choosing the canonical basis in vector space $\mathbb{R}^{L+J}$ we can obtain the associated matrix to $\widetilde{\mathscr{Q}}$. Let us denote by w_j, $j = 1,\ldots,L$ the solution of problem (9.51)–(9.55) corresponding to $\varphi_i = \delta_{ij}$, $i = 1,\ldots,L$ and $I_i = 0$, $i = 1,\ldots,J$. Similarly, let $\widetilde{w}_j$, $j = 1,\ldots,J$ be the solution corresponding to $\varphi_i = 0$, $i = 1,\ldots,L$ and $I_i = \delta_{ij}$, $i = 1,\ldots,J$. We notice that they are, respectively, the unique solutions of the following weak formulations:

Find $w_j \in H^1(\Omega_M)$ with $w_j = \delta_{ij}$ on Γ_i, $i = 1,\ldots,L$, such that,

$$\int_{\Omega_M} \mu \, \mathrm{grad}\, w_j \cdot \mathrm{grad}\, \psi = 0 \quad \forall \psi \in H^1(\Omega_M) \text{ with } \psi_{|\Gamma_i} = 0,\ i = 1,\ldots,L. \tag{9.57}$$

Find $\tilde{w}_j \in \Theta$ with $\tilde{w}_j = 0$ on Γ_i, $i = 1,\ldots,L$ and $[\![\tilde{w}]\!]_{\Sigma_j} = \delta_{ij}$, $i = 1,\ldots,J$, such that,

$$\int_{\tilde{\Omega}_M} \mu \widetilde{\operatorname{grad}} \tilde{w}_j \cdot \operatorname{grad} \psi = 0 \quad \forall \psi \in H^1(\Omega_M) \text{ with } \psi_{|\Gamma_i} = 0, \, i = 1,\ldots,J. \quad (9.58)$$

We notice that the solution of problem (9.51)–(9.55) can be written as the linear combination

$$\tilde{\psi} = \sum_{j=1}^{L} \varphi_j w_j + \sum_{j=1}^{J} I_j \tilde{w}_j.$$

In fact, $\{w_j, \, j = 1,\ldots,L\} \cup \{\tilde{w}_j, \, j = 1,\ldots,J\}$ is a basis of the function space $\mathscr{T}$.

We can split the associated matrix to linear mapping $\widetilde{\mathscr{Q}}$ corresponding to the canonical basis, to be also called $\widetilde{\mathscr{Q}}$, into four blocks:

$$\widetilde{\mathscr{Q}} = \begin{pmatrix} \tilde{q}^{LL} & \tilde{q}^{LJ} \\ \tilde{q}^{JL} & \tilde{q}^{JJ} \end{pmatrix},$$

where

$$\tilde{q}_{ij}^{LL} = \int_{\Gamma_i} \mu \frac{\partial w_j}{\partial \mathbf{v}} \, dA, \quad i,j = 1,\ldots,L,$$

$$\tilde{q}_{ij}^{LJ} = \int_{\Gamma_i} \mu \frac{\partial \tilde{w}_j}{\partial \mathbf{v}} \, dA, \quad i = 1,\ldots,L, \, j = 1,\ldots,J,$$

$$\tilde{q}_{ij}^{JL} = \int_{\Sigma_i} \mu \frac{\partial w_j}{\partial \mathbf{n}_i} \, dA, \quad i = 1,\ldots,J, \, j = 1,\ldots,L,$$

$$\tilde{q}_{ij}^{JJ} = \int_{\Sigma_i} \mu \frac{\partial \tilde{w}_j}{\partial \mathbf{v}} \, dA, \quad i,j = 1,\ldots,J.$$

By using a Green's formula, from (9.51)–(9.55) it is easy to show that

$$\tilde{q}_{ij}^{LL} = \int_{\Omega_M} \mu \operatorname{grad} w_j \cdot \operatorname{grad} w_i \, dV, \quad i,j = 1,\ldots,L,$$

$$\tilde{q}_{ij}^{LJ} = \int_{\tilde{\Omega}_M} \mu \widetilde{\operatorname{grad}} \tilde{w}_j \cdot \operatorname{grad} w_i \, dV, \quad i = 1,\ldots,L, \, j = 1,\ldots,J,$$

$$\tilde{q}_{ij}^{JL} = \int_{\tilde{\Omega}_M} \mu \operatorname{grad} w_j \cdot \widetilde{\operatorname{grad}} \tilde{w}_i \, dV, \quad i = 1,\ldots,J, \, j = 1,\ldots,L,$$

$$\tilde{q}_{ij}^{JJ} = \int_{\tilde{\Omega}_M} \mu \widetilde{\operatorname{grad}} \tilde{w}_j \cdot \widetilde{\operatorname{grad}} \tilde{w}_i \, dV, \quad i,j = 1,\ldots,J.$$

Thus, matrix $\widetilde{\mathscr{Q}}$ is symmetric. Moreover,

$$(\varphi_1,\ldots,\varphi_L, I_1,\ldots,I_J) \widetilde{\mathscr{Q}} (\varphi_1,\ldots,\varphi_L, I_1,\ldots,I_J)^t$$

$$= \int_{\tilde{\Omega}_M} \mu |\widetilde{\operatorname{grad}}(\sum_{j=1}^{L} \varphi_j w_j + \sum_{j=1}^{N} I_j \tilde{w}_j)|^2 \, dV \geq 0$$

9.7 Distributed magnetic circuits: reluctance matrix

Furthermore, if the right-hand side is equal to zero then the function

$$\sum_{j=1}^{L} \varphi_j w_j + \sum_{j=1}^{N} I_j \tilde{w}_j$$

has to be constant in $\widetilde{\Omega}_M$, and then in Ω_M, which immediately implies $I_i = 0$, $i=1,\ldots,J$ and $\varphi_1 = \cdots = \varphi_L$. Thus, the kernel of $\widetilde{\mathscr{D}}$ is given by the one-dimensional space

$$\ker(\widetilde{\mathscr{D}}) = \{(\varphi_1,\ldots,\varphi_L, I_1,\ldots,I_J) \in \mathbb{R}^{L+J} : \varphi_1 = \cdots = \varphi_L, I_i = 0, i = 1,\ldots,J\}.$$

Since $\widetilde{\mathscr{D}}$ is symmetric, the rank of $\widetilde{\mathscr{D}}$ is $L+J-1$ and the image space is the orthogonal complement of $\ker(\widetilde{\mathscr{D}})$ in $\mathbb{R}^{L+J}$, namely,

$$\mathrm{im}(\widetilde{\mathscr{D}}) = \{(\Phi_1,\ldots,\Phi_{L+J}) \in \mathbb{R}^{L+J} : \sum_{j=1}^{L} \Phi_j = 0\}.$$

We notice that, in particular, if $L = 0$ then $\widetilde{\mathscr{D}}$ is one-to-one and hence invertible. Moreover, the restriction of $\widetilde{\mathscr{D}}$ to the quotient space

$$\mathbb{R}^{L+J}/\ker(\widetilde{\mathscr{D}})$$

is a bijection onto the image space so it is invertible. This linear mapping, to be denoted by $\mathscr{D}$ is called *permeance* and its inverse, $\mathscr{F} = \mathscr{D}^{-1}$, is called *reluctance*. Their respective matrices associated with the canonical basis are called in the same way.

Example 9.2. Let Ω_M be a torus with radius a, being R the radius of its section. Let us suppose that μ is constant. Then $L = 0$, $J = 1$ and we have an easy expression for $\tilde{w}_1$ in cylindrical coordinates, namely,

$$\tilde{w}_1(\rho, \theta, z) = \frac{\theta}{2\pi}.$$

Indeed, let us take the cutting surface,

$$\Sigma_1 = \{(\rho, \theta, z) : \theta = 0\}.$$

Then condition (9.51) follows immediately. Furthermore,

$$\mathrm{grad}\,\tilde{w}_1(\rho, \theta, z) = \frac{1}{2\pi\rho}\mathbf{e}_\theta.$$

Since this field is tangent to the torus, condition (9.52) holds. Moreover,

$$\mu \frac{\partial \tilde{w}_1}{\partial \mathbf{n}_1} = \mu \frac{1}{2\pi\rho}\mathbf{e}_\theta \cdot \mathbf{e}_\theta = \mu \frac{1}{2\pi\rho},$$

hence (9.54) is also satisfied. Finally,
$$\mathrm{div}(\mu \,\mathrm{grad}\, \tilde{w}_1) = 0.$$

Since $L = 0$ the permeance matrix is the order one matrix $\mathscr{Q}$ with
$$q_{11} = \int_{\Sigma_1} \mu \frac{1}{2\pi\rho}\, d\rho dz. \tag{9.59}$$

where Σ_1 is the circle with center at $(a,0,0)$ and radius R. One can show that
$$q_{11} = \mu(a - \sqrt{a^2 - R^2}). \tag{9.60}$$

Hence, the reluctance is the order one matrix $\mathscr{F}$ with
$$f_{11} = \frac{1}{q_{11}} = \frac{1}{\mu(a - \sqrt{a^2 - R^2})}.$$

Let us suppose that $R \ll a$. Then (9.60) can be approximated as follows:
$$q_{11} = \mu \frac{a^2 - (a^2 - R^2)}{a + \sqrt{a^2 - R^2}} \simeq \mu \frac{R^2}{2a}$$

and then
$$f_{11} \simeq F_{ap} = \frac{2a}{\mu R^2} = \frac{1}{\mu} \frac{2\pi a}{\pi R^2}, \tag{9.61}$$

which is the usual formula for the reluctance of a thin toroidal core found in electromagnetism books.

Example 9.3. Let us consider a very simple distributed magnetic circuit: the one consisting of a magnetic domain Ω_M with two ports and a magnetic wall (see Fig. 9.4); that is,
$$\partial \Omega_M = \Gamma_0 \cup \Gamma_1 \cup \Gamma_2.$$

Thus, $J = 0$ and $L = 2$.

Assuming there is no current in Ω_M and that the magnetic field is orthogonal to both Γ_1 and Γ_2, the magnetic scalar potential is the solution (unique up to a constant)

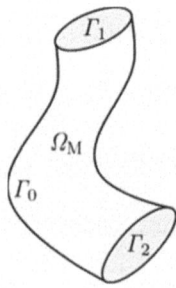

Fig. 9.4. Example of a domain Ω_M with two ports (Γ_1 and Γ_2) and a magnetic wall

of the boundary-value problem (see (9.44)–(9.48)):

Find $\varphi \in \Psi$ such that,

$$\text{div}(\mu \operatorname{grad} \varphi) = 0 \text{ in } \Omega_M, \tag{9.62}$$

$$\mu \frac{\partial \phi}{\partial \mathbf{v}} = 0 \text{ on } \Gamma_0, \tag{9.63}$$

$$\varphi|_{\Gamma_i} = \varphi_i, \, i = 1, 2. \tag{9.64}$$

Let Φ_i be the magnetic flux entering Ω_M through Γ_i, $i = 1, 2$, namely,

$$\Phi_i = \int_{\Gamma_i} \mu \frac{\partial \widetilde{\varphi}}{\partial \mathbf{v}} dA, \, i = 1, 2.$$

Recall that, in this case, mapping $\widetilde{\mathscr{Q}}$ is defined by (see (9.56)),

$$\widetilde{\mathscr{Q}}(\varphi_1, \varphi_2) = (\Phi_1, \Phi_2).$$

As previously established in the general framework, this mapping is not bijective (in fact, since div $\mathbf{B} = 0$, we must have $\Phi_1 = -\Phi_2$). Actually, the kernel consists of couples (φ_1, φ_2) with $\varphi_1 = \varphi_2$. The restriction to the quotient space $\mathbb{R}^2/\ker(\widetilde{\mathscr{Q}})$ is a bijection onto the image space.

Thus, in this case the reluctance matrix is of order one and given by

$$f_{11} = \frac{\varphi_1 - \varphi_2}{\Phi_1}.$$

In the particular case where Ω_M is a cylinder of radius R and height h, with circular bases Γ_i, $i = 1, 2$ and lateral surface Γ_0, the above boundary-value problem can be analytically integrated and the above formula yields (see, for instance, Example 8.1)

$$f_{11} = \frac{h}{\mu \pi R^2}.$$

9.8 Approximate reluctance

Notice that in the examples presented above we can compute the reluctance matrix because the geometry of the magnetic core is very simple and, in the case of the first example, we can build the cutting surface Σ_1 in an easy way. However, the computation of the reluctance matrix in real devices will be more complex and will require the numerical solution of the boundary-value problems introduced above.

On the other hand, in the case of the torus we have computed an approximation for the reluctance, F_{ap}, which actually corresponds to the value given for thin magnetic circuits in classical books of electromagnetism. Next, we will analyze this approximation in more detail in order to generalize it to the case of more complex geometries. To attain this goal, we will focus first on the example of the toroidal core.

Let us consider a thin toroidal magnetic core of n turns carrying intensity I. The radius of the torus is a and the circular cross section has radius R so the area of the later is constant and equal to πR^2. By applying Ampère's law (see Sect. 5.2.1 for a non-magnetic core), we know that the magnetic field inside the core is given by

$$\mathbf{H}(\rho) = H_\theta(\rho)\mathbf{e}_\theta(\rho) = \frac{nI}{2\pi\rho}\mathbf{e}_\theta(\rho)$$

and is null outside. If the section of the core is small compared to its radius, i.e., if $R \ll a$, then we can approximate the magnetic field in the core by its value $\mathbf{H}_{ap}$ at the centerline of the core ($\rho = a$), that is,

$$\mathbf{H}_{ap}(\rho) = \frac{nI}{2\pi a}\mathbf{e}_\theta.$$

By using this approximation and taking into account that each cross section Σ of the core is a circle of constant area $\text{meas}(\Sigma) = \pi R^2$, the magnetic flux through Σ can be approximated by

$$\Phi = \int_\Sigma \mathbf{B}\cdot\mathbf{n}\,dA \approx \int_\Sigma \frac{\mu nI}{2\pi a}\mathbf{e}_\theta\cdot\mathbf{e}_\theta\,dA = \text{meas}(\Sigma)\frac{\mu nI}{2\pi a} = \frac{\mu\pi R^2}{2\pi a}nI.$$

Then, we can conclude that the approximate reluctance F_{ap} given by (9.61) is nothing but the following relation between the magnetomotive force nI and the approximate magnetic flux through surface Σ

$$F_{ap} = nI/(\frac{\mu\pi R^2}{2\pi a}nI) = \frac{2\pi a}{\mu\pi R^2}.$$

We notice that

$$F_{ap} = \frac{2\pi a}{\mu\pi R^2} = \frac{\text{mean length of the core}}{\mu\,\text{meas}(\Sigma)} \qquad (9.65)$$

which is the definition given in books for the reluctance in linear, homogeneous and isotropic magnetic cores with constant permeability μ and constant cross section Σ.

Next, let us obtain an approximate formula similar to (9.65) in more general situations. Let us consider a thin toroidal *magnetic core* made with a linear, homogenous and isotropic material. A winding with n turns and carrying a direct current I is localized at some part of it (see Fig. 9.5). We will assume that the magnetic field is confined within the core so that the *flux leakage* can be neglected; this assumption may be acceptable if the magnetic permeability of the core is very high. Let us consider the centerline l of the conductor which is the line joining the geometrical centers of all its cross sections $\Sigma(l)$.

Since the flux is supposed to be confined in the core and $\mathbf{B}$ is a solenoidal field, from Gauss' theorem we can assume that through any cross section $\Sigma(l)$ of the core

9.8 Approximate reluctance

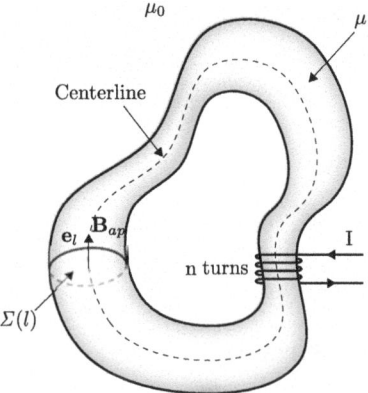

Fig. 9.5. Example of a thin magnetic core

the magnetic flux is constant, i.e.,

$$\Phi = \int_{\Sigma(l)} \mathbf{B} \cdot \mathbf{n} \, dA$$

where $\mathbf{n}$ denotes a unit normal vector to surface $\Sigma(l)$.

If the magnetic circuit is thin enough, then $\mathbf{B}$ at the centerline l can be approximated by an average value $\mathbf{B}_{ap}$ tangent to the centerline l as shown in Fig. 9.5:

$$\mathbf{B}_{ap}(l) := \frac{1}{\text{meas}(\Sigma(l))} \left(\int_{\Sigma(l)} \mathbf{B} \cdot \mathbf{n} \, dA \right) \mathbf{e}_l(l) = \frac{\Phi}{\text{meas}(\Sigma(l))} \mathbf{e}_l(l),$$

$\mathbf{e}_l$ being a unit tangent vector to the centerline l.

On the other hand, if we apply the Ampère's law to l, we have

$$\int_l \mathbf{H} \cdot d\mathbf{l} = \int_S \text{curl}\, \mathbf{H} \cdot \mathbf{m} \, dA = \int_S \mathbf{J} \cdot \mathbf{m} \, dA = n\mathbf{I}, \tag{9.66}$$

S being a surface with boundary l and $\mathbf{m}$ a normal unit vector to this surface.

Thus, by taking into account that $\mathbf{H} = \mathbf{B}/\mu$ and using the approximated value of $\mathbf{B}_{ap}$ given above in (9.66), we obtain (notice that $d\mathbf{l} = \mathbf{e}_l dl$),

$$\int_l \frac{\mathbf{B}_{ap}}{\mu} \cdot d\mathbf{l} = \int_l \frac{\Phi \, dl}{\mu \, \text{meas}(\Sigma(l))} \approx n\mathbf{I}$$

which yields

$$\Phi \approx \frac{n\mathbf{I}}{\int_l \frac{dl}{\mu \, \text{meas}(\Sigma(l))}}. \tag{9.67}$$

172   9 Magnetostatics

This approximate equality establishes a relation between the *magnetomotive force* nI and the quantity

$$F_{ap} = \int_l \frac{dl}{\mu \operatorname{meas}(\Sigma(l))}. \tag{9.68}$$

Thus, F_{ap} is an approximate *reluctance* of the magnetic core (analogous to resistance in electric circuits).

In particular, if the magnetic core has constant cross section Σ and constant permeability μ, Eq. (9.68) leads to the reluctance value (9.65).

9.9 Lumped magnetic circuits

Magnetostatic circuits are similar to direct current circuits. The latter are particular cases of the electrical circuits considered in Chap. 3 when potentials and currents are time independent and there are only resistors and source voltage. The analogy is simple:

Table 9.1. Analogy between electric and magnetic circuits

Electric circuit	Magnetic circuit
Electric potential	Magnetic scalar potential
Current intensity	Magnetic flux
Resistance	Reluctance
Source voltage	Magnetomotive force

As for electrical circuits, a directed graph can be associated to a magnetic circuit. The edges represent reluctances and/or magnetomotive forces. Each edge is traversed by a magnetic flux. The nodes of the graph are the points where edges start and end. They have associated particular values of the magnetic scalar potential. Thus, determining the magnetic flux through branches of complex circuits is similar to the process of determining currents in analogous direct current circuits. In particular, we can use the theory presented in Chap. 3 for this kind of circuits. The definition of nodes and edges in a magnetic circuit is similar to the one of the analogous electrical circuit while the magnetic scalar potential $\widetilde{\varphi}$ plays the role of the electric scalar potential V and the magnetic flux that of the current intensity. As a consequence, by using the notation of Chap. 3, for an edge j with nodes m_{1j} and m_{2j}, where there is a magnetomotive force nI, we have

$$\widetilde{\varphi}_{m_{1j}} - \widetilde{\varphi}_{m_{2j}} = F_j^{int} \Phi_j - \text{nI}$$

F_j^{int} being the (internal) reluctance of the edge.

Similarly, along an edge j where there is only a resistance to the pass of the magnetic flux, i.e. a reluctance F_j, we have

$$\widetilde{\varphi}_{m_{1j}} - \widetilde{\varphi}_{m_{2j}} = F_j \Phi_j.$$

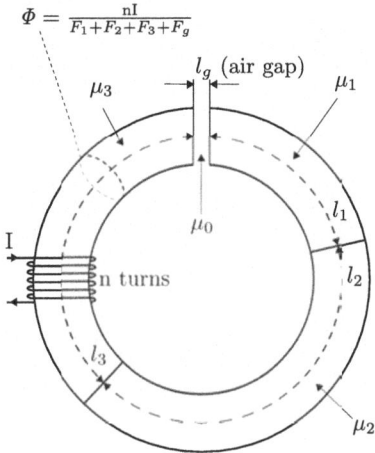

Fig. 9.6. Example of a series magnetic circuit with air-gap

In this way, we could solve magnetic circuits with general topologies by using the standard circuit theory. In particular, the analogy between direct current circuits and magnetic circuits allows us to use the same rules presented for the former to add reluctances in series and in parallel. Thus, for magnetic circuits, the magnetic flux is constant through a series connection and the magnetic potential drop is the same along a parallel connection (see examples below).

Example 9.4. In general, a magnetic circuit will consist of different magnetic materials that may be disposed, for instance, *in series* as in Fig. 9.6, and possibly include a narrow air-gap of length l_g. For this system, by applying the reluctance integral (9.68) to the successive pieces of lengths l_1, l_2, l_3 and l_g where the permeabilities and cross sections $\Sigma_1, \Sigma_2, \Sigma_3$ and Σ_g are constant, we obtain

$$F_{ap} = \frac{l_1}{\mu_1 \operatorname{meas}(\Sigma_1)} + \frac{l_2}{\mu_2 \operatorname{meas}(\Sigma_2)} + \frac{l_3}{\mu_3 \operatorname{meas}(\Sigma_3)} + \frac{l_g}{\mu_0 \operatorname{meas}(\Sigma_g)}$$
$$= F_1 + F_2 + F_3 + F_g$$

which is a similar formula to the one for the equivalent resistance of a series electrical circuit. If there is no flux leakage then the magnetic flux Φ is the same for all elements in the circuit. Thus, by using the previous reluctance in (9.67), we have

$$F_1\Phi + F_2\Phi + F_3\Phi + F_g\Phi - nI = 0, \tag{9.69}$$

which is again analogous to the Kirchhoff's voltage law $(R_1 I + R_2 I + R_3 I_3 + R_g I_g = E)$ for a similar electrical circuit with source voltage E and resistances R_i. This argument can be extended to n reluctances placed in series, where we can speak of an equivalent reluctance equal to the sum of the individual reluctances.

Notice, moreover, that Eq. (9.69) can be alternatively written in terms of the approximate value of **H** in each part of the circuit by using the Ampére's law (9.66),

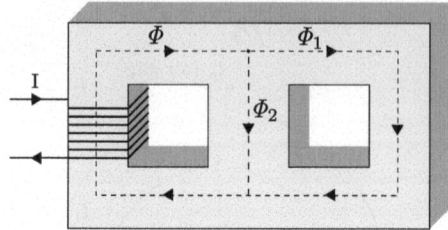

Fig. 9.7. Magnetic core with elements in parallel

namely,

$$\mathbf{H}_{ap}^1 l_1 + \mathbf{H}_{ap}^2 l_2 + \mathbf{H}_{ap}^3 l_3 + \mathbf{H}_{ap}^g l_g = \mathrm{nI},$$

where $\mathbf{H}_{ap}^i$ denotes the approximated value of $\mathbf{H}$ in each part of the circuit.

Example 9.5. Let us consider the magnetic core presented in Fig. 9.7. In this case, the magnetic flux Φ splits into two fluxes Φ_1 and Φ_2 through the branches of the core as the current does in a direct current circuit; namely,

$$\Phi = \Phi_1 + \Phi_2.$$

On the other hand, by using the Ampère's law in each closed loop of the core (analogous to second Kirchhoff's law in the electric circuit), we have:

- left loop: $\quad\quad\quad\quad\quad\quad \mathrm{nI} = \Phi F + F_2 \Phi_2;$
- right loop: $\quad\quad\quad\quad\quad 0 = -F_2 \Phi_2 + F_1 \Phi_1.$

These relationships can be used to obtain the different quantities in a magnetic circuit.

The arguments used in these examples can be extended to circuits with several elements in series or in parallel and can be used to solve the exercises concerning magnetic circuits at the end of the chapter.

9.10 2D Magnetostatic model with J normal to the 2D section

Sometimes the solution of a 3D magnetostatic problem can be approximated by a 2D model. In particular, this is the case when:

- the physical properties are invariant under translation in one privileged direction, let us say the x_3-direction;
- the given current density only has non-null component in the x_3-direction and this component does not depend on x_3, i. e., $\mathbf{J} = J_3(x_1, x_2)\mathbf{e}_3$.

Under these assumptions, the magnetostatic model can be written in a plane transversal to the device, orthogonal to the x_3-axis. Let us denote by $\widehat{\Omega}$ the two-dimensional domain containing the magnetic materials, the support of the current

9.10 2D Magnetostatic model with J normal to the 2D section

density field and the air around. More precisely, $\widehat{\Omega}$ is the intersection of the 3D domain Ω with a plane orthogonal to the x_3-axis.

Firstly, let us assume that all the fields are independent of x_3 and then it is easy to see that the magnetic field and the magnetic induction have only x_1, x_2 components and both are independent of x_3, that is,

$$\mathbf{H} = H_1(x_1,x_2)\mathbf{e}_1 + H_2(x_1,x_2)\mathbf{e}_2,$$

$$\mathbf{B} = B_1(x_1,x_2)\mathbf{e}_1 + B_2(x_1,x_2)\mathbf{e}_2,$$

where $\mathbf{e}_1$ and $\mathbf{e}_2$ denote the first two elements of the cartesian canonical basis. In this 2D case, the formulation in terms of the magnetic vector potential is the most suitable because, according to the above assumptions, we can look for $\mathbf{A}$ independent of the x_3 variable and taking the form

$$\mathbf{A}(x_1,x_2,x_3) = A_3(x_1,x_2)\mathbf{e}_3 \quad \text{in } \widehat{\Omega}.$$

Thus, the main advantage of working in terms of $\mathbf{A}$ is that we have to determine a scalar unknown $A_3(x_1,x_2)$.

Concerning the boundary conditions, if we impose a tangential magnetic field, i.e., $\mathbf{H} \times \mathbf{n} = 0$ on $\partial\Omega$, then

$$\frac{1}{\mu}\mathbf{curl}(A_3\mathbf{e}_3) \times \mathbf{n} = 0 \quad \text{on } \partial\Omega,$$

which is an homogeneous Neumann boundary condition in terms of $\mathbf{A}$.

On the other hand, if the boundary condition on $\partial\Omega$ is $\mu\mathbf{H} \cdot \mathbf{n} = 0$, like in the previous section, then it is enough to consider the Dirichlet condition $A_3 = 0$ on $\partial\widehat{\Omega}$. Thus, we will summarize the 2D magnetostatic model by using homogeneous Dirichlet and Neumann boundary conditions simultaneously. More precisely, let us split the boundary as follows: $\partial\widehat{\Omega} = \widehat{\Gamma}_d \cup \widehat{\Gamma}_n$. Then the 2D magnetostatic model reads:

Given a function J_3, find a field $\mathbf{A} = A_3(x_1,x_2)\mathbf{e}_3$ such that

$$\mathbf{curl}\left(\frac{1}{\mu}\mathbf{curl}(A_3\mathbf{e}_3)\right) = J_3\mathbf{e}_3 \quad \text{in } \widehat{\Omega}, \tag{9.70}$$

$$\frac{1}{\mu}\mathbf{curl}(A_3\mathbf{e}_3) \times \mathbf{n} = 0 \quad \text{on } \widehat{\Gamma}_n, \tag{9.71}$$

$$\mathbf{A} = 0 \quad \text{on } \widehat{\Gamma}_d. \tag{9.72}$$

The weak formulation of the above boundary-value problem can be obtained as in the 3D case by multiplying Eq. (9.70) by a test function $\mathbf{G} = G(x_1,x_2)\mathbf{e}_3$, integrating in $\widehat{\Omega}$ and using a Green's formula. We get the following problem:

Find $\mathbf{A} = A_3\mathbf{e}_3$ with $A_3 = 0$ on $\widehat{\Gamma}_d$ such that

$$\int_{\widehat{\Omega}} \frac{1}{\mu}\mathbf{curl}(A_3\mathbf{e}_3) \cdot \mathbf{curl}(G\mathbf{e}_3) \, dx_1 dx_2 = \int_{\widehat{\Omega}} J_3 G \, dx_1 dx_2, \tag{9.73}$$

for all test function $\mathbf{G}$ with $G|_{\widehat{\Gamma}_d} = 0$.

Since we are considering an isotropic situation (μ is a scalar function rather than a tensor), we have

$$\mathbf{curl}(A_3\mathbf{e}_3) \cdot \mathbf{curl}(G\mathbf{e}_3) = \mathrm{grad}\, A_3 \cdot \mathrm{grad}\, G. \tag{9.74}$$

Thus, the previous formulation can be written in the equivalent form

Find $A_3 \in \mathrm{H}^1(\widehat{\Omega})$ with $A_3 = 0$ on $\widehat{\Gamma}_\mathrm{d}$ such that

$$\int_{\widehat{\Omega}} \frac{1}{\mu} \mathrm{grad}\, A_3 \cdot \mathrm{grad}\, G \, \mathrm{d}x_1 \mathrm{d}x_2 = \int_{\widehat{\Omega}} J_3\, G \, \mathrm{d}x_1 \mathrm{d}x_2, \tag{9.75}$$

for all test function $G \in \mathrm{H}^1(\widehat{\Omega})$ with $G|_{\widehat{\Gamma}_\mathrm{d}} = 0$.

We notice, however, that if μ is a matrix we have to replace $1/\mu$ by the inverse μ^{-1} and equality (9.75) is no longer true.

Remark 9.7. Numerical discretization of the above problem can be done by using continuous piecewise linear elements on a triangular mesh of the domain $\widehat{\Omega}$ (see Appendix E).

9.11 Axisymmetric magnetostatic model with azimuthal current

In some particular situations, the solution of the 3D magnetostatics problem can be approximated by an axisymmetric model. Let us consider a bounded domain Ω containing the magnetic materials, the support of the current density and the air around. Let us assume that the problem in Ω has cylindrical symmetry in the sense that there is a cylindrical coordinate system (ρ, θ, z) such that none of the electromagnetic fields depends on the azimuth θ. Let us denote by $\mathbf{e}_\rho$, $\mathbf{e}_\theta$ and $\mathbf{e}_z$ the orthonormal vectors of the local basis associated to this coordinate system. We also suppose that the only non-null component of the current density field with respect to this basis is the one in the tangential direction $\mathbf{e}_\theta$, namely,

$$\mathbf{J}(\rho, \theta, z) = J_\theta(\rho, z)\, \mathbf{e}_\theta. \tag{9.76}$$

Because of the cylindrical symmetry, we can work on a meridional section of Ω to be denoted by $\widehat{\Omega}$. Let $\widehat{\Gamma}$ be the intersection of $\partial\Omega$ with the half-plane $\{(\rho, 0, z) \in \mathbb{R}^3; \rho > 0\}$. Moreover, let us assume that the boundary of $\widehat{\Omega}$ is the union of $\widehat{\Gamma}$ and Γ_A, the latter being a subset of the symmetry axis, $\rho = 0$.

Firstly, it is easy to see that the magnetic field and the magnetic induction only have $\mathbf{e}_\rho$ and $\mathbf{e}_\theta$ components, that is,

$$\mathbf{H}(\rho, \theta, z) = H_\rho(\rho, z)\mathbf{e}_\rho + H_\theta(\rho, z)\mathbf{e}_\theta,$$
$$\mathbf{B}(\rho, \theta, z) = B_\rho(\rho, z)\mathbf{e}_\rho + B_\theta(\rho, z)\mathbf{e}_\theta.$$

9.11 Axisymmetric magnetostatic model with azimuthal current

In this case, the magnetic vector potential formulation is again the most suitable choice because, according to the above assumptions, we can look for **A** independent of the θ variable and taking the form

$$\mathbf{A}(\rho, \theta, z) = A_\theta(\rho, z) \mathbf{e}_\theta \quad \text{in } \widehat{\Omega}.$$

Concerning boundary conditions, A_θ has to be null on the symmetry axis Γ_A because, otherwise, $\mathbf{B} = \mathbf{curl}\, \mathbf{A}$ would be singular at Γ_A. Let us consider a splitting of the whole boundary of $\widehat{\Omega}$, $\partial \widehat{\Omega} = \Gamma_A \cup \widehat{\Gamma}_d \cup \widehat{\Gamma}_n$. We impose homogeneous Dirichlet and Neumann boundary conditions on $\widehat{\Gamma}_d$ and $\widehat{\Gamma}_n$, respectively, in a similar way as for the 2D model. This means again that $\mathbf{B} \cdot \mathbf{n} = 0$ on $\widehat{\Gamma}_d$ and $\frac{1}{\mu} \mathbf{curl}(A_\theta \mathbf{e}_\theta) \times \mathbf{n} = \mathbf{0}$ on $\widehat{\Gamma}_n$, respectively. Thus, the axisymmetric problem is the following:

Given a function J_θ, find a field $\mathbf{A} = A_\theta(\rho, z)\mathbf{e}_\theta$ satisfying

$$\mathbf{curl}\left(\frac{1}{\mu} \mathbf{curl}(A_\theta \mathbf{e}_\theta)\right) = J_\theta \mathbf{e}_\theta \quad \text{in } \widehat{\Omega}, \tag{9.77}$$

$$\frac{1}{\mu} \mathbf{curl}(A_\theta \mathbf{e}_\theta) \times \mathbf{n} = \mathbf{0} \quad \text{on } \widehat{\Gamma}_n, \tag{9.78}$$

$$A_\theta = 0 \quad \text{on } \Gamma_A \cup \widehat{\Gamma}_d. \tag{9.79}$$

The weak formulation of the above boundary-value problem can be obtained by multiplying Eq. (9.77) by a test field of the form $G(\rho, z)\mathbf{e}_\theta$, integrating in $\widehat{\Omega}$ and using a Green's formula. We obtain the following problem defined in suitable weighted Sobolev spaces (see Sect. C.6):

Find $A_\theta \in \widetilde{H}^1_\rho(\widehat{\Omega})$ such that $A_\theta = 0$ on $\widehat{\Gamma}_d$ and

$$\int_{\widehat{\Omega}} \frac{1}{\mu \rho} \frac{\partial (\rho A_\theta)}{\partial \rho} \frac{\partial (\rho G)}{\partial \rho} \, d\rho dz + \int_{\widehat{\Omega}} \frac{1}{\mu} \frac{\partial A_\theta}{\partial z} \frac{\partial G}{\partial z} \rho \, d\rho dz = \int_{\widehat{\Omega}} J_\theta G \rho \, d\rho dz,$$

for all test functions $G \in \widetilde{H}^1_\rho(\widehat{\Omega})$ with $G|_{\widehat{\Gamma}_d} = 0$.

Remark 9.8. Numerical discretization of this problem can be done by using Lagrangian (nodal) continuous finite element methods on a triangular mesh of $\widehat{\Omega}$; see, for instance [14] for a similar discretization in the framework of the eddy currents model.

Problems

9.1. Let us consider an infinite, straight wire of radius a which carries a static current of given intensity I in the axial direction. The current density is uniformly distributed in any cross section of the wire and the relative magnetic permeability of the wire is equal to one. Compute the flux density **B** inside and outside the wire.
Solution: See Sect. 15.1.1.

9.2. By using cylindrical coordinates, compute the magnetic vector potential inside and outside the wire of the previous problem.
Solution: $\mathbf{A} = A_z \mathbf{e}_z$ with $A_z = \frac{-\mu_0 I \rho^2}{4\pi a^2} + c_1$, $c_1 \in \mathbb{R}$ for $\rho < a$ and $A_z = \frac{-\mu_0 I}{2\pi} \ln(\rho) + c_2$, $c_2 \in \mathbb{R}$ for $\rho > a$.

9.3. Let us consider an infinite thin sheet occupying the $x_2 x_3$-coordinate plane and carrying a constant static surface current density $\mathbf{J}_S = J_S \mathbf{e}_3$ (A/m). Compute the magnetic flux density on both sides of the sheet.
Solution: See Sect. 15.1.3.

9.4. Let us consider an infinitely long solenoid with radius b where the coil carries a static intensity I with n/d turns per unit length. The magnetic induction **B** has been computed in Sect. 5.2.2. Compute the magnetic vector potential **A**.
Solution: $\mathbf{A} = \frac{\mu_0 n I \rho}{2d} \mathbf{e}_\theta$ for $\rho < b$ and $\mathbf{A} = \frac{\mu_0 n I b^2}{2\rho d} \mathbf{e}_\theta$ for $\rho > b$.

9.5. A circular loop of thin wire of radius a carries a static current of intensity I (counterclockwise seen from the top). Let us consider that x_3 is the axis normal to the plane of the circular loop with the origin placed at loop center. Determine the magnetic flux density along the x_3-axis.
Solution: $\mathbf{B} = B_z \mathbf{e}_z = \frac{\mu_0 I a^2}{2(a^2+z^2)^{3/2}} \mathbf{e}_z$.

9.6. A square loop of thin wire is centered at the plane $x_3 = 0$ with sides parallel to the x_1, x_2 axes and carries a static current I flowing counterclockwise looking from the top. The side of the square is equal to $2a$. Use the Biot-Savart law to obtain the magnetic flux density at the center of the square.
Solution: $\mathbf{B}(0,0,0) = \frac{\mu_0 \sqrt{2} I}{\pi a}$.

9.7. A hollow circular non-magnetic conductor of inner radius a and outer radius b carries a static current I in the axial direction. The cylinder has infinite length and is placed in the free space. The current density is assumed to be uniformly distributed in the conductor. Compute the magnetic flux density in the hole ($0 < \rho < a$), in the conducting cylinder ($a < \rho < b$) and outside the conductor ($\rho < b$).
Solution: $\mathbf{B} = 0$ for $\rho < a$, $\mathbf{B} = \frac{\mu_0 I}{2\pi \rho} \frac{\rho^2 - a^2}{b^2 - a^2} \mathbf{e}_\theta$ for $a < \rho < b$ and $\mathbf{B} = 0$ for $\rho > b$.

9.11 Axisymmetric magnetostatic model with azimuthal current

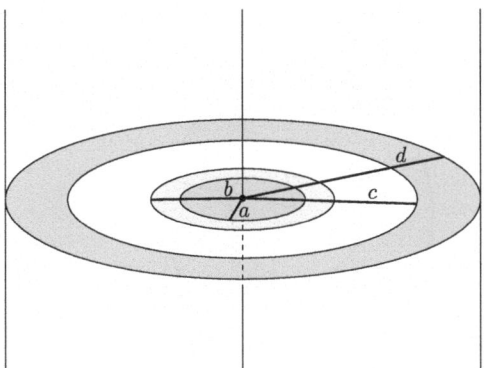

Fig. 9.8. Cross section of the coaxial conductors

9.8. Consider two infinitely long, coaxial circular conductors, $0 < \rho < a$ and $c < \rho < d$ carrying uniformly distributed currents of intensity I upwards and downwards, respectively; see a cross section in Fig. 9.8. The region $a < \rho < b$ is filled with a magnetic material of constant magnetic permeability $\mu = 100\mu_0$ while the region $b < \rho < c$ is air.

Compute the magnetic induction in the air between conductors and in the magnetic region.

Solution: $\mathbf{B} = \frac{100\mu_0 I}{2\pi\rho} \mathbf{e}_\theta$ for $a < \rho < b$ and $\mathbf{B} = \frac{\mu_0 I}{2\pi\rho} \mathbf{e}_\theta$ for $b < \rho < c$.

9.9. Obtain the weak formulation of the 2D magnetostatic problem under the same assumptions considered in Sect. 9.10 but taking into account moreover that the domain $\widehat{\Omega}$ includes a permanent magnet $\widehat{\Omega}_{pm}$ with a linear behavior. The constitutive magnetic law for permanent magnets has the form $\mathbf{B} = \mu \mathbf{H} + \mathbf{B}^r$, where $\mathbf{B}^r$ is the so-called remanent flux density that in this 2D case is assumed to have the form $\mathbf{B}^r = B_1^r(x_1,x_2)\mathbf{e}_1 + B_2^r(x_1,x_2)\mathbf{e}_2$. **Hint:** Use the continuity of $\mathbf{H} \times \mathbf{n}$ through the interface of different materials.

Solution: Find $A_3 \in H^1(\widehat{\Omega})$ with $A_3 = 0$ on $\widehat{\Gamma}_d$ such that

$$\int_{\widehat{\Omega}} \frac{1}{\mu} \operatorname{grad} A_3 \cdot \operatorname{grad} G \, dx_1 dx_2 = \int_{\widehat{\Omega}} J_3 \, G \, dx_1 dx_2$$
$$+ \int_{\widehat{\Omega}_{pm}} \frac{1}{\mu}(-B_2^r \mathbf{e}_1 + B_1^r \mathbf{e}_2) \cdot \operatorname{grad} G \, dx_1 dx_2,$$

for all test function $G \in H^1(\widehat{\Omega})$ with $G|_{\widehat{\Gamma}_d} = 0$ (we have used the notation of Sect. 9.10).

9.10. Consider a toroidal iron core of circular cross section with inner radius 0.05 m and radius of the cross section 0.01 m. A coil of 100 turns and carrying a intensity of 1 A is around the core. The iron core has a relative magnetic permeability equal to 1000. Compute the exact value of the magnetic flux through a cross section of the

180   9 Magnetostatics

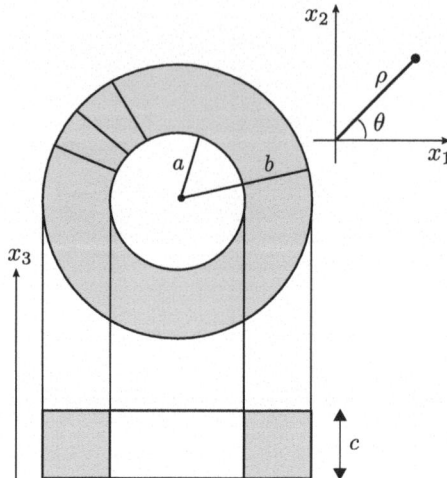

Fig. 9.9. n-turn coil around a magnetic toroidal core with rectangular cross section

core by using the value of the reluctance matrix obtained in Example 9.2 and compare its value with the one obtained with the approximate expression (9.65) for the reluctance.

Solution: Relative error: 1.01 %

9.11. Consider the toroidal core of Fig. 9.9 with rectangular cross section of height c, inner radius equal to a and outer radius equal to b. The coil has n turns and carries a static current I. The material of the core is magnetic with constant magnetic permeability μ.

(a) Compute the magnetic induction inside and outside the magnetic core by following the arguments of Example 9.1.
(b) Compute the exact value of the magnetic flux through a cross section of the core and the exact value of the reluctance.
(c) Compute the approximated value of the magnetic flux by using the concept of approximate reluctance F_{ap} for a thin magnetic circuit given by (9.65).
(d) Compare the values of the exact and approximated flux obtained in (b) and (c) for different ratios between a and b and notice that the difference is very small.

Solution: (a) $\mathbf{B} = \frac{\mu n I}{2\pi \rho} \mathbf{e}_\theta$ for $a < \rho < b$ and $\mathbf{B} = \mathbf{0}$ outside the torus. (b) Exact flux = $\frac{\mu n I c}{2\pi} \ln \frac{b}{a}$, exact reluctance = $\frac{2\pi}{\mu c \ln(\frac{b}{a})}$ H^{-1}. (c) Approximate value of flux = $\frac{\mu n I c}{\pi} \frac{(b-a)}{b+a}$ Wb, $F_{ap} = \frac{\pi(a+b)}{\mu(b-a)c}$ H^{-1}.

To solve the following exercises, consider the theory of thin magnetic circuits.

9.12. Consider a toroidal iron core with a square cross section of area 4e-4 m^2; the inner radius is equal to 0.03 m, the outer radius equal to 0.05 m and there is an air-

9.11 Axisymmetric magnetostatic model with azimuthal current

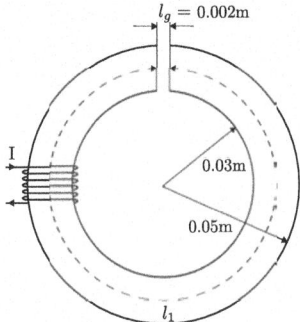

Fig. 9.10. n-turn coil around a magnetic toroidal core with square cross section

gap of 0.002 m (see Fig. 9.10). The coil has 100 turns and 0.5 A. The iron core has a relative magnetic permeability equal to 300.

(a) Compute the reluctance of the magnetic circuit in the core and in the air gap by using the mean lengths l_1 and l_g for each part.
(b) Compute the magnetic flux in the magnetic circuit.
(c) Compute the average value of the magnetic field in the core and in the air gap.

Solution: (a) Reluctance in core: 1.65e6 H^{-1}, reluctance in air-gap: 3.98e6 H^{-1} (b) Φ = 8.8e-6 Wb (c) H_{ap} = 58.87 A/m in core and H_{ap} =1.77e4 A/m in the air-gap.

9.13. A steel ring has a radius of 0.05 m and a cross-sectional area of 4e-4 m^2. A current of 0.5 A flows in a coil wound uniformly around the ring and the flux produced is 1e-4 Wb. If the relative magnetic permeability at this value of current is 200:

(a) Find the approximate reluctance of the steel.
(b) Find the number of turns of the coil.

Solution: (a) 3.125e6 H^{-1} (b) 625 turns.

9.14. A closed magnetic circuit of cast steel contains a 0.06 m length path of cross-sectional area 1e-4 m^2 and a 0.02 m path of a cross-sectional area 5e-5 m^2. A coil of 200 turns is wound around the 0.06 m length of the circuits and carries a current of 0.4 A. Determine the average flux density in the 0.02 m path if the relative permeability of the cast steel is equal to 750.

Solution: B_{ap} = 1.51 T.

9.15. The magnetic core presented in Fig. 9.11 has a constant cross section of area 6e-4 m^2 and is composed by different kinds of steel; the relative magnetic permeability of the steel in regions *bcde*, *be* and *efab* are 4972, 4821 and 2426, respectively. The mean length of each part is the following: $l_{bcde} = l_{efab} = 0.2$ m, $l_{be} = 0.05$ m. The number of turns of the coil is 50. Determine the value of the intensity I in the coil required to establish a flux of 1.54e-4 Wb in the steel region *bcde*.

Solution: 1.76 A.

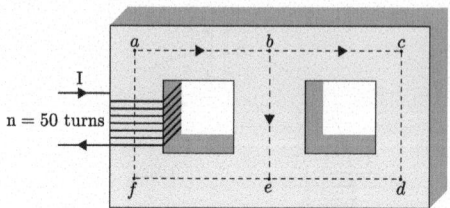

Fig. 9.11. Magnetic core composed by several types of steel

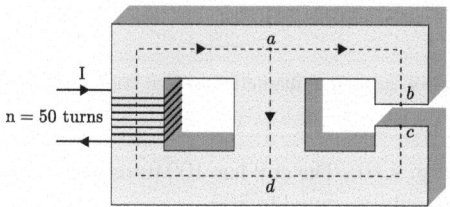

Fig. 9.12. Magnetic core with an air-gap between the points b and c

9.16. The magnetic core presented of in Fig. 9.12 has a constant cross section of area 2.e-2 m^2 and is made of a steel with a relative magnetic permeability equal to 960. The mean length of each part is the following: $l_g = l_{bc} = 0.25$e-3 m, $l_{ab} = l_{cd} = 0.25$ m, $l_{da} = 0.2$ m and $l_{dea} = 0.35$ m. The number of turns of the coil is 200. Determine the value of the intensity I required to establish a flux of 6.e-3 Wb at the air gap.

Solution: 2.96 A.

10
The eddy currents model

In this chapter we deal with the well-known eddy currents model which is obtained from Maxwell's equations by neglecting the electric displacement in the Ampère's law. We will study this model in the time-harmonic regime and in bounded three-dimensional and two-dimensional domains by using different unknowns. At the end of the chapter we give a brief description of the coupling between the eddy currents model and a lumped circuit model.

10.1 The time-harmonic eddy currents model in a bounded domain

The eddy currents model is obtained from Maxwell's equations when the electromagnetic field is slowly time-varying; more precisely, when the current carrying system is small compared with the electromagnetic wavelength associated with the dominant time scale of the problem. In this case fields are propagated instantaneously so we are dealing with a $c \to \infty$ limit, c being the propagation velocity of waves. This leads to neglect field radiation effects. In other words, term $\frac{\partial \mathbf{D}}{\partial t}$ is suppressed in the Ampère's law. Thus, the eddy currents model is

$$\frac{\partial \mathbf{B}}{\partial t} + \operatorname{curl} \mathbf{E} = \mathbf{0}, \tag{10.1}$$

$$\operatorname{curl} \mathbf{H} = \mathbf{J}, \tag{10.2}$$

$$\operatorname{div} \mathbf{B} = 0, \tag{10.3}$$

$$\operatorname{div} \mathbf{D} = \rho_V, \tag{10.4}$$

$$\mathbf{B} = \mu \mathbf{H}, \tag{10.5}$$

$$\mathbf{D} = \varepsilon \mathbf{E}, \tag{10.6}$$

$$\mathbf{J} = \sigma \mathbf{E} + \mathbf{J}_S, \tag{10.7}$$

A. Bermúdez, D. Gómez, P. Salgado: *Mathematical Models and Numerical Simulation in Electromagnetism.* UNITEXT – La Matematica per il 3+2 74
DOI 10.1007/978-3-319-02949-8_10, © Springer International Publishing Switzerland 2014

where $\mathbf{J}_S \in H(\mathrm{div}, \Omega)$ is a divergence-free source current. We also suppose that σ is positive in conductors, null in dielectrics and

$$\mathrm{supp}(\mathbf{J}_S) \cap \mathrm{supp}(\sigma) = \emptyset. \tag{10.8}$$

Since we have neglected the term $\frac{\partial \mathbf{D}}{\partial t}$ in Ampère's law, we do not keep the Gauss' law for electric charge, (10.4), in the conductors. Instead, from (10.2) and (10.7) we get

$$\mathrm{div}(\sigma \mathbf{E}) = 0 \text{ in the conductors.}$$

We are interested in obtaining the magnetic field everywhere but the electric field only in the conductors. If we wanted to get this field also in the dielectrics then Eq. (10.4) would be retained in the model, in the dielectrics.

In what follows we consider the harmonic case. This means that all fields are of the following form

$$\mathbf{G}(x,t) = \mathrm{Re}(e^{i\omega t}\hat{\mathbf{G}}(x)).$$

By replacing in (10.1)–(10.7) we obtain the electromagnetic Helmholtz's equations for the complex phasors $\hat{\mathbf{G}}$ (for the sake of simplicity we will drop the "hat" in the notation of phasors):

$$i\omega \mathbf{B} + \mathbf{curl}\,\mathbf{E} = \mathbf{0}, \tag{10.9}$$
$$\mathbf{curl}\,\mathbf{H} = \mathbf{J}, \tag{10.10}$$
$$\mathrm{div}\,\mathbf{B} = 0, \tag{10.11}$$
$$\mathbf{B} = \mu \mathbf{H}, \tag{10.12}$$
$$\mathbf{J} = \mathbf{J}_S + \sigma \mathbf{E}. \tag{10.13}$$

To solve these equations, we restrict them to a simply connected 3D bounded domain Ω consisting of two parts, Ω_{C} and Ω_{D}, occupied by conductors and dielectrics, respectively. The mathematical framework we are going to analyze covers eddy currents problems posed on different geometrical settings. We sketch a particular case in Fig. 10.1 including several connected components of the conducting domain with different topological properties. The domain Ω is assumed to have a Lipschitz-

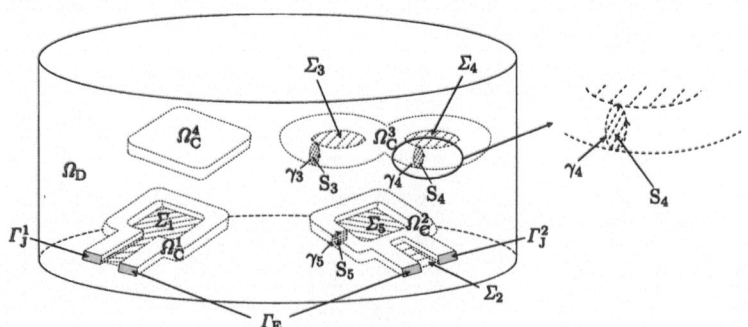

Fig. 10.1. Sketch of the domain

10.1 The time-harmonic eddy currents model in a bounded domain

continuous connected boundary Γ. We denote by Γ_C, Γ_D and Γ_I the open surfaces such that $\bar{\Gamma}_C := \partial\Omega_C \cap \Gamma$ is the outer boundary of the conducting domain, $\bar{\Gamma}_D := \partial\Omega_D \cap \Gamma$ that of the dielectric domain and $\bar{\Gamma}_I := \partial\Omega_C \cap \partial\Omega_D$ the interface between both domains. We also denote by **n** a unit normal vector to a given surface.

As shown in Fig. 10.1, the connected components of the conducting domain are of two types: "inductors" which cross the boundary of Ω, and "workpieces" which have their closure included in Ω. Let us denote $\Omega_C^1, \ldots, \Omega_C^L$ the former and $\Omega_C^{L+1}, \ldots, \Omega_C^M$ the latter.

We assume that the outer boundary of each inductor, $\partial\Omega_C^n \cap \Gamma$ ($n = 1,\ldots,L$), has two disjoint connected components, both being the closure of open surfaces: the current entrance Γ_J^n, where the inductor is connected to an alternate electric current source, and the current exit Γ_E^n. We denote $\Gamma_J := \Gamma_J^1 \cup \cdots \cup \Gamma_J^L$ and $\Gamma_E := \Gamma_E^1 \cup \cdots \cup \Gamma_E^L$. Furthermore, we assume that $\bar{\Gamma}_J^n \cap \bar{\Gamma}_J^m = \emptyset$, $\bar{\Gamma}_E^n \cap \bar{\Gamma}_E^m = \emptyset$, $1 \leq m, n \leq L$, $m \neq n$, and $\bar{\Gamma}_J \cap \bar{\Gamma}_E = \emptyset$.

Let us assume that μ and σ are frequency-independent and there exist constants $\underline{\mu}$, $\overline{\mu}$, $\overline{\sigma}$ and $\underline{\sigma}$ such that

$$0 < \underline{\mu} \leq \mu(x) \leq \overline{\mu}, \quad \text{a.e. } x \in \Omega,$$
$$0 < \underline{\sigma} \leq \sigma(x) \leq \overline{\sigma}, \quad \text{a.e. } x \in \Omega_C \quad \text{and} \quad \sigma \equiv 0 \text{ in } \Omega_D.$$

We have to complete the model with suitable boundary conditions. For the moment, let us consider the following ones:

$$\mathbf{E} \times \mathbf{n} = \mathbf{0} \quad \text{on } \Gamma_E, \tag{10.14}$$
$$\mathbf{E} \times \mathbf{n} = \mathbf{0} \quad \text{on } \Gamma_J, \tag{10.15}$$
$$\mu \mathbf{H} \cdot \mathbf{n} = 0 \quad \text{on } \Gamma. \tag{10.16}$$

Conditions (10.14) and (10.15) mean that the electric current enters and exits domain Ω perpendicularly to the boundary whereas condition (10.16) implies that the magnetic field is tangential to the boundary.

Formal calculations allow us to show that boundary condition (10.16) implies that the tangential component of the electric field **E** is a gradient. Indeed, after integrating $i\omega\mu\mathbf{H}\cdot\mathbf{n}$ on any surface S contained in Γ, by using (10.9), (10.12) and Stokes' theorem, we obtain

$$0 = i\omega \int_S \mu \mathbf{H} \cdot \mathbf{n}\, dS = -\int_S \operatorname{curl} \mathbf{E} \cdot \mathbf{n}\, dS = -\int_{\partial S} \mathbf{E} \cdot \mathbf{t}\, dl = -\int_{\partial S} \mathbf{n} \times \mathbf{E} \times \mathbf{n} \cdot \mathbf{t}\, dl,$$

with **t** being a unit vector tangent to ∂S. Therefore, since Γ is simply connected, we can assert that there exists a sufficiently smooth function V defined in Ω up to a constant, such that $V|_\Gamma$ is a surface potential of the tangential component of **E**, namely, $\mathbf{E} \times \mathbf{n} = -\operatorname{grad} V \times \mathbf{n}$ on Γ. On the other hand, (10.14) and (10.15) imply that V must be constant on each connected component of Γ_J and Γ_E. The complex number $V_n := V|_{\Gamma_E^n} - V|_{\Gamma_J^n}$ is the voltage drop along conductor Ω_C^n.

Many physical applications involve current intensities and voltage drops as boundary data. Let us suppose that the boundary data consist of the voltage drops V_n, for $n = 1,\ldots,\widehat{L}$, and the input current intensities through each surface Γ_J^n, I_n, for $n = \widehat{L}+1,\ldots,L$. We notice that the latter can be written as

$$\int_{\Gamma_J^n} \mathbf{J} \cdot \mathbf{n}\, dA = I_n, \quad n = \widehat{L}+1,\ldots,L.$$

Moreover, Eq. (10.10) yields $\operatorname{div} \mathbf{J} = 0$ and hence, by Gauss' theorem,

$$0 = \int_{\Omega_C^n} \operatorname{div} \mathbf{J}\, dV = \int_{\partial\Omega_C^n} \mathbf{J} \cdot \mathbf{n}\, dA = \int_{\Gamma_J^n} \mathbf{J} \cdot \mathbf{n}\, dA + \int_{\Gamma_E^n} \mathbf{J} \cdot \mathbf{n}\, dA,$$

because $\mathbf{J} \cdot \mathbf{n} = 0$ on $\partial\Omega_C^n \setminus (\Gamma_E^n \cup \Gamma_J^n)$. Thus,

$$\int_{\Gamma_J^n} \mathbf{J} \cdot \mathbf{n}\, dA = -\int_{\Gamma_E^n} \mathbf{J} \cdot \mathbf{n}\, dA, \quad n = 1,\ldots,L. \tag{10.17}$$

These equalities simply mean that the input current intensity coincides with the output one for each conductor Ω_C^n, $n = 1,\ldots,L$.

We summarize the strong problem defined in Ω to be considered in the next section:

$$i\omega \mathbf{B} + \operatorname{curl} \mathbf{E} = \mathbf{0}, \tag{10.18}$$

$$\operatorname{curl} \mathbf{H} = \mathbf{J}, \tag{10.19}$$

$$\operatorname{div} \mathbf{B} = 0, \tag{10.20}$$

$$\mathbf{B} = \mu \mathbf{H}, \tag{10.21}$$

$$\mathbf{J} = \mathbf{J}_S + \sigma \mathbf{E}, \tag{10.22}$$

$$\mathbf{E} \times \mathbf{n} = \mathbf{0} \quad \text{on } \Gamma_E, \tag{10.23}$$

$$\mathbf{E} \times \mathbf{n} = \mathbf{0} \quad \text{on } \Gamma_J, \tag{10.24}$$

$$\mu \mathbf{H} \cdot \mathbf{n} = 0 \quad \text{on } \Gamma, \tag{10.25}$$

$$V = V_J^n \quad \text{on } \Gamma_J^n,\ n = 1,\ldots,\widehat{L}, \tag{10.26}$$

$$V = V_E^n \quad \text{on } \Gamma_E^n,\ n = 1,\ldots,\widehat{L}, \tag{10.27}$$

$$\int_{\Gamma_J^n} \mathbf{J} \cdot \mathbf{n}\, dA = I_n, \quad n = \widehat{L}+1,\ldots,L. \tag{10.28}$$

10.2 Magnetic vector potential/scalar electric potential formulation

We want to formulate and solve the above problem in terms of a magnetic vector potential **A** and an electric scalar potential V.

Firstly, from (10.20), (10.25) and [37, Theorem 1.3.6] we deduce the existence of a vector field $\mathbf{A} \in \mathbf{H}(\mathbf{curl}, \Omega)$ such that

$$\mathbf{curl}\,\mathbf{A} = \mathbf{B} \text{ in } \Omega, \tag{10.29}$$

$$\mathrm{div}\,\mathbf{A} = 0 \text{ in } \Omega, \tag{10.30}$$

$$\mathbf{A} \times \mathbf{n} = \mathbf{0} \text{ on } \Gamma. \tag{10.31}$$

We notice that the latter equality guarantees boundary condition (10.25). Indeed, from Stokes' theorem we have

$$\int_S \mathbf{B} \cdot \mathbf{n}\,\mathrm{d}S = \int_{\partial S} \mathbf{A} \cdot \mathbf{t}\,\mathrm{d}l = \int_{\partial S} \mathbf{n} \times \mathbf{A} \times \mathbf{n} \cdot \mathbf{t}\,\mathrm{d}l = 0 \quad \text{for all surface } S \subset \Gamma.$$

By replacing (10.29) in (10.18) we obtain

$$\mathbf{curl}(i\omega\mathbf{A} + \mathbf{E}) = \mathbf{0} \text{ in } \Omega$$

and then, in particular,

$$i\omega\mathbf{A} + \mathbf{E} = -\,\mathrm{grad}\,V \text{ in } \Omega, \tag{10.32}$$

for some $V \in H^1(\Omega)$.

From (10.23), (10.24) and (10.31) we deduce

$$\mathrm{grad}_\Gamma V := \mathbf{n} \times \mathrm{grad}\,V \times \mathbf{n} = -\mathbf{n} \times \mathbf{E} \times \mathbf{n} = \mathbf{0} \quad \text{on } \Gamma_J \cup \Gamma_E,$$

which implies that V must be constant on each connected component of Γ_J and Γ_E. We notice that the complex number $V_n := V_E^n - V_J^n$ is the voltage drop along conductor Ω_C^n.

From (10.18), (10.21), (10.22) and (10.32) we deduce

$$\sigma(i\omega\mathbf{A} + \mathrm{grad}\,V) + \mathbf{curl}(\frac{1}{\mu}\mathbf{curl}\,\mathbf{A}) = \mathbf{J}_S \text{ in } \Omega. \tag{10.33}$$

We notice that, since $\sigma = 0$ in Ω_D, we only need to compute V in Ω_C.

Finally, from (10.22) and (10.32), boundary conditions (10.28) become

$$\int_{\Gamma_J^n} \sigma(i\omega\mathbf{A} + \mathrm{grad}\,V) \cdot \mathbf{n}\,\mathrm{d}S = -I_n, \quad n = \widehat{L}+1,\ldots,L. \tag{10.34}$$

Summarizing, the problem to be solved reads as follows:

Given a solenoidal field $\mathbf{J}_S$ *with support* Ω_S *included in* Ω_D, *and complex numbers* V_J^n, V_E^n, $n = 1,\ldots,\widehat{L}$ *and* I_n, $n = \widehat{L}+1,\ldots,L$, *find a vector field* **A** *defined in* Ω, *and* a scalar field V defined in Ω_C and null on Γ_J^n, $n = \widehat{L}+1,\ldots,L$, such that

$$\sigma(i\omega \mathbf{A} + \operatorname{grad} V) + \mathbf{curl}(\frac{1}{\mu} \mathbf{curl}\mathbf{A}) = \mathbf{J}_S \text{ in } \Omega, \tag{10.35}$$

$$\operatorname{div} \mathbf{A} = 0 \text{ in } \Omega, \tag{10.36}$$

$$\mathbf{A} \times \mathbf{n} = 0 \text{ on } \Gamma, \tag{10.37}$$

$$\sigma(i\omega \mathbf{A} + \operatorname{grad} V) \cdot \mathbf{n} = 0 \text{ on } \partial\Omega_C^n \setminus (\Gamma_E^n \cup \Gamma_J^n), \ n=1,\ldots,L, \tag{10.38}$$

$$V = V_J^n \text{ on } \Gamma_J^n, \ n=1,\ldots,\widehat{L}, \tag{10.39}$$

$$V = V_E^n \text{ on } \Gamma_E^n, \ n=1,\ldots,\widehat{L}, \tag{10.40}$$

$$\int_{\Gamma_J^n} \sigma(i\omega \mathbf{A} + \operatorname{grad} V) \cdot \mathbf{n} \, dS = -I_n, \ n = \widehat{L}+1,\ldots,L. \tag{10.41}$$

In what follows we write a weak formulation of this problem. Firstly, let us multiply (10.35) by the conjugate of a test function $\mathbf{G} \in \mathbf{H}(\mathbf{curl}, \Omega)$, integrate in Ω and use a Green's formula. We get

$$\int_{\Omega_C} \sigma(i\omega \mathbf{A} + \operatorname{grad} V) \cdot \bar{\mathbf{G}} \, dV + \int_\Omega \frac{1}{\mu} \mathbf{curl}\mathbf{A} \cdot \mathbf{curl}\bar{\mathbf{G}} \, dV = \int_{\Omega_S} \mathbf{J}_s \cdot \bar{\mathbf{G}} \, dV.$$

Next, we notice that, in principle, the traces appearing in (10.41) are not well defined for $\mathbf{A} \in \mathbf{H}(\mathbf{curl}, \Omega)$ and $V \in H^1(\Omega_C)$ so we write a weak formulation of these equations. For this purpose, we first notice that, by taking the divergence operator of (10.35), we deduce

$$\operatorname{div}\left(\sigma(i\omega \mathbf{A} + \operatorname{grad} V)\right) = 0 \text{ in } \Omega_C. \tag{10.42}$$

Now, given two vectors $\vec{V}_J, \vec{V}_E \in \mathbb{R}^{\widehat{L}}$ let us define the affine space

$$\mathscr{L}(\vec{V}_J, \vec{V}_E) = \{V \in H^1(\Omega_C): \ V = V_J^n \text{ on } \Gamma_J^n, \ V = V_E^n \text{ on } \Gamma_E^n, \ n=1,\ldots,\widehat{L},$$
$$V = constant \text{ on } \Gamma_J^n \text{ and on } \Gamma_E^n, \ n = \widehat{L}+1,\ldots,L\}.$$

We denote $\mathscr{L}_0$ the above space corresponding to null vectors $\vec{V}_J$ and $\vec{V}_E$. By multiplying Eq. (10.42) by the conjugate of a test function $Z \in \mathscr{L}_0$, integrating in Ω_C, using a Green's formula, and Eqs. (10.38) and (10.41) we get

$$\int_{\Omega_C} \sigma(i\omega \mathbf{A} + \operatorname{grad} V) \cdot \operatorname{grad} \bar{Z} \, dV = \sum_{n=\widehat{L}+1}^{L} I_n \bar{Z}_n, \tag{10.43}$$

with $Z_n = Z_{|\Gamma_E^n} - Z_{|\Gamma_J^n}$, $n = \widehat{L}+1,\ldots,L$.

Finally, by multiplying the *gauge* condition (10.36) by the conjugate of a test function $\Psi \in H_0^1(\Omega)$ and using a Green's formula we obtain

$$\int_\Omega \mathbf{A} \cdot \operatorname{grad} \bar{\Psi} \, dV = 0.$$

10.2 Magnetic vector potential/scalar electric potential formulation

Let $\mathscr{W}_0$ be the function space

$$\mathscr{W}_0 = \{\mathbf{G} \in \mathbf{H}(\mathbf{curl}, \Omega) : \mathbf{G} \times \mathbf{n} = \mathbf{0} \text{ on } \Gamma\}.$$

Next, we summarize the whole weak problem:

Find $\mathbf{A} \in \mathscr{W}_0$, $V \in \mathscr{L}(\overrightarrow{V}_J, \overrightarrow{V}_E)$ *and* $\varphi \in H_0^1(\Omega)$ *such that*

$$\int_{\Omega_C} \sigma(i\omega \mathbf{A} + \mathrm{grad}\, V) \cdot \bar{\mathbf{G}}\, dV + \int_{\Omega} \frac{1}{\mu} \mathbf{curl}\, \mathbf{A} \cdot \mathbf{curl}\, \bar{\mathbf{G}}\, dV + \int_{\Omega} \mathrm{grad}\, \varphi \cdot \bar{\mathbf{G}}\, dV$$

$$= \int_{\Omega_S} \mathbf{J}_s \cdot \bar{\mathbf{G}}\, dV \quad \forall \mathbf{G} \in \mathscr{W}_0,$$

$$\int_{\Omega_C} \sigma(i\omega \mathbf{A} + \mathrm{grad}\, V) \cdot \mathrm{grad}\, \bar{Z}\, dV = \sum_{n=\widehat{L}+1}^{L} I_n \bar{Z}_n \quad \forall Z \in \mathscr{L}_0,$$

$$\int_{\Omega} \mathbf{A} \cdot \mathrm{grad}\, \bar{\Psi}\, dV = 0 \quad \forall \Psi \in H_0^1(\Omega).$$

Remark 10.1. We notice that in this formulation the scalar potential V only appears under the gradient operator. Since the gradient of V does not change if V is translated by a constant in each connect component Ω_C^n, we can replace the above problem by the following one:

Find $\mathbf{A} \in \mathscr{W}_0$, $V \in \mathscr{M}(\overrightarrow{V})$ *and* $\varphi \in H_0^1(\Omega)$ *such that*

$$\int_{\Omega_C} \sigma(i\omega \mathbf{A} + \mathrm{grad}\, V) \cdot \bar{\mathbf{G}}\, dV + \int_{\Omega} \frac{1}{\mu} \mathbf{curl}\, \mathbf{A} \cdot \mathbf{curl}\, \bar{\mathbf{G}}\, dV + \int_{\Omega} \mathrm{grad}\, \varphi \cdot \bar{\mathbf{G}}\, dV$$

$$= \int_{\Omega_S} \mathbf{J}_s \cdot \bar{\mathbf{G}}\, dV \quad \forall \mathbf{G} \in \mathscr{W}_0,$$

$$\int_{\Omega_C} \sigma(i\omega \mathbf{A} + \mathrm{grad}\, V) \cdot \mathrm{grad}\, \bar{Z}\, dV = \sum_{n=\widehat{L}+1}^{L} I_n \bar{Z}_n \quad \forall Z \in \mathscr{M}_0,$$

$$\int_{\Omega} \mathbf{A} \cdot \mathrm{grad}\, \bar{\Psi}\, dV = 0 \quad \forall \Psi \in H_0^1(\Omega),$$

where $\mathscr{M}(\overrightarrow{V})$ is the affine space

$$\mathscr{M}(\overrightarrow{V}) := \{V \in H^1(\Omega_C) : V = 0 \text{ on } \Gamma_J^n, n = 1, \ldots, L, V = V_n \text{ on } \Gamma_E^n, n = 1, \ldots, \widehat{L},$$

$$V = constant \text{ on } \Gamma_E^n, n = \widehat{L}+1, \ldots, L\}$$

and $\mathscr{M}_0$ is the above space for the null vector $\overrightarrow{V}$.

Remark 10.2. Existence and uniqueness of a solution to this problem can be shown by using the Babuska-Brezzi theory of mixed formulations (see [2]).

Remark 10.3. Numerical discretization can be done by using Nédélec edge finite elements to approximate the vector potential and continuous piecewise linear nodal elements to approximate φ and V.

10.3 A formulation in magnetic field

Problem (10.18)–(10.22) with boundary conditions (10.23)–(10.28) can also be written and solved in terms of the magnetic field.

For this purpose let us multiply (10.32) by the **curl** of the conjugate of a test function $\mathbf{G} \in \mathbf{H}(\mathbf{curl}, \Omega)$ and integrate in Ω. We get

$$\int_\Omega i\omega \mathbf{A} \cdot \mathbf{curl}\,\bar{\mathbf{G}}\,dV + \int_\Omega \mathbf{E} \cdot \mathbf{curl}\,\bar{\mathbf{G}}\,dV = -\int_\Omega \mathrm{grad}\,V \cdot \mathbf{curl}\,\bar{\mathbf{G}}\,dV$$

and using two Green's formulas and boundary condition (10.31)

$$\int_\Omega i\omega\,\mathbf{curl}\,\mathbf{A} \cdot \bar{\mathbf{G}}\,dV + \int_\Omega \mathbf{E} \cdot \mathbf{curl}\,\bar{\mathbf{G}}\,dV = -\int_\Omega \mathrm{grad}\,V \cdot \mathbf{curl}\,\bar{\mathbf{G}}\,dV$$

$$= -\int_\Gamma V\,\mathbf{curl}\,\bar{\mathbf{G}} \cdot \mathbf{n}\,dS. \qquad (10.44)$$

Let us further assume that $\mathbf{curl}\,\mathbf{G} = \mathbf{0}$ in Ω_D. By using (10.22) in (10.44) we get

$$\int_\Omega i\omega \mathbf{B} \cdot \bar{\mathbf{G}}\,dV + \int_{\Omega_\mathrm{C}} \frac{1}{\sigma}\mathbf{curl}\,\mathbf{H} \cdot \mathbf{curl}\,\bar{\mathbf{G}}\,dV = -\int_\Gamma V\,\mathbf{curl}\,\bar{\mathbf{G}} \cdot \mathbf{n}\,dS$$

$$= -\sum_{n=1}^{\widehat{L}} V_\mathbf{J}^n \int_{\Gamma_\mathbf{J}^n} \mathbf{curl}\,\bar{\mathbf{G}} \cdot \mathbf{n}\,dS - \sum_{n=1}^{\widehat{L}} V_\mathbf{E}^n \int_{\Gamma_\mathbf{E}^n} \mathbf{curl}\,\bar{\mathbf{G}} \cdot \mathbf{n}\,dS$$

$$= \sum_{n=1}^{\widehat{L}} V_n \int_{\Gamma_\mathbf{J}^n} \mathbf{curl}\,\bar{\mathbf{G}} \cdot \mathbf{n}\,dS, \qquad (10.45)$$

where we have used that

$$\int_{\Gamma_\mathbf{E}^n} \mathbf{curl}\,\bar{\mathbf{G}} \cdot \mathbf{n}\,dS = -\int_{\Gamma_\mathbf{J}^n} \mathbf{curl}\,\bar{\mathbf{G}} \cdot \mathbf{n}\,dS.$$

This equality follows from the fact that $\mathrm{div}\,\mathbf{curl}\,\mathbf{G} = 0$ and $\mathbf{curl}\,\mathbf{G} \cdot \mathbf{n} = 0$ on $\partial\Omega_\mathrm{D}$.

Let $\vec{\mathbf{I}} \in \mathbb{C}^{L-\widehat{L}}$ be the complex vector of current intensities across conductors Ω_C^n, $n = \widehat{L}+1,\ldots,L$. In order to write a weak formulation of the problem let us introduce the affine space

$$\mathscr{H}(\mathbf{J}_\mathrm{S}, \vec{\mathbf{I}}) := \{\mathbf{H} \in \mathbf{H}(\mathbf{curl}, \Omega) : \mathbf{curl}\,\mathbf{H} = \mathbf{J}_\mathrm{S} \text{ in } \Omega_\mathrm{D},$$

$$\int_{\Gamma_\mathbf{J}^n} \mathbf{curl}\,\mathbf{H} \cdot \mathbf{n}\,dS = -\mathrm{I}_n,\ n = \widehat{L}+1,\ldots,L\}.$$

10.3 A formulation in magnetic field

We denote by $\mathscr{H}_0$ the above space for null $\mathbf{J_S}$ and $\vec{\mathbf{I}}$. We are in a position to write the weak formulation of the problem:

Given a solenoidal field $\mathbf{J_S} \in L^2(\Omega_S)$ *and two complex vectors* $\vec{\mathbf{V}} \in \mathbb{R}^{\hat{L}}$ *and* $\vec{\mathbf{I}} \in \mathbb{R}^{L-\hat{L}}$, *find* $\mathbf{H} \in \mathscr{H}(\mathbf{J_S}, \vec{\mathbf{I}})$ *such that*

$$\int_\Omega i\omega\mu \mathbf{H} \cdot \bar{\mathbf{G}}\, dV + \int_{\Omega_C} \frac{1}{\sigma} \operatorname{curl} \mathbf{H} \cdot \operatorname{curl} \bar{\mathbf{G}}\, dV$$

$$= \sum_{n=1}^{\hat{L}} V_n \int_{\Gamma_J^n} \operatorname{curl} \bar{\mathbf{G}} \cdot \mathbf{n}\, dS \quad \forall \mathbf{G} \in \mathscr{H}_0. \qquad (10.46)$$

Numerical analysis of this problem by a finite element method has been done in [17].

Discretizing the affine space $\mathscr{H}(\mathbf{J_S}, \vec{\mathbf{I}})$ is not easy because of condition $\operatorname{curl} \mathbf{H_S} = 0$ so for numerical solution it is convenient to make first a translation by using a lifting of the current density source $\mathbf{J_S}$. Let us introduce the vector field $\mathbf{H_S} \in H(\operatorname{curl}, \Omega_D)$ satisfying the "magnetostatics" problem

$$\operatorname{curl} \mathbf{H_S} = \mathbf{J_S} \text{ in } \Omega_S, \qquad (10.47)$$

$$\operatorname{curl} \mathbf{H_S} = \mathbf{0} \text{ in } \Omega_D \setminus \Omega_S, \qquad (10.48)$$

$$\operatorname{div}(\mu \mathbf{H_S}) = 0 \text{ in } \Omega_D, \qquad (10.49)$$

$$\mathbf{H_S} \times \mathbf{n} = \mathbf{0} \text{ on } \Gamma_I, \qquad (10.50)$$

$$\mathbf{H_S} \cdot \mathbf{n} = 0 \text{ on } \Gamma_D, \qquad (10.51)$$

$$\int_{\Gamma_I^n} \mu \mathbf{H_S} \cdot \mathbf{n}\, dS = 0, \quad n = 2, \ldots, L. \qquad (10.52)$$

This problem has a unique solution (see [11]). Let us still denote by $\mathbf{H_S}$ its extension by zero to Ω_C. Let us denote by $\mathscr{N}(\vec{\mathbf{I}})$ the affine space $\mathscr{H}(\mathbf{J_S}, \vec{\mathbf{I}})$ for $\mathbf{J_S} = \mathbf{0}$, i.e.,

$$\mathscr{N}(\vec{\mathbf{I}}) := \mathscr{H}(\mathbf{0}, \vec{\mathbf{I}}) = \{\mathbf{H} \in H(\operatorname{curl}, \Omega) : \operatorname{curl} \mathbf{H} = \mathbf{0} \text{ in } \Omega_D,$$

$$\int_{\Gamma_J^n} \operatorname{curl} \mathbf{H} \cdot \mathbf{n}\, dS = -I_n, \; n = \hat{L}+1, \ldots, L\}.$$

Let us define $\hat{\mathbf{H}} := \mathbf{H} - \mathbf{H_S}$. Then $\hat{\mathbf{H}} \in \mathscr{N}(\vec{\mathbf{I}})$ because $\mathbf{H_S} = \mathbf{0}$ in Ω_C. We replace $\mathbf{H}$ in the above problem (10.46) by $\hat{\mathbf{H}} + \mathbf{H_S}$. Then it is straightforward to see that $\hat{\mathbf{H}}$ is a solution to the following problem:

Find $\hat{\mathbf{H}} \in \mathscr{N}(\vec{\mathbf{I}})$ *such that*

$$\int_\Omega i\omega\mu \hat{\mathbf{H}} \cdot \bar{\mathbf{G}}\, dV + \int_{\Omega_C} \frac{1}{\sigma} \operatorname{curl} \hat{\mathbf{H}} \cdot \operatorname{curl} \bar{\mathbf{G}}\, dV$$

$$= -\int_{\Omega_D} i\omega\mu \mathbf{H_S} \cdot \bar{\mathbf{G}}\, dV + \sum_{n=1}^{\hat{L}} V_n \int_{\Gamma_J^n} \operatorname{curl} \bar{\mathbf{G}} \cdot \mathbf{n}\, dS \quad \forall \mathbf{G} \in \mathscr{H}_0. \qquad (10.53)$$

10.4 Magnetic field/scalar magnetic potential formulation

For numerical solution of (10.53) it is convenient to introduce a *scalar magnetic potential* in the dielectric which will be multi-valued if this domain is not simply connected as it is the case in Fig. 10.1 (see again [15]).

In this section we show how the above problem (10.53) can be transformed by replacing the magnetic field in the dielectric domain Ω_D by a (scalar) magnetic potential.

We recall that Ω is assumed to be simply connected with connected boundary Γ. We assume there exist J connected "cut" surfaces $\Sigma_n \subset \Omega_D$, $n = 1,\ldots,J$, such that $\partial \Sigma_n \subset \partial \Omega_D$ and $\widetilde{\Omega}_D := \Omega_D \setminus \bigcup_{n=1}^{J} \Sigma_n$ is pseudo-Lipschitz and simply connected (see, for instance, [5]). We also assume that $\bar{\Sigma}_n \cap \bar{\Sigma}_m = \emptyset$ for $n \neq m$ (see Fig. 10.1). For each inductor, Ω_C^n, $n = 1,\ldots,L$, there exists a cut surface Σ_n such that, necessarily, $\partial \Sigma_n \cap \Gamma_J \neq \emptyset$ (see Fig. 10.1). The remaining cut surfaces, $\Sigma_{L+1},\ldots,\Sigma_J$, are assumed to be contained in the interior of Ω_D (see Fig. 10.1, again).

For each cut surface Σ_n, we assume that there exists a surface $S_n \subset \Omega_C^n$, with $\partial S_n \subset \partial \Omega_C^n$ and such that its boundary γ_n is a simple closed curve which intersects once and only once $\bar{\Sigma}_n$ and does not intersect $\bar{\Sigma}_m$, $m \neq n$. Notice that, for $n = 1,\ldots,L$, we can take $S_n = \Gamma_J^n$. We denote the two faces of each Σ_n by Σ_n^- and Σ_n^+. We choose an orientation for each γ_n by taking its initial and end points on Σ_n^- and Σ_n^+, respectively. We denote by **t** the unit vector tangent to γ_n according to this orientation.

Each function $\widetilde{\Psi} \in H^1(\widetilde{\Omega}_D)$ has in general different traces on each face of Σ_n and we denote by

$$[\![\widetilde{\Psi}]\!]_{\Sigma_n} := \widetilde{\Psi}|_{\Sigma_n^+} - \widetilde{\Psi}|_{\Sigma_n^-}$$

the jump of $\widetilde{\Psi}$ through Σ_n. The gradient of $\widetilde{\Psi}$ in $\mathscr{D}'(\widetilde{\Omega}_D)$ can be extended to $L^2(\Omega_D)^3$ and will be denoted by $\widetilde{\mathrm{grad}\,\Psi}$.

Let $\mathscr{T}$ be the linear subspace of $H^1(\widetilde{\Omega}_D)$ defined by

$$\mathscr{T} := \left\{ \widetilde{\Psi} \in H^1(\widetilde{\Omega}_D) : [\![\widetilde{\Psi}]\!]_{\Sigma_n} = \text{constant}, n = 1,\ldots,J \right\}.$$

Then, for $\widetilde{\Psi} \in H^1(\widetilde{\Omega}_D)$, we have that $\widetilde{\mathrm{grad}\,\Psi} \in H(\mathbf{curl},\Omega_D)$ if and only if $\widetilde{\Psi} \in \mathscr{T}$, in which case $\mathbf{curl}(\widetilde{\mathrm{grad}\,\Psi}) = \mathbf{0}$ (see [5, Lemma 3.11]). We use the following notation: given $\mathbf{G}_C \in L^2(\Omega_C)^3$ and $\mathbf{G}_D \in L^2(\Omega_D)^3$, $(\mathbf{G}_C|\mathbf{G}_D)$ denotes the field $\mathbf{G} \in L^2(\Omega)^3$ defined by $\mathbf{G}|_{\Omega_C} := \mathbf{G}_C$ and $\mathbf{G}|_{\Omega_D} := \mathbf{G}_D$.

Let us introduce the following function spaces:

$$\mathscr{X} = \{\mathbf{H} \in H(\mathbf{curl},\Omega) : \mathbf{curl}\,\mathbf{H} = \mathbf{0} \text{ in } \Omega_D\}$$

and

$$\mathscr{Y} := \left\{ (\mathbf{G},\widetilde{\Psi}) \in H(\mathbf{curl},\Omega_C) \times (\mathscr{T}/\mathbb{C}) : (\mathbf{G}|\widetilde{\mathrm{grad}\,\Psi}) \in H(\mathbf{curl},\Omega) \right\}.$$

Then $(\mathbf{G},\widetilde{\Psi}) \in \mathscr{Y}$ if and only if $(\mathbf{G}|\widetilde{\mathrm{grad}\,\Psi}) \in \mathscr{X}$.

10.4 Magnetic field/scalar magnetic potential formulation

When a magnetic potential $\widetilde{\Psi} \in H^1(\widetilde{\Omega}_D)$ is used, boundary conditions (10.28) can be imposed by fixing its jumps on the cut surfaces. Indeed, if $(\mathbf{G}, \widetilde{\Psi}) \in \mathscr{Y}$ is smooth enough for the following integrals to make sense, we have that

$$\int_{\Gamma_J^n} \operatorname{curl} \mathbf{G} \cdot \mathbf{n} \, dS = \int_{\gamma_n} \mathbf{G} \cdot \mathbf{t} \, dl = \int_{\gamma_n} \widetilde{\operatorname{grad} \Psi} \cdot \mathbf{t} \, dl = [\![\widetilde{\Psi}]\!]_{\Sigma_n}, \qquad (10.54)$$

where we have used Stokes' theorem and the fact that

$$\mathbf{n}_C \times (\mathbf{G} \times \mathbf{n}_C) = \mathbf{n}_D \times (\widetilde{\operatorname{grad} \Psi} \times \mathbf{n}_D)$$

on $\Gamma_I \supset \gamma_n$. Then by using a Green's formula and (10.49) it is straightforward to see that problem (10.53) is equivalent to the following:

Find $(\hat{\mathbf{H}}, \widetilde{\Phi}) \in \mathscr{Y}$ *such that*

$$[\![\widetilde{\Phi}]\!]_{\Sigma_n} = I_n, \quad n = \widehat{L}+1, \dots, L, \qquad (10.55)$$

$$\int_{\Omega_C} i\omega\mu \hat{\mathbf{H}} \cdot \bar{\mathbf{G}} \, dV + \int_{\Omega_D} i\omega\mu \, \widetilde{\operatorname{grad}} \, \widetilde{\Phi} \cdot \widetilde{\operatorname{grad}} \, \bar{\widetilde{\Psi}} \, dV + \int_{\Omega_C} \frac{1}{\sigma} \operatorname{curl} \hat{\mathbf{H}} \cdot \operatorname{curl} \bar{\mathbf{G}} \, dV$$

$$= -i\omega \int_{\Omega_D} \operatorname{div}(\mu \mathbf{H}_S) \cdot \mathbf{n} \bar{\widetilde{\Psi}} \, dS + \sum_{n=1}^{\widehat{L}} V_n \int_{\Gamma_J^n} \operatorname{curl} \bar{\mathbf{G}} \cdot \mathbf{n} \, dS \quad \forall (\mathbf{G}, \widetilde{\Psi}) \in \mathscr{Y}^0,$$
$$(10.56)$$

where

$$\mathscr{Y}^0 := \left\{ (\mathbf{G}, \widetilde{\Psi}) \in \mathscr{Y} \, : \, [\![\widetilde{\Psi}]\!]_{\Sigma_n} = 0, \, n = \widehat{L}+1, \dots, L \right\}.$$

10.4.1 Energy conservation

Let us come back to the weak formulation (10.53). For the sake of simplicity, let us assume $\widehat{L} = L$. Then $\mathscr{N}(\overrightarrow{\mathbf{I}}) = \mathscr{H}_0 = \mathscr{X}$ and we can take $\hat{\mathbf{H}}$ as a test function. We get

$$\int_{\Omega} i\omega\mu \hat{\mathbf{H}} \cdot \bar{\hat{\mathbf{H}}} \, dV + \int_{\Omega_C} \frac{1}{\sigma} \operatorname{curl} \hat{\mathbf{H}} \cdot \operatorname{curl} \bar{\hat{\mathbf{H}}} \, dV$$

$$= -\int_{\Omega_D} i\omega\mu \mathbf{H}_S \cdot \bar{\hat{\mathbf{H}}} \, dV + \sum_{n=1}^{L} V_n \int_{\Gamma_J^n} \operatorname{curl} \bar{\hat{\mathbf{H}}} \cdot \mathbf{n} \, dS.$$

Since $\mathbf{H}_S = \mathbf{0}$ and $\mathbf{E} = \frac{1}{\sigma}\mathbf{J}$ in Ω_C, we easily deduce

$$i\omega \int_{\Omega} \mu |\mathbf{H}|^2 \, dV + \int_{\Omega_C} \mathbf{E} \cdot \bar{\mathbf{J}} \, dV = i\omega \int_{\Omega} \mu \mathbf{H} \cdot \bar{\mathbf{H}}_S \, dV + \sum_{n=1}^{L} V_n \bar{I}_n. \qquad (10.57)$$

In order to give a meaning to the first term on the right-hand side we first notice that from (10.53) we cannot obtain the electric field in Ω_D. However, since we should keep the Faraday's law also in Ω_D we would have

$$i\omega\mu\mathbf{H} + \mathbf{curl}\,\mathbf{E} = \mathbf{0} \text{ in } \Omega_D$$

and then, by using a Green's formula

$$i\omega \int_{\Omega_D} \mu \mathbf{H} \cdot \bar{\mathbf{H}}_S \, dV = -\int_{\Omega_D} \mathbf{curl}\,\mathbf{E} \cdot \bar{\mathbf{H}}_S \, dV = -\int_{\Omega_S} \mathbf{E} \cdot \bar{\mathbf{J}}_S \, dV.$$

Replacing this equality in (10.57) we get

$$i\omega \int_{\Omega} \mu |\mathbf{H}|^2 \, dV + \int_{\Omega_C} \mathbf{E} \cdot \bar{\mathbf{J}} \, dV + \int_{\Omega_S} \mathbf{E} \cdot \bar{\mathbf{J}}_S \, dV = \sum_{i=1}^{L} V_n \bar{I}_n$$

and finally

$$i\omega \int_{\Omega} \mu |\mathbf{H}|^2 \, dV + \int_{\Omega} \mathbf{E} \cdot \bar{\mathbf{J}} \, dV = \sum_{i=1}^{L} V_n \bar{I}_n \qquad (10.58)$$

The first term is the energy stored by the magnetic field; it represents the *reactive power*. The second one is the loss of electromagnetic energy by the Joule effect; it represents the *active power*. The term on the right-hand side is the *power supplied* to conductors Ω_C^n, $n = 1, \ldots, L$.

Actually, since we have to make the average in a cycle, we have (see Appendix D):

- *Reactive power:*

$$\mathscr{P}_r = \frac{1}{2}\omega \int_{\Omega} \mu |\mathbf{H}|^2 \, dx = \frac{1}{2}\mathrm{Im}(\sum_{i=1}^{L} V_i \bar{I}_i). \qquad (10.59)$$

- *Active power:*

$$\mathscr{P}_a = \frac{1}{2} \int_{\Omega} \mathbf{E} \cdot \bar{\mathbf{J}} \, dx = \frac{1}{2}\mathrm{Re}(\sum_{i=1}^{L} V_i \bar{I}_i). \qquad (10.60)$$

10.5 Impedance matrix

We still consider problem (10.53) with $\widehat{L} = L$. Let us notice that after this problem is solved we can obtain the current intensities across boundaries Γ_J^n, $n = 1, \ldots, L$ by using the formula:

$$I_n = \int_{\Gamma_J^n} \mathbf{curl}\,\mathbf{H} \cdot \mathbf{n} \, dS. \qquad (10.61)$$

It is obvious that the mapping from $\mathbb{C}^L$ into itself giving the vector of current intensities $\vec{I}$ from the potential drops vector $\vec{V}$ is affine:

$$\mathscr{L}(\vec{V}) = \vec{I}_S + \mathscr{A}\vec{V}, \qquad (10.62)$$

where $\vec{I}_S$ is the vector of intensities corresponding to the solution of (10.53) for $V_n = 0$, $n = 1,\ldots,\widehat{L}$ and $I_n = 0$, $n = \widehat{L}+1,\ldots,L$.

The order-L matrix $\mathscr{A}$ is called the *admittance matrix*. One can show that it is non-singular. Its inverse $\mathscr{Z} = \mathscr{A}^{-1}$ is called the *impedance matrix*.

In order to compute matrix $\mathscr{A}$ we take $\mathbf{J}_S = \mathbf{0}$ in which case problems (10.46) and (10.53) are the same, $\vec{I}_S = \mathbf{0}$ and then $\mathscr{L}(\vec{V}) = \mathscr{A}\vec{V}$.

For any $\vec{V} \in \mathbb{C}^L$, let us consider the solution $\widehat{\mathbf{H}}$ (=$\mathbf{H}$) of problem (10.53) and compute the vector $\vec{I}$ of current intensities across ports Γ_J^n given by (10.61). Then this $\mathbf{H}$ is also the unique solution to problem (10.53) for $\widehat{L} = 0$ and these current intensities as data (to see this it is enough to notice that $\mathbf{H} \in \mathscr{N}(\vec{I})$ and that the rightmost term in (10.53) disappears for $\widehat{L} = 0$).

From the linearity of the problem we deduce that $\mathbf{H}$ is the linear combination

$$\mathbf{H} = \sum_{j=1}^{L} I_j \mathbf{H}_j, \tag{10.63}$$

where field $\mathbf{H}_j$, $j = 1,\ldots,L$ is the unique solution to problem (10.53) corresponding to $\mathbf{J}_S = \mathbf{0}$ and current intensities $I_i = \delta_{ij}$, $i = 1,\ldots,L$.

By replacing this expression in (10.53) we get

$$\sum_{j=1}^{L} I_j \left(i\omega \int_\Omega \mu \mathbf{H}_j \cdot \bar{\mathbf{G}} dV + \int_{\Omega_C} \frac{1}{\sigma} \operatorname{curl} \mathbf{H}_j \cdot \operatorname{curl} \bar{\mathbf{G}} dV \right)$$
$$= -\sum_{i=1}^{L} V_i \int_{\Gamma_i} \operatorname{curl} \bar{\mathbf{G}} \cdot \mathbf{n} dA \quad \forall \mathbf{G} \in \mathscr{X},$$

and choosing $\mathbf{G} = \mathbf{H}_i$

$$\sum_{j=1}^{L} I_j \left(i\omega \int_\Omega \mu \mathbf{H}_j \cdot \bar{\mathbf{H}}_i dV + \int_{\Omega_C} \frac{1}{\sigma} \operatorname{curl} \mathbf{H}_j \cdot \operatorname{curl} \bar{\mathbf{H}}_i dV \right) = V_i, \quad i = 1,\ldots,L. \tag{10.64}$$

Therefore the terms of the impedance matrix are

$$z_{ij} = i\omega \int_\Omega \mu \mathbf{H}_j \cdot \bar{\mathbf{H}}_i dV + \int_{\Omega_C} \frac{1}{\sigma} \operatorname{curl} \mathbf{H}_j \cdot \operatorname{curl} \bar{\mathbf{H}}_i dV. \tag{10.65}$$

The real part

$$R_i = \int_{\Omega_C} \frac{1}{\sigma} |\operatorname{curl} \mathbf{H}_i|^2 \, dV$$

is called the *resistance* of Ω_C^i while the imaginary part

$$X_i = \omega \int_\Omega \mu |\mathbf{H}_i|^2 \, dV$$

is called the *reactance* of Ω_C^i. The real number

$$L_i = \int_\Omega \mu |\mathbf{H}_i|^2 \, dV$$

is called *self-inductance* of conductor Ω_C^i, while for $i \neq j$, the complex numbers

$$\int_\Omega \mu \mathbf{H}_j \bar{\mathbf{H}}_i \, dV$$

are called *mutual inductances* of the pair of conductors Ω_C^i and Ω_C^j.

Finally, from (10.62) we obtain the inverse relation:

$$\mathscr{L}^{-1}(\vec{\mathbf{I}}) = \vec{\mathbf{V}}_S + \mathscr{Z}\vec{\mathbf{I}},$$

with $\vec{\mathbf{V}}_S := \mathscr{Z}\vec{\mathbf{I}}_S$.

Remark 10.4. We notice that fields $\mathbf{H}_i$ and then impedance matrix $\mathscr{Z}$ are independent of the intensities or the potentials applied to the system. However, they depend on the frequency, on the geometry of the full system and on the electromagnetic properties of the materials the circuits are made of. In this sense, complex number z_{ii}, and then resistance R_i and self-inductance L_i, are not intrinsic properties of the single conductor Ω_C^i. Actually, they also depend on the other conductors Ω_C^j, with $j \neq i$.

10.5.1 Reduced impedances

Sometimes, we know *a priori* that potentials $\vec{\mathbf{V}}$ or intensities $\vec{\mathbf{I}}$ belong to a *given* linear subspace of $\mathbb{C}^L$. In this case a *reduced impedance* matrix can be defined in a natural way.

Let us suppose that $\vec{\mathbf{I}}$ belongs to the space spanned by vectors $\{\vec{K}^1, ..., \vec{K}^{L_r}\}$ with $L_r < L$, where $\vec{K}^1, ..., \vec{K}^{L_r}$ are independent vectors in $\mathbb{C}^L$. Then there exists a vector $\vec{y} \in \mathbb{C}^{L_r}$ such that

$$\vec{\mathbf{I}} = \mathscr{K}\vec{y},$$

where $\mathscr{K}$ is the $L \times L_r$ matrix defined by

$$\mathscr{K} = (\vec{K}^1 | ... | \vec{K}^{L_r}).$$

For any $i \in \{1, ..., L_r\}$, let $\hat{\mathbf{H}}_i$ be the solution to problem (10.46) for $\mathbf{J}_S = \mathbf{0}$ and current intensity vector $\vec{K}^i$. Since we are prescribing the current intensities trough all conductors, $(\hat{L} = L)$, we can determine the potentials by equality

$$\vec{\mathbf{V}}^i = \mathscr{Z}\vec{K}^i.$$

Then, by linearity, the solution to problem (10.46) for current intensities $\vec{\mathbf{I}}$, to be denoted by $\mathbf{H}$, and its respective vector of potentials, $\vec{\mathbf{V}}$, satisfy

$$\mathbf{H} = \sum_{l=1}^{L_r} y_l \hat{\mathbf{H}}_l$$

$$\vec{\mathbf{V}} = \sum_{l=1}^{L_r} y_l \vec{\mathbf{V}}^l.$$

10.5 Impedance matrix

Moreover, (see (10.63)),

$$\hat{\mathbf{H}}_l = \sum_{i=1}^{L} K_i^l \mathbf{H}_i$$

$$\vec{V}^l = \sum_{i=1}^{L} K_i^l \mathscr{L} \vec{e}^i,$$

$\vec{e}^i$ being the i-th vector of the canonical basis in $\mathbb{C}^L$, i.e. $(\vec{e}^i)_j = \delta_{ij}$.

We easily deduce

$$i\omega \int_\Omega \mu \sum_{l=1}^{L_r} y_l \hat{\mathbf{H}}_l \cdot \bar{\hat{\mathbf{H}}}_j \, dx + \int_{\Omega_C} \frac{1}{\sigma} \sum_{l=1}^{L_r} y_l \operatorname{\mathbf{curl}} \hat{\mathbf{H}}_l \cdot \operatorname{\mathbf{curl}} \bar{\hat{\mathbf{H}}}_j \, dx$$

$$+ \sum_{i=1}^{L} \mathsf{V}_i \int_{\Gamma_i} \operatorname{\mathbf{curl}} \bar{\hat{\mathbf{H}}}_j \cdot \mathbf{n} \, dA = 0$$

and then

$$\sum_{l=1}^{L_r} y_l \left[i\omega \int_\Omega \mu \hat{\mathbf{H}}_l \bar{\hat{\mathbf{H}}}_j \, dx + \int_{\Omega_C} \frac{1}{\sigma} \operatorname{\mathbf{curl}} \hat{\mathbf{H}}_l \cdot \operatorname{\mathbf{curl}} \bar{\hat{\mathbf{H}}}_j \, dx \right] = \sum_{i=1}^{L} \mathsf{V}_i \bar{K}_i^j \qquad (10.66)$$

The matrix $\mathscr{Z}_r$ with the following element in position ij

$$i\omega \int_\Omega \mu \hat{\mathbf{H}}_j \bar{\hat{\mathbf{H}}}_i \, dx + \int_{\Omega_C} \frac{1}{\sigma} \operatorname{\mathbf{curl}} \hat{\mathbf{H}}_j \cdot \operatorname{\mathbf{curl}} \bar{\hat{\mathbf{H}}}_i \, dx$$

is called the *reduced impedance* matrix.

Let $\mathscr{K}^* := \overline{\mathscr{K}}^t$ and

$$\vec{v} := \mathscr{K}^* \vec{\mathsf{V}},$$

that is, $v_j = \sum_{i=1}^{L} \mathsf{V}_i \bar{K}_i^j$. Then, from (10.66) we deduce

$$\vec{v} = \mathscr{Z}_r \vec{y}.$$

Moreover,

$$\vec{v} = \mathscr{K}^* \vec{\mathsf{V}} = \mathscr{K}^* \mathscr{L} \vec{\mathsf{I}} = \mathscr{K}^* \mathscr{L} \mathscr{K} \vec{y},$$

because $\vec{\mathsf{I}} = \mathscr{K} \vec{y}$ and $\vec{\mathsf{V}} = \mathscr{L} \vec{\mathsf{I}}$.

Hence,

$$\mathscr{Z}_r = \mathscr{K}^* \mathscr{L} \mathscr{K}. \qquad (10.67)$$

Vectors $\vec{y}$ and $\vec{v}$ are called *reduced* intensities and potentials, respectively. We notice that, since $\mathscr{L}$ is non-singular and $\operatorname{rank}(\mathscr{K}) = L_r$, the space of potentials is the subspace $\operatorname{Im}(\mathscr{K}^*)$.

10.5.1.1 A particular case

We consider a useful particular case: let us assume that the feasible subspace of intensities is defined by

$$\mathscr{P} = \{\vec{\mathrm{I}} \in \mathbb{C}^L : \mathscr{B}\vec{\mathrm{I}} \in \mathfrak{R}\}, \tag{10.68}$$

where $\mathscr{B}$ is a $L_r \times L$ matrix ($L_r \leq L$), with rank($\mathscr{B}$)=L_r and $\mathfrak{R}$ is an R-dimensional linear space in $\mathbb{C}_r^L$ ($R \leq L_r$). Since $\mathscr{L}$ is invertible, the feasible space of potentials will have the same dimension as $\mathscr{P}$, that is, R. Our goal is to compute a reduced impedance matrix relating these two feasible spaces.

However, obtaining a basis for the above space, in other words, computing matrix $\mathscr{K}$, can be cumbersome so we will introduce an $R \times R$ *reduced impedance* matrix for the linear space in (10.68) instead of the one given by (10.67).

Let us consider a vector $\vec{v} \in \mathbb{C}^{L_r}$ and let us solve problem (10.53) for $\mathbf{J}_S = \mathbf{0}$, $\widehat{L} = L$ and potentials given by $\vec{V} = \mathscr{B}^* \vec{v}$. The corresponding intensities are $\vec{\mathrm{I}} = \mathscr{L}^{-1}\vec{V} = \mathscr{L}^{-1}\mathscr{B}^*\vec{v}$.

Conversely, since $\mathscr{B}$ has full-rank, matrix $\mathscr{B}\mathscr{L}^{-1}\mathscr{B}^*$ is invertible. Thus, given a reduced intensity $\vec{y} \in \mathrm{Im}(\mathscr{B})$, the corresponding magnetic field $\mathbf{H}$ is the solution to problem (10.53) for the potentials $\vec{V} = \mathscr{B}^*\mathscr{D}\vec{y}$, where $\mathscr{D} := (\mathscr{B}\mathscr{L}^{-1}\mathscr{B}^*)^{-1}$

However, not all the reduced intensities in $\mathrm{Im}(\mathscr{B})$ are feasible but only those belonging to $\mathfrak{R}$. Let $\{s^1, \ldots, s^R\} \subset \mathbb{C}_r^L$ be a basis of $\mathfrak{R}$ and $\mathscr{S}$ the $L_r \times R$ matrix

$$\mathscr{S} = (s^1|\ldots|s^R).$$

For $\vec{y} \in \mathfrak{R}$, let $\vec{f}$ be such that

$$\vec{y} = \mathscr{S}\vec{f}.$$

Then, the corresponding reduced vector of potentials is

$$\vec{v} = \mathscr{D}\vec{y} = \mathscr{D}\mathscr{S}\vec{f}.$$

Now let us define $\vec{u} \in \mathbb{C}^R$ by

$$\vec{u} = \mathscr{S}^*\vec{v}.$$

Then

$$\vec{u} = \mathscr{S}^*\mathscr{D}\mathscr{S}\vec{f}.$$

In this context, the $R \times R$ matrix

$$\mathscr{L}_r = \mathscr{S}^*\mathscr{D}\mathscr{S} \tag{10.69}$$

will be called the *reduced impedance matrix* while vectors $\vec{f}$ and $\vec{u}$ are called the *reduced current intensities* and the *reduced potentials*, respectively.

We notice that the space of reduced potentials is exactly $\mathrm{Im}(\mathscr{S}^*)$ because $\mathscr{D}$ is invertible and rank($\mathscr{S}$) = R. This implies that $\mathscr{L}_r$ is also invertible. Its inverse matrix $\mathscr{A}_r = (\mathscr{L}_r)^{-1}$ is called *reduced admittance matrix*.

Example 10.1. The goal of this example is to analyze *busbars* like the ones shown in Figs. 10.2 and 10.3.

Fig. 10.2. Busbar system pre-installation. Picture licensed by Creative Commons

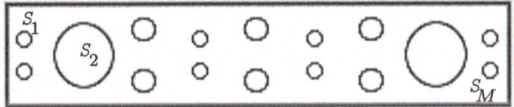

Fig. 10.3. Sketch of a busbar system

We use this system to transport an electric current of intensity y. The M bars are divided into two groups corresponding to the sets of indices $\{1,\ldots,M_1\}$ and $\{M_1+1,\ldots,M\}$.

We choose surfaces $S_1,\ldots,S_M$ "cutting" conductors and corresponding unit vectors $\mathbf{n}$ (see Fig. 10.3). Let I_i be the intensity flowing through S_i in the direction of $\mathbf{n}$, i.e.,

$$I_i = \int_{S_i} \mathbf{J} \cdot \mathbf{n}\, dA$$

for $i = 1,\ldots,M$. We have to impose the constraints

$$\sum_{i=1}^{M_1} I_i - I, \quad \sum_{i=M_1+1}^{M} I_i - -I \tag{10.70}$$

which can be rewritten, in a more compact way, as

$$\mathscr{B}\vec{I} = \vec{y},$$

with

$$\mathscr{B} = \begin{pmatrix} \overset{M_1}{1 \ldots 1} & \overset{M}{0 \ldots 0} \\ 0 \ldots 0 & 1 \ldots 1 \end{pmatrix}$$

and

$$\vec{y} = \begin{pmatrix} I \\ -I \end{pmatrix}.$$

Thus our problem falls into the previous particular case. Specifically, $L_r = 2$ and $R = 1$, hence, in the present case, the reduced impedance matrix is a single complex number that can be computed by using the general formula (10.69). However, it is much simpler to proceed as follows: letting I be any complex number, vector $\vec{y}$ run the one-dimensional space $\mathfrak{R}$ spanned by

$$f_1 = \begin{pmatrix} 1 \\ -1 \end{pmatrix}$$

so $\mathscr{S}$ is the 2×1 matrix

$$\mathscr{S} = \begin{pmatrix} 1 \\ -1 \end{pmatrix}.$$

Let the reduced potential $\vec{u} = (u)$ be given. The corresponding full vector of potentials is

$$\vec{V} = \mathscr{B}^* \mathscr{S} u = (u, \ldots, \overset{M_1}{u}, -u, \ldots, \overset{M}{-u})^t.$$

Hence, in order to compute the reduced impedance we can solve problem (10.53) for this $\vec{V}$. Let $\vec{I} \in \mathbb{C}^M$ be the vector of the corresponding current intensities and

$$I = (\mathscr{B} \vec{I})_1 = \sum_{i=1}^{M_1} I_i,$$

the reduced current intensity. Then, we finally obtain

$$\mathscr{Z}_R = \frac{u}{I}.$$

10.6 2D eddy currents model with J normal to the section

In this section and in the next three ones we will describe some particular situations where the eddy current problem can be solved in a two-dimensional domain. We will restrict ourselves to the time harmonic case and assume linear behavior of all materials. to the time-harmonic case and assume linear behavior in all materials.

As for magnetostatics, sometimes the solution of an eddy currents problem can be approximated by a two-dimensional model which will be different depending on the assumptions on the current density **J**. In particular, in this section we assume that the 3D domain Ω is very long in one of the spatial directions, let us say in the x_3-direction, and furthermore:

- The geometry and the physical parameters are invariant under translation in one privileged direction, let us say the x_3-direction.
- The current density **J** only has non-null component in the x_3-direction and this component does not depend on x_3, i.e., $\mathbf{J}(x) = J_3(x_1, x_2)\mathbf{e}_3$. In particular, the data $\mathbf{J}_S$ is assumed to be of the form $\mathbf{J}_S = J_S(x_1, x_2)\mathbf{e}_3$.

Under these assumptions, we can suppose that none of the fields depends on x_3 and the eddy currents model (10.9)–(10.13) can be written in an orthogonal plane to the x_3-axis. Let us denote by $\widehat{\Omega}$ the two-dimensional domain containing the region to be modelled. Let $\widehat{\Omega}_C$ be the subset of $\widehat{\Omega}$ occupied by the conductors and $\widehat{\Omega}_D := \widehat{\Omega} \setminus \overline{\widehat{\Omega}_C}$

10.6 2D eddy currents model with J normal to the section

that occupied by the surrounding dielectrics (air, for instance). Moreover, let us assume that conducting domain $\widehat{\Omega}_C$ is composed by M mutually disjoint connected components $\widehat{\Omega}_n$, for $n = 1, \ldots, M$, i.e., $\widehat{\Omega}_C := \widehat{\Omega}_1 \cup \cdots \cup \widehat{\Omega}_M$ which are strictly contained in $\widehat{\Omega}$, i.e., $\widehat{\overline{\Omega}}_n \subset \widehat{\Omega}$. On the other hand, let $\widehat{\Omega}_S$ denote the two-dimensional section of the support of $\mathbf{J}_S$, which is assumed to be contained in $\widehat{\Omega}_D$ like in the three-dimensional case.

From the above assumptions, it is easy to see that the magnetic field $\mathbf{H}$ and the magnetic induction $\mathbf{B}$ have only components on the $x_1 x_2$-plane, that is,

$$\mathbf{H}(x) = H_1(x_1, x_2)\mathbf{e}_1 + H_2(x_1, x_2)\mathbf{e}_2, \tag{10.71}$$
$$\mathbf{B}(x) = B_1(x_1, x_2)\mathbf{e}_1 + B_2(x_1, x_2)\mathbf{e}_2. \tag{10.72}$$

Similar to the magnetostatic case, there exists a magnetic vector potential $\mathbf{A}$ independent of x_3 and that does not have either x_1 or x_2 components, i.e., of the form $\mathbf{A} = A_3(x_1, x_2)\mathbf{e}_3$.

We are interested in solving the eddy currents model in $\widehat{\Omega}$, by assuming that we know the current density field in $\widehat{\Omega}_S$ while the current intensities or the potential drops are given in some connected components of the conducting domain. Next, we will explain how to introduce the two latter sources in the two-dimensional model and their specific meaning. Firstly, notice that from Ohm's law, $\mathbf{J} = \sigma \mathbf{E}$, the electric field in each conductor has to be of the form

$$\mathbf{E}(x) = E_3(x_1, x_2)\mathbf{e}_3. \tag{10.73}$$

However, as it is argued in [22], the electric field should have a more general form outside the conductors. Indeed, from Faraday's law

$$i\omega \mathbf{B} + \mathbf{curl}\,\mathbf{E} = 0,$$

a scalar potential V exists such that

$$i\omega \mathbf{A} + \mathbf{E} = -\operatorname{grad} V.$$

Given the shape of $\mathbf{A}$, we deduce from this equality that

$$E_1 = -\frac{\partial V}{\partial x_1}, \tag{10.74}$$

$$E_2 = -\frac{\partial V}{\partial x_2}, \tag{10.75}$$

$$i\omega A_3 + E_3 = -\frac{\partial V}{\partial x_3}. \tag{10.76}$$

If $\mathbf{E}$ also had the form (10.73) outside conductors, i.e., if $E_1 \equiv 0$ and $E_2 \equiv 0$ then V would only be dependent on x_3, in contradiction with the fact that conductors may have different electric potentials.

Anyway, in conductors, Eq. (10.76) and the fact that the left-hand side does not depend on x_3 while the right-hand side is independent of x_1 and x_2, lead to

$$\frac{\partial V}{\partial x_3} = \widehat{V}_n \in \mathbb{C}, \quad \text{in } \widehat{\Omega}_n, \ n = 1, \ldots, M.$$

Notice that constant $\widehat{V}_n$ represents the potential drop per unit length in direction x_3, in conductor $\widehat{\Omega}_n$. Hence, from the previous equation and (10.76) one deduces

$$i\omega\sigma A_3 + \sigma E_3 = -\sigma\widehat{V}_n \quad \text{in } \widehat{\Omega}_n, \quad n = 1, \ldots, M. \tag{10.77}$$

By integrating this equation on each $\widehat{\Omega}_n$ we get

$$i\omega \int_{\widehat{\Omega}_n} \sigma A_3(x_1, x_2)\, dx_1 dx_2 + \int_{\widehat{\Omega}_n} \sigma E_3(x_1, x_2)\, dx_1 dx_2 = -\widehat{V}_n \int_{\widehat{\Omega}_n} \sigma\, dx_1 dx_2 \tag{10.78}$$

and hence, from Ohm's law,

$$i\omega \int_{\widehat{\Omega}_n} \sigma A_3(x_1, x_2)\, dx dy + I_n = -\widehat{V}_n \int_{\widehat{\Omega}_n} \sigma\, dx_1 dx_2,$$

being $I_n := \int_{\widehat{\Omega}_n} J_3(x_1, x_2)\, dx_1 dx_2$ the current intensity through $\widehat{\Omega}_n$. Taking into account the previous discussion we will assume that, for each conductor $\widehat{\Omega}_n$, we know either the potential drop per unit length in the x_3-direction $\widehat{V}_n$, $(n = 1, \ldots, \widehat{M})$, or the current intensity I_n, $(n = \widehat{M} + 1, \ldots, M)$.

On the other hand, on boundary $\partial\widehat{\Omega}$ we will consider homogeneous Dirichlet and Neumann boundary conditions like in the 2D magnetostatics model. Thus, the problem to be solved is the following:

Given complex numbers $\widehat{V}_n$, $n = 1, \ldots, \widehat{M}$, I_n, $n = \widehat{M} + 1, \ldots, M$, and function J_S in $\widehat{\Omega}_S$, find $A_3 : \widehat{\Omega} \to \mathbb{C}$ and a complex vector $\overrightarrow{V} \in \mathbb{C}^{M-\widehat{M}}$ satisfying

$$i\omega\sigma A_3\, \mathbf{e}_3 + \mathbf{curl}\left(\frac{1}{\mu}\mathbf{curl}(A_3\, \mathbf{e}_3)\right) = -\sigma\widehat{V}_n \quad \text{in } \widehat{\Omega}_n, \ n = 1, \ldots, \widehat{M}, \tag{10.79}$$

$$i\omega \int_{\widehat{\Omega}_n} \sigma A_3(x_1, x_2)\, dx_1 dx_2 + I_n = -\widehat{V}_n \int_{\widehat{\Omega}_n} \sigma\, dx_1 dx_2 \quad \text{in } \widehat{\Omega}_n, \ n = \widehat{M}+1, \ldots, M, \tag{10.80}$$

$$\mathbf{curl}\left(\frac{1}{\mu}\mathbf{curl}(A_3\, \mathbf{e}_3)\right) = J_S \mathbf{e}_3 \quad \text{in } \widehat{\Omega}_D, \tag{10.81}$$

$$A_3 = 0 \quad \text{on } \widehat{\Gamma}_d. \tag{10.82}$$

Notice that Eq. (10.80) allows writing $\widehat{V}_n$ in terms of data I_n, for $n = \widehat{M}+1, \ldots, M$. Thus, this expression for $\widehat{V}_n$ could be used in (10.79). However, from the numerical point of view this option is not interesting because it leads to a dense matrix. Therefore, the numerical solution will be performed by imposing the current intensities in a weak form. By using this procedure, the values of $\widehat{V}_n$, for $n = \widehat{M}+1,\ldots,M$ are additional unknowns of the problem which play the role of *Lagrange multipliers* associated with the corresponding intensities. Therefore, the weak formulation reads as follows:

Given $\vec{V} = (\widehat{V}_1,\ldots,\widehat{V}_{\widehat{M}}) \in \mathbb{C}^{\widehat{M}}$, $\vec{I} = (I_{\widehat{M}+1},\ldots,I_M) \in \mathbb{C}^{M-\widehat{M}}$, and function J_S in $\widehat{\Omega}_S$, find $A_3 \in H^1(\widehat{\Omega})$ such that $A_3 = 0$ on $\widehat{\Gamma}_d$ and a complex vector $\vec{V} \in \mathbb{C}^{M-\widehat{M}}$ satisfying

$$i\omega \int_{\widehat{\Omega}_C} \sigma A_3 \bar{G}\, dx_1 dx_2 + \int_{\widehat{\Omega}} \frac{1}{\mu} \operatorname{grad} A_3 \cdot \operatorname{grad} \bar{G}\, dx_1 dx_2 + \sum_{n=\widehat{M}+1}^{M} \widehat{V}_n \int_{\widehat{\Omega}_n} \sigma \bar{G}\, dx_1 dx_2$$

$$= -\sum_{n=1}^{\widehat{M}} \widehat{V}_n \int_{\widehat{\Omega}_n} \sigma \bar{G}\, dx_1 dx_2 + \int_{\widehat{\Omega}_S} J_S \bar{G}\, dx_1 dx_2 \qquad \forall \psi \in H^1(\widehat{\Omega}), \text{ null on } \widehat{\Gamma}_d,$$

$$\sum_{n=\widehat{M}+1}^{M} \left(\int_{\widehat{\Omega}_n} \sigma A_3\, dx_1 dx_2\right) \bar{W}_n - \frac{i}{\omega} \sum_{n=\widehat{M}+1}^{M} \int_{\widehat{\Omega}_n} \sigma \widehat{V}_n \bar{W}_n\, dx_1 dx_2$$

$$= \frac{i}{\omega} \sum_{n=\widehat{M}+1}^{M} I_n \bar{W}_n \qquad \forall \mathbf{W} \in \mathbb{C}^{M-\widehat{M}}.$$

Remark 10.5. Recall that for magnetically anisotropic materials, i.e., when μ is a matrix, equality (9.74) is not valid so the same is true for the above formulation.

Remark 10.6. Numerical discretization of this problem can be done by using Lagrangian (nodal) continuous finite elements on a triangular mesh of $\widehat{\Omega}$.

10.7 2D eddy currents model with J lying on the section

There are electromagnetic devices in which the geometry and physical properties are invariant under translation in the privileged x_3-direction and the current density **J** has the form $\mathbf{J}(x) = J_1(x_1,x_2)\mathbf{e}_1 + J_2(x_1,x_2)\mathbf{e}_2$. In this section, we focus on this case and assume moreover that $\Omega_S = \emptyset$.

In these cases, it may be interesting to write the eddy currents model in an orthogonal plane to the x_3-axis by including conductors only. Let us denote by $\widehat{\Omega}_C$ the conducting domain, where the electric field has the form

$$\mathbf{E}(x) = E_1(x_1,x_2)\mathbf{e}_1 + E_2(x_1,x_2)\mathbf{e}_2$$

and then

$$\mathbf{H}(x) = H_3(x_1,x_2)\mathbf{e}_3,$$
$$\mathbf{B}(x) = B_3(x_1,x_2)\mathbf{e}_3.$$

10 The eddy currents model

From these properties it is straightforward to obtain the following equation for the scalar unknown H_3:

$$i\omega\mu(H_3\mathbf{e}_3) + \mathbf{curl}\left(\frac{1}{\sigma}\mathbf{curl}(H_3\mathbf{e}_3)\right) = \mathbf{0} \text{ in } \widehat{\Omega}_C. \tag{10.83}$$

This model has two important advantages: it is defined only in the conducting domain and it has only one scalar unknown. However, it is necessary to define suitable boundary conditions on $\partial\widehat{\Omega}_C$, which is not always an easy task. In particular, let us suppose that $\partial\widehat{\Omega}_C = \widehat{\Gamma}_d \cup \widehat{\Gamma}_n$. We distinguish two kind of boundary conditions to obtain a suitable weak formulation of (10.83):

$$\mathbf{H} \times \mathbf{n} = \mathbf{g} \quad \text{on } \widehat{\Gamma}_d, \tag{10.84}$$

$$\mathbf{J} \times \mathbf{n} = \mathbf{0} \quad \text{on } \widehat{\Gamma}_n. \tag{10.85}$$

In some cases the value of $\mathbf{g} = g_1\mathbf{e}_1 + g_2\mathbf{e}_2$ can be obtained from physical data such as current intensities entering the domain or surface current densities (see for instance [9]). However, the computation of $\mathbf{g}$ may be an important restriction to the use of this model. Actually, condition (10.84) leads to a Dirichlet boundary condition in terms of H_3, i.e., $H_3 = H_d$ for some H_d. The second condition, (10.85), is suitable for those parts of the boundary where we can assume that current enters perpendicularly into the domain. Moreover, this condition will be a natural boundary condition for the weak formulation of Eq. (10.83). Summarizing, the corresponding 2D eddy currents model reads as follows:

Given a function H_d defined on $\widehat{\Gamma}_d$, find a field $\mathbf{H} = H_3(x_1,x_2)\mathbf{e}_3$ such that

$$i\omega\mu(H_3\mathbf{e}_3) + \mathbf{curl}\left(\frac{1}{\sigma}\mathbf{curl}(H_3\mathbf{e}_3)\right) = \mathbf{0} \text{ in } \widehat{\Omega}_C, \tag{10.86}$$

$$\frac{1}{\sigma}\mathbf{curl}(H_3\mathbf{e}_3) \times \mathbf{n} = \mathbf{0} \text{ on } \widehat{\Gamma}_n, \tag{10.87}$$

$$\mathbf{H} = H_d\mathbf{e}_3 \text{ on } \widehat{\Gamma}_d. \tag{10.88}$$

To obtain a weak formulation of this problem, let us multiply (10.86) by the conjugate of a test function $\mathbf{G}(x) = G(x_1,x_2)\mathbf{e}_3 \in \mathbf{H}(\mathbf{curl},\Omega)$ with $\mathbf{G} \times \mathbf{n} = \mathbf{0}$ on $\widehat{\Gamma}_d$, integrate in Ω and use a Green's formula. We get

$$\int_{\widehat{\Omega}_C} i\omega\mu\mathbf{H} \cdot \bar{\mathbf{G}}\,dx_1dx_2 + \int_{\widehat{\Omega}_C} \frac{1}{\sigma}\mathbf{curl}\,\mathbf{H} \cdot \mathbf{curl}\,\bar{\mathbf{G}}\,dx_1dx_2 = 0.$$

Let us notice that we have again $\mathbf{curl}(H_3\mathbf{e}_3) \cdot \mathbf{curl}(G\mathbf{e}_3) = \text{grad}\,H_3 \cdot \text{grad}\,G$. Thus, the previous formulation can be written in the equivalent form:

Given a function H_d defined on $\widehat{\Gamma}_d$, find $H_3 \in H^1(\widehat{\Omega}_C)$ with $H_3 = H_d$ on $\widehat{\Gamma}_d$ such that

$$\int_{\widehat{\Omega}_C} i\omega\mu H_3\bar{G}\,dx_1dx_2 + \int_{\widehat{\Omega}_C} \frac{1}{\sigma}\text{grad}\,H_3 \cdot \text{grad}\,\bar{G}\,dx_1dx_2 = 0,$$

for all test function $G \in H^1(\widehat{\Omega}_C)$ with $G|_{\widehat{\Gamma}_d} = 0$.

Remark 10.7. Numerical discretization of this problem can be done by using Lagrangian (nodal) continuous finite elements on a triangular mesh of $\widehat{\Omega}_C$.

10.8 Axisymmetric eddy currents model with azimuthal J

Sometimes the solution of the 3D eddy currents problem can be approximated by an axisymmetric model. This happens, for instance, if we are interested in computing the eddy currents in a cylindrical workpiece surrounded by an helical coil carrying an alternating current. Figure 10.4 illustrates a typical example corresponding to an induction heating furnace.

In this situation, in order to have a domain with cylindrical symmetry, we replace the coil by several superimposed rings with toroidal geometry. On the other hand, in order to solve the electromagnetic model in a bounded domain, we introduce a sufficiently large three dimensional cylinder, Ω, containing the coil and the workpiece. Because of the cylindrical symmetry, we can work on a meridional section of Ω to be denoted by $\widehat{\Omega}$ (see Fig. 10.5). Let $\widehat{\Gamma}$ be the intersection of $\partial \Omega$ with the half-plane

Fig. 10.4. Sketch of a cylindrical workpiece surrounded by an helical coil

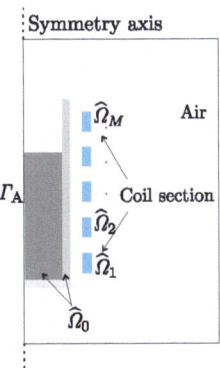

Fig. 10.5. Sketch of the meridional section $\widehat{\Omega}$

$\{(\rho,0,z) \in \mathbb{R}^3; \rho > 0\}$. Moreover, let us assume that the boundary of $\widehat{\Omega}$ is the union of $\widehat{\Gamma}$ and Γ_A, the latter being a subset of the symmetry axis.

Let $\widehat{\Omega}_0$ be the radial section of the workpiece and $\widehat{\Omega}_1, \widehat{\Omega}_2, \ldots, \widehat{\Omega}_M$ be the radial sections of the turns of the coil. We assume $\widehat{\Omega}_0, \ldots, \widehat{\Omega}_M$ are connected and mutually disjoint. We also assume $\widehat{\Omega}_n \cap \Gamma_A = \emptyset$, $n = 1, \ldots, M$. Let $\widehat{\Omega}_C := \widehat{\Omega}_0 \cup \widehat{\Omega}_1 \cup \cdots \cup \widehat{\Omega}_M$ denote the section of the domain occupied by all the conductors and $\widehat{\Omega}_D := \widehat{\Omega} \setminus \overline{\widehat{\Omega}_C}$ that of the surrounding air. Finally, let $\widehat{\Omega}_S$ denote the radial section of the support of $\mathbf{J}_S$.

In order to exploit the cylindrical symmetry of the problem, we will consider a cylindrical coordinate system (ρ, θ, z) with the z-axis coinciding with the symmetry axis of the device. Hereafter we denote by $\mathbf{e}_\rho$, $\mathbf{e}_\theta$ and $\mathbf{e}_z$ the local unit vectors in the corresponding coordinate directions. Now, cylindrical symmetry allows us to consider that no field depends on azimuth θ. We further assume that the current density field has non-zero component only in the tangential direction $\mathbf{e}_\theta$, namely

$$\mathbf{J}(\rho, \theta, z) = J_\theta(\rho, z) \mathbf{e}_\theta. \tag{10.89}$$

In particular, $\mathbf{J}_S$ is assumed to be of the form $\mathbf{J} = J_S(\rho, z) \mathbf{e}_\theta$.

The previous assumptions lead to

$$\mathbf{H}(\rho, \theta, z) = H_\rho(\rho, z) \mathbf{e}_\rho + H_z(\rho, z) \mathbf{e}_z \quad \text{in } \widehat{\Omega}, \tag{10.90}$$

$$\mathbf{B}(\rho, \theta, z) = B_\rho(\rho, z) \mathbf{e}_\rho + B_z(\rho, z) \mathbf{e}_z \quad \text{in } \widehat{\Omega}, \tag{10.91}$$

$$\mathbf{E}(\rho, \theta, z) = E_\theta(\rho, z) \mathbf{e}_\theta \quad \text{in } \widehat{\Omega}_C. \tag{10.92}$$

Therefore, the arguments developed in [14] allows us to conclude that only the azimuthal component of the magnetic vector potential, hereafter denoted by A_θ, does not vanish, i.e.,

$$\mathbf{A}(\rho, \theta, z) = A_\theta(\rho, z) \mathbf{e}_\theta. \tag{10.93}$$

Hence, the relation $\mathbf{B} = \mathbf{curl}\, \mathbf{A}$ and the expression of the **curl** operator in cylindrical coordinates (see (B.49)) lead to

$$B_\rho(\rho, z) = -\frac{\partial A_\theta}{\partial z} \quad \text{and} \quad B_z(\rho, z) = \frac{1}{r} \frac{\partial (\rho A_\theta)}{\partial \rho} \quad \text{in } \widehat{\Omega}. \tag{10.94}$$

On the other hand, from Faraday's law and Ohm's law we deduce

$$\mathbf{curl}\left(\left(i\omega A_\theta + \sigma^{-1} J_\theta \right) \mathbf{e}_\theta \right) = 0 \quad \text{in } \widehat{\Omega}_C,$$

from which it follows that

$$\frac{\partial}{\partial z} \left(i\omega A_\theta + \sigma^{-1} J_\theta \right) = 0 \quad \text{in } \widehat{\Omega}_C,$$

$$\frac{\partial}{\partial \rho} \left(\rho \left(i\omega A_\theta + \sigma^{-1} J_\theta \right) \right) = 0 \quad \text{in } \widehat{\Omega}_C.$$

10.8 Axisymmetric eddy currents model with azimuthal **J**

Hence, there exist constants $C_n \in \mathbb{C}$, $n = 0, \ldots, M$, such that

$$i\omega A_\theta + \sigma^{-1} J_\theta = \frac{C_n}{\rho} \quad \text{in } \widehat{\Omega}_n \tag{10.95}$$

(recall that $\widehat{\Omega}_n$ are the connected components of $\widehat{\Omega}_C$).

Next, from Ampère's equation, Ohm's law, (10.91), (10.94) and (10.89) we have

$$\mathbf{curl}\left(-\frac{1}{\mu}\frac{\partial A_\theta}{\partial z}\mathbf{e}_\rho + \frac{1}{\mu\rho}\frac{\partial(\rho A_\theta)}{\partial \rho}\mathbf{e}_z\right) = J_\theta \mathbf{e}_\theta.$$

Thus, taking into account (10.95), we obtain for $n = 0, \ldots, M$,

$$-\left(\frac{\partial}{\partial \rho}\left(\frac{1}{\mu\rho}\frac{\partial(\rho A_\theta)}{\partial \rho}\right) + \frac{\partial}{\partial z}\left(\frac{1}{\mu}\frac{\partial A_\theta}{\partial z}\right)\right) + i\omega\sigma A_\theta = \frac{\sigma}{\rho} C_n \quad \text{in } \widehat{\Omega}_n, \tag{10.96}$$

whereas

$$-\left(\frac{\partial}{\partial \rho}\left(\frac{1}{\mu\rho}\frac{\partial(\rho A_\theta)}{\partial \rho}\right) + \frac{\partial}{\partial z}\left(\frac{1}{\mu}\frac{\partial A_\theta}{\partial z}\right)\right) = J_S \quad \text{in } \widehat{\Omega}_D. \tag{10.97}$$

Notice that any simply-connected axisymmetric region must have a non-empty intersection with the z-axis, where ρ is zero. Since the current density cannot be infinite, Eq. (10.95) implies that C_n is zero in those regions $\widehat{\Omega}_n$ corresponding to 3D simply connected Ω_n. In particular, $C_0 = 0$ in the workpiece. On the other hand, from a physical point of view, constants C_n for $n = 1, \ldots, M$ are equal to $\widehat{V}_n/(2\pi)$, $\widehat{V}_n$, $n = 1, \ldots, M$ being the potential drops between the ends of the turns of the coil.

Taking into account the previous discussion, we will assume that the following data are given:

- the potential drops $\widehat{V}_n$, for $n = 1, \ldots, M_V$,
- the current intensities I_n flowing through each cylindrical ring, i.e.,

$$\int_{\widehat{\Omega}_n} J_\theta \, d\rho dz = I_n, \quad \text{for } n = \widehat{M}+1, \ldots, M, \tag{10.98}$$

and finally,

- the current density $J_S(\rho, z)$ in $\widehat{\Omega}_S$.

Concerning boundary conditions, A_θ has to be null in order to prevent unphysical singularities on the symmetry axis Γ_A. Moreover, let us consider that the whole boundary of $\widehat{\Omega}$ is split as $\partial\widehat{\Omega} = \Gamma_A \cup \widehat{\Gamma}_d \cup \widehat{\Gamma}_n$, in order to define homogeneous Dirichlet and Neumann boundary conditions on $\widehat{\Gamma}_d$ and $\widehat{\Gamma}_n$ as in the 2D model. This means that $\mathbf{B} \cdot \mathbf{n} = 0$ on $\widehat{\Gamma}_d$ and $\frac{1}{\mu}\mathbf{curl}(A_\theta \mathbf{e}_\theta) \times \mathbf{n} = \mathbf{0}$ on $\widehat{\Gamma}_n$, respectively. Finally, we summarize the axisymmetric problem as follows:

Given complex numbers $\widehat{V}_n$, $n = 1,\ldots,\widehat{M}$, I_n, $n = \widehat{M}+1,\ldots,M$, and a complex function J_S in $\widehat{\Omega}_S$, find $A_\theta : \widehat{\Omega} \to \mathbb{C}$ and a complex vector $\overrightarrow{V} \in \mathbb{C}^{M-\widehat{M}}$ satisfying

$$i\omega \sigma A_\theta \mathbf{e}_\theta + \mathbf{curl}\left(\frac{1}{\mu}\mathbf{curl}(A_\theta \mathbf{e}_\theta)\right) = \sigma \frac{\widehat{V}_n}{2\pi\rho} \quad \text{in } \widehat{\Omega}_n,\, n = 1,\ldots,\widehat{M}, \tag{10.99}$$

$$\mathbf{curl}\left(\frac{1}{\mu}\mathbf{curl}(A_\theta \mathbf{e}_\theta)\right) = J_S \mathbf{e}_\theta \quad \text{in } \widehat{\Omega}_D, \tag{10.100}$$

$$i\omega \int_{\widehat{\Omega}_n} \sigma A_\theta(\rho,z)\,d\rho dz + I_n = \frac{V_n}{2\pi}\int_{\widehat{\Omega}_n} \frac{\sigma}{\rho}\,d\rho dz \quad \text{in } \widehat{\Omega}_n,\, n = \widehat{M}+1,\ldots,M, \tag{10.101}$$

$$A_\theta = 0 \quad \text{on } \widehat{\Gamma}_d \cup \Gamma_A. \tag{10.102}$$

Let us notice that for $n = \widehat{M}+1,\ldots,M$,

$$\widehat{V}_n = \frac{2\pi}{d_n}\left(I_n + i\omega \int_{\widehat{\Omega}_n} \sigma A_\theta\,d\rho dz\right), \tag{10.103}$$

where

$$d_n := \int_{\widehat{\Omega}_n} \frac{\sigma}{\rho}\,d\rho dz.$$

Although we can eliminate the potential drops in (10.99) by using (10.103), this option is not interesting from the numerical point of view because it leads to a dense matrix. Thus, an alternative consists in writing constraint (10.98) in a weak form and introducing the potential drops $\widehat{V}_n$, for $n = 1,\ldots,M$, as additional unknowns of the problem which play the role of Lagrange multipliers associated with the corresponding intensities (see, for instance, [14]). Then, the weak formulation reads as follows:

Given $\overrightarrow{\widehat{V}} = (\widehat{V}_1,\ldots,\widehat{V}_{\widehat{M}}) \in \mathbb{C}^{\widehat{M}}$, $\overrightarrow{I} = (I_{\widehat{M}+1},\ldots,I_M) \in \mathbb{C}^{M-\widehat{M}}$, and function J_S in $\widehat{\Omega}_S$, find $A_\theta \in \widetilde{H}^1_\rho(\widehat{\Omega})$ such that $A_\theta = 0$ on $\widehat{\Gamma}_d$ and a complex vector $\overrightarrow{V} \in \mathbb{C}^{M-\widehat{M}}$ satisfying

$$i\omega \int_{\widehat{\Omega}_C} \sigma A_\theta \bar{G}\rho\,d\rho dz + \int_{\widehat{\Omega}} \frac{1}{\mu\rho}\frac{\partial(\rho A_\theta)}{\partial\rho}\frac{\partial(\rho \bar{G})}{\partial\rho}\,d\rho dz + \int_{\widehat{\Omega}} \frac{1}{\mu}\frac{\partial A_\theta}{\partial z}\frac{\partial \bar{G}}{\partial z}\rho\,d\rho dz$$

$$-\frac{1}{2\pi}\sum_{n=1}^{M} \widehat{V}_n \int_{\widehat{\Omega}_n} \sigma \bar{G}\,d\rho dz = \frac{1}{2\pi}\sum_{n=M+1}^{\widehat{M}} V_n \int_{\widehat{\Omega}_n} \sigma \bar{G}\,d\rho dz + \int_{\widehat{\Omega}_S} J_S \bar{G}\rho\,d\rho dz,$$

$$\forall \psi \in \widetilde{H}^1_\rho(\widehat{\Omega}) \text{ null on } \widehat{\Gamma}_d.$$

$$\sum_{n=1}^{M}\left(\int_{\Omega_n} \sigma A_\theta\,d\rho dz\right)\bar{W}_n\,d\rho dz - \frac{i}{2\pi\omega}\sum_{n=1}^{M}\int_{\widehat{\Omega}_n}\sigma\frac{\widehat{V}_n}{\rho}\bar{W}_n\,d\rho dz = \frac{i}{\omega}\sum_{n=1}^{M} I_n \bar{W}_n,\, \forall \mathbf{W} \in \mathbb{C}^M.$$

10.8 Axisymmetric eddy currents model with azimuthal J

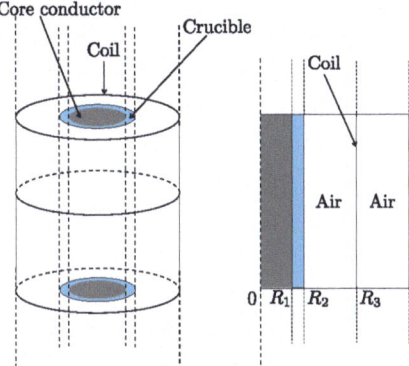

Fig. 10.6. Sketch of the domain with a crucible of one layer ($N = 2$)

Remark 10.8. Numerical discretization of this problem can be done by using Lagrangian (nodal) continuous finite elements on a triangular mesh of $\widehat{\Omega}$; see, for instance [14].

Next, we present an example with cylindrical geometry and azimuthal **J**, for which it is possible to obtain an analytical solution of the eddy currents model in the whole space in terms of the magnetic vector potential.

Example 10.2. Let us consider an infinite cylinder consisting of a core conductor surrounded by a crucible and an extremely thin coil. The crucible itself may consist of several concentric layers made with different materials. The multi-turn coil is modeled as a continuous single coil with a uniform current intensity. Let $R_1, \ldots, R_N$ be the respective radii of the core and crucible materials and R_{N+1} the radius of the coil (see Fig. 10.6 for $N = 2$). We also define $R_0 = 0$.

In order to compute the analytical solution, we assume that the electrical conductivity and magnetic permeability are constant in each material, i.e,

$$\sigma = \sigma_j, \quad \mu = \mu_j \quad \text{if } R_{j-1} < \rho < R_j, \quad j = 1, \ldots, N.$$

We will provide the analytical solution, $\mathbf{A} = A(\rho)\mathbf{e}_\theta$, for any number of layers. Indeed, let us denote

$$A_j = A|_{(R_{j-1}, R_j)}, \quad j = 1, \ldots, N+1,$$

$$A_{ext} = A|_{(R_{N+1}, +\infty)}.$$

We will assume that electric current flows in the coil in the $\mathbf{e}_\theta$-direction and that it is uniformly distributed in the $\mathbf{e}_z$-direction. Then, taking into account that we only have one connected component, Eqs. (10.96)–(10.97) become

$$-\frac{1}{\mu_j} \frac{d}{d\rho} \left(\frac{1}{\rho} \frac{d(\rho A_j)}{d\rho} \right) + i\omega \sigma_j A_j = 0 \quad \text{if } R_{j-1} < \rho < R_j, j = 1, \ldots, N, \quad (10.104)$$

10 The eddy currents model

$$-\frac{1}{\mu_0}\frac{d}{d\rho}\left(\frac{1}{\rho}\frac{d(\rho A_{N+1})}{d\rho}\right) = 0 \quad \text{if } R_N < \rho < R_{N+1}, \tag{10.105}$$

$$-\frac{1}{\mu_0}\frac{d}{d\rho}\left(\frac{1}{\rho}\frac{d(\rho A_{ext})}{d\rho}\right) = 0 \quad \text{if } \rho > R_{N+1}, \tag{10.106}$$

with boundary conditions

$$A_1(\rho) \text{ is bounded as } \rho \to 0, \tag{10.107}$$

$$A_{ext}(\rho) = O\left(\frac{1}{\rho}\right) \text{ as } \rho \to \infty. \tag{10.108}$$

Equation (10.108) follows from the Biot-Savart law in the case of an infinitely long conductor.

Now we must add interface conditions at the boundaries of the conductors and at the inductor position. On the one hand, since $\mathbf{B} = \mathbf{curl\,A}$ and $\mathbf{A} = A(\rho)\mathbf{e}_\theta$, we have that A must be continuous. Then we set

$$A_j(R_j) = A_{j+1}(R_j), \quad j = 1, \ldots, N, \tag{10.109}$$

$$A_{N+1}(R_{N+1}) = A_{ext}(R_{N+1}). \tag{10.110}$$

Moreover, we have

$$\frac{1}{\mu_j}\frac{1}{\rho}\frac{d(\rho A_j)}{d\rho}(R_j) = \frac{1}{\mu_{j+1}}\frac{1}{\rho}\frac{d(\rho A_{j+1})}{d\rho}(R_j), \quad j = 1, \ldots, N. \tag{10.111}$$

Finally, if we denote by I the intensity per unit length in the induction coil, we have

$$\frac{1}{\mu_0}\frac{1}{\rho}\left(\frac{d(\rho A_{N+1})}{d\rho} - \frac{d(\rho A_{ext})}{d\rho}\right)(R_{N+1}) = I. \tag{10.112}$$

Consequently, the model consists of Eqs. (10.104)–(10.106) with boundary conditions (10.107)–(10.108) and interface conditions (10.109)–(10.112).

In order to solve Eq. (10.104) for $j = 1, \ldots, N$ we perform the change of variable $x = \rho \gamma_j$ in (R_{j-1}, R_j), where $\gamma_j = \sqrt{i \omega \mu_j \sigma_j}$. Then we get

$$x^2 \frac{d^2 \tilde{A}_j}{dx^2} + x \frac{d\tilde{A}_j}{dx} - (x^2 + 1)\tilde{A}_j = 0 \quad \text{in } (R_{j-1}, R_j), \tag{10.113}$$

where $\tilde{A}_j(x) = A_j(x/\gamma_j)$. Equation (10.113) is a Bessel's equation, the general solution of which is given by

$$\tilde{A}_j(x) = \alpha_j \mathscr{I}_1(x) + \beta_j \mathscr{K}_1(x),$$

where $\mathscr{I}_1$ and $\mathscr{K}_1$ are the modified Bessel function of the first and second kind, respectively.

10.8 Axisymmetric eddy currents model with azimuthal **J**

Simple integration of Eqs. (10.105) and (10.106) yields

$$A_{N+1} = \mu_0 \frac{\rho}{2} \alpha_{N+1} + \frac{\beta_{N+1}}{\rho},$$

$$A_{ext} = \mu_0 \frac{\rho}{2} \alpha_{ext} + \frac{\beta_{ext}}{\rho}.$$

By using boundary conditions (10.107)–(10.108) we first deduce that $\beta_1 = 0$ for A_1 to be bounded as $\rho \to 0$ and $\alpha_{ext} = 0$. Besides, from (10.112) we obtain that

$$\alpha_{N+1} = I.$$

On the other hand, interface conditions (10.110) imply

$$\alpha_1 \mathscr{I}_1(\gamma_1 R_1) - \alpha_2 \mathscr{I}_1(\gamma_2 R_1) - \beta_2 \mathscr{K}_1(\gamma_2 R_1) = 0,$$

$$\alpha_j \mathscr{I}_1(\gamma_j R_j) + \beta_j \mathscr{K}_1(\gamma_j R_j) - \alpha_{j+1} \mathscr{I}_1(\gamma_{j+1} R_j) - \beta_{j+1} \mathscr{K}_1(\gamma_{j+1} R_j) = 0,$$
$$\text{for } j = 2, \ldots, N-1,$$

$$\alpha_N \mathscr{I}_1(\gamma_N R_N) + \beta_N \mathscr{K}_1(\gamma_N R_N) - \frac{\mu_0}{2} R_N \alpha_{N+1} - \frac{\beta_{N+1}}{R_N} = 0,$$

$$\frac{\mu_0}{2} R_{N+1} \alpha_{N+1} + \frac{\beta_{N+1}}{R_{N+1}} - \frac{\beta_{ext}}{R_{N+1}} = 0.$$

Finally, from (10.111) we deduce, for $j = 1$,

$$\alpha_1 \left\{ \mathscr{I}_1(\gamma_1 R_1) + R_1 \gamma_1 \mathscr{I}'_1(\gamma_1 R_1) \right\} = \alpha_2 \left\{ \mathscr{I}_1(\gamma_2 R_1) + R_1 \gamma_2 \mathscr{I}'_1(\gamma_2 R_1) \right\}$$
$$+ \beta_2 \left\{ \mathscr{K}_1(\gamma_2 R_1) + R_1 \gamma_2 \mathscr{K}'_1(\gamma_2 R_1) \right\},$$

for $j = 2, \ldots, N-1$,

$$\alpha_j \left\{ \mathscr{I}_1(\gamma_j R_j) + R_j \gamma_j \mathscr{I}'_1(\gamma_j R_j) \right\} + \beta_j \left\{ \mathscr{K}_1(\gamma_j R_j) + R_j \gamma_j \mathscr{K}'_1(\gamma_j R_j) \right\} =$$
$$\alpha_{j+1} \left\{ \mathscr{I}_1(\gamma_{j+1} R_j) + R_j \gamma_{j+1} \mathscr{I}'_1(\gamma_{j+1} R_j) \right\} + \beta_{j+1} \left\{ \mathscr{K}_1(\gamma_{j+1} R_j) + R_j \gamma_{j+1} \mathscr{K}'_1(\gamma_{j+1} R_j) \right\},$$

and for $j = N$,

$$\alpha_N \left\{ \mathscr{I}_1(\gamma_N R_N) + R_N \gamma_N \mathscr{I}'_1(\gamma_N R_N) \right\} + \beta_N \left\{ \mathscr{K}_1(\gamma_N R_N) + R_N \gamma_N \mathscr{K}'_1(\gamma_N R_N) \right\} = \mu_0 R_N \alpha_{N+1}.$$

Thus we obtain a linear system of order $2N+2$ for unknowns α_1, β_{ext} and α_j, β_j, $j = 2, \ldots, N+1$.

10.9 Axisymmetric eddy currents model with J lying on the meridional section

There are eddy currents problems where the assumption of cylindrical symmetry is still suitable but the current density **J** has the form $\mathbf{J}(\rho,\theta,z) = J_\rho(\rho,z)\mathbf{e}_\rho + J_z(\rho,z)\mathbf{e}_z$, that is, lies on a meridional section of the 3D domain Ω. Here, we deal with this case and assume moreover that $\Omega_S = \emptyset$.

In these cases, it may be again interesting to write the eddy currents model in a meridional section of the conducting domain which will be denoted by $\widehat{\Omega}_C$. From Ohm's law, the electric field has the following form in conductors

$$\mathbf{E}(\rho,\theta,z) = E_\rho(\rho,z)\mathbf{e}_\rho + E_z(\rho,z)\mathbf{e}_z,$$

which leads to

$$\mathbf{H}(\rho,\theta,z) = H_\theta(\rho,z)\mathbf{e}_\theta,$$
$$\mathbf{B}(\rho,\theta,z) = B_\theta(\rho,z)\mathbf{e}_\theta.$$

Hence, we can obtain the following equation defined in $\widehat{\Omega}_C$ for the scalar unknown H_θ:

$$i\omega\mu(H_\theta \mathbf{e}_\theta) + \mathbf{curl}\left(\frac{1}{\sigma}\mathbf{curl}(H_\theta \mathbf{e}_\theta)\right) = 0. \tag{10.114}$$

This model has two important advantages: it is defined only in the conducting domain and has one scalar unknown. However, it is necessary to define suitable boundary conditions on $\partial\widehat{\Omega}_C$, which is not always an easy task. Let us consider $\partial\widehat{\Omega}_C = \Gamma_A \cup \widehat{\Gamma}_d \cup \widehat{\Gamma}_n$. Firstly, H_θ has to be null on the symmetry axis Γ_A in order to avoid unphysical singularities. Secondly, on the other parts of the boundary we will consider similar conditions to those of the 2D case, namely,

$$\mathbf{H} \times \mathbf{n} = \mathbf{g} \quad \text{on } \widehat{\Gamma}_d, \tag{10.115}$$

$$\mathbf{J} \times \mathbf{n} = 0 \quad \text{on } \widehat{\Gamma}_n. \tag{10.116}$$

As we have advanced above for the 2D model, in some cases the value of $\mathbf{g} = (g_1 \mathbf{e}_1 + g_2 \mathbf{e}_2)$ may be obtained from physical data such us current intensities entering the domain or surface currents (see for instance [6]). However, in general, the computation of $\mathbf{g}$ may be an important restriction to the use of this model. Anyway, (10.115) leads to a Dirichlet boundary condition in terms of H_θ, i.e., $H_\theta = H_d$ for a given H_d. The second condition, (10.116), is suitable for those parts of the boundary where we can assume that the current enters perpendicularly. It will be a natural boundary condition in the weak formulation of Eq. (10.114). Thus, the corresponding axisymmetric model reads as follows:

Given a function H_d defined on $\widehat{\Gamma}_d$, find a field $\mathbf{H} = H_\theta(\rho,z)\mathbf{e}_\theta$ such that

$$i\omega\mu(H_\theta \mathbf{e}_\theta) + \mathbf{curl}\left(\frac{1}{\sigma}\mathbf{curl}(H_\theta \mathbf{e}_\theta)\right) = 0 \text{ in } \widehat{\Omega}_C, \tag{10.117}$$

$$\frac{1}{\sigma}\mathbf{curl}(H_\theta \mathbf{e}_\theta) \times \mathbf{n} = \mathbf{0} \text{ on } \widehat{\Gamma}_n, \tag{10.118}$$

$$\mathbf{H} = H_d \mathbf{e}_\theta \text{ on } \widehat{\Gamma}_d. \tag{10.119}$$

The corresponding weak formulation reads as follows:

Given a function H_d defined on $\widehat{\Gamma}_d$, find $H_\theta \in \widetilde{H}^1_\rho(\widehat{\Omega}_C)$ with $H_\theta = H_d$ on $\widehat{\Gamma}_d$ such that

$$\int_{\widehat{\Omega}_C} i\omega\mu H_\theta \bar{G} \rho \, d\rho dz + \int_{\widehat{\Omega}_C} \frac{1}{\sigma\rho} \frac{\partial(\rho H_\theta)}{\partial \rho} \frac{\partial(\rho \bar{G})}{\partial \rho} \, d\rho dz + \int_{\widehat{\Omega}_C} \frac{1}{\sigma} \frac{\partial H_\theta}{\partial z} \frac{\partial \bar{G}}{\partial z} \rho \, d\rho dz = 0,$$

for all test function $G \in \widetilde{H}^1_\rho(\widehat{\Omega}_C)$ with $G|_{\widehat{\Gamma}_d} = 0$.

Remark 10.9. Numerical discretization of this problem can be done by using Lagrangian (nodal) continuous finite elements on a triangular mesh of $\widehat{\Omega}_C$ (see, for instance [14]).

10.10 Coupling lumped and distributed models

Sometimes we want to make a distributed model of an electromagnetic device that is part of a large electrical circuit whose remaining elements are represented by lumped models. Thus, the whole model will be a coupling between distributed and lumped parameter models.

We will assume that the coupling of the distributed circuit with the lumped circuit is made through the *ports* of the former that are identified to nodes of the latter. Moreover, we recall that at the ports of a distributed circuit we should know either the potential or the current intensity (see Sect. 10.1). Similarly, at the nodes of a lumped circuit (see Chap. 3) we have to prescribe either the potential or the current intensity exchanged with the "outside world". Thus, coupling a distributed model with a circuit can be simply made by imposing equal voltages and current intensities at the coupled *ports/nodes*.

Let us consider a lumped circuit as those considered in Chap. 3. More precisely, we refer to model (3.13)–(3.15). Coupled with this circuit we have a distributed one, Ω, including P ports Γ_n^P, $n = 1, \ldots, P$, modelled by the eddy currents problem,

$$\frac{\partial \mathbf{B}}{\partial t} + \mathbf{curl}\, \mathbf{E} = \mathbf{0}, \tag{10.120}$$

$$\mathbf{curl}\, \mathbf{H} = \mathbf{J}, \tag{10.121}$$

$$\text{div}\, \mathbf{B} = 0, \tag{10.122}$$

$$\mathbf{B} = \mu \mathbf{H}, \tag{10.123}$$

$$\mathbf{J} = \mathbf{J}_S + \sigma \mathbf{E}, \tag{10.124}$$

$$\mu \mathbf{H} \cdot \mathbf{n} = 0 \quad \text{on } \partial\Omega, \tag{10.125}$$

$$V = V_n^P \text{ on } \Gamma_n^P, \, n = 1, \ldots, P. \tag{10.126}$$

We notice that we are assuming that the current enters perpendicularly to the ports so that the electric potential is constant on each of them. For instance, if the topology of the domain is the one assumed in Sect. 10.1, the number of ports P will be equal to $2L$ with L the number of inductors and $\Gamma^P = \Gamma_J^1 \cup \ldots \Gamma_J^L \cup \Gamma_E^1 \cup \ldots \Gamma_E^L$.

Let us suppose that these ports are connected to P nodes of a lumped circuit. We denote by $\mathscr{B}_P$ the $P \times N$ matrix extracting from the vector of all potentials at the nodes of the lumped circuit, $\vec{V} \in \mathbb{R}^N$, the sub-vector whose components are the potentials at the P ports. The $N - P$ remaining nodes will be called *boundary nodes*. At each boundary node either the potential or the current intensity has to be given. Let us denote by $\mathscr{B}_d$ (respectively, $\mathscr{B}_u$) the matrix that extracts from the vector of all potentials at the nodes the ones corresponding to the boundary nodes where the potentials (respectively, the current intensities) are prescribed. Let

$$\vec{V}^P = \mathscr{B}_P \vec{V}, \quad \vec{V}^u = \mathscr{B}_u \vec{V} \text{ and } \vec{V}^d = \mathscr{B}_d \vec{V},$$

then we have

$$\vec{V}(t) = \mathscr{B}_P^t \vec{V}^P(t) + \mathscr{B}_u^t \vec{V}^u(t) + \mathscr{B}_d^t \vec{V}^d(t).$$

Similarly,

$$\vec{\Psi}(t) = \mathscr{B}_P^t \vec{\Psi}^P(t) + \mathscr{B}_u^t \vec{\Psi}^d(t) + \mathscr{B}_d^t \vec{\Psi}^u(t),$$

where $\vec{\Psi}^P$ is the vector of current intensities introduced into the lumped circuit through the ports, $\vec{\Psi}^u$ are the (unknown) current intensities at the boundary nodes where the potentials are given and $\vec{\Psi}^d$ are the (given) current intensities at the boundary nodes where the potentials are unknown. By replacing the above expressions in (3.13)–(3.15) we obtain

$$\mathscr{D}(\vec{I})(t) + \mathscr{A}^t \vec{V}(t) = \vec{E}(t), \tag{10.127}$$

$$\mathscr{A} \vec{I}(t) + \mathscr{B}_P^t \vec{\Psi}^P(t) + \mathscr{B}_d^t \vec{\Psi}^u(t) = -\mathscr{B}_u^t \vec{\Psi}^d(t), \tag{10.128}$$

$$\mathscr{B}_d \vec{V}(t) = \vec{V}^d(t). \tag{10.129}$$

Comparing these equations to (3.13)–(3.15) we have the new unknowns $\vec{\Psi}^P$ that have to be computed by coupling these equations to the distributed circuit. More specifically, we have

$$\Psi_n^P(t) = \int_{\Gamma_n^P} \mathbf{J} \cdot \mathbf{n} \, dA, \quad n = 1, \ldots, P, \tag{10.130}$$

with $\mathbf{n}$ being a unit normal vector to Γ_n^P that points outward from Ω. Summarizing, the problem to be solved is the following:

Let us suppose that the data are: the current density $\mathbf{J}_S$, *the electromotive forces* $\vec{E}$, *the potentials* $\vec{V}^d$, *and the intensities* $\vec{\Psi}^d$. *Find the electromagnetic field in* Ω, *the potentials at the nodes of the lumped circuit,* $\vec{V}$, *the current intensities along the edges of the lumped circuit,* $\vec{I}$, *the current intensities through the ports,* $\vec{\Psi}^P$, *and the current intensities* $\vec{\Psi}^u$ *satisfying Eqs.* (10.120)–(10.126) *and* (10.127)–(10.130)

with
$$V_n^P = (\mathscr{B}_P \vec{V})_n, \quad n = 1, \ldots, P. \tag{10.131}$$

Remark 10.10. In order to solve this problem, we can use different formulations for the model of the distributed circuit as, for instance, the one involving the magnetic vector potential and the electric potential considered above. Next, this formulation has to be discretized in space by using, for instance, finite element methods. Finally, a time discretization of the whole coupled problem has to be introduced. If an implicit scheme is chosen (say, for instance, the Euler implicit method), a linear numerical system has to be solved at each time step.

Problems

10.1. Let us consider an infinite cylinder of radius R_c composed by a conducting material with constant electrical conductivity σ and constant magnetic permeability μ. An alternating current goes through the conductor along its axis, i.e., $\mathbf{J} = J_3 \mathbf{e}_z$; the intensity through each cross section of the cylinder is given by $I(t) = I_0 \cos(\omega t)$, where I_0 is the complex amplitude and ω the angular frequency. Compute the complex amplitude of magnetic field, $\mathbf{H}$, solution of the eddy currents model inside and outside the cylinder.

Solution: See Sect. 16.1.1.

10.2. By using the solution $\mathbf{H}$ of the previous problem, compute the current density inside the conductor and plot it by using a graphic software such us MATLAB to appreciate the skin effect for different values of μ, σ and f. Choose, for instance, $f = 50, 500, 1000$ Hz, $\sigma = 1.\text{e}3, 1.\text{e}6 \, (\Omega\text{m})^{-1}$ and $\mu = \mu_0, 1000\mu_0$ Hm^{-1}.

Solution: $\mathbf{J}(\rho,\theta,z) = \frac{I_0 \gamma}{2\pi R_c} \frac{\mathscr{I}_0(\gamma \rho)}{\mathscr{I}_1(\gamma R_c)} \mathbf{e}_\theta$ (we use the notation in Sect. 16.1.1).

10.3. In the same framework of the previous problem, compute the electric field inside the conductor. Notice that in this case $\mathbf{E} \times \mathbf{n} = -\text{grad}V \times \mathbf{n}$ on the boundary of the conductor. Compute the value of V by using cylindrical coordinates and taking into account that it is constant in each section $z = z_0$.

Solution: $V = -\frac{I_0 \gamma}{2\pi R_c \sigma} \frac{\mathscr{I}_0(\gamma R_c)}{\mathscr{I}_1(\gamma R_c)} z$ (we use the notation in Sect. 16.1.1).

10.4. Let us consider an infinite round wire of radius R_c composed by a conducting material with constant electrical conductivity σ and constant magnetic permeability μ. The wire is placed in a uniform external magnetic field parallel to it and with positive direction. The magnetic field is time-harmonic with angular frequency ω. The amplitude of the magnetic field intensity on the surface of the wire is H_0. Determine the distribution of the complex amplitude of the magnetic field inside the wire.

Solution: $\mathbf{H}(\rho,\theta,z) = H_0 \frac{\mathscr{I}_0(\gamma \rho)}{\mathscr{I}_0(\gamma R_c)} \mathbf{e}_\theta$.

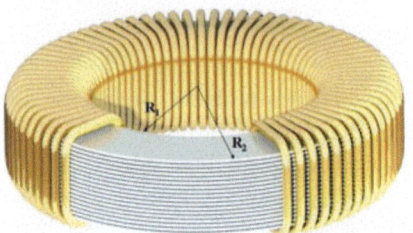

Fig. 10.7. Toroidal laminated core

10.5. Let us consider a toroidal laminated core consisting of N sheets of rectangular section and thickness d (see Fig. 10.7). Let us denote by R_1 and R_2 the internal and external radius of the core, respectively.

Let n be the number of turns of the coil and I the current intensity which is assumed to be alternating and having the form $I(t) = I_0 \cos(\omega t)$. Consider that the coil is infinitely thin so that it can be modelled as a surface current of surface density (A/m) given by

$$\mathbf{J}_S(R_1, z, t) = \frac{\mathrm{n}I(t)}{2\pi R_1} \mathbf{e}_z,$$

$$\mathbf{J}_S(\rho, L, t) = \frac{\mathrm{n}I(t)}{2\pi \rho} \mathbf{e}_\rho,$$

$$\mathbf{J}_S(R_2, z, t) = -\frac{\mathrm{n}I(t)}{2\pi R_2} \mathbf{e}_z,$$

$$\mathbf{J}_S(\rho, 0, t) = -\frac{\mathrm{n}I(t)}{2\pi \rho} \mathbf{e}_\rho,$$

on the interior, upper, exterior and lower surfaces, respectively. Let us neglect the thickness of the dielectric between each two sheets so that $L = \mathrm{n}$. Exploit the cylindrical symmetry of the problem to obtain the eddy currents model to be solved in each sheet by assuming that the normal component of the current density has to be null on the internal surfaces between sheets, because they are isolated. Consider that the core has a linear magnetic behavior, $\mathbf{B} = \mu \mathbf{H}$, and write the time-harmonic eddy currents model in the radial section $\widehat{\Omega}$ of each sheet in terms of the complex amplitude of the azimuthal component of the magnetic field.

Solution: Find $H_\theta(\rho, z)$ such that,

$$i\omega\mu H_\theta - \frac{\partial}{\partial \rho}\left(\frac{1}{\sigma\rho}\frac{\partial(\rho H_\theta)}{\partial \rho}\right) - \frac{\partial}{\partial z}\left(\frac{1}{\sigma}\frac{\partial H_\theta}{\partial z}\right) = 0 \text{ in } \widehat{\Omega},$$

$$H_\theta = \frac{\mathrm{n}I}{2\pi\rho} \text{ on } \partial\widehat{\Omega}.$$

11
An introduction to nonlinear magnetics. Hysteresis

This chapter is an introduction to nonlinear magnetic materials. Nonlinearity means that the relation between the magnetic field intensity and the magnetic induction is nonlinear. We also consider hysteresis, a behavior appearing in many ferromagnetic materials for which the magnetic induction at each particular point depends not only on its present magnetic intensity but also on the past history of its magnetic state. We introduce a well-known model for magnetic hysteresis: the *classical Preisach model* which is carefully described. We also address the identification problem of the Preisach operator for a specific ferromagnetic material by means of the so-called Everett function.

11.1 Magnetic behavior of materials

We refer to Sect. 6.3 where the concept of *magnetization* has been introduced. Magnetic behavior of materials is classified in *diamagnetic*, *magnetic* and *ferromagnetic*. The diamagnetism is characterized by a small negative magnetic susceptibility (see Sect. 6.3.3). This means that these materials are weakly magnetized in the opposite direction to the applied magnetic field. Paramagnetic materials have small and positive magnetic susceptibility so they are also weakly magnetized but in the same direction as the applied magnetic field.

Ferromagnetic materials are characterized by their strong remanent magnetization even in the absence of an applied magnetic field. In most cases they exhibit hysteresis which means that the value of the magnetic induction is a function not only of the present value of the magnetic field but also of the past magnetic history of the volume element under consideration. Thus, building a mathematical model of the magnetic constitutive law is a very difficult task and numerical simulation of devices involving ferromagnetic materials is still quite a challenge. For the sake of simplicity, in this chapter we will assume an *isotropic* magnetic behavior, i.e., magnetic induction and magnetic field are aligned at any position and at any time (see Sect. 6.63).

A. Bermúdez, D. Gómez, P. Salgado: *Mathematical Models and Numerical Simulation in Electromagnetism*. UNITEXT – La Matematica per il 3+2 74
DOI 10.1007/978-3-319-02949-8_11, © Springer International Publishing Switzerland 2014

11.1.1 Magnetic hysteresis

Ferromagnetic materials are very sensitive to be magnetized. These materials are made up of small regions known as *magnetic domains*. Domains are very small regions in the material structure, where all the dipoles are paralleled in the same direction. In each domain, all of the atomic dipoles are coupled together in a preferential direction (see Fig. 11.1, left). In other words, the domains are like small permanent magnets oriented randomly in the material.

Ferromagnetic materials become magnetized when the magnetic domains within the material are aligned (see Fig. 11.1, right). This can be done by placing the material in a strong external magnetic field or by passing electrical current through the material. Some or all of the domains can become aligned. The more the aligned domains, the stronger the magnetic field in the material. When all of the domains are aligned, the material is said to be magnetically saturated. When a material is magnetically saturated, no additional amount of external magnetization force will cause an increase in its internal level of magnetization. After removing this external field, most of domains come again to random positions, but a few of them still remain in their changed position. Because of these unchanged domains the substance becomes slightly magnetized permanently. A similar process takes place if we consider a material magnetically saturated but in the opposite direction. The phenomenon which causes B to lag behind H, so that the magnetization curve for increasing and decreasing fields is not the same, is called *hysteresis* and the loop traced out by the magnetization curve is called a *hysteresis cycle* or *hysteresis loop*.

Figure 11.2 shows an example of a hysteresis loop. In this loop we represent the relationship between the induced magnetic flux density B and the magnetic field H. It is often referred to as the B-H loop.

The loop is generated by measuring the magnetic flux of a ferromagnetic material while the magnetic field is changed. We start at the *demagnetized* state, that is, when a ferromagnetic material has never been previously magnetized or has been thoroughly demagnetized.

Let us suppose a demagnetized ferromagnetic material which is subjected to a monotonically increasing magnetic field starting from zero. Then the couples $(H(t), B(t))$ describe the curve number 1 shown in Fig. 11.2. Thus, the magnetic

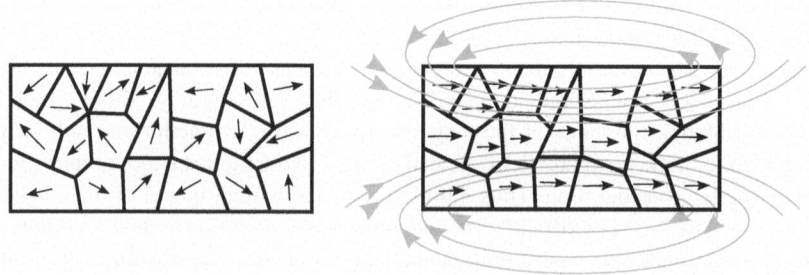

Fig. 11.1. Randomly oriented domains (left) and aligned domains (right)

induction also increases up to a maximum value B_m at which saturation is attained. This curve is called *initial* (or *normal*) *magnetization curve*.

Next, let us decrease *monotonically* the magnetic field from the saturation value H_m to the opposite saturation value $-H_m$. Then points $(H(t), B(t))$ do not trace back the above initial curve but follow curve number 2 in Fig. 11.2 until the magnetic field attains the value $-H_m$. If we increase again the magnetic field, then points $(H(t), B(t))$ describe curve number 3. More generally, if the magnetic field oscillates between two extreme and opposite values H_m and $-H_m$ monotonically (i.e. $H(t)$ does not have any local extremes apart from the global ones) then the couples $(H(t), B(t))$ follow alternatively curves 2 and 3 in the indicated sense, i.e. they travel along the so-called hysteresis *major loop*.

Two important quantities are related with ferromagnetic materials: the *remanence* and the *coercive field*. Remanence represents the magnetization after applying a large magnetic field and then removing it. Thus, it corresponds to the remanent magnetic induction denoted by B_r in Fig. 11.2. In its turn, the coercive field is the intensity of the magnetic field needed to bring the magnetization from the remanent value to zero, i.e., the value H_c in Fig. 11.2.

According to these parameters, ferromagnetic materials can be classified in *soft* and *hard* magnetic materials. Soft magnetic materials have small coercive fields so they are easy to magnetize and their hysteresis loops are thin. On the contrary, hard magnetic materials have large coercive fields (see Fig. 11.3).

A similar behavior is observed when the magnetic field oscillates monotonically between two extreme values H_1 and H_2 with $-H_m < H_1 < H_2 < H_m$. Then points $(H(t), B(t))$ describe a *minor loop* as the one shown in Fig. 11.4.

If $H_2 = -H_1 = H_e > 0$ then the center of the loop is at $(0, B_i)$ where B_i is the initial magnetic induction which is not null when the material is initially magnetized (see Fig. 11.5). Otherwise, $B_i = 0$ so the minor loops are centered at the origin $(0,0)$. In this case, for H_e varying along the interval $[0, H_m]$ a continuum of minor loops is obtained (see Fig. 11.6) the tips of which describe the initial magnetization curve.

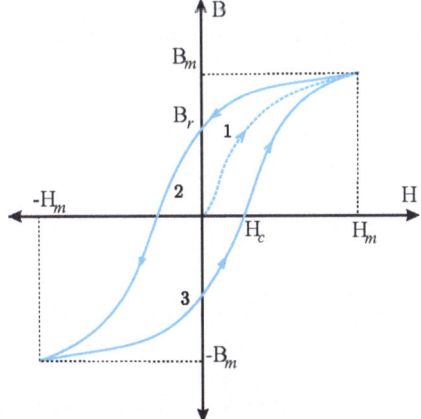

Fig. 11.2. Magnetic hysteresis

220   11 An introduction to nonlinear magnetics. Hysteresis

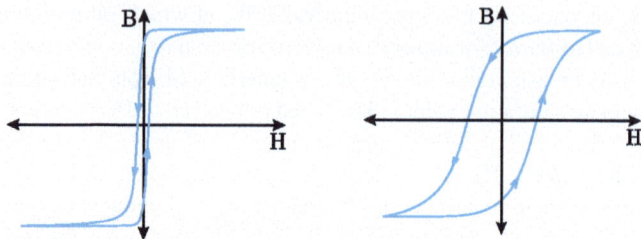

Fig. 11.3. Hysteresis loops for soft (left) and hard (right) materials

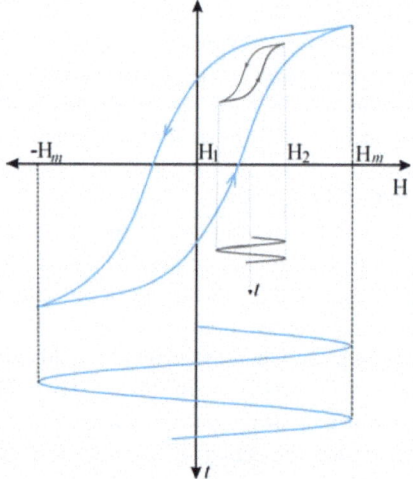

Fig. 11.4. Major and minor loops

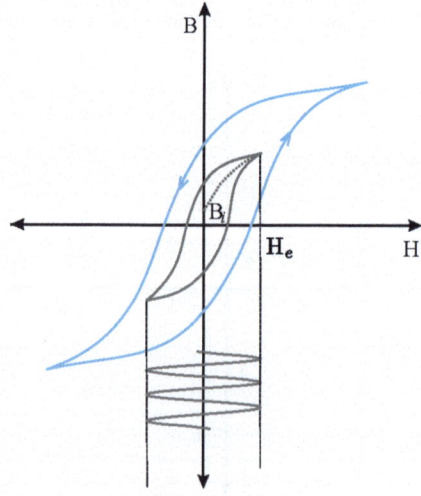

Fig. 11.5. Minor loop

11.1 Magnetic behavior of materials

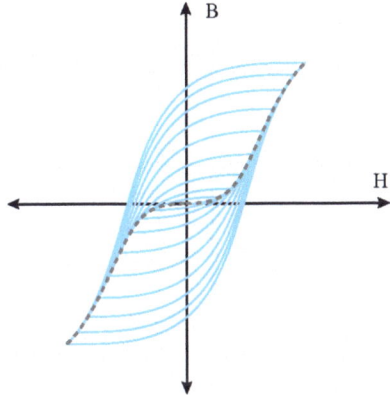

Fig. 11.6. Family of minor loops centered at the origin

For more general time evolution of the magnetic field, the curves described by points $(H(t), B(t))$ along time are also contained in the major loop of the hysteresis diagram but become much more complicated. Instead of looping we get branching at points corresponding to local extreme values of $H(t)$ (i.e., local maxima or minima). These points in the H–B plane are called *reversal points*. Indeed, after an extreme value is attained by the magnetic field, points $(H(t), B(t))$ does not trace back the same curve in reversed sense but a new one emerging from the corresponding reversal point with a different slope. This curve is called *reversal branch*. In general reversal branches depend on the past magnetic history.

Implicit in the above discussion is the fact that hysteresis is a *rate-independent* phenomenon. This means that it does not depend on the rate at which the input (i.e., the magnetic field) varies but only on the sequence of its attained values along time during the evolution of the system. More precisely, branches of such hysteresis nonlinearities are determined only by the past extremum values of input, while the speed (or particular manner) of input variations between extremum points has no influence on branching.

A simpler particular situation is the so-called hysteresis with *local memory*. In this case, the state at time t is characterized by the present values $(H(t), B(t))$ and its subsequent evolution only depends on the sense (increasing or decreasing) of the magnetic field (see Fig. 11.7). More specifically, to each point inside the major hysteresis loop we can associate exactly two branches arising from it: the so-called ascending and descending curves. One corresponding to increasing values of the magnetic field and another one to decreasing values. The two branches together constitute the graph of a monotonic univalued function giving $B(s)$ from $H(s)$ for $s > t$ until a new local extreme value is attained by the magnetic field, in other words, up to the next reversal point.

This is not true for hysteresis with *nonlocal memory*, in which case, at any reachable point in the diagram there is an infinity of curves that may represent the future behavior of the system. Each of these curves depends on a particular past history, namely, on a particular sequence of past extremum values of input.

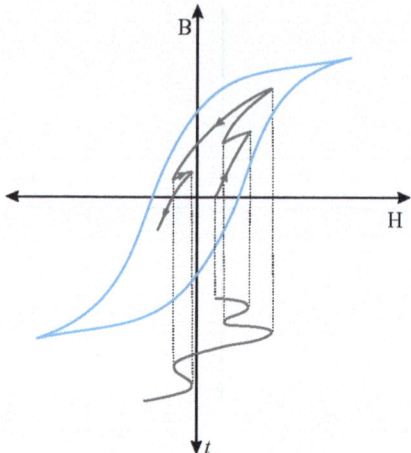

Fig. 11.7. Local memory hysteresis

11.1.2 Energy balance

For hysteretic ferromagnetic materials it will not be possible to define a magnetic energy density as a function of the present values of the magnetic field and the magnetic induction (see Sect. 6.8). Moreover, the presence of hysteresis implies that (for periodic solutions of Maxwell's equations) the electromagnetic energy dissipated during a cycle $[t_1,t_2]$ not only consists of the losses by Joule effect (i.e., the eddy current losses). Indeed, the energy balance (6.58) also involves the integral

$$\int_\Omega \int_{t_1}^{t_2} \frac{\partial \mathbf{B}}{\partial t}(x,t) \cdot \mathbf{H}(x,t) \, dt \, dV$$

which, in general, is not null. The scalar field

$$\int_{t_1}^{t_2} \frac{\partial \mathbf{B}}{\partial t}(x,t) \cdot \mathbf{H}(x,t) \, dt$$

gives the so-called *density of hysteresis* losses at point x during $[t_1,t_2]$.

According to Proposition B.3, in the case of isotropic hysteresis (i.e., if at any time vectors $\mathbf{H}(x,t)$ and $\mathbf{B}(x,t)$ have the same direction, see (6.63)) the value of this integral is equal to the "algebraic sum" of the areas of the regions enclosed by the closed curve

$$t \in [t_1,t_2] \longrightarrow (\mathrm{H}(x,t), \mathrm{B}(x,t)). \tag{11.1}$$

Recall that the area of one of such regions is taken to be positive if its boundary (which consists of pieces of the above curve) is walked counter-clockwise, and negative otherwise.

As previously stated, an important and simpler case arises when field $\mathrm{H}(x,t)$ oscillates periodically in time and *monotonically* (i.e., without local extremes) between two opposite values $H_e(x)$ and $-H_e(x)$. Then the closed curve (11.1) does not intersects itself, i.e., it is a loop (in general a minor loop).

Under the previous circumstances, the hysteresis loss density at each point x during a cycle is just the area of the loop, $\mathscr{A}(B_e(x))$, which can be measured for each particular ferromagnetic material by experiments. Precisely,

$$\mathscr{W}_h(x) = \mathscr{A}(B_e(x)) \ (J/m^3). \tag{11.2}$$

Thus, the power *density of the hysteresis loss* at point x, averaged in a cycle, is

$$\mathscr{P}_h(x) = f \mathscr{A}(B_e(x)) \ (W/m^3), \tag{11.3}$$

being $f = 1/(t_2 - t_1)$ the frequency.

Hence, the power of the hysteresis losses in a region Ω, averaged in a cycle, is given by

$$f \int_\Omega \mathscr{A}(B_e(x)) \, dV \ (W).$$

Notice that this power depends linearly on frequency.

Steinmetz (see [65]) proposed to approximate the loop area by the formula,

$$\mathscr{A}(B) = k_h B^\eta, \tag{11.4}$$

for some constants k_h and η depending on each particular material. The exponent η is called *Steinmetz coefficient* and varies between 1.5 y 3.

Moreover, according to the previous formulas and in order to determine the loss density, we need to know the maximum values of H and B at each point x of the considered region. For this purpose, Maxwell's equations have to be solved (more precisely, the eddy currents model) which, in its turn, makes necessary to dispose of a mathematical model for hysteresis. The complexity of existing models together with the difficulty of making experimental measurements of the parameters involved in them should explain the fact that many of the commercial packages for electromagnetism do not include any model to compute the electromagnetic field including hysteresis.

An alternative, which is suitable for soft magnetic materials, consists in neglecting the hysteresis by using a (nonlinear in general) univalued relation between the magnetic field and the magnetic induction given by the normal magnetization curve of the material. Thus, one computes the electromagnetic field by solving numerically a transient nonlinear eddy currents model.

Obviously this numerical simulation does not allow us to compute directly the hysteresis losses because this phenomena is not considered in the model. Nevertheless, the previous discussions, more precisely the formula (11.3), makes possible to get an approximation of them from the electromagnetic field computed by numerical simulation, by using the hysteresis diagram of the material.

11.2 Mathematical modelling of magnetic hysteresis

Different models have been proposed to represent the magnetic hysteresis phenomenon. At the macroscopic level, the most popular is the *classical Preisach model* [62]. This model is based on some hypotheses concerning the physical mechanisms of magnetization, and for this reason was primarily known in the area of magnetics. It was not until fifty years later when a group of Russian mathematicians developed the model into an abstract mathematical frame of *hysteresis operators* which can be applied to a wide variety of hysteresis phenomena [45].

11.2.1 The classical Preisach model

The classical Preisach model was first suggested to describe ferromagnetism (see, for instance, [52, 67]). Nowadays it is recognized as a fundamental tool for describing a wide range of hysteresis phenomena in different subjects as physics, mechanics or superconductivity, among others. Here we briefly recall the classical definition and some properties of this operator.

The classical Preisach model is constructed from an infinite set of hysteresis operators called *relay operators*. A relay operator is represented by elementary rectangular loops with "up" and "down" switching values. Given any couple $\rho = (\rho_1, \rho_2) \in \mathbb{R}^2$, with $\rho_1 < \rho_2$, the corresponding relay operator h_ρ, depicted in Fig. 11.8, is defined as follows: for any $u \in C([0,T])$ and $\xi \in \{1, -1\}$, $h_\rho(u, \xi)$ is a function from $[0,T]$ to $\mathbb{R}$ such that (see, for instance, [52, 67])

$$h_\rho(u, \xi)(0) := \begin{cases} -1 & \text{if } u(0) \leq \rho_1, \\ \xi & \text{if } \rho_1 < u(0) < \rho_2, \\ 1 & \text{if } u(0) \geq \rho_2. \end{cases}$$

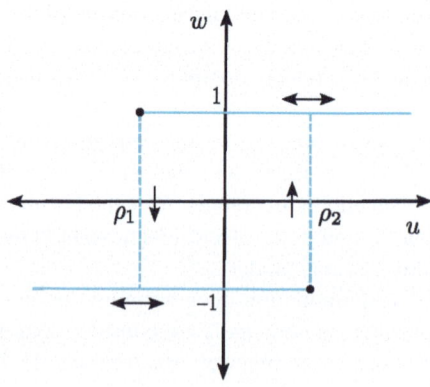

Fig. 11.8. Scalar relay

11.2 Mathematical modelling of magnetic hysteresis

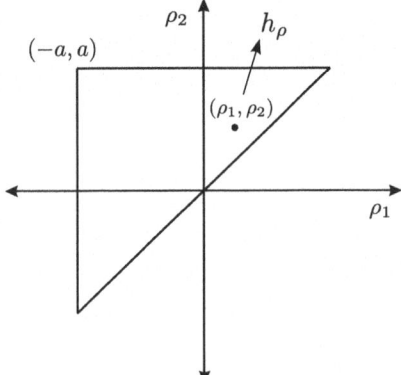

Fig. 11.9. Preisach triangle

Then, for any $t \in (0,T]$, let us set $X_u(t) := \{\tau \in (0,t] : u(\tau) = \rho_1 \text{ or } \rho_2\}$ and define

$$h_\rho(u,\xi)(t) := \begin{cases} h_\rho(u,\xi)(0) & \text{if } X_u(t) = \emptyset, \\ -1 & \text{if } X_u(t) \neq \emptyset \text{ and } u(\max X_u(t)) = \rho_1, \\ 1 & \text{if } X_u(t) \neq \emptyset \text{ and } u(\max X_u(t)) = \rho_2. \end{cases}$$

We notice that $h_\rho = \pm 1$ with up-switch at ρ_2 and down-switch at ρ_1. The value of the relay operator remains at the last value (± 1) until u takes the value of one opposite switch, that is, switch to value $+1$ when u attains the value ρ_2 from below, and to -1 when it attains ρ_1 from above. This operator is the most simple model of discontinuous hysteresis.

Let us now introduce the *Preisach triangle* $\mathscr{T}_a := \{\rho = (\rho_1,\rho_2) \in \mathbb{R}^2 : -a \leq \rho_1 \leq \rho_2 \leq a\}$ (see Fig. 11.9). Let us denote by Y the family of Borel measurable functions (see, for instance, [33]) $\mathscr{T}_a \to \{-1,1\}$ and by ξ a generic element of Y. Let us define the *Preisach operator*

$$\mathscr{F} : C([0,T]) \times Y \longrightarrow L^\infty(0,T),$$
$$(u,\xi) \longmapsto [\mathscr{F}(u,\xi)](t) = \int_{\mathscr{T}_a} [h_\rho(u,\xi(\rho))](t)p(\rho)d\rho, \quad (11.5)$$

with $0 < p \in L^1(\mathscr{T}_a)$, usually known as *Preisach function*. The Preisach model can be interpreted as the sum of a family of relays, distributed with a certain density $p(\rho)$.

11.2.2 Geometric interpretation

The understanding of the Preisach operator is considerably facilitated by its geometric interpretation. This interpretation is based on the fact that there is a one-to-one correspondence between relay operators h_ρ and points (ρ_1,ρ_2) of the Preisach triangle $\mathscr{T}_a$.

11 An introduction to nonlinear magnetics. Hysteresis

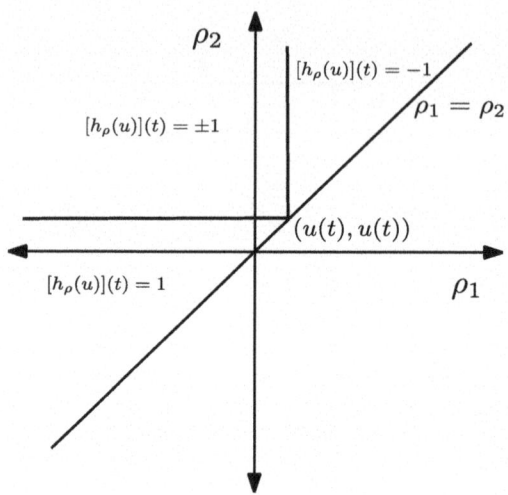

Fig. 11.10. Function $h_\rho(u, \xi(\rho))$

We notice that, given $u \in C([0,T])$ and ξ, each relay $h_\rho(u, \xi(\rho))$ is such that, for any $t \in [0,T]$

$$\begin{cases} \text{if } u(t) \leq \rho_1, & \text{then } [h_\rho(u, \xi(\rho))](t) = -1 \\ \text{if } u(t) \geq \rho_2, & \text{then } [h_\rho(u, \xi(\rho))](t) = 1 \\ \text{if } \rho_1 < u(t) < \rho_2, & \text{then } [h_\rho(u, \xi(\rho))](t) \text{ depends on } u|_{[0,t]} \text{ and } \xi(\rho). \end{cases} \quad (11.6)$$

That is, for a given $u(t)$, all the relays h_ρ such that $\rho_1 \geq u(t)$ are "switched down". Similarly the relays h_ρ such that $\rho_2 \leq u(t)$ are "switched up" (see Fig.11.10).

Now, to understand the geometrical interpretation of the Preisach operator, we consider a simple setting. First we assume that $u(t)$ at some instant of time t_0 has a value less than $-a$. Notice that, from the particular choice of u, all the relays are well defined in $\mathcal{T}_a$ without the need of giving an "initial state" ξ, for $t > t_0$. Given that, $u(t_0) \leq -a \leq \rho_1$ for all $(\rho_1, \rho_2) \in \mathcal{T}_a$, then from (11.6) it follows that all the relay operators $[h_\rho(u)](t) := [h_\rho(u, \xi)](t_0) = -1$ in $\mathcal{T}_a$. Now, we consider that u increases monotonically. From the definition of the relay operator, the relays will only change to a positive state. Thus, triangle $\mathcal{T}_a$ is subdivided into two sets (one possibly empty):

$$S_u^-(t) = \{(\rho_1, \rho_2) \in \mathcal{T}_a : [h_\rho(u)](t) = -1\} \quad (11.7)$$

and

$$S_u^+(t) = \{(\rho_1, \rho_2) \in \mathcal{T}_a : [h_\rho(u)](t) = 1\}. \quad (11.8)$$

Since the change to a positive state of the relay h_ρ depends only on the value of ρ_2, we obtain that $L_u(t) := \partial S_u^-(t) \cap \partial S_u^+(t)$ is orthogonal to the ρ_2-axis and moves up.

11.2 Mathematical modelling of magnetic hysteresis

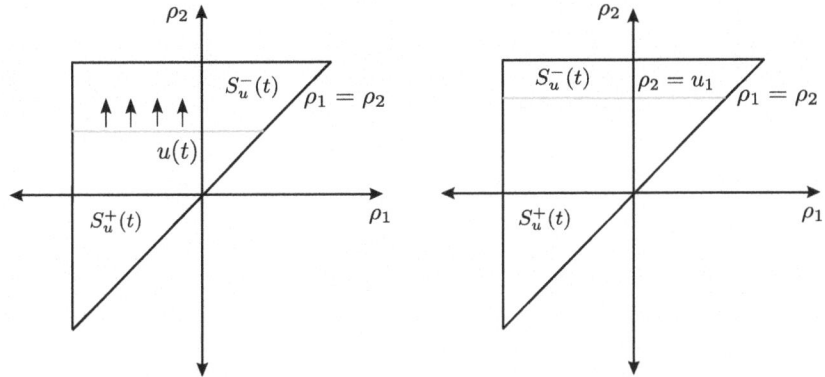

Fig. 11.11. $L_u(t)$: $u(t)$ is increasing (left) and attains a maximum at u_1 (right)

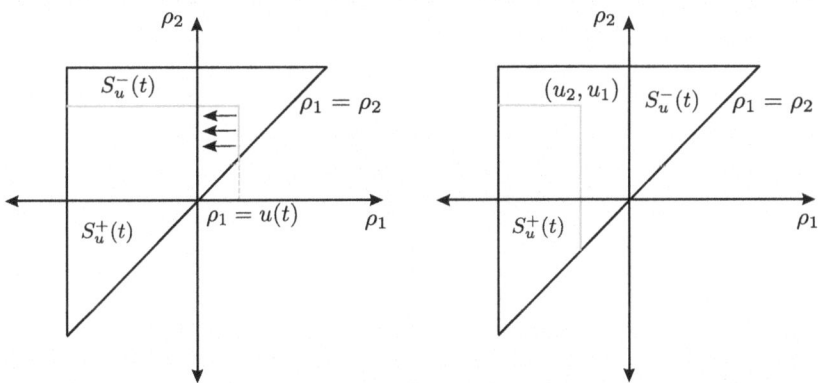

Fig. 11.12. $L_u(t)$: $u(t)$ is decreasing from u_1 (left) and attains a minimum at u_2 (right)

This subdivision is made by the line $\rho_2 = u(t)$ (see Fig. 11.11, left). Function u increases until it reaches some maximum value $-a < u_1 < a$ at time t_1 (see Fig. 11.11, right).

Now, we assume that $u(t)$ decreases monotonically. Then, the relays will only change to a negative state. Since the change to a negative state of the relay h_ρ depends only on the value of ρ_1, we obtain that the line $\rho_1 = u(t)$ moves from right to left (see Fig. 11.12, left). Function u decreases until it reaches at time t_2 some value $u_2 > -a$. At this point, the interface $L_u(t)$ between $S_u^+(t)$ and $S_u^-(t)$ has now two segments, the horizontal and vertical ones depicted in Fig. 11.12 (right).

Now, we assume that $u(t)$ increases again until it reaches at time t_3 some maximum value u_3 that is less than u_1. Geometrically, this increment produces a new horizontal segment in $L_u(t)$ which moves up. This motion is terminated when the maximum u_3 is reached. This is shown in Fig. 11.13 (left). Finally we assume that $u(t)$ decreases until it reaches at time t_4 some minimum value $u_4 > u_2$. This variation results in the formation of a new vertical line in $L_u(t)$ that moves from right to left

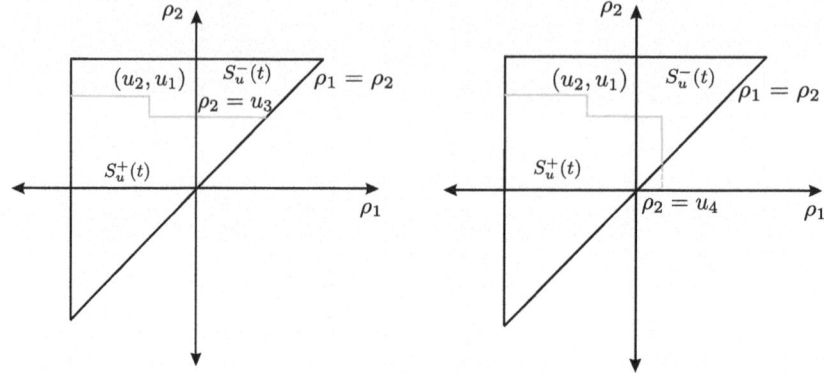

Fig. 11.13. $L_u(t)$: $u(t)$ attains a maximum at u_3 (left) and a minimum at u_4 (right)

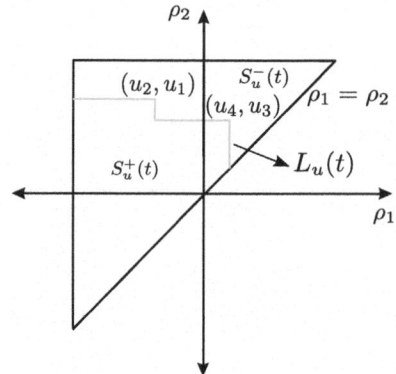

Fig. 11.14. Staircase line L_u at $t = t_4$

as it is shown in Fig. 11.13 (right). At this point, $L_u(t)$ has two vertices (u_2, u_1) and (u_4, u_3) (see Fig. 11.14).

A similar figure can be obtained if we consider a function $v \in C([0,T])$ such that, unlike $u(t)$, at some instant of time t_0 has a value that is greater than a. We assume that $v(t)$ decreases to $v_1 > -a$, then increases to $v_2 \leq a$, next decreases to v_3 and finally increases to v_4 as depicted in Fig. 11.15 (right). $L_v(t)$ is illustrated in Fig. 11.15 (left); we notice that the first line of $L_v(t)$ is a vertical line, because $v(t)$ decreases from a value greater than a.

We can summarize the above analysis as follows; for a given $u \in C([0,T])$, and any instant t of time, the triangle $\mathcal{T}_a$ is subdivided into two sets: $S_u^+(t)$ consisting of points (ρ_1, ρ_2) for which the corresponding relay operators $h_\rho(u)$ are positive (in the "up" position), and $S_u^-(t)$ consisting of points (ρ_1, ρ_2) for which the corresponding relay operators $h_\rho(u)$ take negative values (in the "down" position). The interface $L_u(t)$ between $S_u^+(t)$ and $S_u^-(t)$ is a staircase line whose vertices have coordinates (ρ_1, ρ_2) coinciding respectively with the local minimum and maximum values of u at previ-

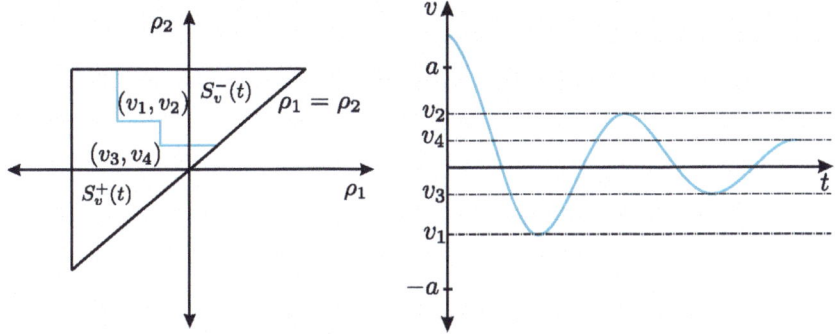

Fig. 11.15. Staircase line $L_v(t)$ (left) and input $v(t)$ (right)

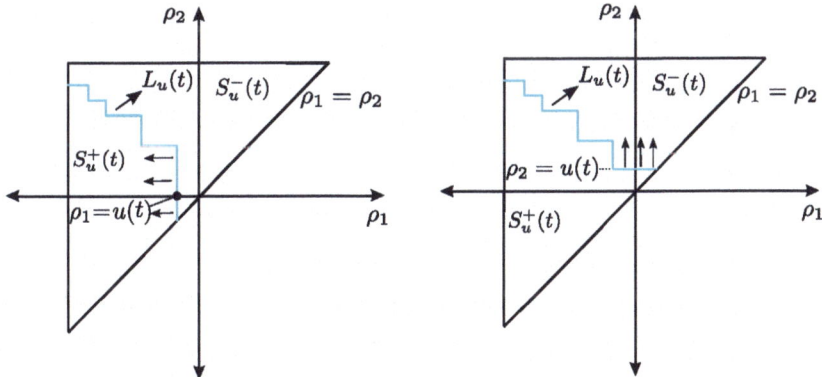

Fig. 11.16. Staircase line $L_u(t)$ moving right to left (left) and moving up (right)

ous instants of time. At time t, the staircase line $L_u(t)$ is attached to the line $\rho_1 = \rho_2$ in the current value of u, namely, $L_u(t)$ intersects the line $\rho_1 = \rho_2$ in $(u(t), u(t))$. $L_u(t)$ moves when $u(t)$ changes, intersects the line $\rho_1 = \rho_2$ horizontally and it moves up as $u(t)$ increases. Otherwise, $L_u(t)$ intersects the line $\rho_1 = \rho_2$ vertically and it moves from right to left as $u(t)$ decreases (see Fig. 11.16).

Hence, from the latter we notice that, at any instant of time t, the integral in (11.5) can be subdivided into two integrals, over $S_u^+(t)$ and $S_u^-(t)$, respectively:

$$w_u(t) := [\mathscr{F}(u)](t) = \int_{\mathscr{T}_a} [h_\rho(u)](t) p(\rho) \mathrm{d}\rho$$

$$= \int_{S_u^+(t)} [h_\rho(u)](t) p(\rho) \mathrm{d}\rho + \int_{S_u^-(t)} [h_\rho(u)](t) p(\rho) \mathrm{d}\rho.$$

We recall that, because of the particular choice of the values of u we do not need an "initial state" ξ. Moreover, from (11.7), (11.8) and the latter equation we obtain that

$$w_u(t) = \int_{S_u^+(t)} p(\rho)\,d\rho - \int_{S_u^-(t)} p(\rho)\,d\rho. \tag{11.9}$$

Remark 11.1. To compute the Preisach operator in $(0,T]$, in general it is enough to know $u(0)$, the Preisach function p and the history of u represented by $S_u^-(t)$ and $S_u^+(t)$, which contain the minimum information to compute (11.9). Notice that for $t = 0$ the above sets are deduced from the "initial state" ξ.

From expression (11.9), it follows that $[\mathscr{F}(u)](t)$ depends on the particular subdivision of the limiting triangle, $\mathscr{T}_a$, into $S_u^+(t)$ and $S_u^-(t)$. Therefore, it depends on the shape of the interface $L_u(t)$ which, in its turn, is determined by the extremum values of $u(t)$ at previous instants of time. It turns out that not all extremum input values are stored by the model. In fact, given the dependence of the staircase line $L_u(t)$ we can see that the Preisach operator has a wiping-out property. This property states that each time the input reaches a local maximum $u(t)$, $L_u(t)$ erases, or "wipes out" the previous vertices whose ρ_2 value is lower than the current value $u(t)$. As a result, all previous maximum values recorded in $L_u(t)$ having a value lower than the current maxima are taken out. Similarly, each time an input reaches a local minimum $u(t)$, the memory curve erases all previous vertices whose ρ_1 value is higher than the current $u(t)$ value. To illustrate this property, we consider a simple setting. Let $u \in C([0,T])$ be characterized by a finite decreasing sequence $\{u_1, u_3, u_5, u_7\}$ of local maxima and an increasing sequence $\{u_2, u_4, u_6, u_8\}$ of local minima, with $-a < u_i < a, i = 1,\ldots,8$ (see Fig. 11.17). Now, let us assume that $u(t)$ is monotonically increasing until it reaches u_9, such that $u_3 < u_9 < u_1$. This increase of $u(t)$ results in the formation of a new line in $L_u(t)$ which intersects the line $\rho_1 = \rho_2$ horizontally and moves up until the maximum value u_9 is reached. Then we obtain a modified staircase line $L_u(t)$ where all vertices whose ρ_2-coordinates were below u_9 have been wiped out (see Fig. 11.18).

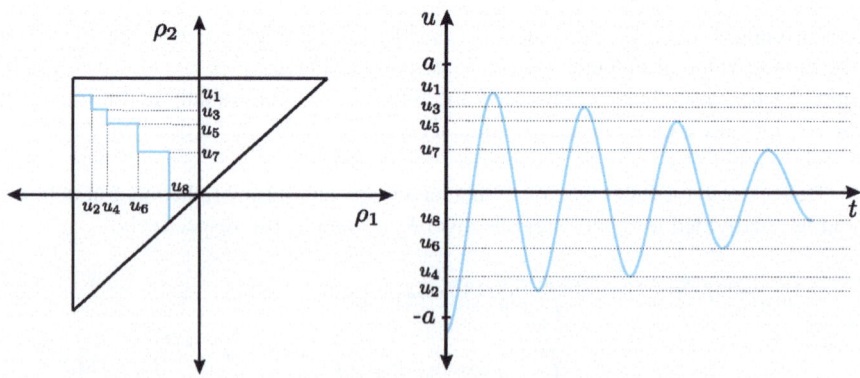

Fig. 11.17. Initial staircase line L_u (left) and function u (right)

11.2 Mathematical modelling of magnetic hysteresis

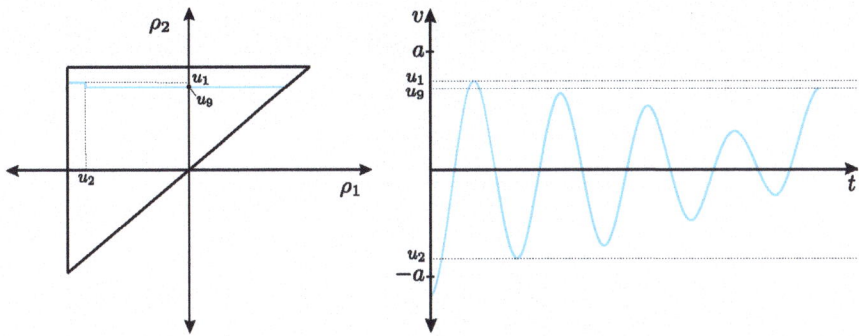

Fig. 11.18. L_u for increasing u until u_9 (left) and function u (right)

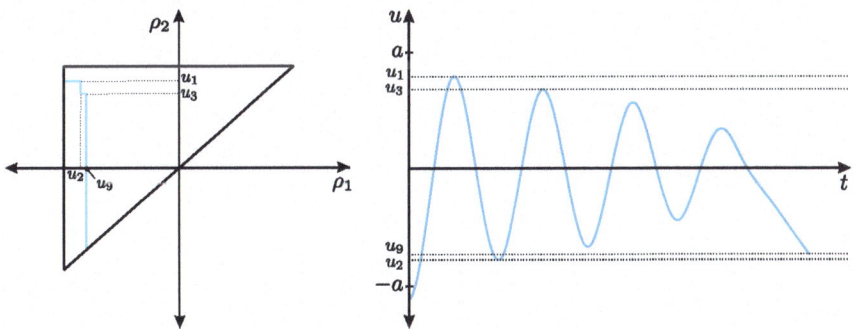

Fig. 11.19. L_u for decreasing u until u_9 (left) and function u (right)

Similarly, instead of assuming that $u(t)$ is monotonically increasing, let us suppose that it decreases until it reaches u_9, such that $u_2 < u_9 < u_4$. Function u and the corresponding staircase line $L_u(t)$ are depicted in Fig. 11.19.

Another important property of the Preisach operator is referred to as the *congruency property*. This property states that as the input is cycled between two extremum values, the minor loop traced will have the same shape, independently of history (see Fig. 11.20). However, the position of the minor loop along the output axis will be determined by the history of past input variations (for further details, see [52]).

11.2.3 Identification problem

When using the Preisach operator to model an electromagnetic system it is necessary to find the density function p that characterizes the material. Therefore, the identification of density p is an important step for the effective use of this model in real applications.

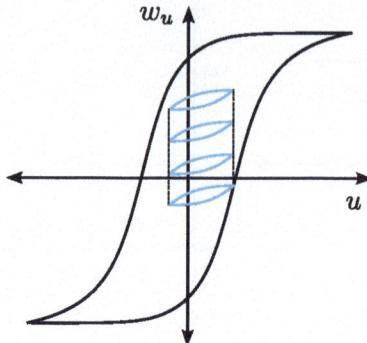

Fig. 11.20. Congruence property

There are many different analytical expressions to represent the Preisach distribution, e.g., the *Factorized-Lorentzian* distribution function,

$$p(\rho_1,\rho_2) := N\left(\left(1+\left(\frac{\rho_2-\beta}{\gamma\beta}\right)^2\right)\left(1+\left(\frac{\rho_1+\beta}{\gamma\beta}\right)^2\right)\right)^{-1},$$

or the *Gauss–Gauss* distribution function,

$$p(\rho_1,\rho_2) := N\exp\left(-\frac{\left(\frac{\rho_2-\rho_1}{2}-\beta\right)^2-\left(\frac{\rho_2+\rho_1}{2}\right)^2}{2\gamma^2\beta^2}\right).$$

Parameter N is the so-called normalization factor, and β and γ are adjustable parameteres which can be determined with only a few measurements. Examples of the above distribution functions are shown in Figs. 11.21 and 11.22.

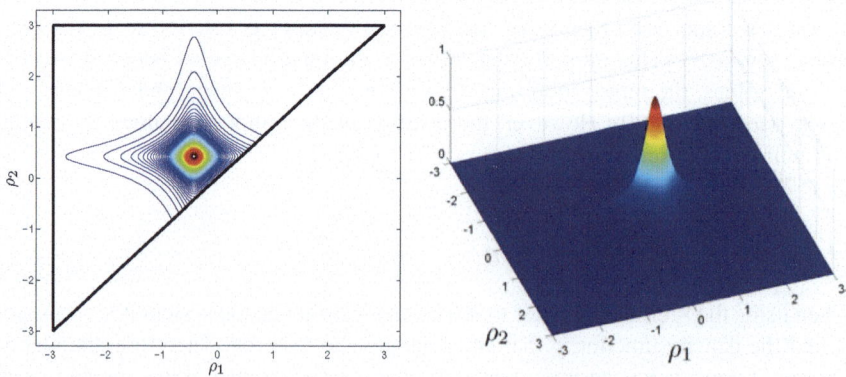

Fig. 11.21. Factorized-Lorentzian distribution function with $\gamma = 0.614152$ and $\beta = 0.427471$

11.2 Mathematical modelling of magnetic hysteresis

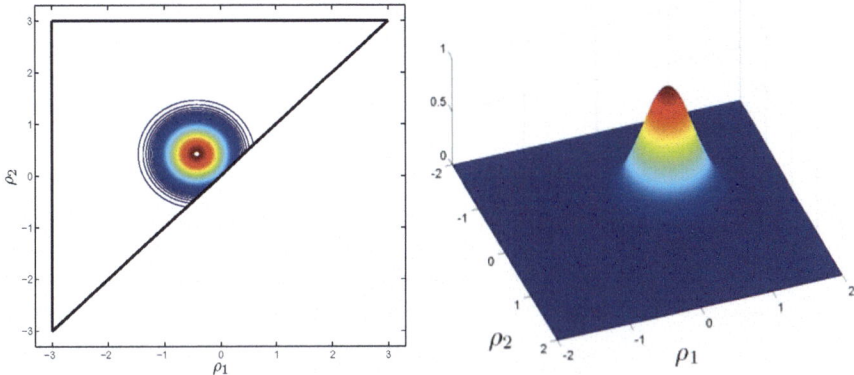

Fig. 11.22. Gauss-Gauss distribution function with $\gamma = 0.582933$ and $\beta = 0.425094$

11.2.4 Everett function

Once we have the density function p and an "initial state" ξ, given $u \in C([0,T])$ we can compute $w_u(t) := [\mathscr{F}(u,\xi)](t)$ by means of (11.5). Based on this feature, Mayergoyz [52] uses another approach for the numerical implementation of the Preisach model that does not require the Preisach function p, but a function E called *Everett function* which describes the effect of p on the hysteresis operator. To obtain the Everett function, the so-called first order *transition curves* are required. To define a first order transition curve, we consider a function $u \in C([0,T])$ such that at time t_0, $u(t_0) \leq -a$. Then, u is monotonically increased until it reaches some value ρ_2' at time t_1. We denote by $w_{\rho_2'} = w_u(t_1)$. A first order transition curve is formed when the above monotonic increase of u is followed by a subsequent monotonic decrease, namely, from ρ_2', u decreases monotonically until it reaches some value ρ_1' at time t_2; let us introduce $w_{\rho_2',\rho_1'} =: w_u(t_2)$ (see Figs. 11.23 and 11.24).

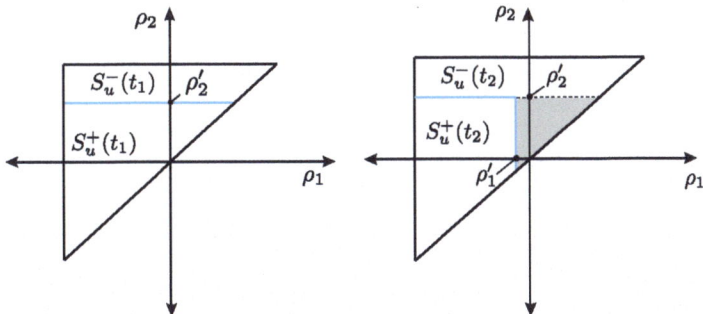

Fig. 11.23. Staircase line $L_u(t)$ at time t_1 (left) and at time t_2 (right)

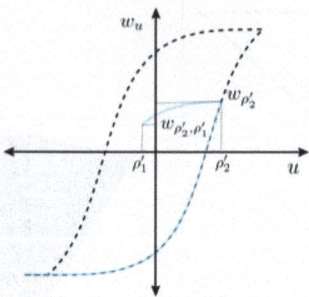

Fig. 11.24. First order transition curve

Then we define the *Everett function* $E : \mathcal{T}_a \to \mathbb{R}$ by

$$E(\rho_1', \rho_2') = \frac{w_{\rho_2'} - w_{\rho_2', \rho_1'}}{2}. \tag{11.10}$$

From (11.9), we notice that

$$w_{\rho_2', \rho_1'} - w_{\rho_2'} = \left(\int_{S_u^+(t_2)} p(\rho)\,d\rho - \int_{S_u^-(t_2)} p(\rho)\,d\rho \right)$$
$$- \left(\int_{S_u^+(t_1)} p(\rho)\,d\rho - \int_{S_u^-(t_1)} p(\rho)\,d\rho \right) = -2 \int_{\mathcal{T}(\rho_1', \rho_2')} p(\rho)\,d\rho,$$

with $\mathcal{T}(\rho_1', \rho_2')$ the triangle such that its hypotenuse is part of the line $\rho_1 = \rho_2$, while the remaining vertex has coordinates (ρ_1', ρ_2'). This is so because $S_u^+(t_2) = S_u^+(t_1) \setminus \mathcal{T}(\rho_1', \rho_2')$ and $S_u^-(t_2) = S_u^-(t_1) \cup \mathcal{T}(\rho_1', \rho_2')$ (see Fig. 11.23, right). Therefore, we obtain the following relation between the Preisach function p and the Everett function

$$E(\rho_1, \rho_2) = \int_{\mathcal{T}(\rho_1, \rho_2)} p(\rho)\,d\rho \qquad \forall (\rho_1, \rho_2) \in \mathcal{T}_a. \tag{11.11}$$

To take into account this relation in the computation of the Preisach operator, first we rewrite (11.9). By adding and subtracting the integral of p over $S_u^+(t)$, the expression (11.9) can be represented in the form

$$w_u(t) = 2 \int_{S_u^+(t)} p(\rho)\,d\rho - \int_{\mathcal{T}_a} p(\rho)\,d\rho.$$

Moreover, from (11.11) and the definition of the limiting triangle, namely, $\mathcal{T}_a = \mathcal{T}(-a, a)$ it follows that,

$$w_u(t) = 2 \int_{S_u^+(t)} p(\rho)\,d\rho - E(-a, a). \tag{11.12}$$

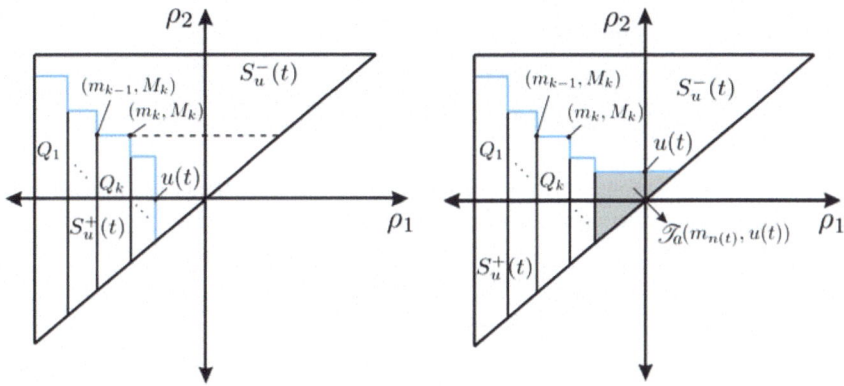

Fig. 11.25. Staircase line for a decreasing input (left) and a increasing input (right)

By assuming that the Preisach function p is known, in order to obtain $w_u(t)$ we can compute the first term on the right-hand side of (11.12). For this purpose we consider two cases: increasing and decreasing arguments. For decreasing arguments, we subdivide $S_u^+(t)$ into n trapezoids $Q_k(t)$ (see Fig. 11.25, left). We can perform this subdivision because, for decreasing arguments, the staircase line $L_u(t)$ intersects the line $\rho_1 = \rho_2$ vertically. Then, by taking into account the wiped-out property we have

$$\int_{S_u^+(t)} p(\rho)\mathrm{d}\rho = \sum_{k=1}^{n(t)} \int_{Q_k(t)} p(\rho)\mathrm{d}\rho, \qquad (11.13)$$

where $n(t)$ is the number of local maxima of u up to time t.

Each trapezoid $Q_k(t)$ depends on the local maximum M_k and on the local minima m_k and m_{k-1}. Notice that, for $k = 1$, m_0 is equal to $-a$. Moreover, each trapezoid can be represented as the difference of two triangles $\mathcal{T}(m_{k-1}, M_k)$ and $\mathcal{T}(m_k, M_k)$, and then

$$\int_{Q_k(t)} p(\rho)\mathrm{d}\rho = \int_{\mathcal{T}(m_{k-1}, M_k)} p(\rho)\mathrm{d}\rho - \int_{\mathcal{T}(m_k, M_k)} p(\rho)\mathrm{d}\rho. \qquad (11.14)$$

Now, from (11.11), it follows that

$$E(m_{k-1}, M_k) = \int_{\mathcal{T}(m_{k-1}, M_k)} p(\rho)\mathrm{d}\rho \quad \text{and} \quad E(m_k, M_k) = \int_{\mathcal{T}(m_k, M_k)} p(\rho)\mathrm{d}\rho.$$

Then, from the latter and (11.14), we rewrite (11.13) in terms of the Everett function as follows

$$\int_{S_u^+(t)} p(\rho)\mathrm{d}\rho = \sum_{k=1}^{n(t)} \left(E(m_{k-1}, M_k) - E(m_k, M_k) \right). \qquad (11.15)$$

Finally, from (11.15) and (11.12), we obtain

$$w_u(t) = 2 \sum_{k=1}^{n(t)} (E(m_{k-1},M_k) - E(m_k,M_k)) - E(-a,a).$$

Since we consider u monotonically decreasing, we obtain that the last minimum value $m_{n(t)}$ is equal to the current value of u, namely, $m_{n(t)} = u(t)$. Then

$$w_u(t) = -E(-a,a) + 2 \sum_{k=1}^{n(t)-1} (E(m_{k-1},M_k) - E(m_k,M_k)) \quad (11.16)$$

$$+ 2 \left(E(m_{n(t)-1},M_{n(t)}) - E(u(t),M_{n(t)}) \right). \quad (11.17)$$

Because of the decomposition of S_u^+ into trapezoids (see Fig. 11.25), this expression is valid only for u being monotonically decreasing. If $u(t)$ is monotonically increasing, then staircase line $L_u(t)$ intersects the line $\rho_1 = \rho_2$ horizontally. Hence, we may decompose S_u^+ into trapezoids and a triangle (see Fig. 11.25, right). It follows that

$$\int_{S_u^+(t)} p(\rho) d\rho = \sum_{k=1}^{n(t)-1} (E(m_{k-1},M_k) - E(m_k,M_k)) + E(m_{n(t)-1},M_{n(t)}). \quad (11.18)$$

In this case, the last maximum value $M_{n(t)}$ is equal to the current value of u, namely, $M_{n(t)} = u(t)$. Hence, from (11.18) we write (11.12) for a monotonically increasing u as follows:

$$w_u(t) = -E(-a,a) + 2 \sum_{k=1}^{n(t)-1} (E(m_{k-1},M_k) - E(m_k,M_k)) + 2E(m_{n(t)-1},u(t)).$$

$$(11.19)$$

From (11.16) and (11.19) we obtain the following expression to compute the Preisach operator in terms of the Everett function

$$w_u(t) = \begin{cases} -E(-a,a) + 2 \sum_{k=1}^{n(t)-1} (E(m_{k-1},M_k) - E(m_k,M_k)) \\ + 2 \left(E(m_{n(t)-1},M_{n(t)}) - E(u(t),M_{n(t)}) \right) \quad \text{for } u \text{ decreasing,} \\ -E(-a,a) + 2 \sum_{k=1}^{n(t)-1} (E(m_{k-1},M_k) - E(m_k,M_k)) \\ + 2E(m_{n(t)-1},u(t)) \quad \text{for } u \text{ increasing.} \end{cases}$$

As an example, we compute $w_u(t)$ by using the Preisach function p given by the Factorized-Lorentzian distribution with parameters $N = 1$, $\beta = 0.8$ and $\gamma = 0.6$ (see Fig. 11.26). Also, the Preisach triangle $\mathcal{T}_a$ is characterized by $a = 5$. Figures 11.27 to 11.29 illustrate two different examples where the $w_u - u$ loop, the final staircase line $L_u(t)$ and the input functions $u(t)$ are represented.

11.2 Mathematical modelling of magnetic hysteresis   237

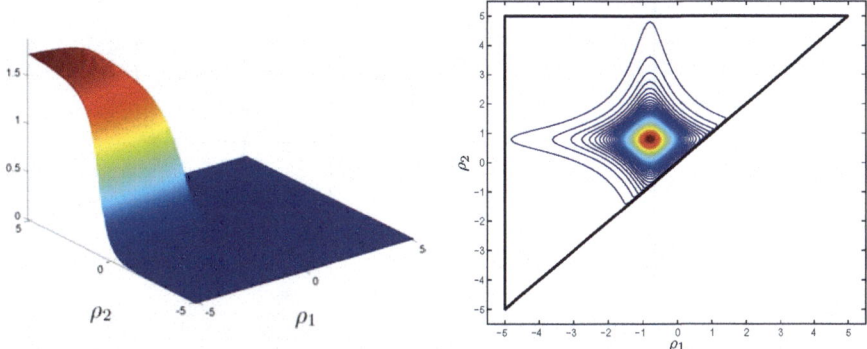

Fig. 11.26. Everett function (left) and isolines of the Factorized-Lorentzian distribution (right)

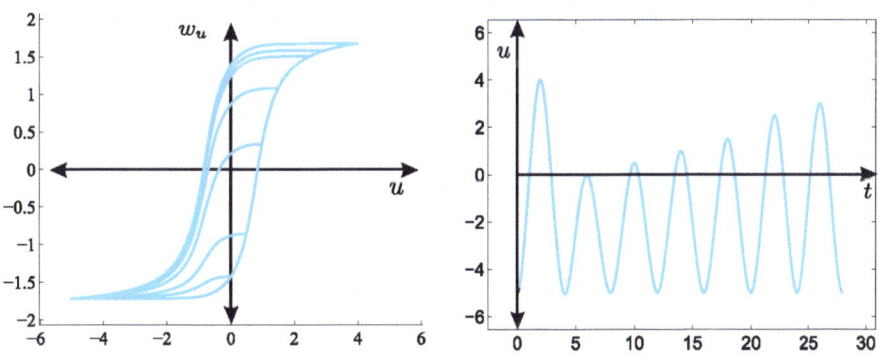

Fig. 11.27. $w_u - u$ curve (left) and input function u (right)

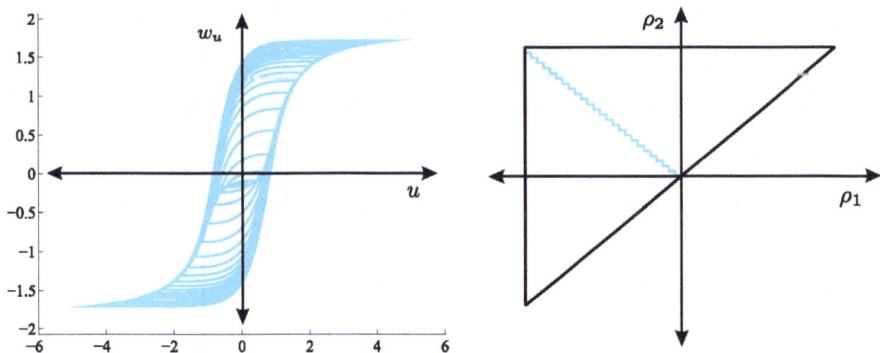

Fig. 11.28. $w_u - u$ curve (left) and staircase function corresponding to the input function u in Fig. 11.29 (right)

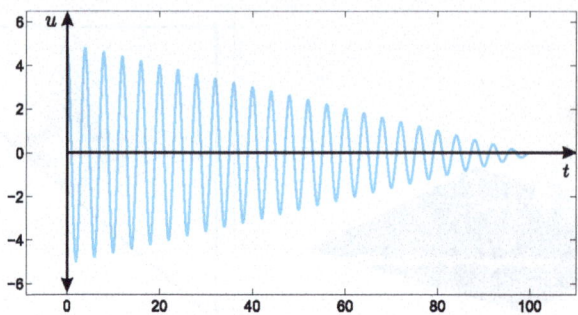

Fig. 11.29. Input function u corresponding to the staircase function in Fig. 11.28 (right)

Remark 11.2. In the previous examples we consider different inputs u, such that $u(0) \geq a$ or $u(0) \leq -a$. Clearly, in both cases $S_u^+(0)$ and $S_u^-(0)$ are determined and because of that there is not need to consider additional information to compute $w_u(t), t \geq 0$. In particular, we have $w_u(0) = E(-a,a)$ if $u(0) \geq 0$, and $-E(-a,a)$ if $u(0) \leq -a$. In the case of $-a < u(0) < a$ in order to compute $w_u(0)$ we must have an "initial state". Depending on this state we obtain different values of $w_u(0)$. For instance, if we consider the "initial states" given by the staircase lines $L_u^1(0)$, $L_u^2(0)$ and $L_u^3(0)$ depicted in Fig. 11.30 (left) then we obtain the different values w_u^1, w_u^2 and w_u^3, respectively (Fig. 11.30, right).

From a practical point of view, the Everett function is given at various points throughout the limiting Preisach triangle $\mathcal{T}_a$ as depicted in Fig. 11.31. The value of the Everett function at each point can be obtained experimentally from the first order transition curves (cf. (11.10)). Using this discretization, the values of the Everett function on $\mathcal{T}_a$ are calculated by interpolation.

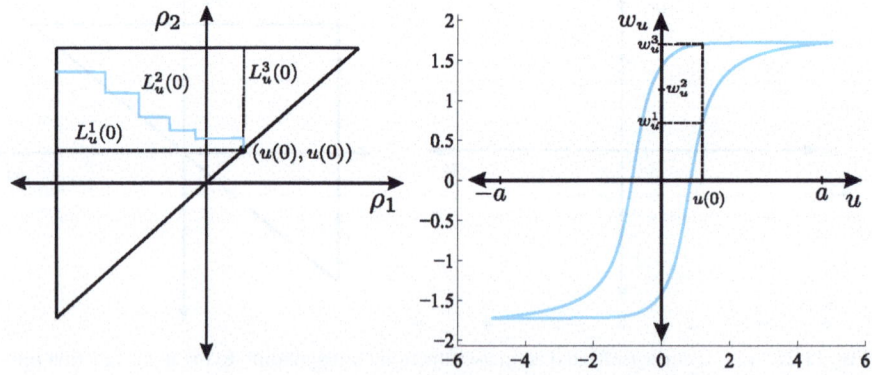

Fig. 11.30. Staircase function (left) and $w_u - u$ curve (right)

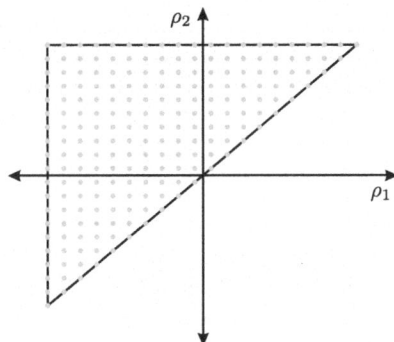

Fig. 11.31. Discretization of the Preisach triangle

11.2.5 Numerical computation of the electromagnetic field: the transient eddy currents model with hysteresis

Let us consider the transient eddy currents model introduced at the beginning of Chap. 10 and suppose we are dealing with ferromagnetic materials with hysteresis. Then we have to replace the linear constitutive equation (10.5) with

$$\mathbf{B} = \mu_0 (\mathbf{H} + \mathbf{M}),$$

where $\mathbf{M}$ is the magnetization field and μ_0 is the magnetic permeability of the vacuum. Moreover, the relation between $\mathbf{M}$ and $\mathbf{H}$ exhibits a history-dependent behavior and must be represented by a suitable constitutive law accounting for hysteresis. We synthetically represent this dependence in the form

$$\mathbf{M} = \vec{\mathscr{F}}(\mathbf{H}),$$

where $\vec{\mathscr{F}}$ is a *vector hysteresis operator* (see [52]). This dependence is nonlocal in time but local in space. If the magnetic behavior is isotropic, $\mathbf{H}$ and $\mathbf{M}$ are aligned at each (x,t) and so the above described Preisach model can be used.

Thus, the constitutive magnetic law writes as follows

$$\mathbf{B} = \mu_0 \left(\mathbf{H} + \vec{\mathscr{F}}(\mathbf{H}) \right), \tag{11.20}$$

which leads to solve a transient nonlinear model which could be written in terms of different unknowns as we have seen in Chap. 10 for the harmonic case. Numerical solution requires as a first step to use a suitable time-discretization scheme (Euler implicit, for instance) and take into account that at time step $n+1$ we know the previous history $(\mathbf{H}^0, \mathbf{B}^0), \cdots, (\mathbf{H}^n, \mathbf{B}^n)$. After time discretization, the semidescrized model involves nonlinear partial differential equations which will be written in terms of $\mathbf{H}$, $\mathbf{A} - V$, etc. depending on the used formulation. In particular, we refer the reader to [66] for the numerical solution of an axisymmetric formulation in terms of the magnetic field which is only defined in the conducting domain.

11.3 Excess eddy current losses. The dynamic Preisach model

Up to now we have considered two mechanisms for dissipating electromagnetic energy into heat: the Joule effect depending on the current density flowing in the material and leading to the so-called *classical eddy current losses*, and the magnetic hysteresis leading to the *hysteresis losses* (see 11.1.2). However, there is experimental evidence that these two mechanisms do not explain the whole electromagnetic losses in a ferromagnetic material. Actually, there are other *dynamic losses* apart from the classical eddy current ones: the so called *excess eddy current losses*. In order to build a mathematical model taking into account these losses, we recall that classical (or static) hysteresis is a *rate-independent* phenomenon so this goal cannot be achieved with a *sensu stricto* hysteresis operator like the one arising from the above studied classical Preisach model. Indeed, let us recall that only the extreme values, maxima and minima, influence future values of the magnetic induction. Thus, the model is not able to consider the speed with which these extreme values occur and then it cannot reflect the dependence with frequency or field waveform. Actually, the classical Preisach model is a quasi-static model in that it calculates the magnetic respponse to an excitation whose frequency approaches zero. Therefore, we need a dynamic model to be able to compute the excess eddy current losses. At present there are several extensions of the classical Preisach model to behave like a dynamic model. They are generically called *dynamic Preisach models*. The study of these models is beyond the scope of this book but the interested reader can find this subject in, for instance, Chap. 2 of [52].

Part III
Numerical Solution of Maxwell's Equations

In this part of the book we will focus on the numerical simulation of electromagnetic problems. According to Parts I and II, they are splitted into five chapters devoted, respectively, to electrical circuits, electrostatics, direct current, magnetostatics and eddy currents problems.

The aims are:

- To illustrate some of the theoretical concepts developed along the book facilitating its understanding by means of its visualization.
- To numerically solve problems with analytical solution so that students can compare both the numerical and the analytical solutions. This is also intended to train the students in assessing the validity and accuracy of the underlying numerical techniques in commercial codes. This is particularly important since the colorful maps produced by the computer can mislead students into trusting all of the results (correct or erroneous) offered by the simulation.
- To numerically solve technological applications or relevant industrial problems whose analytical solution is not possible and that are far away from simple theoretical idealizations.
- To motivate the student showing the power of modelling that allows us to understand and analyze complex electromagnetic systems.

In all cases we use MaxFEM software package to obtain the numerical solution and we also develop the calculus needed to obtain the analytical solution for those problems where we are able to find it.

12
Electrical circuits with MaxFEM

In this chapter we deal with the numerical solution of some linear circuit problems. In particular, we are interested in the determination of voltage drops and currents associated with the circuit elements, given the values of the power sources feeding the system. To do that, the numerical method described in Chap. 3 has been implemented in a Fortran code that allows us to solve linear circuits consisting of several connected components. The code also allows to compute the exchanged currents with the outside in the case they exist. Linear circuits with sources both harmonic and purely transient can be solved.

12.1 Some examples with harmonic source voltage

To begin with, we will consider some examples where the source voltage is a harmonic function of time. In this case, the model of the circuit can be reduced to a complex linear system of equations.

12.1.1 Magnetically coupled circuits. A simple transformer

Two circuits are said to be *magnetically coupled* if the interaction between them takes place through a magnetic field. In that case, there exists a *mutual inductance* between them, since a changing current in one produces, by electromagnetic induction, an electromotive force in the other.

The typical example is a transformer. Basically, a transformer is a device composed of two or more coils linked by magnetic lines of force, used to transfer electrical energy from one circuit to another by altering the voltage. The simplest case consists of two electrical conductors called *primary*, which takes electrical power from the source, and *secondary*, which is the the coil where the current is induced (see Fig. 12.1). An alternating current in the primary winding creates a varying magnetic flux through the secondary winding. This varying magnetic flux induces a current in the secondary winding.

A. Bermúdez, D. Gómez, P. Salgado: *Mathematical Models and Numerical Simulation in Electromagnetism.* UNITEXT – La Matematica per il 3+2 74
DOI 10.1007/978-3-319-02949-8_12, © Springer International Publishing Switzerland 2014

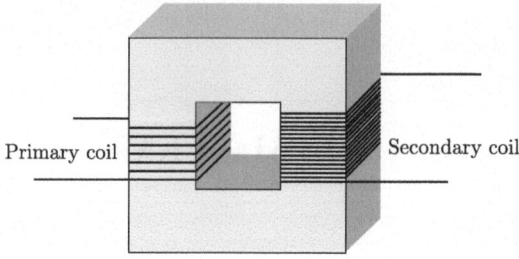

Fig. 12.1. Sketch of a typical step up transformer

In an ideal transformer, the ratio of the primary and secondary voltages is equal to the ratio of the number of turns in their windings, that is, the voltage per turn is the same for both windings. This leads to the most common use of a transformer: to convert electrical energy at one voltage to energy at a different voltage by means of windings with different numbers of turns. For instance, in a *step up transformer* the number of turns in secondary coil is larger than the number of turns in primary coil, thus increasing the voltage from the primary to the secondary. The opposite case is the *step down transformer*s as the ones commonly used in electrical power plants and distribution systems. Indeed, because electric power often has to travel long distances, it is transformed or "stepped down" repeatedly into more manageable states, as it gets closer to its destination.

The physical behavior of a transformer may be represented by an equivalent circuit model which can be more or less complicated depending on the transformer properties. For simplicity, in this example we will consider the circuit depicted in Fig. 12.2. The primary winding is connected to a source voltage providing an alternating current $E(t) = 50\cos(t)$ V. This winding induces current in the secondary one in such a way that the inductance matrix is given by

$$\mathscr{L}^S = \begin{pmatrix} 1 & 10^2 \\ 10^2 & 10^4 \end{pmatrix}.$$

As we have seen in Chap. 3, the system of equations resulting from the application of Kirchhoff's voltage and current laws depends on the graph of the circuit, that is to

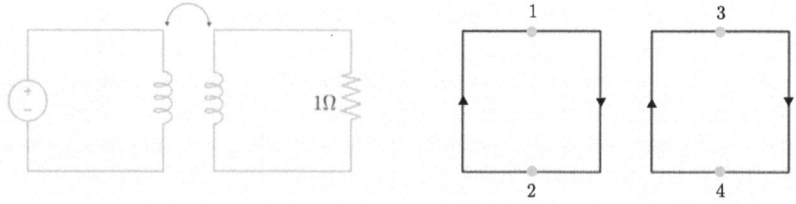

Fig. 12.2. Electric circuit associated to the transformer (left) and its graph (right)

12.1 Some examples with harmonic source voltage

Table 12.1. Data for circuit depicted in Fig. 12.2

Edge	From node	To node	Element type	Value (Unit)
1	2	1	Power generator	$E(t) = 50\cos(t)$ (V)
2	3	4	Resistance	1 (Ω)
3	1	4	Magnetically coupled inductor	$\mathscr{L}^S$
4	4	3	Magnetically coupled inductor	$\mathscr{L}^S$

say, on the way the circuit elements are interconnected. Consequently, the first thing to do is to identify the graph of the circuit, which has been sketched in Fig. 12.2.

For the sake of clarity, we have chosen to represent the circuits topology with directed graphs or digraphs in which all of its edges indicate a direction from one node to another. The directions can be arbitrarily chosen. Moreover, and as stated in Chap. 3, all the graphs have been made in accordance with the fact that each edge only can contain one circuit element.

In Table 12.1 we have summarized all the data concerning this circuit. In particular, the first three columns indicate the number of each edge and the nodes it connects. The other data in the row correspond to the type of element which is present in that edge and its value. In particular, we will assume that the power generators appearing along this chapter do not have internal resistance.

Let us recall that the system to solve is (see Sect. 3.4)

$$\mathscr{D}(\omega)\widetilde{\mathbb{I}} + \mathscr{A}^t \widetilde{\mathbb{V}} = \widetilde{\mathbb{E}},$$
$$\mathscr{A}\widetilde{\mathbb{I}} = -\widetilde{\psi}.$$

In this case, $\omega = 1$, $\widetilde{\psi} = \mathbf{0}$ and $\widetilde{\mathbb{E}}$ has null components except for the first one which is equal to 50. Moreover,

$$\mathscr{D}(\omega) = \begin{pmatrix} 0 & 0 & 0 & 0 \\ 0 & 1 & 0 & 0 \\ 0 & 0 & 0 & 0 \\ 0 & 0 & i & 10^2 i \\ 0 & 0 & 10^2 i & 10^4 i \end{pmatrix} \text{ and } \mathscr{A}^t = \begin{pmatrix} 1 & -1 & 0 & 0 \\ 0 & 0 & -1 & 1 \\ -1 & 1 & 0 & 0 \\ 0 & 0 & 1 & -1 \end{pmatrix}.$$

By replacing these values in the system above we obtain the following complex linear system:

$$i\widetilde{\mathbb{I}}_p + 10^2 i\widetilde{\mathbb{I}}_s = 50,$$
$$10^2 i\widetilde{\mathbb{I}}_p + (10^4 i + 1)\widetilde{\mathbb{I}}_s = 0,$$

where $\widetilde{\mathbb{I}}_p$ and $\widetilde{\mathbb{I}}_s$ denote the complex amplitudes of the currents traversing the primary and the secondary windings, respectively. This system can be analytically solved. By doing so, we obtain that the loop currents $I_p(t) = \text{Re}(\widetilde{\mathbb{I}}_p e^{i\omega t})$ and $I_s(t) = \text{Re}(\widetilde{\mathbb{I}}_s e^{i\omega t})$

Table 12.2. Intensity phasor values per edge for circuit in Fig. 12.2

Edge	Intensity phasor (A)
1	500000 − 50i
2	−5000 + 0i
3	500000 − 50i
4	−5000 + 0i

Table 12.3. Voltage values per node for circuit in Fig. 12.2

Node	Voltage (V)
1	0
2	−50
3	0
4	5000

are given by:

$$I_p(t) = 500000 \cos\left(t - \frac{7\pi}{219000}\right),$$
$$I_s(t) = -5000 \cos(t).$$

Let us compare these solutions with the ones obtained by numerically solving the system. They have been summarized in Tables 12.2 and 12.3.

In Fig. 12.3 we have represented, from top to bottom, the source voltage, the intensities traversing each edge and the voltages per node obtained from the numerical computation. We have also represented the exact solutions I_p and I_s in order to compare them with the numerical ones. As expected, the intensities I_1 and I_3 traversing the primary winding agree with I_p; the same occurs with currents I_2 and I_4 traversing the secondary winding which agree with I_s. Notice also that this circuit acts as a step up transformer since the voltage supplied to the primary winding is multiplied by 10^2 in the secondary.

12.1.2 A circuit with three connected components

In this section we will consider the circuit illustrated in Fig. 12.4. It consists of three connected components each containing different circuit elements. The graph we have considered to solve the problem has been depicted in Fig. 12.5.

Similar to the previous example, we have summarized in Table 12.4 the edges, the nodes they connect, the element type appearing in the edges and the physical parameters assigned to them. Notice that in this case we have two sets of magnetically coupled sub-circuits: the first one involving edges 6 and 9, and the second one involving edges 8 and 12 (see Table 12.4 and Figs. 12.4 and 12.5). The corresponding

12.1 Some examples with harmonic source voltage

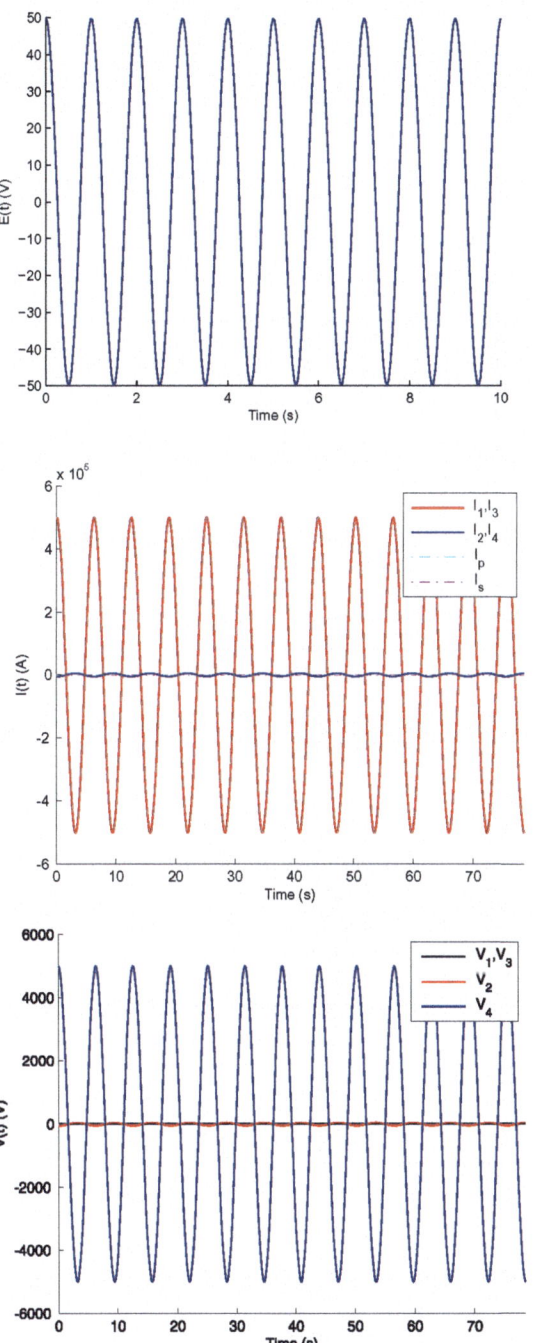

Fig. 12.3. From top to bottom, source voltage, intensity per edge and voltage per node for the circuit of Fig. 12.2

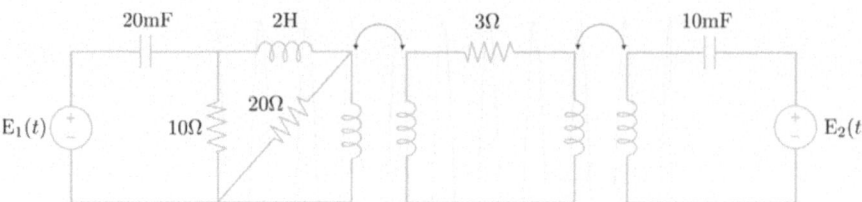

Fig. 12.4. Circuit for example in Sect. 12.1.2

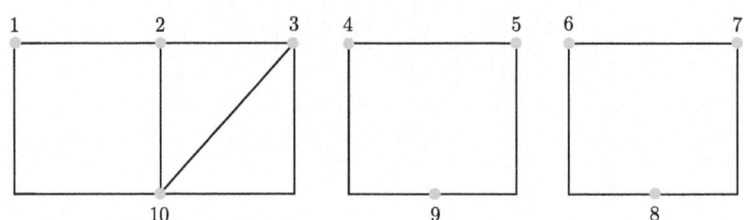

Fig. 12.5. Graph considered for the circuit in Fig. 12.4

Table 12.4. Data for the circuit depicted in Fig. 12.4

Edge	From node	To node	Element type	Value (Unit)
1	1	2	Capacitor	0.02 (F)
2	2	10	Resistance	10 (Ω)
3	10	1	Power generator	$E_1(t) = 4\cos(5t - \frac{\pi}{6})$ (V)
4	2	3	Inductor	2 (H)
5	3	10	Resistance	20 (Ω)
6	3	10	Magnetically coupled inductor	$\mathscr{L}_1^S$
7	4	5	Resistance	3 (Ω)
8	5	9	Magnetically coupled inductor	$\mathscr{L}_2^S$
9	9	4	Magnetically coupled inductor	$\mathscr{L}_1^S$
10	6	7	Capacitor	0.01 (F)
11	7	8	Power generator	$E_2(t) = 2\cos(5t)$ (V)
12	8	6	Magnetically coupled inductor	$\mathscr{L}_2^S$

inductance matrices are given by

$$\mathscr{L}_1^S = \begin{pmatrix} 2 & 3 \\ 3 & 6 \end{pmatrix} \quad \text{and} \quad \mathscr{L}_2^S = \begin{pmatrix} 6 & 3 \\ 3 & 2 \end{pmatrix}.$$

The frequency of the harmonic signals is 0.79 Hz. Figure 12.6 shows the voltage supplied by the power generators placed on edges 3 and 11, the intensities per edge and the voltages per node along the time. The corresponding values have been summarized in Table 12.5 and Table 12.6, respectively.

12.1 Some examples with harmonic source voltage

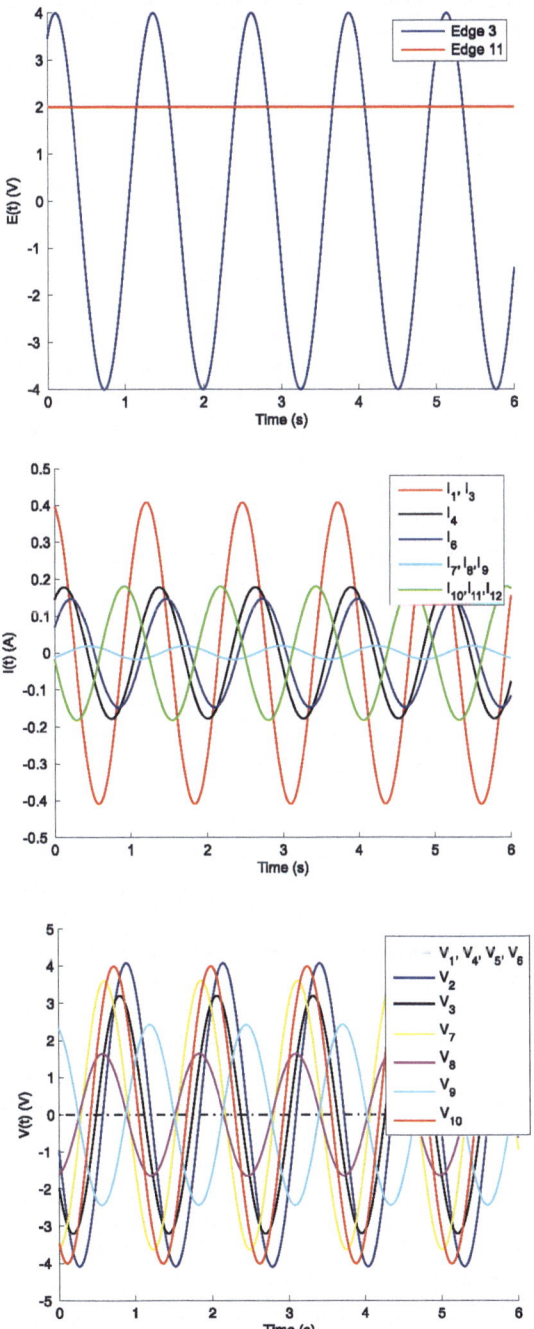

Fig. 12.6. From top to bottom, source voltage, intensity per edge and voltage per node for the circuit of Fig. 12.4

Table 12.5. Intensity phasor per edge for circuit in Fig. 12.4

Edge	Intensity phasor (A)
1	$0.397 + 0.094i$
2	$0.252 + 0.197i$
3	$0.397 + 0.094i$
4	$0.145 - 0.103i$
5	$0.075 + 0.026i$
6	$0.071 - 0.129i$
7	$-0.012 - 0.013i$
8	$-0.012 - 0.013i$
9	$-0.012 - 0.013i$
10	$-0.019 + 0.180i$
11	$-0.019 + 0.180i$
12	$-0.019 + 0.180i$

Table 12.6. Voltage phasor per node for circuit in Fig. 12.4

Node	Voltage (V)
1	$0.000 + 0.000i$
2	$-0.943 + 3.973i$
3	$-1.973 + 2.521i$
4	$0.000 + 0.000i$
5	$0.000 + 0.000i$
6	$0.000 + 0.000i$
7	$-3.600 - 0.371i$
8	$-1.600 - 0.371i$
9	$2.336 + 0.689i$
10	$-3.464 + 2.000i$

12.1.3 Three-phase alternating current

Circuits in which the alternating current sources operate at the same frequency and amplitude but different phases are known as *polyphase*. In particular, a three-phase system is produced by a generator consisting of three sources having the same amplitude and frequency but out of phase with each other by 120°. It is the most prevalent and most economical polyphase system. Electric power, for instance, is generated, transmitted and distributed in three-phase form. In the aluminium industry, where 48 phases are required for melting purposes, they can be provided by manipulating a three phases supplier (see [1]). Moreover, when single-phase power is required, it can be merely tapped off from a basic three-phase system, which is cheaper. Other advantage is that the instantaneous power in a three-phase system can be constant (not pulsating) and this results in uniform power transmission and less vibrations in the feeded device.

12.1 Some examples with harmonic source voltage

In this section we will consider the three-phase system depicted in Fig. 12.7. As in the previous sections, we will compute the currents traversing the edges and the voltage drops between the nodes of the system, assuming the circuit element values specified in Table 12.7. The graph we have considered is shown in Fig. 12.8.

In Fig. 12.3 we have represented, from top to bottom, the source voltage, the intensities traversing each edge and the voltages per node obtained from the numerical computation. Notice that, in particular, the source voltages have the same amplitude but they are out of phase by 120°. The frequency of the harmonic signals is 1 Hz.

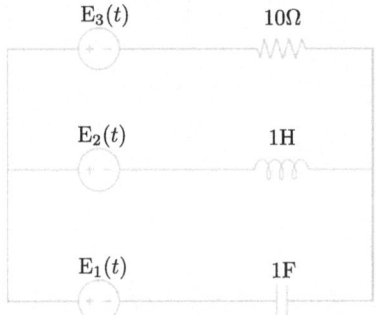

Fig. 12.7. A three-phase system

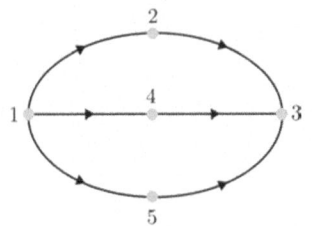

Fig. 12.8. Graph considered for the three-phase system in Fig. 12.7

Table 12.7. Data for the circuit depicted in Figure 12.7

Edge	From node	To node	Element type	Value (Unit)
1	1	2	Power generator	$E_1(t) = 200\cos(2\pi t)$ (V)
2	2	3	Resistance	10 (Ω)
3	1	4	Power generator	$E_2(t) = 200\cos(2\pi t - \frac{2\pi}{3})$ (V)
4	4	3	Inductor	1 (H)
5	1	5	Power generator	$E_3(t) = 200\cos(2\pi t + \frac{2\pi}{3})$ (V)
6	5	3	Capacitor	1 (F)

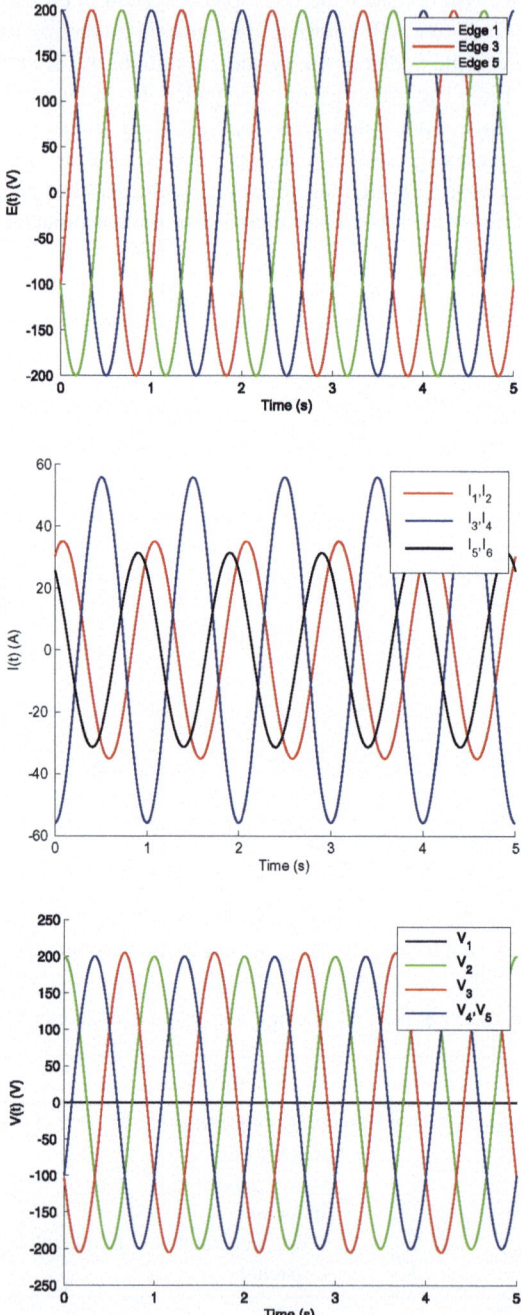

Fig. 12.9. From top to bottom, source voltage, intensity per edge and voltage per node for the circuit in Fig. 12.7

Table 12.8. Intensity phasor per edge for the circuit depicted in Fig. 12.7

Edge	Intensity phasor (A)
1	30.289 − 17.726i
2	30.289 − 17.726i
3	−55.779 − 0.461i
4	−55.779 − 0.461i
5	25.489 + 18.187i
6	25.489 + 18.187i

Table 12.9. Voltage values per node for the circuit depicted in Fig. 12.7

Node	Voltage (V)
1	0.000 + 0.000i
2	200.000 + 0.000i
3	−102.895 + 177.262i
4	−100.000 − 173.205i
5	−100.000 + 173.205i

12.2 Some examples with transient source voltage

In this section we will consider some examples where the power source is purely transient. The same electrical circuit is supplied with different power sources. In particular, the cases of a linear voltage source, a rectangular pulse train, a sawtooth wave and PWM (*Pulse Width Modulation*) excitations have been considered.

Let us consider a simple circuit as the one depicted in Fig. 12.10. As associated graph we have considered the one in Fig. 12.11. The characteristics associated to the circuit elements have been summarized in Table 12.10. As usual, the first five columns indicate the number of each edge, the nodes it connects, the type of element in that edge and its value. Moreover, as we deal with a transient problem, we need to know the initial values of the intensity in the coils and the capacitor and the initial charge in the capacitor, which have been specified in the two last columns. The

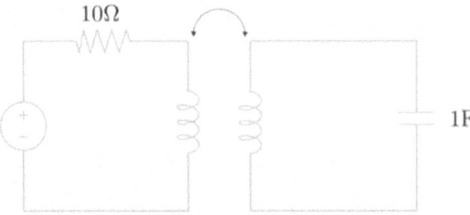

Fig. 12.10. Electrical circuit for Example 12.2

254   12 Electrical circuits with MaxFEM

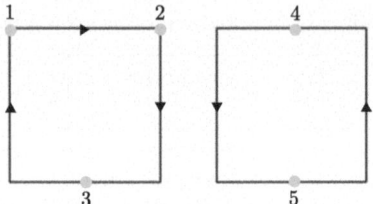

Fig. 12.11. Graph associated to the circuit in Fig. 12.10

Table 12.10. Data for circuit depicted in Figure 12.10

Edge	From node	To node	Element type	Value (Unit)	I(0) (A)	Q(0) (C)
1	1	2	Resistance	10 (Ω)		
2	2	3	Magnetically coupled inductor	$\mathscr{L}^S$	0	
3	3	1	Power generator	$E(t)$ (V)		
4	4	5	Magnetically coupled inductor	$\mathscr{L}^S$	0	
5	5	4	Capacitor	1 (F)	0	0

inductance matrix is given by

$$\mathscr{L} = \begin{pmatrix} 5 & 1 \\ 1 & 3 \end{pmatrix}.$$

Firstly, let us consider that the circuit is supplied with a voltage source which grows linearly from 0 to 100 in 10 seconds. It has been depicted in Fig. 12.12 together with the resulting intensity current. In the next example, a sawtooth wave is applied. In Fig. 12.13 we have represented both the wave and the resulting current. Next, Fig. 12.14 shows the obtained current under a rectangular pulse train excitation. In all these cases, a time discretization of the interval $[0, 20]$ (seconds) with 2000 time steps has been considered. Finally, a PWM voltage as the ones that usually feed electrical machines has been employed; these kind of signals are discontinuous with a large number of discontinuities in each period, thus requiring the use of very small time steps and thereby a greater computational effort. In this case, we have considered 2000 time steps in the interval $[0, 0.08]$ (four cycles). The signal and the resulting current in the circuit are shown in Fig. 12.15.

12.2.1 Results for the linear voltage source

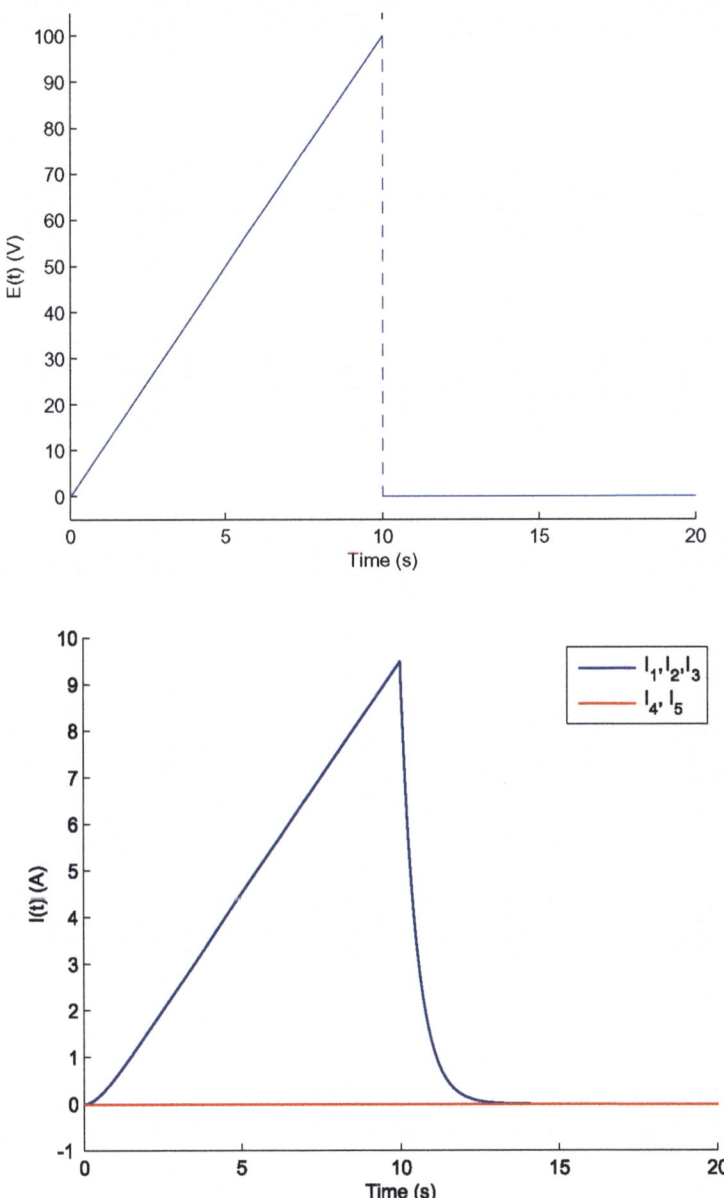

Fig. 12.12. Linear voltage source (top) and resulting current intensity (bottom) traversing the circuit in Fig. 12.10

12.2.2 Results for the sawtooth wave

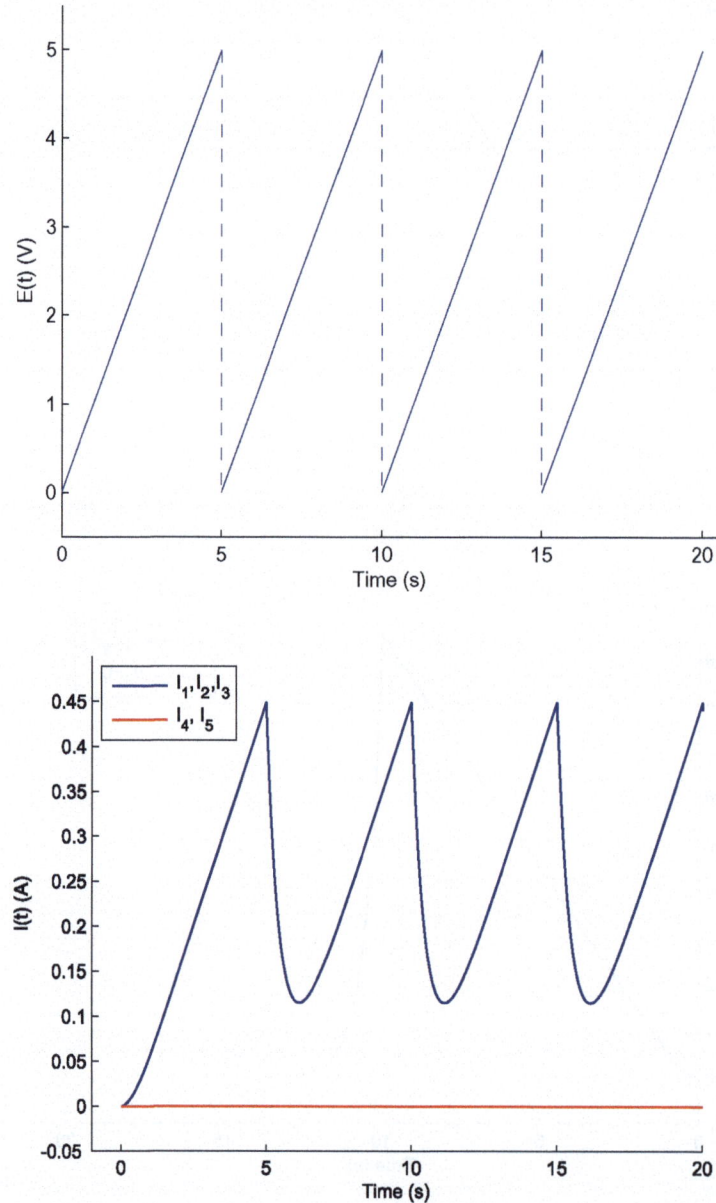

Fig. 12.13. Linear voltage source (top) and resulting current intensity (bottom) traversing the circuit in Fig. 12.10

12.2.3 Results for the rectangular pulse train

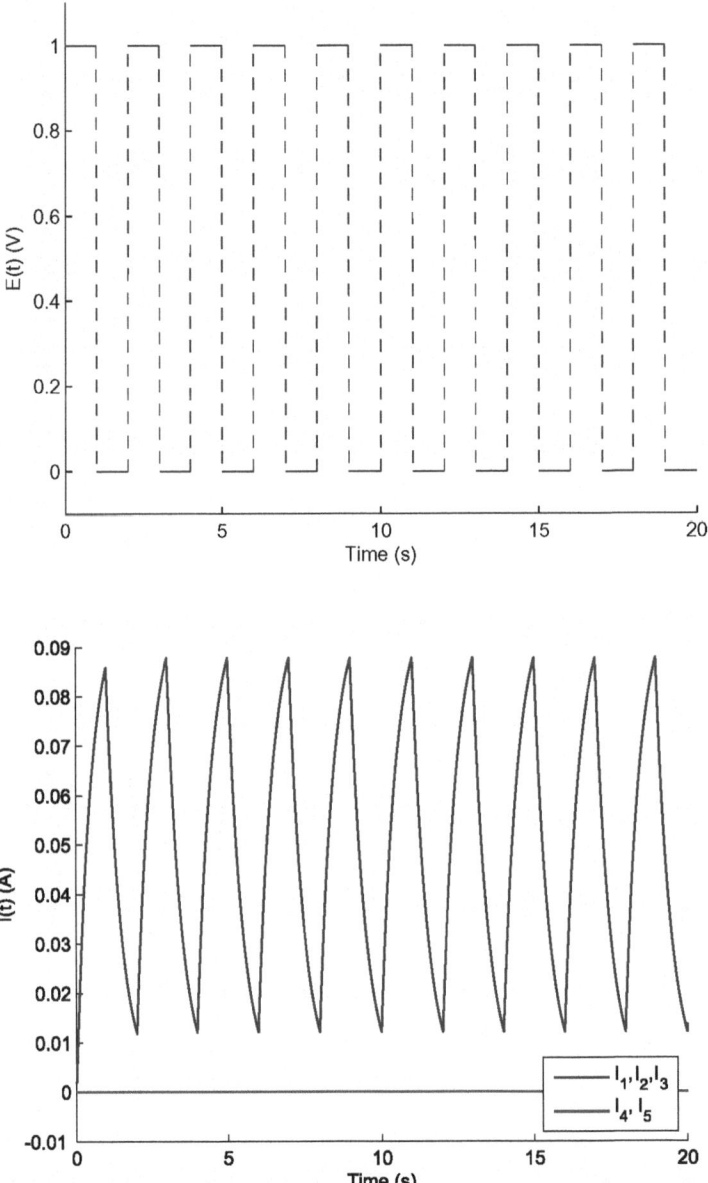

Fig. 12.14. Rectangular pulse train (top) and resulting current intensity (bottom) traversing the circuit in Fig.12.10

12.2.4 Results for the PWM signal

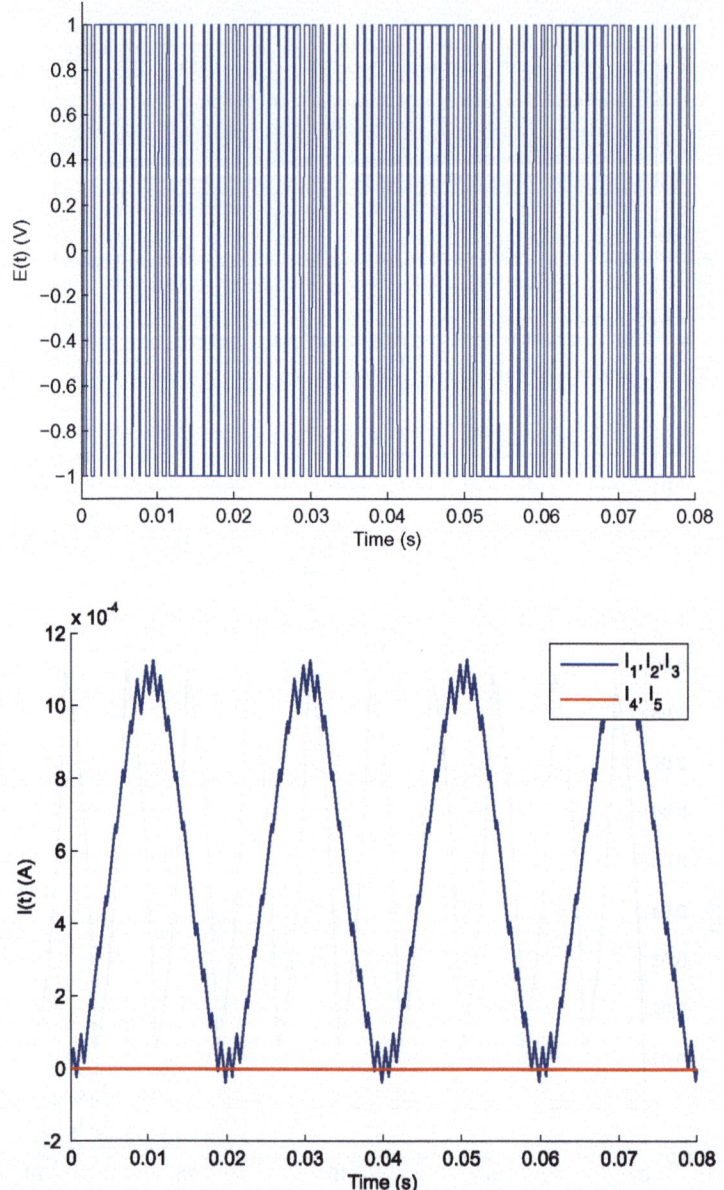

Fig. 12.15. PWM source (top) and resulting current intensity (bottom) traversing the circuit in Fig. 12.10

12.3 Some other typical configurations

In this section we will present some other examples corresponding to typical configurations in electrical circuits.

12.3.1 Bridge circuit

There exist a wide variety of bridge circuits for measuring all kinds of electrical values, for example the impedance of components. In this example we will consider the circuit proposed in Problem 3.3. In Fig. 12.16 we have depicted a digraph associated to this circuit. The element values have been summarized in Table 12.11 below, 1 Hz being the operating frequency. The results obtained from the numerical simulation have been represented in Fig. 12.17. Notice that, as expected, the current intensity is the same along the edges 1 and 7, and that the current along edge 2 (respectively, edge 3) is very similar to the one along edge 5 (respectively, edge 6). Finally, the current along the central edge (edge 4) is small in comparison to the ones along the other edges.

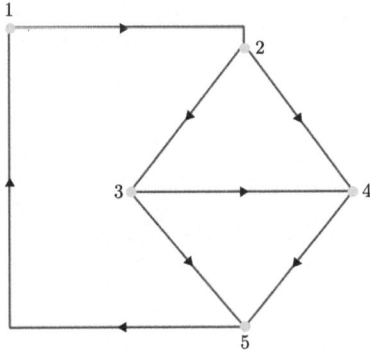

Fig. 12.16. Graph associated to the circuit of Fig. 3.3

Table 12.11. Data for the circuit depicted in Fig. 3.3

Edge	First node	Second node	Element type	Value (Unit)
1	1	2	Resistance	10 (Ω)
2	2	3	Resistance	5 (Ω)
3	2	4	Resistance	3 (Ω)
4	3	4	Resistance	1 (Ω)
5	3	5	Resistance	7 (Ω)
6	4	5	Resistance	4 (Ω)
7	5	1	Power generator	$E(t) = 15\cos(2\pi t)$ (V)

Table 12.12. Intensity phasor per edge for the circuit depicted in Fig. 3.3

Edge	Intensity phasor (A)
1	1.040
2	0.389
3	0.651
4	0.010
5	0.379
6	0.661
7	1.040

Table 12.13. Voltage values per node for the circuit depicted in Fig. 3.3

Node	Voltage (V)
1	0.000
2	−10.402
3	−12.346
4	−12.346
5	−15.000

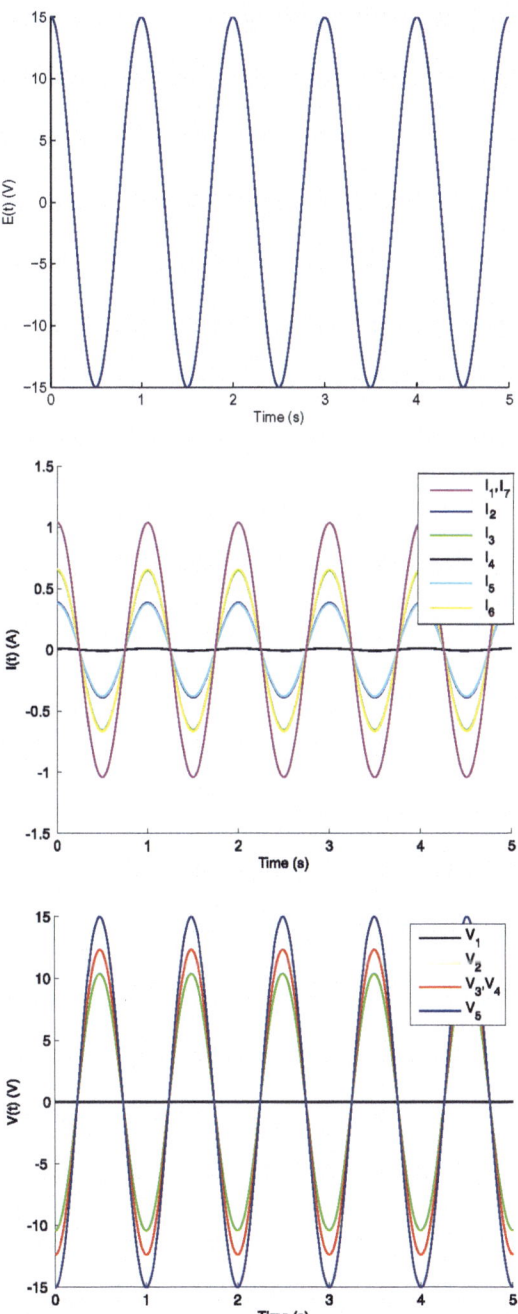

Fig. 12.17. From top to bottom, source voltage, intensity per edge and voltage per node for the circuit in Fig. 3.3

12.3.2 Star delta connection circuit

In this section we will consider the star delta connection circuit introduced in Problem 3.5 and represented in Fig. 3.5. In Fig. 12.18 we have depicted a possible associated digraph. The element values have been summarized in Table 12.14 below. The operating frequencey is 1 Hz, and the results obtained from the numerical simulation have been represented in Figs. 12.19 and 12.20. In this case we have considered non-null input current in certain circuit nodes; more precisely, we suppose that a current of complex amplitud 2 A enters into the circuit through the first node and that a current of complex amplitude 1 A exits the circuit through the third node, both having a frequency of 1 Hz. In Fig. 12.20 we can see that this causes a current of complex amplitude 1 A to flow outside the circuit through the fourth node.

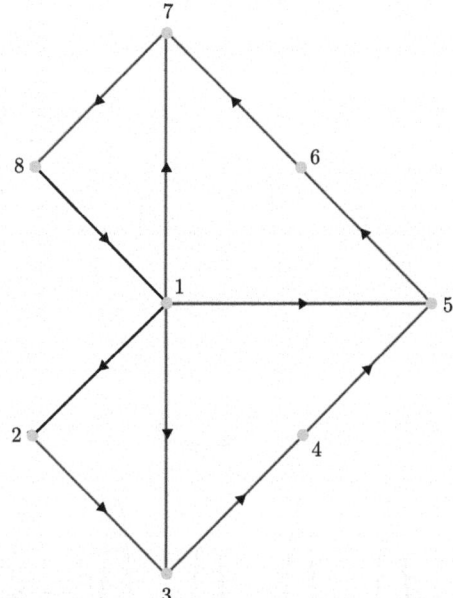

Fig. 12.18. Graph associated to the circuit of Fig. 3.5

12.3 Some other typical configurations 263

Table 12.14. Data for circuit depicted in Fig. 3.5

Edge	From node	To node	Element type	Value (Unit)
1	1	2	Resistance	5 (Ω)
2	2	3	Power generator	$E_1(t) = 12\cos(2\pi t)$ (V)
3	3	4	Power generator	$E_2(t) = 6\cos(2\pi t)$ (V)
4	4	5	Resistance	5 (Ω)
5	5	6	Power generator	$E_3(t) = 9\cos(2\pi t)$ (V)
6	6	7	Resistance	5 (Ω)
7	7	8	Resistance	5 (Ω)
8	8	1	Power generator	$E_4(t) = 10\cos(2\pi t)$ (V)
9	1	7	Resistance	5 (Ω)
10	1	5	Resistance	5 (Ω)
11	1	3	Power generator	$E_5(t) = 4\cos(2\pi t)$ (V)

Table 12.15. Intensity phasor per edge for circuit depicted in Fig. 3.5

Edge	Intensity phasor (A)
1	1.600
2	1.600
3	1.950
4	1.950
5	1.900
6	1.900
7	1.950
8	1.950
9	−0.050
10	−0.050
11	1.350

Table 12.16. Voltage values per node for circuit depicted in Fig. 3.5

Node	Voltage (V)
1	0.000
2	−8.000
3	4.000
4	10.000
5	0.250
6	9.250
7	− 0.250
8	−10.000

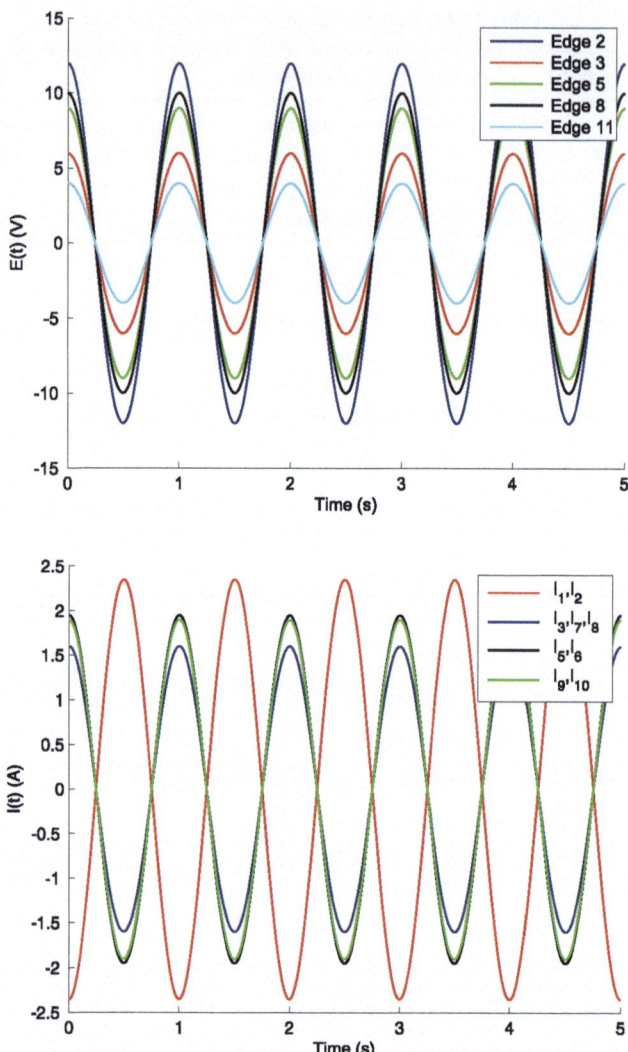

Fig. 12.19. Voltage source and intensity currents for circuit in Fig. 3.5

12.3 Some other typical configurations 265

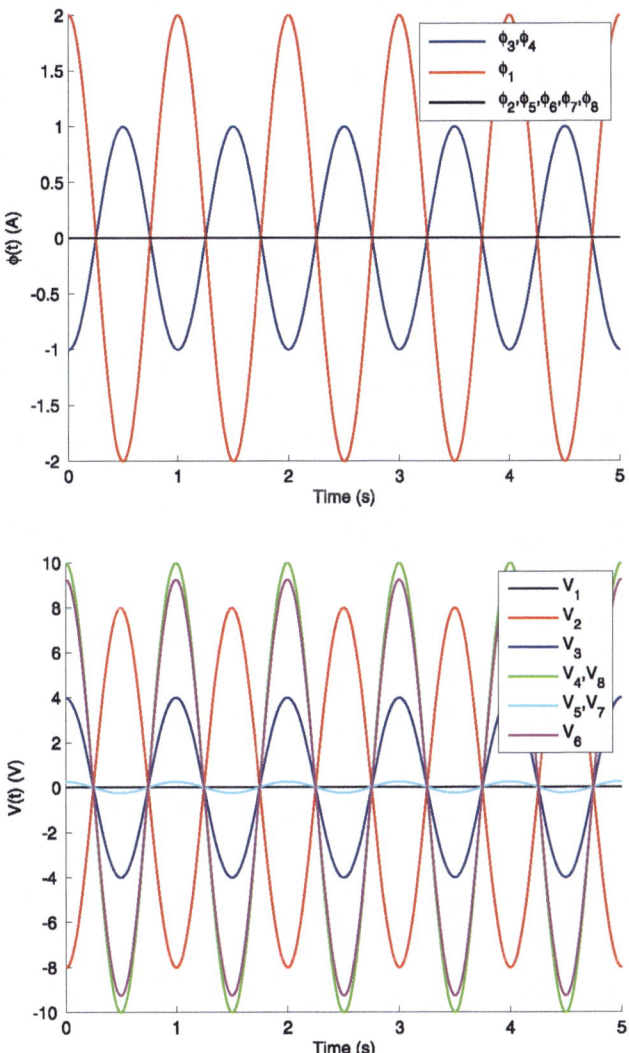

Fig. 12.20. Intensity currents exchanged with the outside and voltage per node for circuit in Fig. 3.5

13
Electrostatics with MaxFEM

In this chapter we deal with the numerical solution of some electrostatics problems by using MaxFEM. The aim is to show a set of numerical simulations to illustrate the concepts of electric field and electric potential. The chapter begins with a review of some classical problems with analytical solution appearing in the vast majority of textbooks. These are limited to situations where the geometry is especially simple and maps onto a standard coordinate system, more specifically, cartesian, cylindrical or spherical coordinates. Some of these analytical examples deal with situations that are never encountered in practice (for instance, those where the geometry is "infinite" in some direction). However, they appear in the classical literature in electromagnetism since they are excellent approximations to some real systems. The second part proposes problems oriented to physical applications where a numerical approach is necessary or simpler than an analytical one. We refer the reader to Chap. 7 and Sect. 7.4 for information about the mathematical model and the weak formulation of electrostatics problems, respectively.

13.1 Some classical problems with analytical solution

As mentioned in Chap. 7, electrostatics problems are essentially of two types: problems stated in the whole space and boundary-value problems. For problems lying on the the first type, a specification of the charge in the whole space is needed and the problem of finding the electric field reduces to compute a convolution integral (see Sects. 5.1 and 7.1).

In this section we deal with the numerical solution of some classical examples for which the analytical solution can be computed.

A. Bermúdez, D. Gómez, P. Salgado: *Mathematical Models and Numerical Simulation in Electromagnetism.* UNITEXT – La Matematica per il 3+2 74
DOI 10.1007/978-3-319-02949-8_13, © Springer International Publishing Switzerland 2014

13.1.1 Point charge at the origin

Let us numerically compute the electrostatic potential V and the electric field **E** generated by a positive point charge $Q = 1.\text{e-}10$ C located at the origin of coordinates. As stated in Sect. 5.1.2, the electromagnetic field can be analytically deduced from Gauss' law and it is given by

$$\mathbf{E}(x) = \frac{1}{4\pi\varepsilon_0} \frac{Q}{r^2(x)} \mathbf{e}_r(x).$$

Thus, the electric field intensity of a point charge is a vector field directed along imaginary straight lines emerging from the charge and directed away from it or toward it depending on whether the charge is positive or negative. We have also computed the electrostatic potential (see (5.5)) which is given by

$$V(x) = \frac{1}{4\pi\varepsilon_0} \frac{Q}{r(x)}.$$

In order to use the finite element method we need to construct a bounded domain containing the charge and large enough as to assume that the potential on the surface of this domain is equal to zero. In this case, we have considered a sphere centered at $(0,0,0)$ and radius 1 meter.

Figure 13.1 left shows the electric field which, as expected, is radial outward from the positive charge. The equipotential surfaces are concentric spherical shells. The circles in Fig. 13.1 right represent equipotential lines. Finally, Fig. 13.2 illustrates the variation of the modulus of the electric field and the potential depending on the distance to the charge.

Table 13.1. Data for the numerical test

Radius of the computational domain:	1 m
Point charge density, Q_p:	1.e-10 C
Electric permittivity, ε_0:	8.8542e-12 F/m

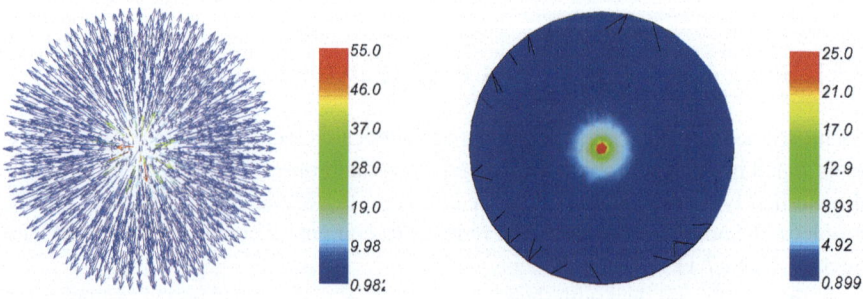

Fig. 13.1. Electric field created by a positive point charge (left) and equipotential regions (right)

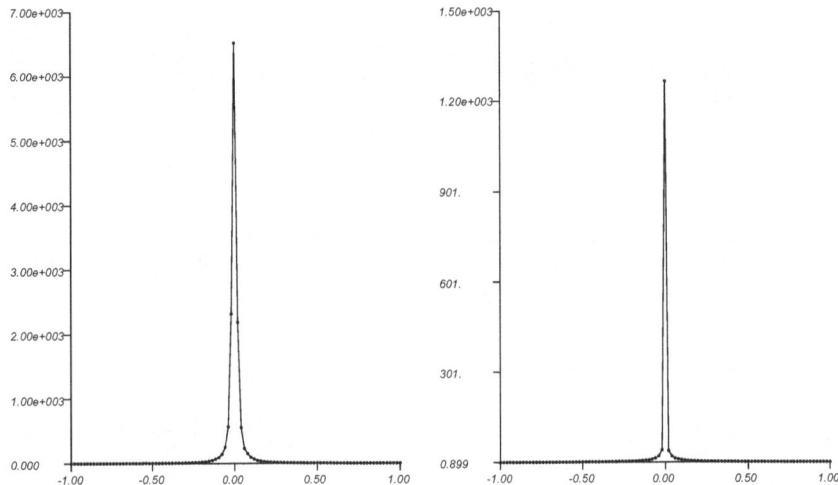

Fig. 13.2. Point charge located at the origin. Modulus of the electric field (left) and potential (right) along the diameter of the computational domain

13.1.2 Electric field of n point charges

In a similar way as we have done for a single point charge, we can determine the electrostatic potential and the electric field due to an arbitrary distribution of charges $Q_1 \ldots Q_n$. In free space, the superposition principle applies and then we have that the potential at a point x located at distance $r_k(x)$ from charge Q_k, for $k = 1, \ldots, n$ is given by

$$V(x) = \sum_{k=1}^{n} \frac{Q_k}{4\pi\varepsilon_0 r_k(x)}.$$

In a similar way the electric field due to n point charges is simply the superposition of the electric fields due to each point charge, i.e.,

$$\mathbf{E}(x) = \sum_{k=1}^{n} \mathbf{E}_k(x) = \frac{1}{4\pi\varepsilon_0} \sum_{k=1}^{n} \frac{Q_k}{|\mathbf{r}(x) - \mathbf{r}_k|^2} \frac{\mathbf{r}(x) - \mathbf{r}_k}{|\mathbf{r}(x) - \mathbf{r}_k|},$$

where $\mathbf{r}_k$, $k = 1, \ldots, n$ is the position vector of charge Q_k.

To illustrate this fact, let us determine the electrostatic field and the electric potential created by a pair of electric charges of equal magnitude but opposite sign located at the x_1-axis. The computational domain is an ellipsoid (see Fig. 13.3) centered at the origin of the coordinate system and aligned with the axes, where the length of its principal semi-axes a, b and c are specified in Table 13.2. A null Dirichlet boundary condition has been imposed on the ellipsoid surface.

Let us consider two point charges $-Q = 10^{-10}$ C and $-Q = -10^{-10}$ C located at points $p = (2,0,0)$ and $q = (-2,0,0)$. Figures 13.4 and 13.5 show the electric field and the equipotential surfaces, respectively. Finally, Fig. 13.6 shows the potential on the x_1-axis.

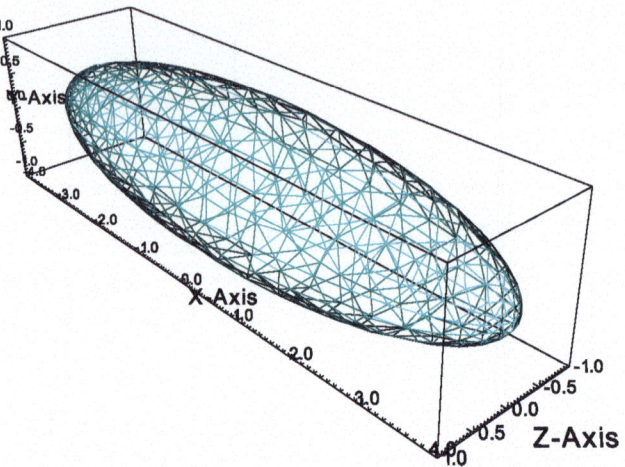

Fig. 13.3. Two opposite charged points. Computational domain

Table 13.2. Two point charges. Data for the numerical test

Length of the semi-axis a:	4 m
Length of the semi-axis b:	1 m
Length of the semi-axis c:	1 m
Point charge value, Q_p:	$\pm$ 1.e-10 C
Location of point charge Q:	(2,0,0)
Location of point charge $-Q$:	(-2,0,0)
Vacuum electric permittivity, ε_0:	8.8542e-12 F/m

Fig. 13.4. Two opposite charged points. Electric field

13.1 Some classical problems with analytical solution   271

Fig. 13.5. Two opposite charged points. Equipotential surfaces

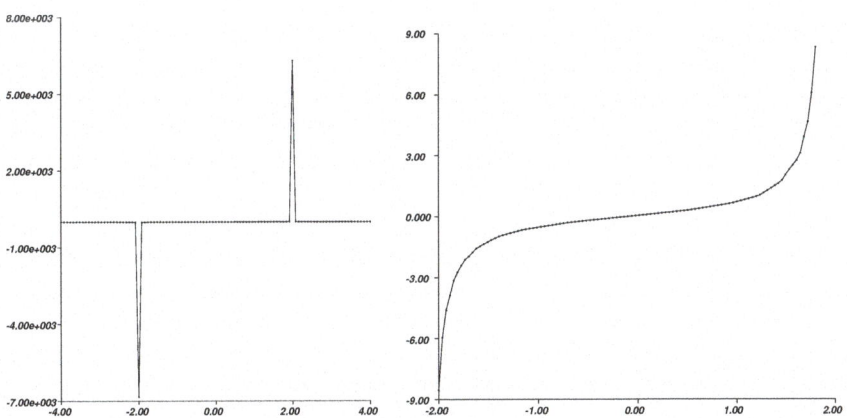

Fig. 13.6. Two opposite charged points. Potential on the x_1-axis (left). Zoom between the charged points (right)

13.1.3 A uniformly charged spherical volume

Let us consider a ball of radius r_0 centered at the origin with a uniform volumetric charge density ρ_V C/m^3 and total charge Q. As we have shown in Sect. 5.1.3, in this case the electric field is given by

$$\mathbf{E}(x) = \begin{cases} \dfrac{\rho_V r_0^3}{3\varepsilon_0 r^2(x)} \mathbf{e}_r(x), & \text{if } r(x) \geq r_0, \\[1em] \dfrac{\rho_V r(x)}{3\varepsilon_0} \mathbf{e}_r(x) & \text{if } r(x) < r_0. \end{cases}$$

Table 13.3. Ball with a uniform volumetric charge. Data for the numerical test

Radius of the charged ball, r_0:	1 m
Radius of the computational domain:	6 m
Volumetric charge density, ρ_V:	1.e-10 C/m^3
Vacuum electric permittivity, ε_0:	8.8542e-12 F/m

and the electrostatic potential is

$$V(x) = \begin{cases} \dfrac{\rho_V r_0^3}{3\varepsilon_0 \, r(x)} & \text{if } r(x) \geq r_0, \\[2ex] \dfrac{\rho_V (3r_0^2 - r^2(x))}{6\varepsilon_0} & \text{if } r(x) < r_0. \end{cases}$$

For the numerical resolution with MaxFEM, a ball of center (0,0,0) and radius 6 m containing a charged ball of radius 1 m is considered as computational domain. The radius of the outer ball has been taken large enough as to consider an homogeneous Dirichlet boundary condition on its surface. The relative permittivity is considered to be constant and equal to 1. The volumetric charge density ρ_V is 1.e-10 C/m^3. These data have been summarized in Table 13.3.

Figure 13.7 left shows the electric field which, as expected, is radially outward from the center of the sphere. The equipotential surfaces are concentric spherical shells as the ones represented in Fig. 13.7 (right). As predicted by theory, the electric field-strength is proportional to r inside the sphere, but falls off like $1/r^2$ outside the sphere. Figure 13.8 (left) illustrates this fact. Notice that the electric field outside the ball is identical to that of a point charge Q located at the center of the sphere.

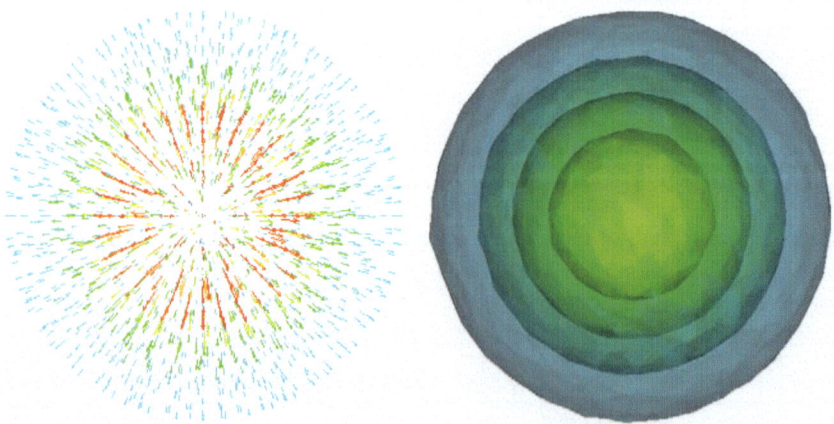

Fig. 13.7. Electric field (left) and equipotential surfaces (right) created by a sphere with a uniform volumetric charge

13.1 Some classical problems with analytical solution 273

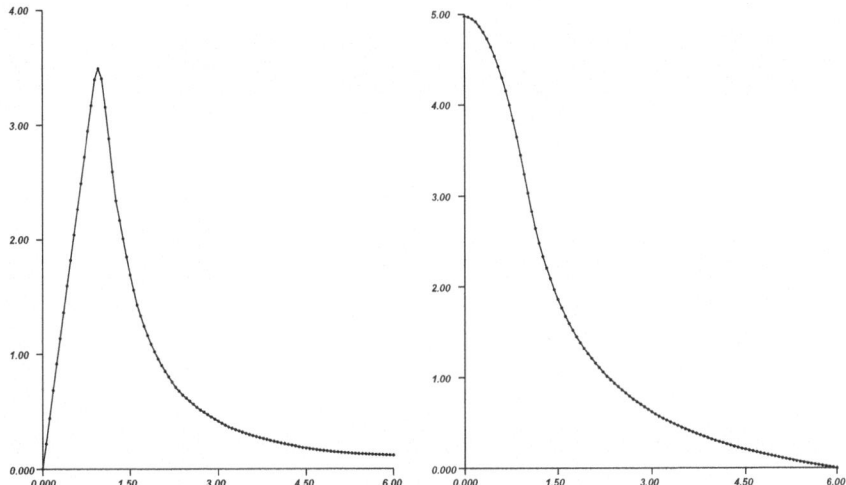

Fig. 13.8. A uniformly charged ball. Modulus of the electric field (left) and potential (right). Plots over the radius of the computational domain

Figure 13.8 (right) shows the potential plotted over the radius of the computational domain.

Remark 13.1. The reader can think about what happens in the case of a sphere with a uniform charge density ρ_S on its surface (see Problem 7.1).

13.1.4 A sphere with a non-uniform volumetric charge density

In this example let us consider a non-conducting sphere of radius r_0 centered at the origin that carries a volumetric charge density proportional to the distance from the center, i.e., $\rho_V = \rho_0 r$, where ρ_0 is a constant; $\rho_0 = 0$ for $r > r_0$.

Now, the charge distribution is not uniform but has spherical symmetry, i.e, ρ_V varies with the distance from the origin of the coordinate system, r, but not with the angle. As a consequence, the electric field **E** must be of the form $\mathbf{E}(x) = E_r(r(x))\mathbf{e}_r(x)$, with $\mathbf{e}_r(x) = \mathbf{r}(x)/r(x)$ and, in a similar manner as in example 13.1.3, it can be obtained by applying the Gauss' law. Let us first choose B as the ball centered at the origin of radius $r \geq r_0$ containing the sphere. Then, Eq. (4.22) leads to

$$\int_{\partial B} \varepsilon_0 \mathbf{E} \cdot \mathbf{n}\, dA = \varepsilon_0 E_r(r) 4\pi r^2 = \int_B \rho_V(r)\, dV.$$

Now, in order to compute the charge contained in B we must integrate the charge density $\rho_V(r)$ over the spherical volume B but since ρ_V is not constant it cannot be taken outside the integral as in the previous example (see Sect. 5.1.3). As ρ_0 is zero for $r > r_0$, we have

$$\int_B \rho_V(r)\, dV = \int_0^{2\pi}\!\int_0^{\pi}\!\int_0^{r_0} \rho_0 r r^2 \sin\phi\, dr d\phi d\theta = \rho_0 \pi r_0^4.$$

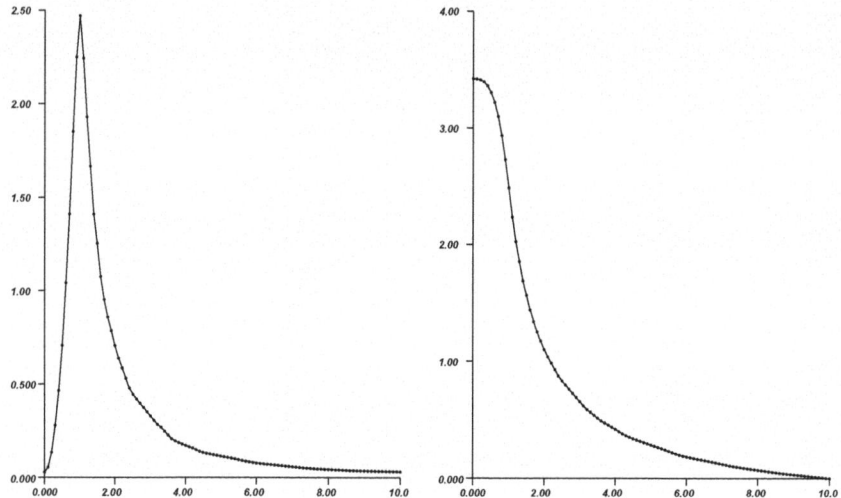

Fig. 13.9. Sphere with non-uniform volumetric charge density. Modulus of the electric field (left) and potential (right). Plots over the radius of the computational domain

Thus,

$$E_r(r) = \frac{\rho_0 r_0^4}{4\varepsilon_0 r^2} \quad r \geq r_0,$$

that is, for $r \geq r_0$, E_r behaves like a point charge. For $r < r_0$, B will enclose less than the total charge; indeed,

$$\int_B \rho_V(r) \, dV = \pi \rho_0 r^4$$

and then

$$E_r(r) = \frac{\rho_0 r^2(x)}{4\varepsilon_0}, \quad r < r_0,$$

that is, E_r varies as r^2 inside the charge distribution. The potential can be obtained by simple integration and is given by

$$V(x) = \begin{cases} -\dfrac{\rho_0 r^3(x)}{12\varepsilon_0} + \dfrac{\rho_0 r_0^3}{3\varepsilon_0} & \text{if } r(x) < r_0, \\ \dfrac{\rho_0 r_0^4}{4\varepsilon_0 r(x)} & \text{if } r(x) \geq r_0. \end{cases}$$

For numerical computation with MaxFEM we have considered a sphere of radius 1 m inside a ball of radius 10 m which is the computational domain. This radius is assumed to be large enough as to impose an homogeneous Dirichlet boundary condition on the boundary of the outer ball. The value ρ_0 has been taken equal to 1.e-10, as specified in Table 13.4. In Fig. 13.9 we have represented the modulus of the electromagnetic field and the potential along the diameter of the computational domain. As expected, these fields vary in the form predicted by the above analytical formulas.

Table 13.4. Ball with a non-uniform volumetric charge. Data for the numerical test

Radius of the charged sphere, r_0:	1 m
Radius of the computational domain:	10 m
Value of the constant ρ_0:	1.e-10 C/m^3
Vacuum electric permittivity, ε_0:	8.8542e-12 F/m

13.1.5 A uniformly charged infinite line

In Sect. 5.1.4 we have computed the electric field and the electric potential due to an infinitely long, straight line in free space with a uniform line charge density ρ_l. Let us recall that the electric field is given by

$$\mathbf{E}(x) = \frac{\rho_l}{2\pi\varepsilon_0 \rho(x)} \mathbf{e}_\rho(x),$$

and the electric potential by

$$V(x) = \frac{\rho_l}{2\pi\varepsilon_0} \ln\left(\frac{1}{\rho(x)}\right). \tag{13.1}$$

To illustrate this, let us numerically compute the electrostatic potential and the electric field generated in the $x_1 x_2$-plane by a uniformly charged infinite line lying along the x_3-axis.

Notice that, as the line is infinitely long, the line charge density in the 3D problem corresponds to a point charge in 2D. Taking this fact into account, we have solved a 2D problem and considered a circle of center (0,0) and radius 2 meters as computational domain (see Table 13.5). In this case, the analytical solution (13.1) is used to compute the exact Dirichlet boundary condition on the boundary of this domain. The line charge density is $\rho_l = 1.\text{e-}10$ C/m.

Figure 13.10 shows the equipotential lines in a transversal plane to the x_3-axis (right) and the electric field (left). As expected, the electric field is radially directed and only depends on the radial coordinate ρ. The equipotential surfaces are always perpendicular to the electric field lines. In this case, the equipotential surfaces are cylinders with axis the x_3-axis; as a consequence, the potential distribution is the same on each transversal plane to the x_3-axis. In Fig. 13.11 we have represented the modulus of the electric field and the electric potential along the radius of the computational domain. Notice that the potential tends to the value imposed as boundary condition obtained from formula (13.1).

Table 13.5. Infinitely long charged line. Data for the numerical test

Radius of the computational domain:	2 m
Value of ρ_l:	1.e-10 C/m
Vacuum electric permittivity, ε_0:	8.8542e-12 F/m

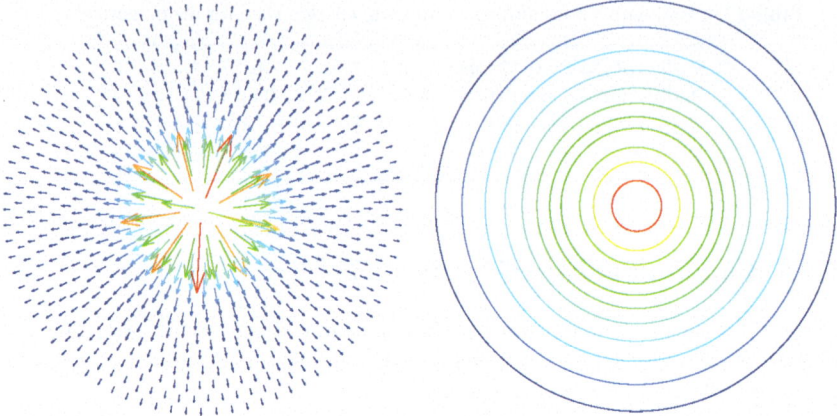

Fig. 13.10. Electric field (left) and equipotential lines (right) created for the infinite charged line in a transversal plane to the x_3-axis.

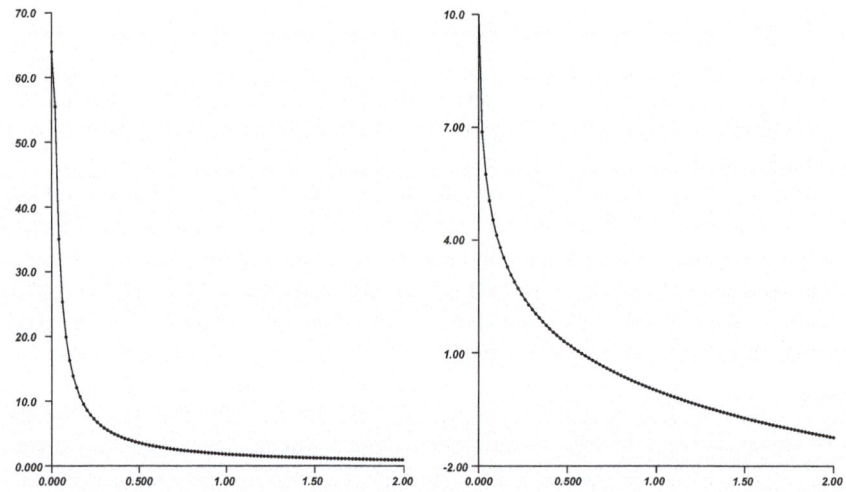

Fig. 13.11. Modulus of the electric field (left) and potential (right) due to a infinite charged line. Plots over the radius of the computational domain

13.1.6 A uniformly charged segment

Let us compute the electric field generated by a charged segment of length $2L$ centered at the origin and lying along the x_3-axis as illustrated in Fig. 13.12. The segment is assumed to be uniformly charged with line charge density ρ_l C/m.

As explained in Sect. 7.1, electric field **E** for line charge distributions can be computed from expression (7.6). In this case, and assuming ε is constant, solving Eq. (7.6)

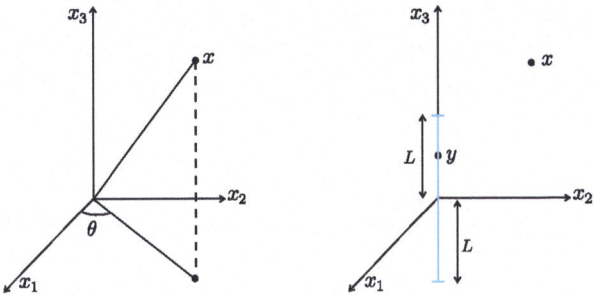

Fig. 13.12. Charged segment

reduces to compute the integral

$$\mathbf{E}(x) = \frac{\rho_l}{4\pi\varepsilon} \int_{-L}^{L} \frac{x-y}{|x-y|^3} dz(y).$$

Let x be the point where we want to compute the electric field and y be the source point. If o denotes the origin of the affine space, by using cylindrical coordinates we can write

$$x = o + \rho(x)\mathbf{e}_\rho(x) + z(x)\mathbf{e}_z,$$
$$y = o + \rho(y)\mathbf{e}_\rho(y) + z(y)\mathbf{e}_z.$$

We notice that $\rho(y) = 0$ because y is on the x_3-axis. Then,

$$x - y = \rho(x)\mathbf{e}_\rho(x) + (z(x) - z(y))\mathbf{e}_z,$$

and

$$|x-y| = \sqrt{\rho(x)^2 + (z(x) - z(y))^2}.$$

Thus,

$$\mathbf{E}(x) = \frac{\rho_l}{4\pi\varepsilon} \left\{ \left[\rho(x) \int_{-L}^{L} \frac{1}{\left(\sqrt{\rho(x)^2 + (z(x) - z(y))^2}\right)^3} dz(y) \right] \mathbf{e}_\rho(x) \right.$$

$$\left. + \left[\int_{-L}^{L} \frac{z(x) - z(y)}{\left(\sqrt{\rho(x)^2 + (z(x) - z(y))^2}\right)^3} dz(y) \right] \mathbf{e}_z \right\}$$

$$= \frac{\rho_l}{4\pi\varepsilon} \left\{ \left[\frac{L - z(x)}{\rho(x)\sqrt{\rho(x)^2 + (z(x) - L)^2}} + \frac{L + z(x)}{\rho(x)\sqrt{\rho(x)^2 + (z(x) + L)^2}} \right] \mathbf{e}_\rho(x) \right.$$

$$\left. + \left[\frac{1}{\left(\sqrt{\rho(x)^2 + (z(x) - L)^2}\right)} - \frac{1}{\left(\sqrt{\rho(x)^2 + (z(x) + L)^2}\right)} \right] \mathbf{e}_z \right\}.$$

(13.2)

13 Electrostatics with MaxFEM

Table 13.6. Charged segment. Data for the numerical test

Radius of the cylindrical computational domain:	1 m
Height of the cylindrical computational domain:	3 m
Half length of the segment, L:	1 m
Value of ρ_l:	1.e-10 C/m
Vacuum electric permittivity, ε_0:	8.8542e-12 F/m

On the other hand, from (7.4) we can easily deduce the following expression for the electric potential:

$$V(x) = -\frac{\rho_l}{4\pi\varepsilon} \ln\left(\frac{\sqrt{\rho(x)^2 + (L-z(x))^2} - (L-z(x))}{\sqrt{\rho(x)^2 + (L+z(x))^2} + (L+z(x))}\right). \quad (13.3)$$

For numerical resolution, a cylinder with the charged segment centered on its axis has been considered as computational domain. We have set $L = 1$ m and a cylinder of radius 1 meter and height 3 meters. The line charge density is $\rho_l = 1.\text{e-}10$ C/m. In this case, since we know an analytical expression for the electric potential, this expression can be used to compute an exact Dirichlet boundary condition on the surface of the cylinder. This is not the case in real situations where the computational domain must be considered large enough as to suppose that V vanishes on its boundary. Here we have chosen the other option in order to consider a smaller computational domain.

Figure 13.13 shows a cut of the computational domain where the charged segment can be appreciated. Figure 13.14 shows the electric field and Fig. 13.15 the equipotential surfaces, which are ellipsoids of revolution. The corresponding electric fields lines are hyperboloids of revolution. Finally, in Fig. 13.16 we have represented the modulus of the electric field as a function of the distance to the cylinder axis.

Notice that the limit of expression (13.2) as the line-charge length L tends to infinity is

$$\mathbf{E}(x) = \frac{\rho_l}{2\pi\varepsilon\rho(x)}\mathbf{e}_\rho(x), \quad (13.4)$$

which agrees with formula (5.10).

Electrostatic forces and potentials are key issues in determining the interactions between biomolecules (see, for instance, [43]); in particular, in explaining how the DNA (Deoxyribonucleic acid) interacts with various proteins transcribing genetic information. DNA has a basically linear form and it carries a negative charge, so from the point of view of computation it can be considered as a uniformly line charge density. Scientists use the double-stranded DNA to study the electric transport systems in single molecules. This allows them to structurally manipulate DNA properties, such as the energy level structure, for DNA damage repair.

13.1 Some classical problems with analytical solution 279

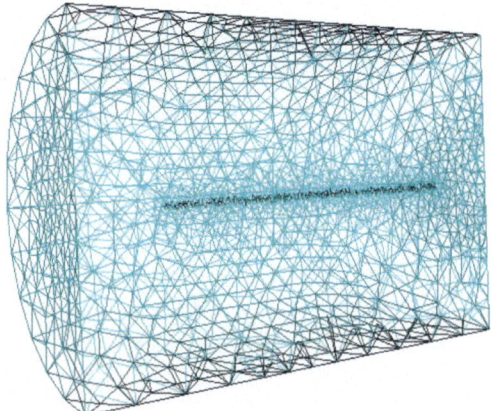

Fig. 13.13. Charged segment: computational domain

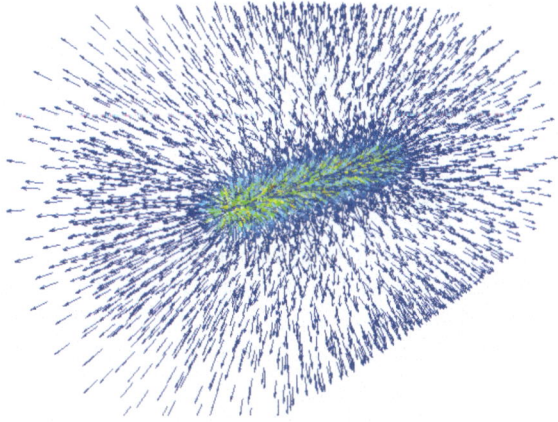

Fig. 13.14. Electric field created by a uniformly charged segment

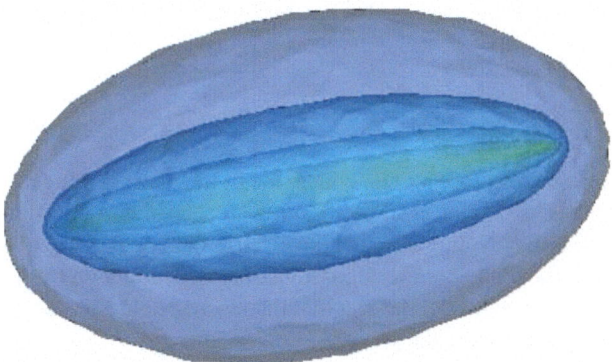

Fig. 13.15. Equipotential surfaces created by a uniformly charged segment

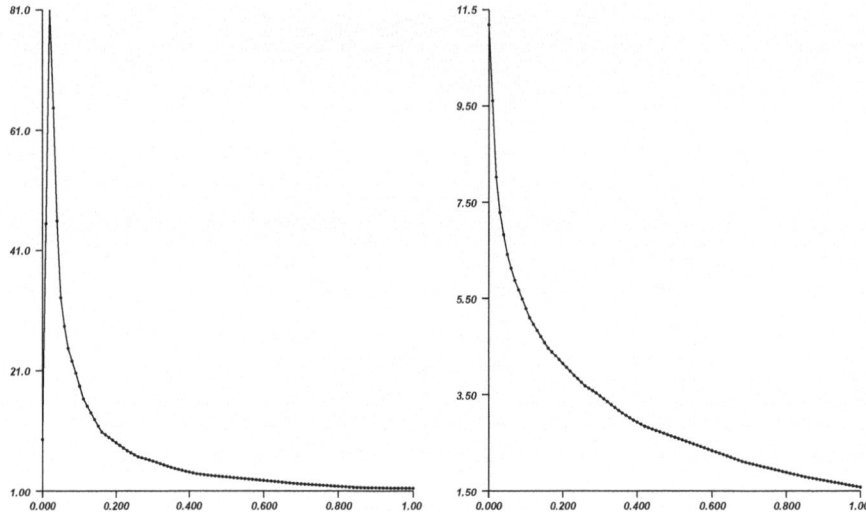

Fig. 13.16. Modulus of the electric field (left) and potential (right) due to a charged segment. Plots over the radius of the computational domain

13.1.7 A uniformly charged infinite plane

In this section we are going to compute the electric field and the electrostatic potential due to an infinite charged plane as the one studied in Sect. 5.1.5. We recall that the plane is perpendicular to the x_1-axis at $x_1 = 0$ (see Fig. 5.3) and uniformly charged with charge per unit area ρ_S. As deduced in Sect. 5.1.5 the electric field is given by

$$\mathbf{E}(x) = \begin{cases} \dfrac{\rho_S}{2\varepsilon_0} \mathbf{e}_1 & \text{if } x_1 > 0, \\ -\dfrac{\rho_S}{2\varepsilon_0} \mathbf{e}_1 & \text{if } x_1 < 0. \end{cases}$$

and the electrostatic potential

$$V(x) = -\dfrac{\rho_S}{2\varepsilon_0} |x_1| + constant.$$

By symmetry, the electric field on either side of the plane is a function of x_1 only, normal to the plane, and pointing away from/towards the plane depending on whether ρ_S is positive/negative. Notice also that the electric field does not get weaker as you move away from the plane but it is a constant only depending on the surface charge density. This is consistent with the fact that the electric field lines do not get further apart as we move away from the plane.

In what follows we will use MaxFEM in order to reproduce these facts. In this case, as the plane is infinite we cannot consider an artificial domain enclosing it and large enough as to consider the field vanishes on its boundary, so other arguments

13.1 Some classical problems with analytical solution

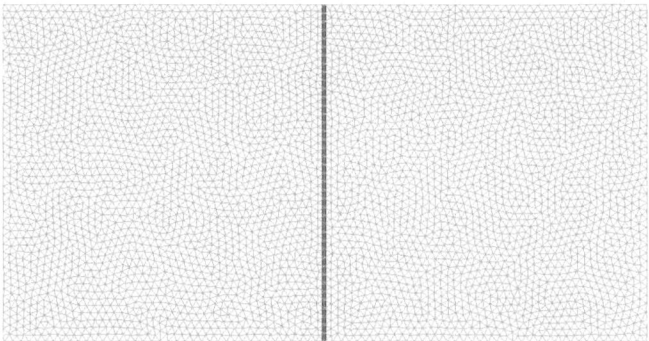

Fig. 13.17. Infinite charged plane. Computational domain

must be employed. First, and due to the symmetry of the problem, let us consider the (infinite) line of charge which results from the intersection between the charged plane and the x_1x_2-plane. As computational domain we will consider the rectangle $\mathscr{R} = \{(x_1,x_2) \in \mathbb{R}^2, -a \leq x_1 \leq a, 0 \leq x_2 \leq b\}$ which is cut in half by the charged line (in dark blue) as sketched in Fig. 13.17.

Let us see that it is possible to impose boundary conditions on the boundary of $\mathscr{R}$ which allows us to solve the problem in this bounded two-dimensional domain. Indeed, it is immediate to check that

$$\mathbf{E} \cdot \mathbf{n} = 0 \quad \text{on} \quad \Gamma_n = \{(x_1,x_2), x_2 = 0 \text{ or } x_2 = b\}, \tag{13.5}$$

$$\mathbf{E} \times \mathbf{n} = \mathbf{0} \quad \text{on} \quad \Gamma_d = \{(x_1,x_2), x_1 = a \text{ or } x_1 = -a\}. \tag{13.6}$$

Since $\mathbf{E} = -\operatorname{grad} V$, Eq. (13.5) leads us to consider homogeneous Neumann boundary conditions on Γ_n. On the other hand, Eq. (13.6) implies that $V = C$ on Γ_d, with C an arbitrary constant. For numerical computation, we have considered $C = 100$, $\rho_S = 1.\text{e-}6$ C/m^2 and $a = b = 1$ as summarized in Table 13.7. Figure 13.18 shows the electric field in any plane perpendicular to the infinite charged plane. As expected, it is normal to the plane and it points outward the plane (since the charge is positive). In Fig. 13.19 we have made a 3D representation of the first component of the electric field. Figure 13.20 (left) shows the potential isolines in any plane perpendicular to the infinite charged plane. The equipotential surfaces are planes parallel to the charged plane. Finally, Fig. 13.20 (right) represents a plot of the potential over the the line determined by the points $(-1, 0.5)$ and $(1, 0.5)$.

Table 13.7. Infinite charged plane. Data for the numerical test

Length of the rectangle,	2 m
Height of the rectangle,	1 m
Surface charge density, ρ_S:	1.e-6 C/m^2
Vacuum electric permittivity, ε_0:	8.8542e-12 F/m

282   13 Electrostatics with MaxFEM

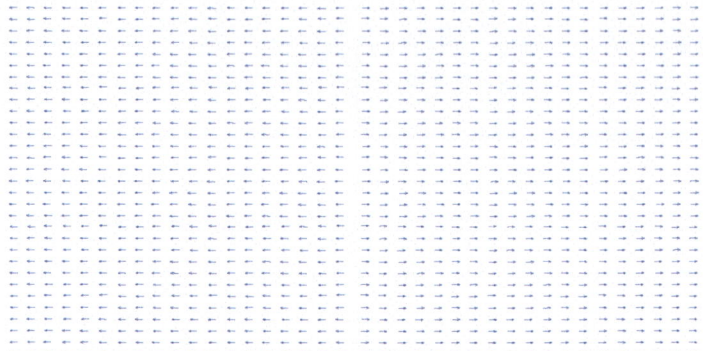

Fig. 13.18. Electric field created in any plane perpendicular to the infinite charged plane

Fig. 13.19. Infinite charged plane. First component of the electric field

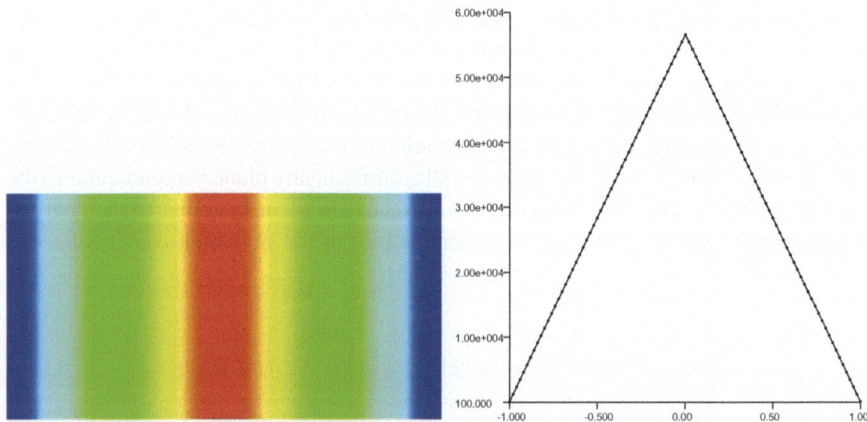

Fig. 13.20. Equipotential regions in any plane perpendicular to the infinite charged plane (left). Potential over the line determined by the points $(-1, 0.5)$ and $(1, 0.5)$ (right)

13.1.8 An infinitely long cylinder uniformly charged on its surface

In this example we consider an infinitely long cylindrical surface of radius a in free space with a uniform surface charge density ρ_S.

By symmetry, the electric field **E** at a point x must be a function only of the normal distance $\rho(x)$ of x to the axis of the cylinder (let us assume it is the x_3-axis), that is

$$\mathbf{E}(x) = E_\rho(\rho(x))\mathbf{e}_\rho(x).$$

Let us compute the electric field at a point x. In order to apply the Gauss' law let Ω be a cylinder coaxial to the charged one, with radius $\rho(x)$ and height h. On the bases of the cylinder, electric field **E** and unit normal vector **n** are perpendicular so their scalar product is zero, while on the lateral surface S the two vectors are parallel. If $\rho(x) \geq a$, we obtain

$$\int_S \varepsilon_0 E_\rho(\rho)\mathbf{e}_\rho \cdot \mathbf{n}\, dA = E_\rho(\rho)2\pi\rho h = \rho_S 2\pi a h,$$

so that

$$\mathbf{E}(x) = \rho_S \frac{a}{\varepsilon_0 \rho(x)} \mathbf{e}_\rho(x) \quad \text{for } \rho(x) \geq a.$$

On the contrary, if $\rho(x) < a$ then the charge inside the cylinder is zero and

$$E_\rho(\rho(x)) = 0 \quad \text{for } \rho(x) < a.$$

By simple integration, and taking as zero-potential reference point any x with $\rho(x) = a$, we obtain that

$$V(x) = \begin{cases} \rho_S \dfrac{a}{\varepsilon_0} \ln\left(\dfrac{a}{\rho(x)}\right) & \text{for } \rho(x) \geq a, \\ 0 & \text{for } \rho(x) \leq a. \end{cases}$$

For numerical resolution of the problem, the circle of center $(0,0)$ and radius 0.5 m with a uniform line charge density on its boundary has been considered. Notice that this circle represents any transversal section of a cylinder coaxial with the x_3-axis and radius 0.5 m. Then, the line charge density corresponds to a surface charge density $\rho_S = 1.\text{e-}10$ C/m² in the 3D problem (see Table 13.8).

On the other hand, and in order to apply the finite element method, a circle of radius 10 m concentric with the charged one has been considered as computational domain. In this case, the analytical solution for the electric potential is used to compute the exact Dirichlet boundary condition on the boundary of this domain. Figures 13.21 and 13.22 show the electric field and the potential, respectively. From Fig. 13.21 (left) we can deduce that the electric fields lines point radially outward and are parallel to

Table 13.8. Infinitely long charged cylinder. Data for the numerical test

Radius of the cylinder:	0.5 m
Radius of the computational domain:	10 m
Surface charge density, ρ_S:	1.e-10 C/m²
Vacuum electric permittivity, ε_0:	8.8542e-12 F/m

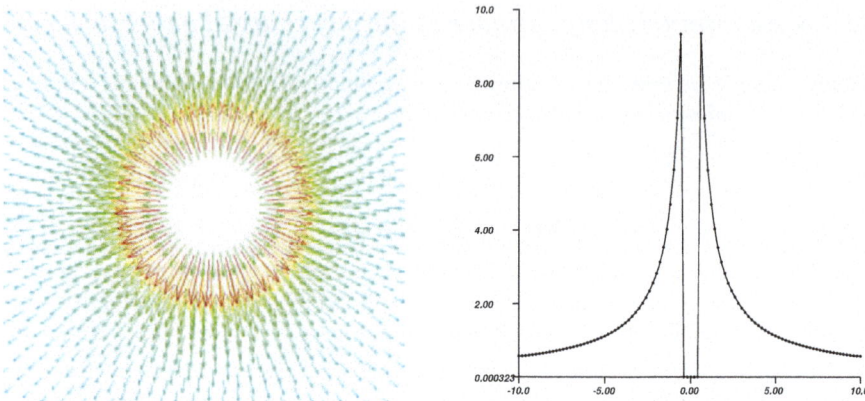

Fig. 13.21. An infinitely long cylinder uniformly charged on its surface. Detail of the electric field (left) and plot over the diameter of the computational domain (right)

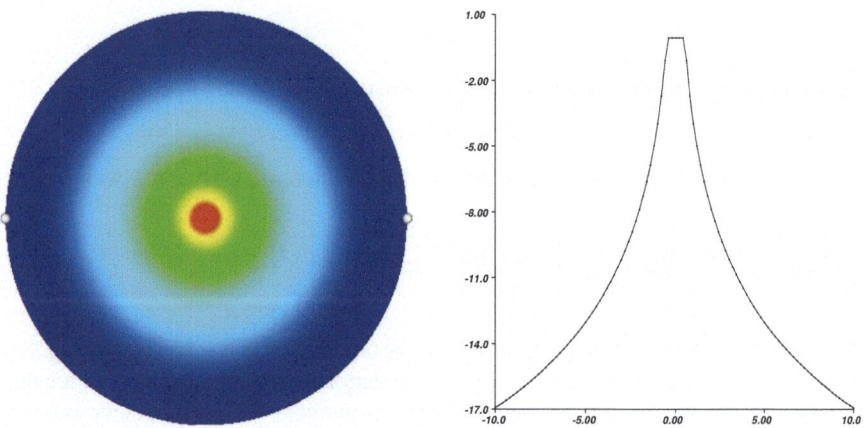

Fig. 13.22. An infinitely long cylinder uniformly charged on its surface. Equipotential regions (left) and plot over the diameter of the computational domain (right)

the two end caps. On the right side of Figs. 13.21 and 13.22 we can see the modulus of the electric field and the potential over the diameter of the computational domain, respectively. Notice that the potential is continuous but the electric field is not.

13.1.9 An infinitely long solid cylinder uniformly charged

Let us consider an infinitely long, solid cylinder of radius $\rho = a$ in free space containing a static, uniform volume charge density ρ_V. Because of the cylinder symmetry one expects the electric field to be only dependent on the distance ρ to the axis of the cylinder. In a similar way as we proceed for the cylindrical surface (see Sect. 13.1.8) we can apply the Gauss' law to compute the electric field at a point x of the space. In

13.1 Some classical problems with analytical solution   285

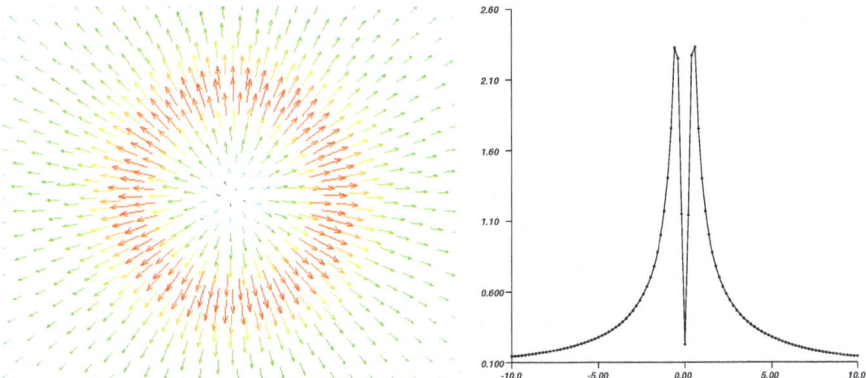

Fig. 13.23. An infinitely long cylinder uniformly charged with a volumetric charge density. Detail of the electric field (left) and plot of its modulus over the diameter of the computational domain (right)

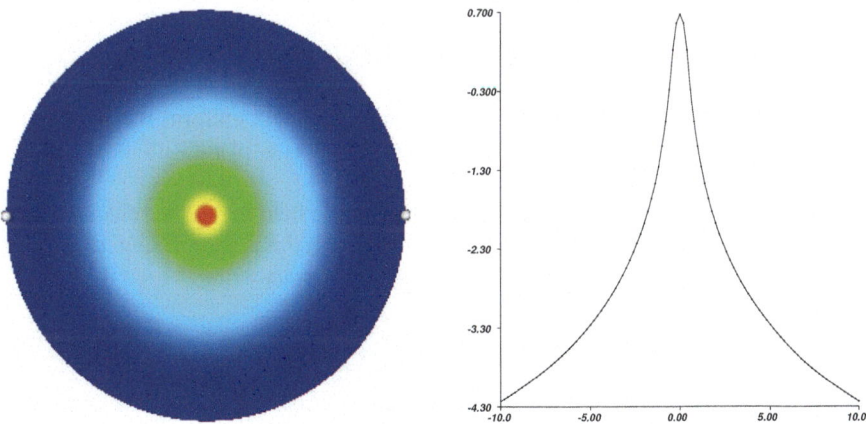

Fig. 13.24. An infinitely long cylinder uniformly charged with a volumetric charge density. Equipotential regions (left) and plot over the diameter of the computational domain (right)

this case, for $\rho(x) \geq a$ the following equation holds true

$$\mathbf{E}(x) = \rho_V \frac{a^2}{2\varepsilon_0 \rho(x)} \mathbf{e}_\rho(x) \quad \text{for } \rho(x) \geq a.$$

However, and contrary to example 13.1.8, for $\rho(x) < a$ the electric field is not equal to zero. In this case we have

$$\mathbf{E}(x) = \rho_V \frac{\rho(x)}{2\varepsilon_0} \mathbf{e}_\rho(x) \quad \text{for } \rho(x) < a.$$

Then, the potential is given by

$$V(x) = \begin{cases} \rho_V \dfrac{a^2}{2\varepsilon_0} \ln\left(\dfrac{a}{\rho(x)}\right) & \text{for } \rho(x) \geq a, \\ -\rho_V \dfrac{\rho(x)^2}{4\varepsilon_0} + \dfrac{\rho_V}{4\varepsilon_0} a^2 & \text{for } \rho(x) < a. \end{cases}$$

In order to compute the numerical solution, the same domain as in example 13.1.8 has been considered, the only difference being that now a volume charge density $\rho_V = 1.\text{e-}10$ C/m^3 is considered. Again, the analytical solution for the electric potential is used to compute the exact Dirichlet boundary condition on the boundary of this domain.

According to the above expressions, the electric field increases within the cylinder with the distance to the axis as shown in Fig. 13.24 above, whereas it decreases outside the cylinder as this distance increases. Notice that the exterior electric field is the same as that expected if the same total charge per unit length were concentrated as a line charge along the x_3-axis (see Sect. 13.1.5).

13.1.10 A uniformly charged circular disk

Let us consider a circular disc of radius a which carries a uniform surface charge density ρ_S in C/m^2. Let us suppose the disk lies on the x_1x_2-plane with its axis along the x_3-axis as shown in Fig. 13.25.

As explained in Sect. 7.1 we can obtain the electric field for surface charge distributions from expression (7.7). Let x be a point on the x_3-axis and y a source point on the disk. If o denotes the origin of the affine space, we can write

$$x = o + z(x)\mathbf{e}_z,$$
$$y = o + \rho(y)\mathbf{e}_\rho(y).$$

Then

$$x - y = z(x)\mathbf{e}_z - \rho(y)\mathbf{e}_\rho(y),$$
$$|x - y| = \sqrt{z(x)^2 + \rho(y)^2}.$$

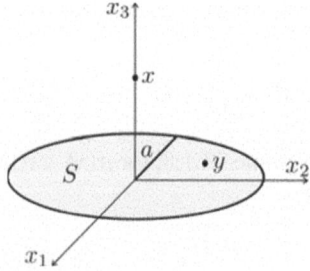

Fig. 13.25. Circular disk on the x_1x_2-plane

Thus

$$\mathbf{E}(x) = \frac{\rho_S}{4\pi\varepsilon} \int_S \frac{z(x)\mathbf{e}_z - \rho(y)\mathbf{e}_\rho(y)}{\left(\sqrt{z(x)^2 + \rho(y)^2}\right)^3} dA_y$$

$$= \frac{\rho_S}{4\pi\varepsilon} \left[\int_0^{2\pi} d\theta \int_0^a \frac{z(x)\rho(y)d\rho(y)}{\left(\sqrt{z(x)^2 + \rho(y)^2}\right)^3} \right] \mathbf{e}_z \qquad (13.7)$$

$$= \frac{\rho_S}{2\varepsilon} \left[\text{sign}(z(x)) - \frac{z(x)}{\sqrt{z(x)^2 + a^2}} \right] \mathbf{e}_z,$$

because, for symmetry reasons, the integral of the term with $\mathbf{e}_\rho$ must be null.

As the radius of the disk tends to infinity we have

$$\lim_{a \to \infty} \mathbf{E}(x) = \begin{cases} \dfrac{\rho_S}{2\varepsilon} \mathbf{e}_z & \text{if } z(x) > 0, \\ -\dfrac{\rho_S}{2\varepsilon} \mathbf{e}_z & \text{if } z(x) < 0, \end{cases}$$

which is consistent with formula (5.13) for an infinite planar charge.

In order to numerically solve the problem, we have considered a disk of radius 1 m centered at the origin, lying on the x_1x_2-plane and enclosed by an artificial boundary corresponding to the surface of a cylinder of radius 5 m and height 10 m around the x_3-axis. This boundary is assumed to be enough far away as to consider that V vanishes on it. Figure 13.26 shows the computational domain and the electric field obtained from the numerical simulation with MaxFEM. A zoom of the field around the disk is shown in Fig. 13.27. Finally, in Fig. 13.28 we have plotted the modulus of the z component of the electric field over the segment delimited by points $(0, 0, -5)$

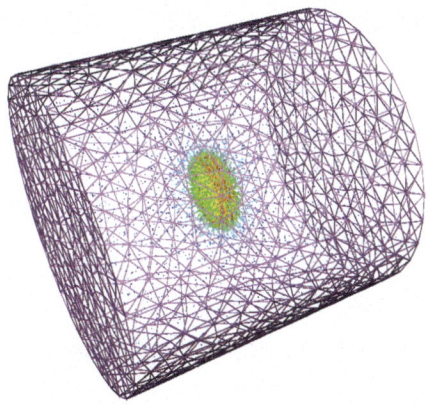

Fig. 13.26. Electric field around the charged disk in the computational domain

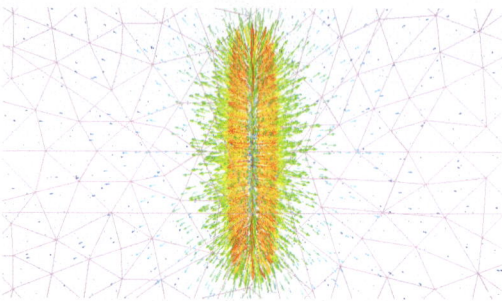

Fig. 13.27. Zoom of the electric field around the charged disk (x_2x_3-view)

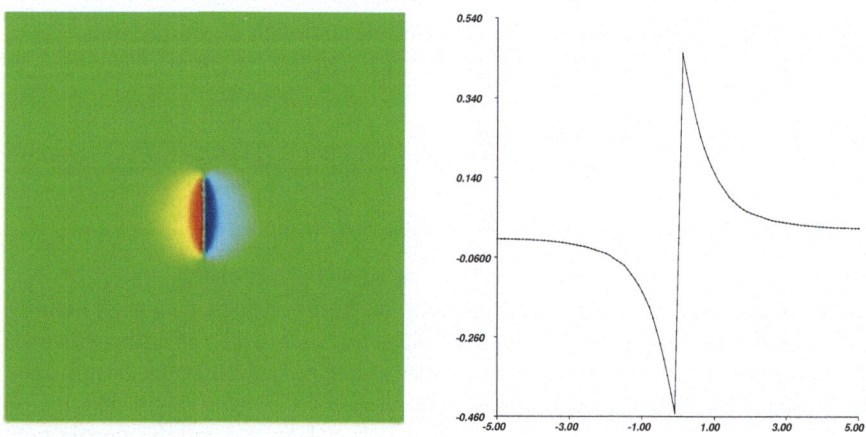

Fig. 13.28. E_z component of the electric field (left) and plot of E_z over the segment delimited by points $(0,0,-5)$ and $(0,0,5)$

Table 13.9. Charged disk. Data for the numerical test

Radius of the disk:	1 m
Radius of the cylindrical computational domain:	5 m
Height of the cylindrical computational domain:	10 m
Surface charge density, ρ_S:	8.85e-12 C/m²
Vacuum electric permittivity, ε_0:	8.8542e-12 F/m

and $(0,0,5)$ on the x_3-axis. The values obtained with MaxFEM agree which those obtained from exact formula (13.7).

13.2 Some problems arising in physical applications

Contrary to the examples in the previous section, in most practical situations solution of electrostatic problems cannot be analytically computed. In these cases, only electrostatic conditions (charge or potential) at some boundaries of the domain are known and it is desired to compute **E** and V throughout the region. Such problems are referred to as boundary-value problems for the Poisson's or Laplace's equations and must be solved, in general, by using numerical techniques. In this section we will show some examples.

13.2.1 The coaxial cable

A coaxial cable as that shown in Fig. 13.29 (left) consists of two metallic conductors with cylindrical symmetry, sharing a common cylinder axis and separated by some kind of insulator. Many coaxial cables also have an insulating outer jacket. Coaxial cables are widely used as transmission lines, for instance in computer network connections or to transport television signals. In an ideal coaxial cable Gauss' law shows that the electromagnetic field is zero except between the conducting cylindrical surfaces, where it is equal to the field produced by the inner cylinder. This is an advantage because power losses appearing in other types of transmission lines are avoided.

In this example, we will consider a long coaxial cable consisting of two coaxial conductors separated by an insulator with null charge density, as depicted in Fig. 13.29 (right). Let $V_b - V_a$ be the potential difference between the inner and outer conductors and let us compute the electric field in the dielectric. As the dimension in the orthogonal direction to the figure is much larger than the other two, we will assume that the cable is infinite in that direction, let us say x_3, and that the geometry and the material properties are translational invariant in x_3. Under these assumptions, it is possible to reduce the computations to a 2D cut orthogonal to x_3.

If ε is constant, the problem can be analytically solved in a simple way by using polar coordinates. Indeed, for symmetry reasons the solution only depends on the radial coordinate, i. e., $V(x) = V(\rho(x))$. Then, the equation to be solved reads

$$-\varepsilon \frac{1}{\rho} \frac{\partial}{\partial \rho} \left(\rho \frac{\partial V}{\partial \rho} \right) = 0,$$

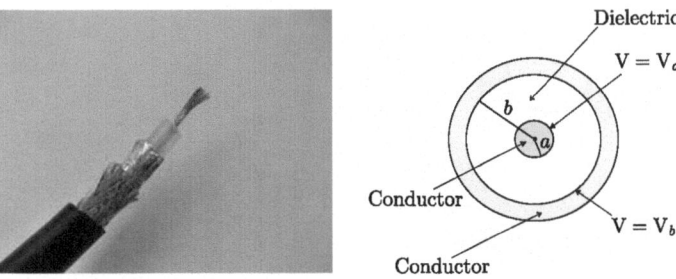

Fig. 13.29. Coaxial cable (left; picture licensed by Creative Commons) and sketch of its transversal section (right)

with boundary conditions $V(a) = V_a$ and $V(b) = V_b$. It is easy to show that the electric potential at distance ρ from the inner conductor is given by

$$V(x) = V_a - \frac{(V_a - V_b)\ln(\rho(x)/a)}{\ln(b/a)} \qquad a \leq \rho \leq b, \tag{13.8}$$

and, as a consequence, the electric field in the dielectric is

$$\mathbf{E}(x) = \frac{(V_a - V_b)}{\rho(x)\,\ln(b/a)}\mathbf{e}_\rho. \tag{13.9}$$

We have numerically solved the above problem for the values specified in Table 13.10. Figures 13.30 and 13.31 show the electric field and the potential, respectively.

Finally, we have considered a coaxial cable where the relative electric permittivity of the dielectric is not constant but depends on the distance to the center of the cable, namely $\varepsilon_r = \rho^2$. Figure 13.32 represents the modulus of the electric field (left) and the potential (right), along the diameter of the computational domain. Figure 13.33 shows the modulus of the electric field and the isopotential curves on the computational domain.

Table 13.10. Coaxial cable. Data for the 2D numerical test

Inner radius, a:	0.0010 m
Outer radius, b:	0.0047 m
Relative electric permittivity, ε_r:	1.78
Vacuum electric permittivity, ε_0:	8.8542e-12 F/m
Potential on $\rho = a$:	0 V
Potential on $\rho = b$:	3000 V

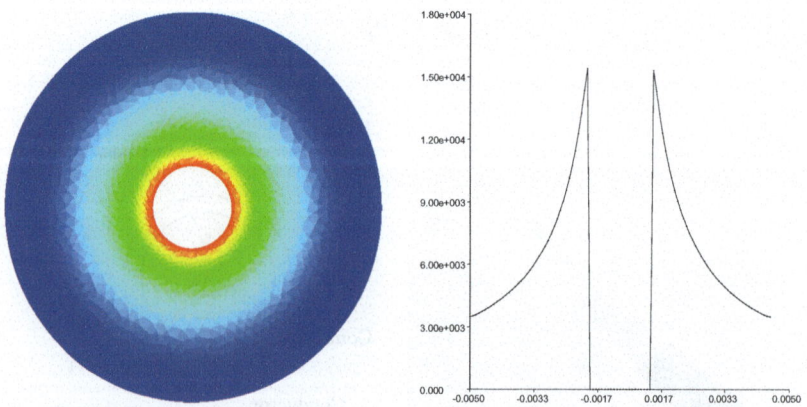

Fig. 13.30. Coaxial cable with $\varepsilon_r = 1.78$. Electric field. Modulus (left) and plot over the diameter of the computational domain (right)

13.2 Some problems arising in physical applications 291

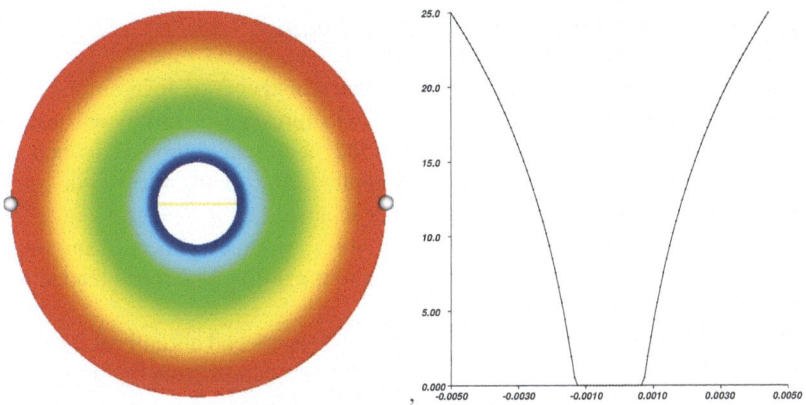

Fig. 13.31. Coaxial cable with $\varepsilon_r = 1.78$. Equipotential lines (left) and plot over the diameter of the computational domain.

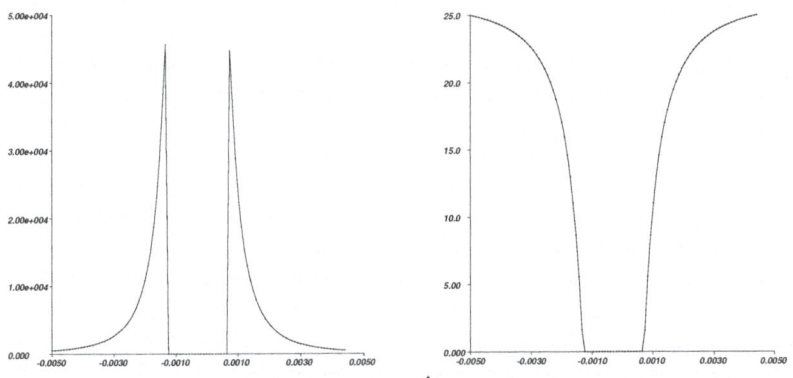

Fig. 13.32. Coaxial cable withe with $\varepsilon_r = \rho^2$. Modulus of the electric field (left) and potential (right) along the diameter of the cable

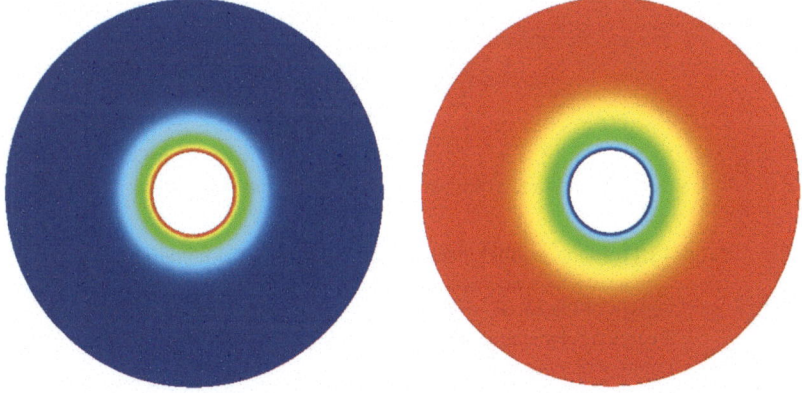

Fig. 13.33. Coaxial cable with $\varepsilon_r = \rho^2$. Modulus of the electric field (left) and potential (right) along the diameter of the cable

13.2.2 A planar capacitor with parallel plates

A capacitor is an electrical device that stores charge and energy. In its simplest form, it consists of two conductors (usually called armatures) separated by an insulating (dielectric) material. The energy is stored by the electric field produced when a potential difference between the armatures is applied. The simplest capacitor consists of two parallel and oppositely charged plates separated by a dielectric with permittivity ε, as the one sketched in Fig. 13.34. Air, mica, ceramics, paper, oil and vacuum are usually employed as dielectrics according to the utility which is intended to the device.

The electric field created due to the capacitor does not end abruptly at the edge of the plates. There is some field outside the plates that curves from one to the other, that is, there are the so-called *fringing fields*. To illustrate this phenomenon, let us consider a parallel plate plane capacitor whose characteristics have been specified in Table 13.11. Moreover, let us assume that the dimension of the plates in the x_3-direction is much higher than in the other two spatial directions in such a way that a 2D numerical computation can be performed.

If the plates are supposed to be close enough, the *fringe effect* can be neglected and the field between them can be approximated by the field between two parallel plates infinite in the two directions x_1 and x_3. Although this is a one dimensional case that can be analytically solved, we are going to perform 2D numerical computations in order to highlight the difference with the shape of the electric field when fringe effect is considered. In this case the computational domain consists only of a rectangular box representing the transversal section of the capacitor (see Fig. 13.35, right). The (known) potential is imposed as Dirichlet boundary condition on the plates; on the other two sides, a homogeneous Neumann boundary condition holds. Figure 13.36 illustrates this case. The field between the plates is uniform and the field lines are parallel and perpendicular to each plate, so fringe effect is ignored.

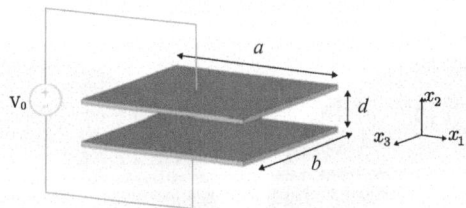

Fig. 13.34. Sketch of a planar capacitor of parallel plates

Table 13.11. Plane parallel plate capacitor. Data for the numerical test

Length of the capacitor plates, a:	0.010 m
Distance between the plates, d:	0.006 m
Relative electric permittivity, ε_r:	1
Vacuum electric permittivity, ε_0:	8.8542e-12 F/m
Potential drop, V_0:	100 V

13.2 Some problems arising in physical applications

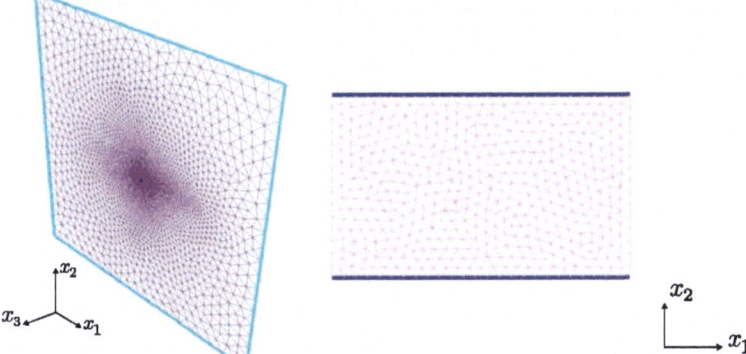

Fig. 13.35. Parallel plates plane capacitor. Computational domain to show the fringe effect (left; isoview) and neglecting the fringe effect (right)

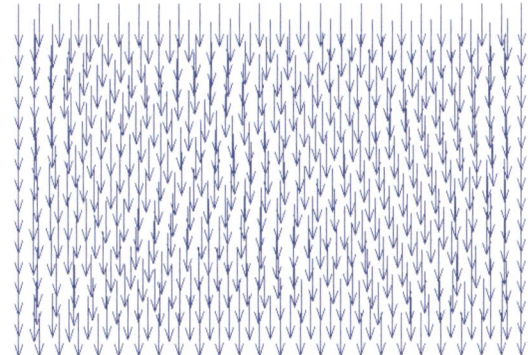

Fig. 13.36. Ideal parallel plates plane capacitor. Fringing fields are ignored

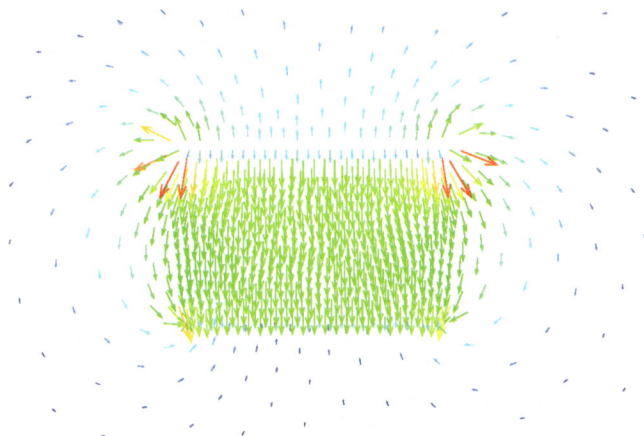

Fig. 13.37. Parallel plates plane capacitor. Fringe effect can be observed

Table 13.12. Circular parallel plates capacitor. Data for the numerical test

Radius of the capacitor plates, a:	0.010 m
Distance between the capacitor plates, d:	0.006 m
Radius of the computational domain:	0.050 m
Height of the computational domain:	0.015 m
Relative electric permittivity, ε_r:	1
Vacuum electric permittivity, ε_0:	8.8542e-12 F/m
Potential drop, V_0:	100 V

On the contrary, to show the fringe effect, an artificial boundary enclosing this domain must be considered. This boundary is assumed to be far enough as to impose a null Dirichlet boundary condition. In this case, we have considered a square of side 100 mm surrounding the capacitor as shown in Fig. 13.35 (left). Figure 13.37 shows the electric field obtained from the numerical simulation and we can appreciate the fringe effect. In particular, this causes the real capacitance to be larger than the one obtained for infinite plates. You have more electric field because of the fringing fields. Both simulations have been performed with MaxFEM and setting $a = 10$ mm, $d = 6$ mm (see Fig. 13.34). The potential drop between the plates is 100 V and the dielectric is air.

The fringe effect can also be observed for a planar capacitor with circular parallel plates as illustrated in Fig. 13.38. In this case a 3D computation has been performed. Figure 13.39 shows the computational domain considered in the simulation, which is assumed to be big enough as to impose a homogeneous Dirichlet boundary condition. In Fig. 13.40 we have represented the electric potential in a cut view of the computational domain.

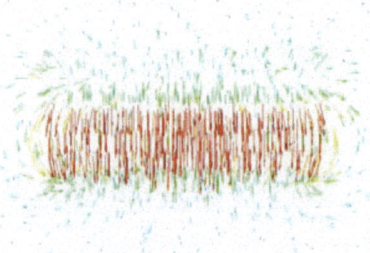

Fig. 13.38. Circular parallel plates capacitor. Fringe effect

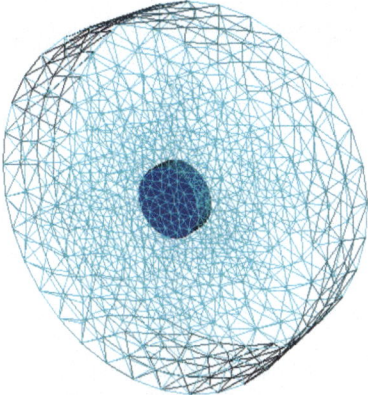

Fig. 13.39. Circular parallel plates capacitor. Computational domain. The capacitor appears in blue

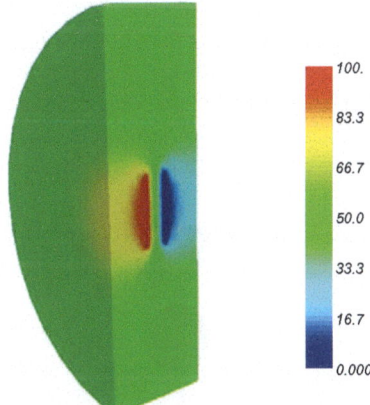

Fig. 13.40. Circular parallel plates capacitor. Electric potential

13.2.3 Planar capacitor with discontinuities in the dielectric

Above a particular electric field, known as the *dielectric strength*, the dielectric in a capacitor becomes conductive. The voltage at which this occurs is called the *breakdown voltage*. The maximum energy that can be stored safely in a capacitor is limited by the breakdown voltage. For air dielectric capacitors the breakdown field strength is of the order 2 to 5 MV/m; for mica the breakdown is 100 to 300 MV/m, for oil 15 to 25 MV/m to show some examples, but it can be much less if other materials are used (see [34]).

In this example we will study the effects on the electric field produced by a small air bubble present in the dielectric of a plane capacitor. The importance of the study lies on the fact that when a critical electric field is exceeded, a dielectric breakdown may occur that can lead to a failure of the device. This is the case, for instance, in high-voltage transformers containing oil as insulating dielectric.

296   13 Electrostatics with MaxFEM

Table 13.13. Plane parallel plates capacitor. Data for the numerical test

Length of the plates, a:	0.010 m
Width of the plates, b:	0.005 m
Distance between the plates, d:	0.006 m
Dielectric relative electrical permittivity, ε_r:	200
Bubble relative electrical permittivity, ε_r:	1
Vacuum electric permittivity, ε_0:	8.8542e-12 F/m
Potential drop, V_0:	100 V

Let us consider a planar capacitor of parallel plates as the one sketched in Fig. 13.34, assuming that there is a spherical air bubble inside.

We have set $a = 10$ mm, $b = 5$ mm, $d = 6$ mm (see Fig. 13.34). The potential drop between the plates is 100 V. The dielectric is a material with relative electrical permittivity equal to 200. These data are summarized in Table 13.13.

Figure 13.41 shows the computational domain. We have restricted ourselves to the 3D box formed by the capacitor with the dielectric inside, thus neglecting the fringe effect. As boundary conditions we have imposed the (known) potential on the plates of the capacitor (Dirichlet condition) and a homogeneous Neumann condition on the other sides. Figures 13.42 to 13.46 compare the results obtained with the

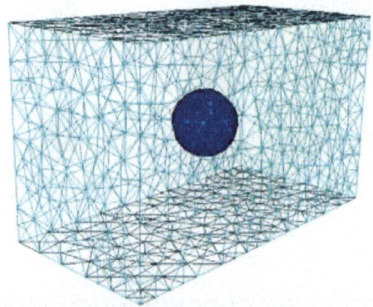

Fig. 13.41. Mesh of the computational domain with the bubble inside

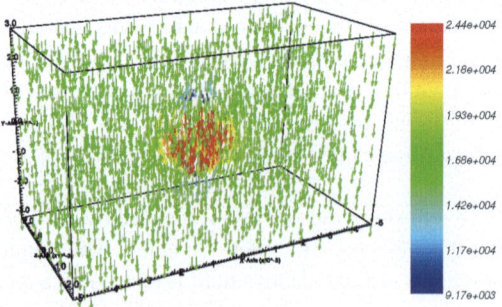

Fig. 13.42. Electric field around the bubble

13.2 Some problems arising in physical applications   297

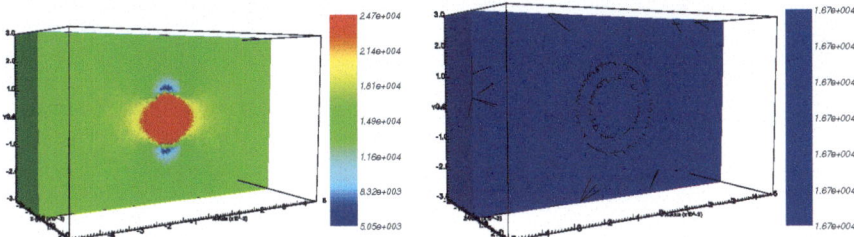

Fig. 13.43. Modulus of the electric field in a $x_3 = C$ plane. Left, with bubble; right, without bubble

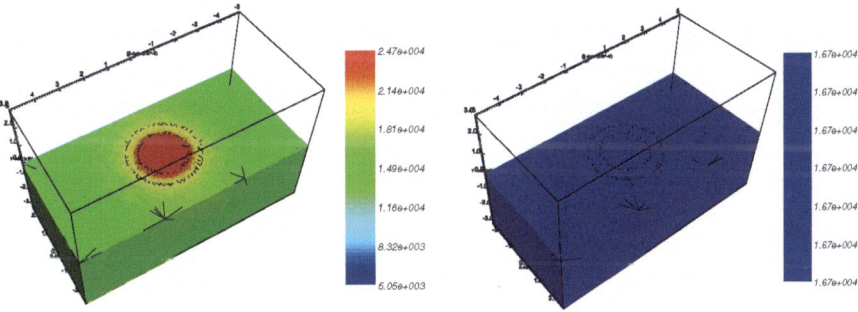

Fig. 13.44. Modulus of the electric field in a $x_2 = C$ plane. Left, with bubble; right, without bubble

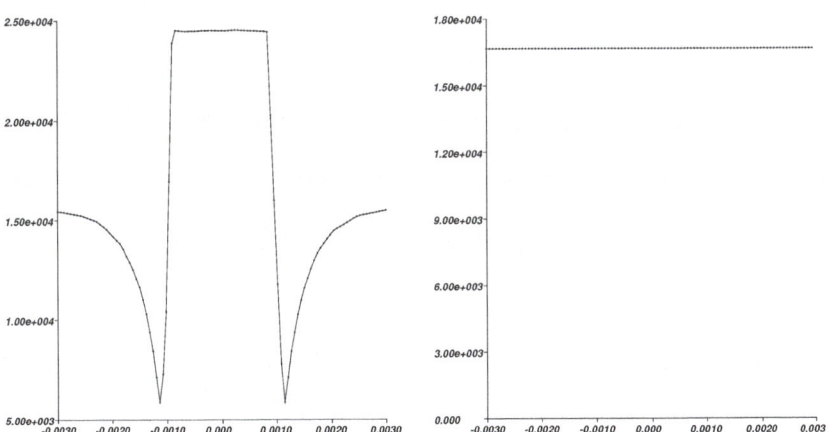

Fig. 13.45. Modulus of the electric field over the line defined by points $(0, -0.03, 0)$, $(0, 0.03, 0)$. Left, with bubble; right, without bubble

MaxFEM code depending on the presence or absence of the bubble. In particular, we can observe how the modulus of the electric field becomes higher as the bubble air is considered (see Fig. 13.45).

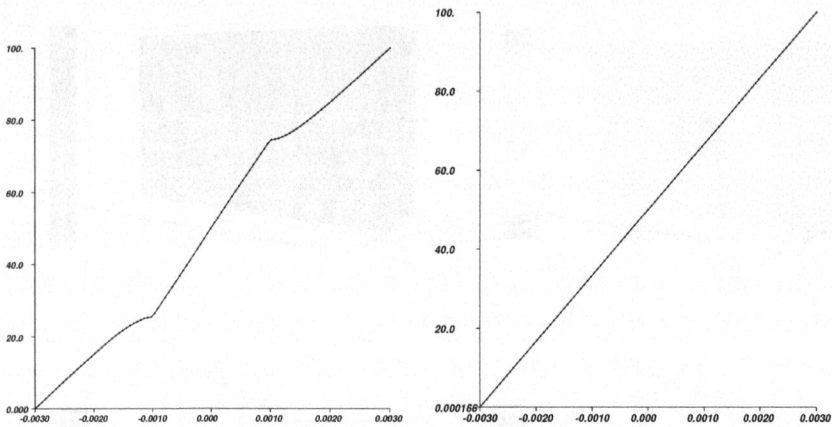

Fig. 13.46. Electric potential in the dielectric with (left) or without (right) bubble

14
Direct current with MaxFEM

In ths chapter we deal with the numerical solution of some direct current problems.

14.1 Current traversing a copper bar

Let us consider a conducting copper bar as the one sketched in Fig. 14.1. We assume that the potential is constant on each end of the bar, the potential drop between them being equal to 0.1 V. As a consequence, a current flowing through the conductor is produced which does not depend on time. Thus, the problem states in the framework of direct current problems studied in Chap. 8. The computational domain has been sketched in Fig. 14.1. As there is no kind of symmetry we must solve the problem in a 3D domain.

We will assume that the conductor is made of copper. The electrical conductivity and the dimensions considered for the bar has been summarized in Table 14.1.

Figures 14.2 and 14.3 show the current density field and the electric potential obtained from MaxFEM simulations.

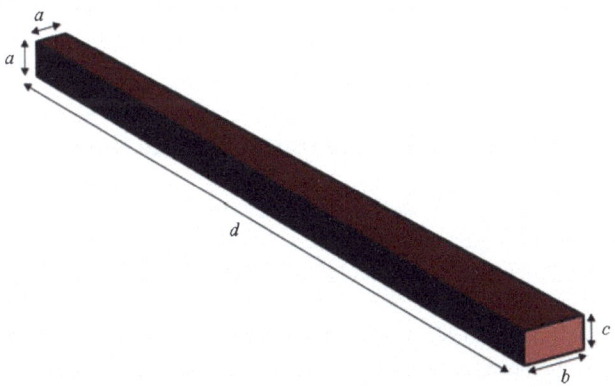

Fig. 14.1. Computational domain

A. Bermúdez, D. Gómez, P. Salgado: *Mathematical Models and Numerical Simulation in Electromagnetism.* UNITEXT – La Matematica per il 3+2 74
DOI 10.1007/978-3-319-02949-8_14, © Springer International Publishing Switzerland 2014

Table 14.1. Geometrical and physical data

Width and height of the right end, a:	0.002 m
Height of the left end, b:	0.004 m
Width of the left end, c:	0.002 m
Length of the bar, d:	0.050 m
Electrical conductivity of the bar, σ:	58.e6 $(\Omega m)^{-1}$

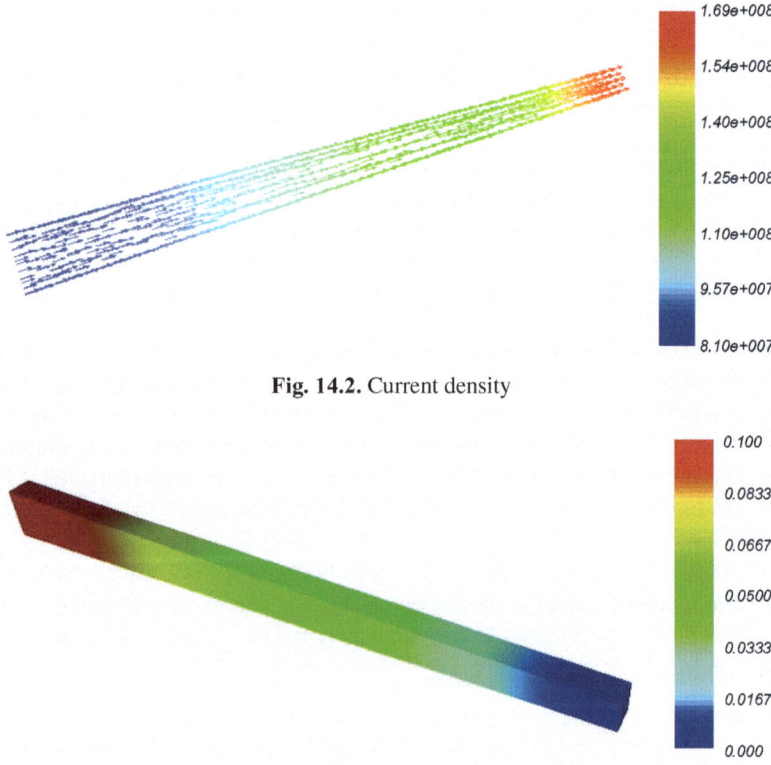

Fig. 14.2. Current density

Fig. 14.3. Potential

14.2 An electrolytic cell for aluminium production

In this section we deal with an industrial application motivated by the mathematical modelling of an aluminium electrolytic cell. We will give a brief description of the problem and refer the reader to works [13, 20, 32] and references therein for further details.

Aluminium is produced by reduction of alumina (Al_2O_3) dissolved in an electrolytic bath based on molten cryolite (NA_3AlF_6). This complex process, called *Hall-Héroult*, takes place in a rectangular cell usually lined with thermally insulating refractory materials (see [40] and Fig. 14.4). Inside it there are several prebaked carbon

14.2 An electrolytic cell for aluminium production

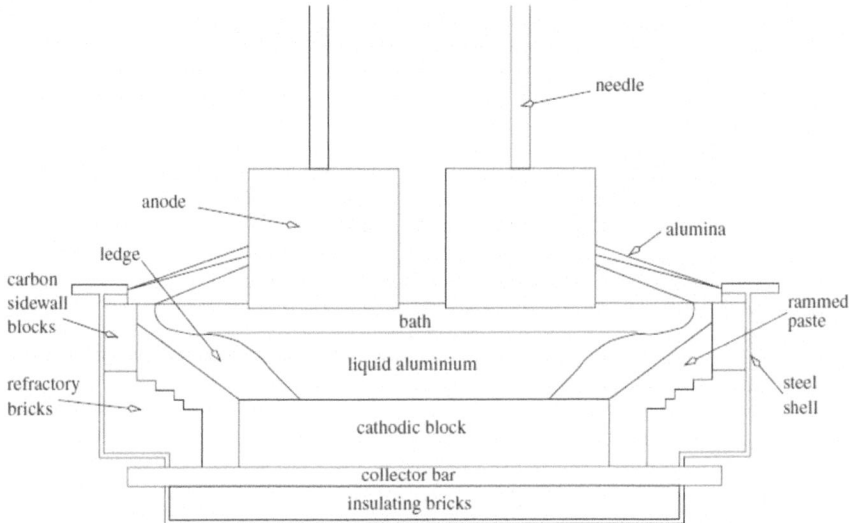

Fig. 14.4. Cross-section of an electrolytic cell. Figure courtesy of M.C. Muñiz

cathode blocks with embedded steel current collector bars surrounding the aluminium and the electrolytic bath, which are the liquid parts of the cell. The voltage drop between the anode and the cathode causes an increasing of the temperature due to the Joule effect which melts the aluminium. As the aluminium is forming, a solidified bath layer, the so-called ledge, protects the side wall of the cathode from corrosive electrolyte and reduces the heat loss from the cell. This ledge also reduces heat loss from the cathode, and works as a heat sink when an extra power is supplied to the cell. Moreover, the electromagnetic force causes the liquid metal and bath to move. The profile of the ledge strongly influences the horizontal current component so it plays a major role in the hydrodynamic behavior of the cell. Indeed, the electro-magnetic force due to this current component may induce oscillations of the aluminium bath interphase leading to a loss of cell voltage stability and current efficiency in commercial cells. This is why one of the objectives of cell sidewall design is to promote the formation of a good ledge profile to give stable, efficient cell operation and long sidewall life.

The overall process is very complex and involves many physical problems which are coupled: thermoelectrical and magneto-hydrodynamical phenomena, electrochemical reactions, phase equilibria and so on.

In this section we only deal with a simplified electromagnetic model. In particular, we will be interested in the computation of the electrostatic potential V in the cell. Since the current does not depend on time, the problem can be modelled by using the direct current problem (8.3)–(8.5) presented in Chap. 8.

As the length of the cell is greater than the two other dimensions, we will assume that the cell is infinite in that direction (let us say x_3). We also assume that the geometry is invariant in that spatial direction. Then we can reduce the problem to a two

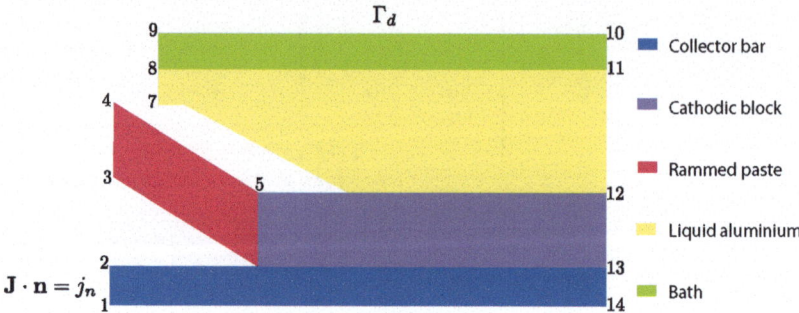

Fig. 14.5. Transversal section of the electrolytic cell. Distribution of materials and points defining the computational domain

dimensional domain. For symmetry reasons, we only consider the left half vertical section of the cell as depicted in Fig. 14.5. We notice that we have included the liquid aluminium in the domain. To determine this domain it is necessary to solve a coupled thermoelectrical model (see [20, 32]); thus, in this case, we have approximated the liquid domain from thermal results obtained in [20].

Concerning the boundary conditions, we consider a homogeneous Dirichlet condition $V = 0$ on the boundary Γ_d (see Fig. 14.5). On the rest of the boundaries, a Neumann boundary condition is considered which is null except in the cathodic bar where we impose the inward current flow $\mathbf{J} \cdot \mathbf{n}$ as illustrated in Fig. 14.5.

Table 14.2 summarizes the electrical conductivity of the materials composing the cell and also the value of the current density traversing the cathodic bar. In Table 14.3 we have included the coordinates of the points defining the cell geometry (see Fig. 14.5).

Figures 14.6 and 14.7 show the potential and the current density on the computational domain, respectivly.

We have also made some simulations considering a three dimensional domain which is more realistic. In this case, we do not consider the liquid aluminium but we take into account the genuine 3D geometry of the cathodic bar (see Fig. 14.10). The depth is considered to be 0.135 m at the widest part. In this case, as computational domain we have considered the collector bar, the cathodic block, the rammed paste

Table 14.2. Electrical conductivities and current density

Bath:	1000 $(\Omega m)^{-1}$
Cathodic block:	27000 $(\Omega m)^{-1}$
Collector bar:	1670000 $(\Omega m)^{-1}$
Cast iron:	9382600 $(\Omega m)^{-1}$
Liquid aluminium:	76900000 $(\Omega m)^{-1}$
Rammed paste:	17800 $(\Omega m)^{-1}$
RMS current density:	4577 (A/m^2)

14.2 An electrolytic cell for aluminium production

Table 14.3. Coordinates (in meters) of the points defining the computational domain

Point	x_1	x_2
1	0	0
2	0	0.15
3	0.02	0.50
4	0.02	0.79
5	0.60	0.44
6	1	0.44
7	0.20	0.79
8	0.20	0.93
9	0.20	1.06
10	2	1.06
11	2	0.93
12	2	0.44
13	2	0.15
14	2	0.15
15	0.60	0.15
16	0.30	0.79

Fig. 14.6. Distribution of the electric potential on the 2D section of the cell

Fig. 14.7. Modulus of the current density

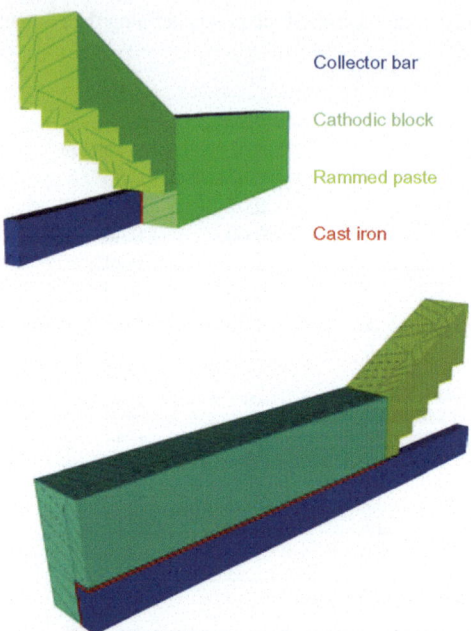

Fig. 14.8. Two different views of the 3D geometry and material distribution of the electrolytic cell

Fig. 14.9. Potential distribution

and a new cast iron part whose electrical conductivity is also included in Table 14.2. Moreover, the inward current density through the cathodic bar is 4577 A/m^2.

In Fig. 14.9 we have shown the distribution of the electric potential in the computational domain. Finally, in Fig. 14.10 we have represented the current density distribution.

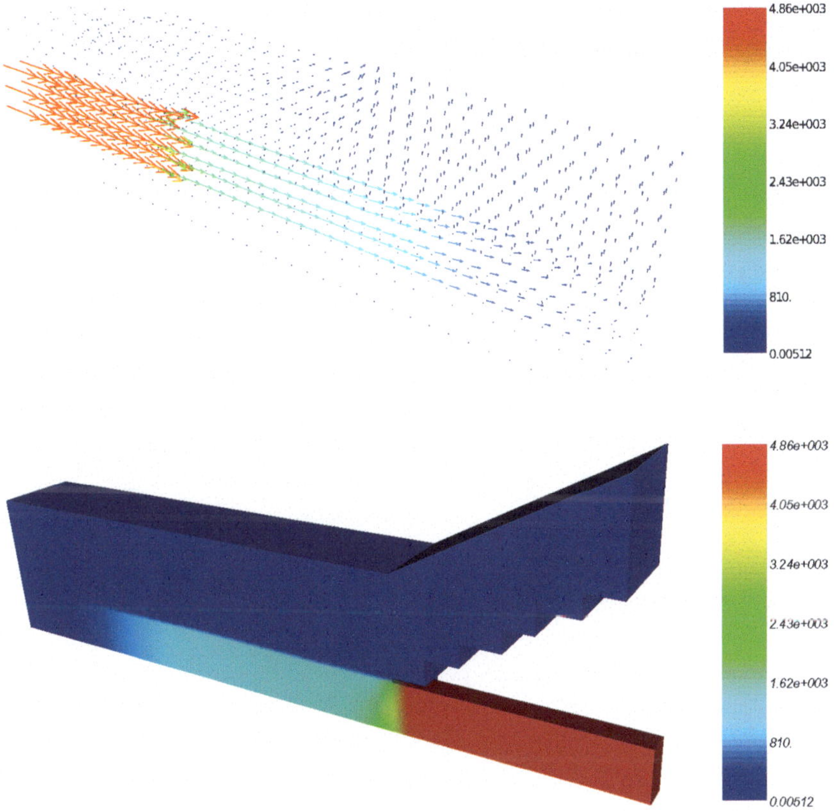

Fig. 14.10. Current density distribution

15
Magnetostatics with MaxFEM

In this chapter we solve several magnetostatics problems by using MaxFEM. For some of the problems we will provide the analytical and the numerical solution. The examples include linear and nonlinear magnetic materials.

15.1 Problems with analytical solution

15.1.1 An infinite cylinder carrying a static current intensity

In this section we present an example with cylindrical geometry, for which it is possible to obtain an analytical solution of the magnetostatics model (9.1)–(9.3) defined in the whole space.

Let us consider an infinite, straight wire of radius a which carries a static current of given intensity I in the axial direction (see Fig. 15.1). We assume the current density is uniform in the wire, i.e.,

$$\mathbf{J} = \frac{\mathrm{I}}{\pi a^2} \mathbf{e}_3.$$

The relative magnetic permeability of the wire is equal to one, i.e., $\mu = \mu_0$, and in order to compute the analytical solution, we consider a cylindrical coordinate system (ρ, θ, z) with the z-axis coinciding with the axis of the cylinder.

Since $\frac{1}{\mu_0} \mathbf{curl}\,\mathbf{B} = \mathbf{J}$, then $\mathbf{B} = B_\theta(\rho) \mathbf{e}_\theta$ by symmetry reasons. On the other hand, the Ampère's law (5.15) applied to a circle l of radius ρ (see Fig. 15.1) yields

$$\int_l \frac{B_\theta}{\mu_0} \mathbf{e}_\theta \cdot \mathbf{dl} = \int_l \frac{B_\theta}{\mu_0} \mathbf{e}_\theta \cdot \mathbf{e}_\theta dl = \mathrm{I},$$

if $\rho > a$. Therefore

$$\frac{2\pi\rho}{\mu_0} B_\theta = \mathrm{I},$$

and then

$$\mathbf{B} = \frac{\mu_0 \mathrm{I}}{2\pi\rho} \mathbf{e}_\theta.$$

A. Bermúdez, D. Gómez, P. Salgado: *Mathematical Models and Numerical Simulation in Electromagnetism*. UNITEXT – La Matematica per il 3+2 74
DOI 10.1007/978-3-319-02949-8_15, © Springer International Publishing Switzerland 2014

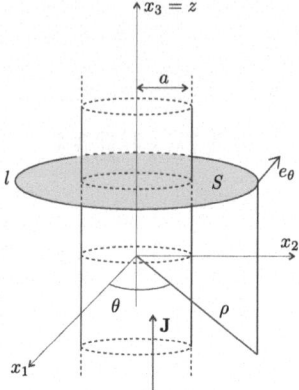

Fig. 15.1. Infinite cylinder carrying a static current. Surface S and curve l used in calculus

If $\rho < a$, we can use the same argument for a circle l of radius $\rho < a$ to obtain

$$\int_l \frac{B_\theta(\rho)}{\mu_0} \mathbf{e}_\theta \cdot \mathrm{d}\mathbf{l} = \int_l \frac{B_\theta(\rho)}{\mu_0} \mathbf{e}_\theta \cdot \mathbf{e}_\theta \mathrm{d}l = \int_S \mathbf{J} \cdot \mathbf{n} \, \mathrm{d}A = \frac{I}{\pi a^2} \pi \rho^2$$

and hence

$$\mathbf{B} = \frac{\mu_0 I \rho}{2\pi a^2} \mathbf{e}_\theta.$$

On the other hand, the magnetic field is given by $\mathbf{H} = H_\theta \mathbf{e}_\theta = \frac{B_\theta}{\mu_0} \mathbf{e}_\theta$, i.e.,

$$\mathbf{H} = \begin{cases} \dfrac{\rho I}{2\pi a^2} \mathbf{e}_\theta & \text{if } 0 \leq \rho \leq a, \\ \dfrac{I}{2\pi \rho} \mathbf{e}_\theta & \text{if } \rho \geq a. \end{cases}$$

This problem can be approximated by using a finite element method in a bounded domain Ω with suitable boundary conditions. In particular, let us consider a piece of the cylinder with height L, to be denoted by Ω. Moreover, we notice that the magnetic field only has θ-component so it satisfies the boundary condition $\mu \mathbf{H} \cdot \mathbf{n} = 0$ on $\partial \Omega$. Since this condition is satisfied on the boundary of the conducting domain we can approximate the solution of the problem by using the discretization of weak formulation (9.26)–(9.27) implemented in the MaxFEM code without considering air around. We could consider air around the cylinder in the computational domain if we were interested in approximating the magnetic field there.

To solve the problem in MaxFEM we have used the geometrical parameters and physical properties detailed in Table 15.1. Figure 15.2 shows the vector field $\mathbf{H}$ and its modulus in the cylinder.

This problem also fits in the framework of the 2D magnetostatic model (9.70)–(9.72) proposed in Sect. 9.10 to approximate the electromagnetic field in a bounded 2D domain because $\mathbf{J}$ is normal to any cross section of the cylinder.

15.1 Problems with analytical solution   309

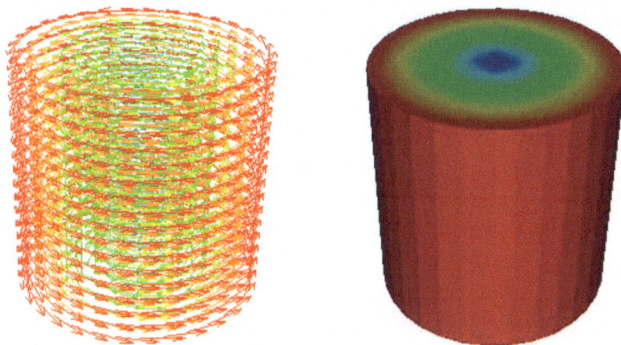

Fig. 15.2. Magnetic field in the cylinder: vector field (left) and modulus (right). Results obtained with the MaxFEM 3D magnetostatic code

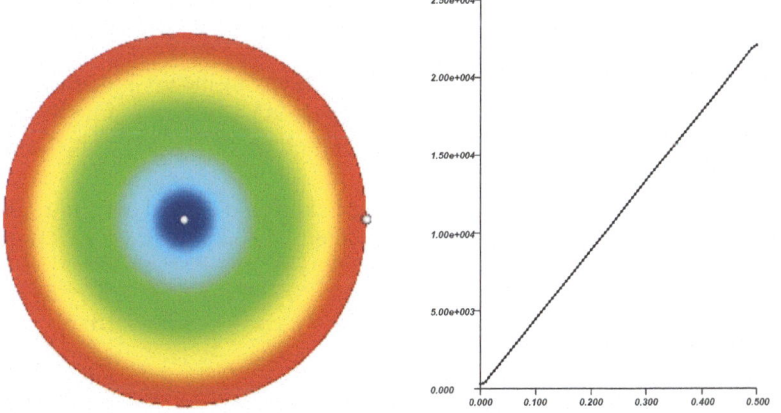

Fig. 15.3. Modulus of the magnetic field in a cross section of the domain (left) and $|\mathbf{H}|$ vs. radius (right) Results obtained with the MaxFEM 2D magnetostatic code

Table 15.1. Geometrical data and physical parameters for the numerical test

Radius of the conducting domain, a:	0.5 m
Height of the domain, L:	1 m
Relative magnetic permeability of the domain, μ_r:	1
Current intensity, I:	70000 A

Let us consider a circle of radius a as the two-dimensional computational domain $\widehat{\Omega}$. The boundary condition $A_3 = 0$ on the surface of the circle $\widehat{\Omega}$ guarantees that $\mu \mathbf{H} \cdot \mathbf{n} = 0$ and therefore there is no need again to consider the air around. Fig. 15.3 shows the modulus of the magnetic field in a cross section of the domain and its plot versus the radius obtained with the 2D magnetostatic code in MaxFEM.

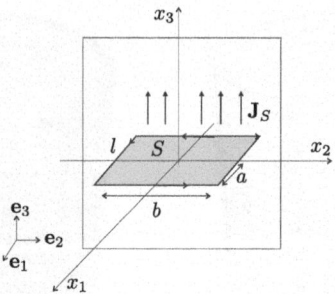

Fig. 15.4. Sheet on the x_2x_3-plane. Surface S and curve l

Notice that the numerical solution **H** only has θ-component and varies linearly with the radius ρ inside the conductor as the analytical solution predicts.

15.1.2 Magnetic field created by an infinite plane current

Let us consider a finite thin sheet finite thin sheet occupying the x_2x_3-coordinate plane and carrying a static surface current density $\mathbf{J}_S = J_S \mathbf{e}_3$ (A/m) (see Fig. 15.4). We are going to compute the magnetic field solution of problem (9.1)–(9.3) on both sides of the sheet.

Firstly, for symmetry reasons the magnetic field would depend only on x_1 and, in cartesian coordinates, only one component would be non-null. Since $\mathbf{curl}\,\mathbf{B} = \mu_0 \mathbf{J}$ and

$$\mathbf{curl}\,\mathbf{B} = \left(\frac{\partial B_3}{\partial x_2} - \frac{\partial B_2}{\partial x_3}\right)\mathbf{e}_1 + \left(\frac{\partial B_1}{\partial x_3} - \frac{\partial B_3}{\partial x_1}\right)\mathbf{e}_2 + \left(\frac{\partial B_2}{\partial x_1} - \frac{\partial B_1}{\partial x_2}\right)\mathbf{e}_3,$$

we have

$$\frac{\partial B_3}{\partial x_1} = 0, \quad \frac{\partial B_2}{\partial x_1} = \mu_0 J_S,$$

and then

$$\mathbf{B}(x) = B_2(x_1)\mathbf{e}_2.$$

Let S be the rectangle in Fig. 15.4 and l its boundary. From Ampère's law,

$$\int_l \frac{B_2(x_1)}{\mu_0} \mathbf{e}_2 \cdot d\mathbf{l} = \int_S J_S \mathbf{e}_3 \cdot \mathbf{e}_3 = J_S b,$$

and then

$$B_2(a)b + B_2(-a)(-1)b = \mu_0 J_S b. \tag{15.1}$$

Since $|B_2(a)| = |B_2(-a)|$ for symmetry reasons, the above Eq. (15.1) is only compatible with

$$B_2(a) = -B_2(-a),$$

and then

$$B_2(a) = \frac{\mu_0 J_S}{2}.$$

Thus,
$$\mathbf{B}(x) = \begin{cases} \mu_0 \dfrac{J_S}{2} \mathbf{e}_2 & \text{if } x_1 > 0, \\ -\mu_0 \dfrac{J_S}{2} \mathbf{e}_2 & \text{if } x_1 < 0, \end{cases}$$

and it is easy to see that
$$\mathbf{B}^+ \times \mathbf{n}^+ + \mathbf{B}^- \times \mathbf{n}^- = \mu_0 J_S \mathbf{e}_3.$$

This problem fits into the framework of the 2D magnetostatic model (9.70)–(9.72) by using suitable boundary conditions. In particular, it has been numerically solved with MaxFEM by considering $\widehat{\Omega} = \{(x_1, x_2) : a \leq x_1 \leq a,\ 0 \leq x_2 \leq b\}$ and taking into account that the current density is normal to any plane $x_3 = constant$. The values of a and b are arbitrary and the surface current density J_S is imposed on the surface $x_1 = 0$. Concerning boundary conditions, $A_3 = C$, $C \in \mathbb{R}$ on $x_1 = -a$ and $x_1 = a$ guarantees that $\mathbf{B} \cdot \mathbf{n} = \mathbf{0}$ on these planes; C may be chosen arbitrarily. On the other hand, notice that $\mathbf{H} \times \mathbf{n} = \mathbf{0}$ on $x_2 = 0$ and $x_2 = b$ which leads to a homogeneous Neumann boundary condition in terms of A_3.

To solve the problem we have used the geometrical parameters and physical properties detailed in Table 15.2. Figure 15.5 shows the vector field $\mathbf{B}$ in the domain $\widehat{\Omega}$ which clearly corresponds to the analytical solution presented above. Indeed, it only has x_2-component with the same value on both sides of the sheet and opposite direction.

Table 15.2. Geometrical data and physical parameters used in the solution by MaxFEM

Length of the domain in x_1-direction, $2a$:	2 m
Height of the domain, b:	1 m
Relative magnetic permeability of the domain, μ_r:	1
Surface current density, J_S:	1.e9 Am^{-1}

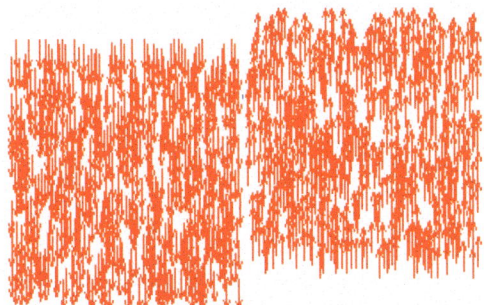

Fig. 15.5. Flux density in a plane transversal to the sheet. Results obtained with the MaxFEM 2D magnetostatic code

15.1.3 Two infinite coaxial conductors carrying a static current

Let us consider the magnetostatic problem in a cylindrical domain, consisting of two coaxial and infinite conductors separated by a magnetic material, as shown in Fig. 15.6. The conductors carry uniformly distributed currents **J** of the same intensity, but in opposite directions, that is,

$$\mathbf{J}(\rho) = \begin{cases} \dfrac{I}{\pi a^2} \mathbf{e}_z & \text{if } 0 \leq \rho \leq a, \\ 0 & \text{if } a \leq \rho \leq b, \\ -\dfrac{I}{\pi(c^2 - b^2)} \mathbf{e}_z & \text{if } b < \rho < c. \end{cases}$$

In order to obtain an analytical solution, let us assume that the relative magnetic permeability of the conductors is one, while the magnetic material has a linear behavior $\mathbf{B} = \mu \mathbf{H}$, with $\mu = \mu_r \mu_0$, being $\mu_r > 1$.

By using similar arguments to those developed in Sect. 15.1.1 in a cylindrical coordinate system, we can show that the magnetic field is given by

$$\mathbf{H} = \begin{cases} \dfrac{\rho I}{2\pi a^2} \mathbf{e}_\theta & \text{if } 0 \leq \rho \leq a, \\ \dfrac{I}{2\pi \rho} \mathbf{e}_\theta & \text{if } a \leq \rho \leq b, \\ \left\{ -\dfrac{\rho I}{2\pi(c^2 - b^2)} + \dfrac{1}{\rho}\left[\dfrac{I}{2\pi} + \dfrac{b^2 I}{2\pi(c^2 - b^2)}\right] \right\} \mathbf{e}_\theta & \text{if } b < \rho < c. \end{cases}$$

This problem fits into the framework of the 3D magnetostatic formulation (9.26)–(9.27) implemented in MaxFEM and has been solved in a bounded cylinder Ω

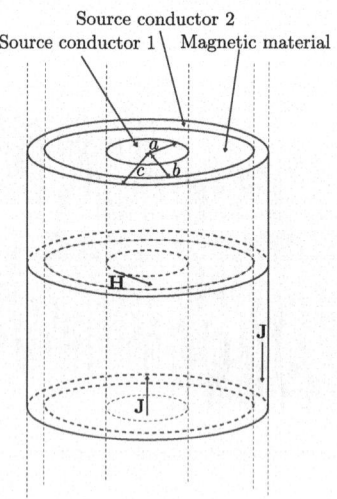

Fig. 15.6. Two infinite coaxial conductors carrying opposite currents in the axial direction

15.1 Problems with analytical solution

Table 15.3. Geometrical data and physical parameters

Radius of the inner conducting domain, a:	0.5 m
Radius of the magnetic material, b:	1 m
Outer radius of the conducting domain, c:	1.25 m
Height of the domain, L:	0.5 m
Relative magnetic permeability of the conducting domain, μ_r:	1
Relative magnetic permeability of the magnetic domain, μ_r:	10000
Current intensity in each conducting domain, I:	70000 A

of height L which contains the conducting cylinders and the magnetic material. Notice that boundary condition $\mu \mathbf{H} \cdot \mathbf{n} = 0$ is satisfied on $\partial \Omega$. To solve the problem in MaxFEM we have used the geometrical parameters and physical properties detailed in Table 15.3.

This problem also fits into the framework of the 2D magnetostatic model (9.70)–(9.72) and can be numerically solved by MaxFEM as the current density is normal to any cross section $\widehat{\Omega}$ of the 3D domain. In this case, $\widehat{\Omega}$ will be taken as the circle of radius c because boundary condition $A_3 = 0$ on the surface of the cylinder guarantees that $\mu \mathbf{H} \cdot \mathbf{n} = 0$ without the need of considering air around. Figure 15.7 shows the modulus of the magnetic field in $\widehat{\Omega}$ and its plot versus the radius. This solution approximates the analytical solution presented above.

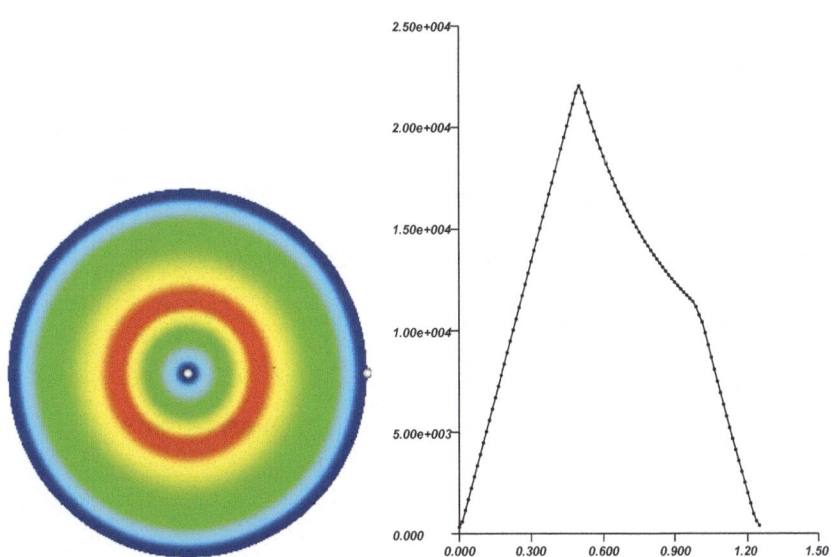

Fig. 15.7. Modulus of the magnetic field in a cross section of the domain (left) and $|\mathbf{H}|$ vs. radius (right). Results obtained with the MaxFEM 2D magnetostatic code

15.1.4 An infinite cylinder surrounded by an infinitely thin coil

Let us consider a cylindrical domain composed by a magnetic core surrounded by an infinitely thin coil (see Fig. 15.8). We consider again a cylindrical coordinate system being the z-axis the axis of the cylindrical domain. Moreover, the core and the coil are assumed to be infinite in the z-direction.

Let us suppose axisymmetry of the current sources and also that they are supported on the core-air interphase. Thus, the coil is modeled as a surface conductor in 3D and then as a curve in 2D. More precisely, let the inner coil be located on the surface S_1, $\rho = R_1$, and the outer one on the surface S_2, $\rho = R_2$. The surface current density on these surfaces is imposed by means of the equation

$$[\mathbf{H} \times \mathbf{n}] = \mathbf{J}_S = J_S \mathbf{e}_z, \qquad (15.2)$$

where $[.]$ denotes the jump across the corresponding surface. Let us suppose that $\mathbf{J}_S$ is given by $\mathbf{J}_S(\rho, \theta, z) = J_S(\rho)\mathbf{e}_z$, for $\rho = R_1, R_2$, with

$$J_S(R_1) = \frac{nI}{2\pi R_1}$$

and

$$J_S(R_2) = -\frac{nI}{2\pi R_2},$$

n being the "number of turns of the coil" and I the static current intensity in the coil.

Under the assumptions above, all the fields are independent of the azimuthal variable and the magnetic field has the form $\mathbf{H} = H_\theta(\rho)\mathbf{e}_\theta$. Thus, from the Ampère's law

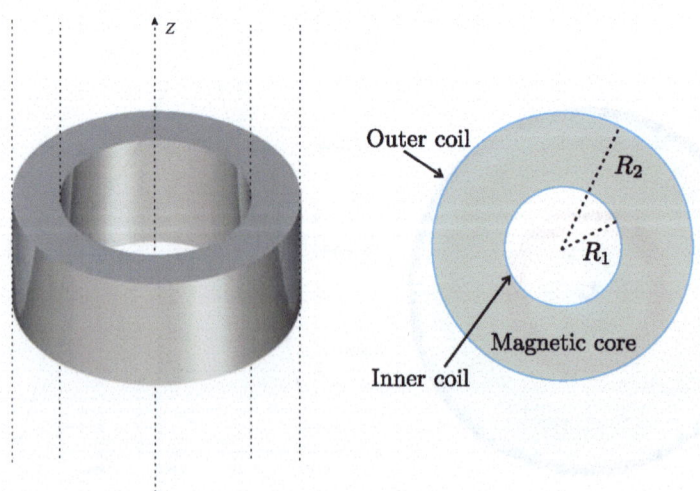

Fig. 15.8. Infinite cylindrical core (left) and sketch of the cross section (right)

the magnetic field H_θ can be obtained as

$$H_\theta(\rho) = \begin{cases} 0, & \text{if } 0 \leq \rho \leq R_1, \\ \dfrac{nI}{2\pi\rho}, & \text{if } R_1 \leq \rho \leq R_2, \\ 0, & \text{if } \rho \geq R_2. \end{cases}$$

Notice that the magnetic field H_θ does not depend on the magnetic properties of the core.

For the sake of completeness, we will also obtain the analytical expression of the magnetic vector potential in the linear case and in a particular nonlinear one.

We recall that, for linear materials, the constitutive law relating the magnetic field to the flux density reads $\mathbf{B} = \mu \mathbf{H}$, where μ is the magnetic permeability, while in nonlinear media μ depends on $|\mathbf{H}|$, i.e.,

$$\mathbf{B} = \mu(|\mathbf{H}|)\mathbf{H}. \tag{15.3}$$

In the present case, we have,

$$B_\theta(\rho) = \mathscr{B}(H_\theta(\rho)),$$

with $\mathscr{B}(H_\theta) = \mu(|H_\theta|)H_\theta$.

Taking into account that $\mathbf{B} = \mathbf{curl}\,\mathbf{A}$, we can choose the magnetic vector potential of the form

$$\mathbf{A}(\rho,\theta,z) = A_z(\rho)\mathbf{e}_z. \tag{15.4}$$

Hence,

$$\mathbf{curl}\,\mathbf{A} = -\frac{\partial A_z}{\partial \rho}\mathbf{e}_\theta \tag{15.5}$$

and equations $\mathbf{curl}\,\mathbf{H} = 0$ in the air and core, and $[\mathbf{H} \times \mathbf{n}] = \mathbf{J}_S$ become

$$\frac{\partial}{\partial \rho}\left(\frac{\rho}{\mu}\frac{\partial A_z}{\partial \rho}\right) = 0 \quad \text{in the core and in the air,} \tag{15.6}$$

$$\left[\frac{1}{\mu}\frac{\partial A_z}{\partial \rho}\right] = J_S \quad \text{on the surface of the core.} \tag{15.7}$$

On the other hand, since $\mathbf{B} = \mathbf{curl}\,\mathbf{A}$, from (15.5) we deduce

$$A_z(\rho) = A_z(\infty) + \int_\rho^\infty B_\theta(s)\,ds = \int_\rho^\infty B_\theta(s)\,ds. \tag{15.8}$$

Therefore,

$$A_z(\rho) = \begin{cases} \displaystyle\int_{R_1}^{R_2} B_\theta(s)\,ds & \text{if } 0 \leq \rho \leq R_1, \\ \displaystyle\int_\rho^{R_2} B_\theta(s)\,ds & \text{if } R_1 \leq \rho \leq R_2, \\ 0 & \text{if } \rho \geq R_2. \end{cases}$$

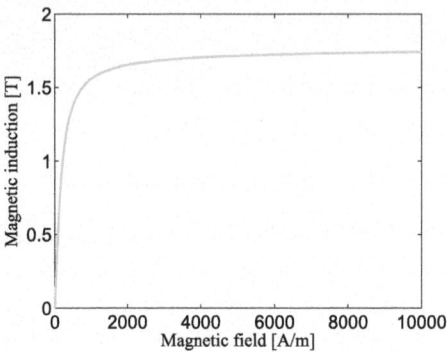

Fig. 15.9. Curve $\mathscr{B}(H_\theta)$ for the magnetic core

Then, in a linear case, the analytical expression of A_z is the following

$$A_z(\rho) = \begin{cases} \dfrac{\mu nI}{2\pi} \log(\dfrac{R_2}{R_1}) & \text{if } 0 \leq \rho \leq R_1, \\ \dfrac{\mu nI}{2\pi} \log(\dfrac{R_2}{\rho}) & \text{if } R_1 \leq \rho \leq R_2, \\ 0 & \text{if } \rho \geq R_2. \end{cases}$$

For the nonlinear case let us suppose function $\mathscr{B}$ is given by

$$\mathscr{B}(H_\theta) = \mu_0 H_\theta + \frac{2J_s}{\pi} \arctan\left(\frac{\pi(\mu_r - 1)\mu_0 H_\theta}{2 J_s}\right), \tag{15.9}$$

where $\mu_0 = 4\pi 10^{-7}$ Hm^{-1}, $\mu_r = 5000$ and $J_s = 1.75$ T. This curve, which has been represented in Fig. 15.9, is very similar to the first magnetization curve of laminated steel. By denoting $\alpha = \frac{2J_s}{\pi}$ and $\gamma = (\mu_r - 1)\frac{\mu_0}{\alpha}$, the previous expression can be written as

$$\mathscr{B}(H_\theta) = \mu_0 H_\theta + \alpha \arctan(\gamma H_\theta). \tag{15.10}$$

Then, if $R_1 \leq \rho \leq R_2$,

$$B_\theta(\rho) = \mathscr{B}(H_\theta(\rho)) = \mathscr{B}\left(\frac{nI}{2\pi\rho}\right) = \mu_0 \frac{nI}{2\pi\rho} + \alpha \arctan\left(\frac{\gamma nI}{2\pi\rho}\right). \tag{15.11}$$

Let us compute A_z. First, let us denote,

$$\beta := \frac{\gamma nI}{2\pi}.$$

Then, we have

$$A_z(\rho) = \mu_0 \frac{nI}{2\pi} \log(\frac{R_2}{R_1})$$
$$+ \alpha \left(\frac{\beta}{2} \log(\frac{\beta^2 + R_2^2}{\beta^2 + R_1^2}) + R_2 \arctan(\frac{\beta}{R_2}) - R_1 \arctan(\frac{\beta}{R_1}) \right),$$

for $0 \leq \rho \leq R_1$,

$$A_z(\rho) = \mu_0 \frac{nI}{2\pi} \log(\frac{R_2}{\rho})$$
$$+ \alpha \left(\frac{\beta}{2} \log(\frac{\beta^2 + R_2^2}{\beta^2 + \rho^2}) + R_2 \arctan(\frac{\beta}{R_2}) - \rho \arctan(\frac{\beta}{\rho}) \right),$$

for $R_1 \leq \rho \leq R_2$ and

$$A_z(\rho) = 0, \quad \text{for } \rho \geq R_2.$$

This problem fits into the framework of the 2D magnetostatic model implemented in MaxFEM by using surface current sources. Thus, it has been solved in a circle $\widehat{\Omega}$ containing the magnetic core and air around.

In this example, the current density of the coil is provided by means of the total current intensity. The device is surrounded by air in order to impose the boundary condition $A_z = 0$ on the outer boundary of the whole domain.

The geometrical data and physical parameters used in the numerical simulation have been summarized in Table 15.4. Figure 15.10 shows the modulus of the magnetic field in the ferromagnetic core which is a good approximation of the analytical solution obtained above.

Table 15.4. Geometrical data and physical parameters used in MaxFEM

Inner radius of the magnetic core, R_1:	1 m
Outer radius of the magnetic core, R_2:	1.401 m
Relative magnetic permeability of the vacuum, μ_r:	1
Current intensity, I:	3000 A (inner coil), −3000 A (outer coil)
Number of turns of the coil, n:	1

Fig. 15.10. Modulus of the magnetic field in the ferromagnetic core. Results obtained with the 2D magnetostatic code

15.2 Problems arising in physical applications

15.2.1 2D magnetostatic fields in an electromagnetic contactor

The objective of this example is to simulate an electromagnetic contactor used to establish or break electrical circuits. In general, this device is composed by a fixed part and a movable part (armature). The fixed part consists of a ferromagnetic core, an eventual permanent magnet and a coil. The movable part is composed by a metallic blade.

In this example we will simulate the magnetic behavior of a contactor where the coil is supplied by a time independent current density and magnet is not considered. Moreover, the movable part is placed at a fixed position.

We will consider a cross-section of the device by using a two-dimensional domain $\widehat{\Omega}$ as illustrated in Fig. 15.11; it includes a coil, a ferromagnetic core and the armature. The distance between the armature and the core is usually called *air-gap*.

The armature and the ferromagnetic core are made with magnetic steel which has a nonlinear behavior described by the **B** − **H** curve represented in Fig. 15.9. As we anticipated, due to the fixed position of the armature, the air-gap will have a fixed value.

The coil is made with a non-magnetic material, that is, it has the same magnetic permeability as the free space. The source current of the coil is static and equal to 2.e-5 Am^{-2}.

Notice that there is an important concentration of magnetic flux in the corners of the core and in the part of the armature close to the core. Notice moreover, that the arrows of the flux density are aligned with the core. It is very useful to analyze the numerical solution in this kind of devices. The magnetic forces depend on the size of the air-gap, on the current density carried by the coil and on the presence of magnets which enforce the magnetic fields.

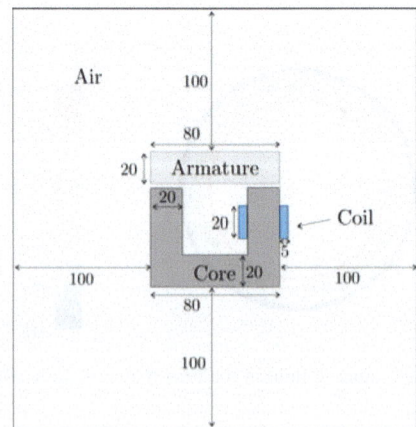

Fig. 15.11. 2D section of the contactor. Dimensions in mm

15.2 Problems arising in physical applications   319

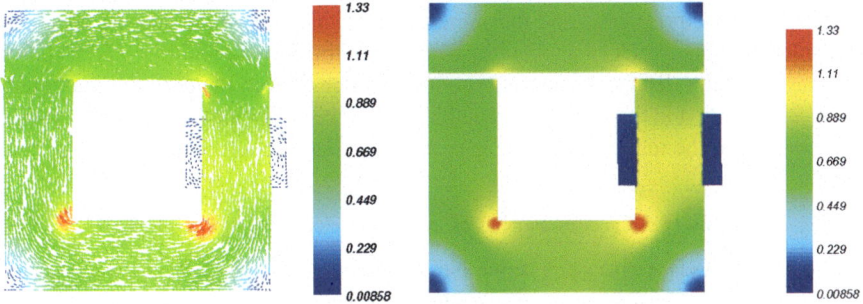

Fig. 15.12. Magnetic flux density in the contactor: vector field (left) and modulus (right)

15.2.2 A cylindrical electromagnet

The objective of this example is to compute the static magnetic fields in an axisymmetric electromagnet. The geometry consists of a ferromagnetic cylindrical core, surrounded by a toroidal coil with a rectangular cross section. A meridian section of the core and the coil is presented in Fig. 15.13 which also shows the limits of the box Ω surrounding the device. The artificial boundary of Ω is supposed to be far enough as to assume the boundary condition $\mu \mathbf{H} \cdot \mathbf{n} = 0$ there.

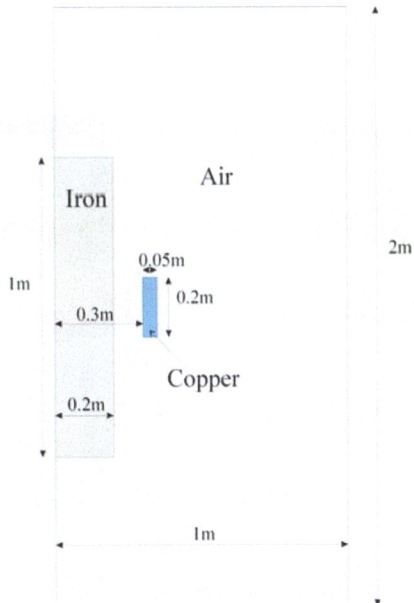

Fig. 15.13. Electromagnet. Axial section of the domain. Dimensions in m

The magnetic properties and the source current intensity in the coil are specified in Table 15.5. Figure 15.14 shows the 3D mesh of the domain and Fig. 15.15 shows the magnetic flux density in the set coil-core. Notice that, in the coil, the maximum values are reached in the zone placed at the level of the coil. Finally, Fig. 15.16 shows the modulus of the magnetic flux density in a meridian section of the magnetic core.

Table 15.5. Electromagnet. Physical parameters used in MaxFEM

Relative magnetic permeability of coil, μ_r:	1
Relative magnetic permeability of the core, μ_r:	100
Current intensity in the coil, I:	1 A

Fig. 15.14. Electromagnet. 3D mesh of the domain

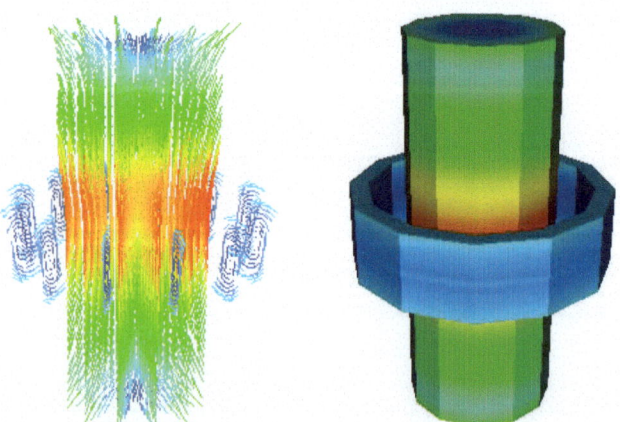

Fig. 15.15. Magnetic flux density in the set coil-core: vector field (left) and modulus (right)

15.2 Problems arising in physical applications   321

Fig. 15.16. Magnetic flux density (modulus) in a radial section of the magnetic core

15.2.3 3D magnetostatic fields in a C-magnetic core

The objective of this example is to compute the static magnetic fields in a ferromagnetic core with a copper coil wrapped around its central part (see Fig. 15.17). The core is a cylinder bent to obtain a C-shape. Figure 15.18 shows a x_2x_3-section of the domain with the dimensions in meters. The example is inspired in one presented in [28].

Fig. 15.17. C-magnetic core. 3D mesh of the domain

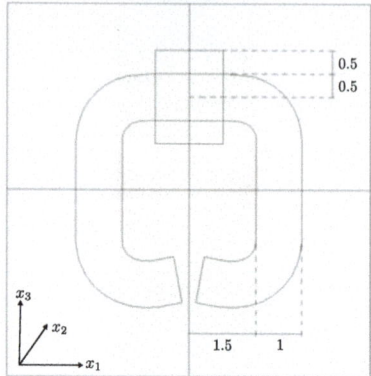

Fig. 15.18. C-magnetic core. Dimensions in m

The device is surrounded by air and on the boundary of the whole domain we consider the boundary condition $\mu \mathbf{H} \cdot \mathbf{n} = 0$. The distance between the core and the box boundary is equal to one meter. We provide a circular current $\mathbf{J}$ in the coil given by the following expression in cartesian coordinates (see Fig. 15.19):

$$\mathbf{J} = 1.\mathrm{e}7 \left(-\frac{(x_3 - 2)}{x_2^2 + (x_3 - 2)^2} \mathbf{e}_2 + \frac{x_2}{x_2^2 + (x_3 - 2)^2} \mathbf{e}_3 \right).$$

The magnetic properties of all materials are specified in Table 15.6. Figure 15.20 shows the magnetic flux density in vector form in the set coil-core. Notice that the magnetic flux density is aligned with the core and the highest values are reached in the core due to its high magnetic permeability.

Fig. 15.19. Current density in the coil

15.2 Problems arising in physical applications   323

Table 15.6. Physical parameters used in MaxFEM

Relative magnetic permeability of coil, μ_r:	1
Relative magnetic permeability of the core, μ_r:	4000

Fig. 15.20. Magnetic flux density in the set core-coil (right)

16
Eddy currents with MaxFEM

In this chapter we solve several examples governed by the time-harmonic eddy currents model by using MaxFEM. For some of the problems we will provide the analytical and the numerical solution. We exploit that some problems can be approximated by 2D or axisymmetric models but some others will require a genuine 3D model.

16.1 Problems with analytical solution

16.1.1 An infinite cylinder carrying an alternating current

In this section we present an example with cylindrical geometry, for which it is possible to obtain an analytical solution of the 3D eddy currents model (10.9)–(10.13) defined in the whole space.

Let us consider an infinite cylinder of radius R_c composed by a conducting material. We consider an alternating current $\mathbf{J}$, the intensity of which is given, going through the conductor along its axis. More precisely, this current is assumed to be axisymmetric with intensity $I(t) = I_0 \cos(\omega t)$, where I_0 is the amplitude and ω the angular frequency.

We assume that the electrical conductivity of the conductor, σ, is constant and that the relative magnetic permeability is one, i.e., $\mu = \mu_0$.

Under the above assumptions, we can obtain an analytical solution of the problem by using a cylindrical coordinate system (ρ, θ, z) with the z-axis coinciding with the axis of the cylinder. We denote by $\mathbf{e}_\rho$, $\mathbf{e}_\theta$, and $\mathbf{e}_z$ the unit vectors in the corresponding coordinate directions.

Because of the assumed conditions on $\mathbf{J}$, only the z-component of the electric field $\mathbf{E} = \frac{1}{\sigma}\mathbf{J}$ does not vanish in the conductor. Moreover, it depends on the radial coordinate ρ, but it is independent of the other two coordinates z and θ. Consequently, only the θ-component of the magnetic field $\mathbf{H} = \frac{i}{\omega\mu}\operatorname{curl}\mathbf{E}$ is not null and it only depends on coordinate ρ. Then, taking into account the expression of the **curl** operator in cylindrical coordinates (see (B.49)), we have $\mathbf{H}(\rho,\theta,z) = H_\theta(\rho)\mathbf{e}_\theta$, with H_θ

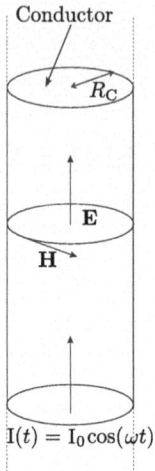

Fig. 16.1. Sketch of an infinite conductor cylinder

satisfying the equation

$$i\omega\mu H_\theta(\rho) - \frac{d}{d\rho}\left\{\frac{1}{\sigma\rho}\frac{d}{d\rho}[\rho H_\theta(\rho)]\right\} = 0, \qquad \rho \in (0, R_c),$$

and the boundary conditions

$$|H_\theta(0)| < \infty \tag{16.1}$$

$$H_\theta(R_c) = \frac{I_0}{2\pi R_c}. \tag{16.2}$$

To solve this problem, we perform the change of variable $x = \gamma\rho$, where $\gamma = \sqrt{i\omega\mu\sigma} \in \mathbb{C}$. We obtain the equation

$$x^2\frac{d^2}{dx^2}\tilde{H}_\theta(x) + x\frac{d}{dx}\tilde{H}_\theta(x) - (x^2+1)\tilde{H}_\theta(x) = 0, \qquad x \in (0, \gamma R_c),$$

where $\tilde{H}_\theta(x) = H_\theta(x/\gamma)$. This is a Bessel's equation, the solution of which is given by

$$\tilde{H}_\theta(x) = \alpha \mathscr{I}_1(x) + \beta \mathscr{K}_1(x), \tag{16.3}$$

where $\mathscr{I}_1$ and $\mathscr{K}_1$ are the *modified Bessel functions* of order 1 of first and second kind, respectively, while α and β are complex constants to be determined by using (16.1)–(16.2). Indeed, since $\mathscr{I}_1$ is bounded when $x \to 0$ while $\mathscr{K}_1$ is unbounded, we deduce that $\beta = 0$ from condition (16.1). On the other hand, taking into account boundary condition (16.2), we obtain that the magnetic field in the conductor is given by

$$\mathbf{H}(\rho,\theta,z) = \frac{I_0}{2\pi R_c}\frac{\mathscr{I}_1(\gamma\rho)}{\mathscr{I}_1(\gamma R_c)}\mathbf{e}_\theta, \qquad \rho \in (0, R_c), \ \theta \in [0, 2\pi], \ z \in \mathbb{R}.$$

16.1 Problems with analytical solution

On the other hand, the magnetic field created by an infinite circular cylindrical conductor of radius R_c carrying an axially aligned and symmetric current of intensity I_0, is computed using the Ampère's law (see for instance Sect.15.1.1). In cylindrical coordinates it is given by $\mathbf{H}(\rho,\theta,z) = H_\theta(\rho)\mathbf{e}_\theta$, with

$$H_\theta(\rho) = \frac{I_0}{2\pi\rho}, \quad z \in \mathbb{R}.$$

Once more, the magnitude of H_θ depends only on the radial coordinate ρ.

Notice that one can obtain the analytical magnetic induction from the magnetic field, $\mathbf{B} = \mu\mathbf{H} = \mu H_\theta(\rho)\mathbf{e}_\theta$, the electric field inside the conductor, $\mathbf{E} = \frac{1}{\sigma}\mathbf{curl}\,\mathbf{H}$, and the current density $\mathbf{J} = \sigma\mathbf{E}$.

This problem fits into the framework of the 3D time-harmonic model (10.18)–(10.28) to approximate the electromagnetic fields in a bounded domain by using a finite element method. In particular, let us consider a bounded domain Ω containing a conductor Ω_C and a dielectric Ω_D. More precisely, Ω_C and Ω_D are coaxial cylinders of radius R_C and R_D and bounded sections with height L of their respective infinite cylinders (see Fig. 16.2).

To impose boundary conditions, we assume that the current enters the conductor through the bottom of the inner cylinder, $\Gamma_J = \Gamma_J^1$, and leaves it through Γ_E, (see Fig. 16.2). Notice that in this case we only use current intensities as data and $\mathbf{J}_S = \mathbf{0}$. Since the current density goes through Ω_C in the axial direction, the electric field automatically satisfies the boundary condition $\mathbf{E} \times \mathbf{n} = \mathbf{0}$ on $\Gamma_J \cup \Gamma_E$. Moreover, we notice that the magnetic field, which has only θ-component, also satisfies the boundary condition $\mu\mathbf{H} \cdot \mathbf{n} = 0$ on $\partial\Omega$. Thus, we can approximate the solution by using the numerical code implemented in MaxFEM which solves problem (10.18)–(10.28) in terms of the magnetic field. The numerical method is described in detail in [10].

To solve the problem we have used the geometrical parameters and physical properties detailed in Table 16.1.

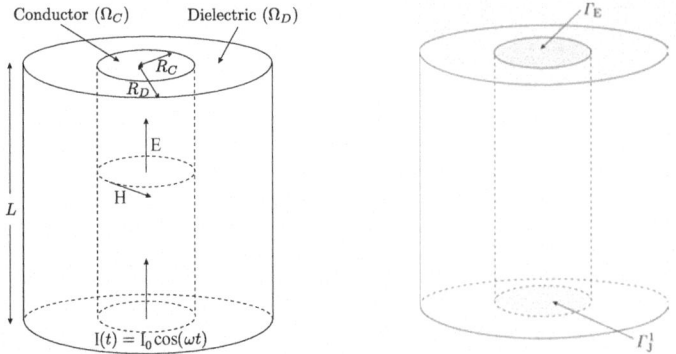

Fig. 16.2. Sketch of the bounded domain Ω (left). Description of the boundary of the conductor domain: Γ_J and Γ_E (right)

Table 16.1. Geometrical data and physical parameters used in the solution by MaxFEM

Radius of the conducting domain, R_c:	1 m
Radius of the whole domain, R_D:	2 m
Height of the domain, L:	0.5 m
Electrical conductivity of the conductor, σ:	240000 $(\Omega m)^{-1}$
Relative magnetic permeability of the whole domain:	1
Current input intensity (RMS value and phase):	62000 A, null phase
Frequency, f:	50 Hz

Figure 16.3 shows the RMS current density in the inner conductor and its variation with respect to the radius, where we can appreciate the skin effect that concentrates the current density on the surface of the conductor. Moreover, the numerical solution agrees quite well with the analytical one.

We emphasize that although this problem has cylindrical symmetry it cannot be solved by means of the axisymmetric code implemented in MaxFEM because it requires the current density to be normal to any meridian section of the domain; a situation that has not yet been included in the code. However, it could be solved by using the two-dimensional model by defining suitable boundary conditions on Γ for the magnetic vector potential which is the main unknown in this case (see a similar example below).

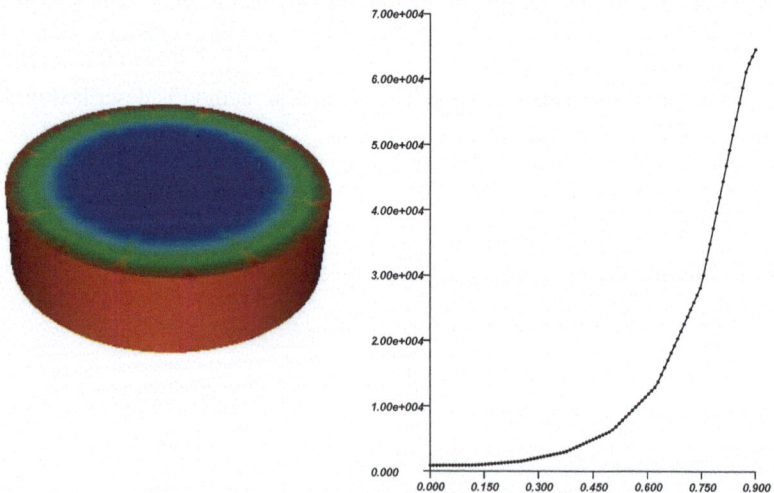

Fig. 16.3. Magnetic field in the whole domain

16.1.2 Two infinite coaxial conductors carrying an alternating current

In this section we state a similar problem to the previous one, but with two coaxial conductors.

Let us consider a cylindrical domain of infinite length which consists of two coaxial different conductors separated by a dielectric. Let us denote by R_{C_1} the radius of the inner conductor and by R_{C_2} the one of the outer cylinder (see Fig. 16.4).

Concerning the physical data, the electrical conductivity and the magnetic permeability can be different in both conducting domains. They will be denoted by σ_1 and μ_1 (respectively σ_2 and μ_2) in the inner conductor (respectively in the outer conductor).

We consider an alternating current **J**, the intensity of which is given, going through the inner conductor along its axis. More precisely, this current is assumed to be axisymmetric with an intensity $I(t) = I_0 \cos(\omega t)$. We also assume that the current intensity through any horizontal section of the second conductor vanishes.

Next, we obtain the analytical solution by using a cylindrical coordinate system and following similar steps to those developed in the previous section. First, because of the assumed conditions on **J**, only the z-component of the electric field $\mathbf{E} = \frac{1}{\sigma}\mathbf{J}$ is not null in the conductor and it depends only on the radial coordinate ρ, $\mathbf{E}(\rho,\theta,z) = E_z(\rho)\mathbf{e}_z$. Consequently, only the θ-component of the magnetic field $\mathbf{H} = \frac{i}{\omega\mu}\,\mathbf{curl}\,\mathbf{E}$ does not vanish and it also depends only on coordinate ρ. Then, taking into account the expression of the **curl** operator in cylindrical coordinates, we have

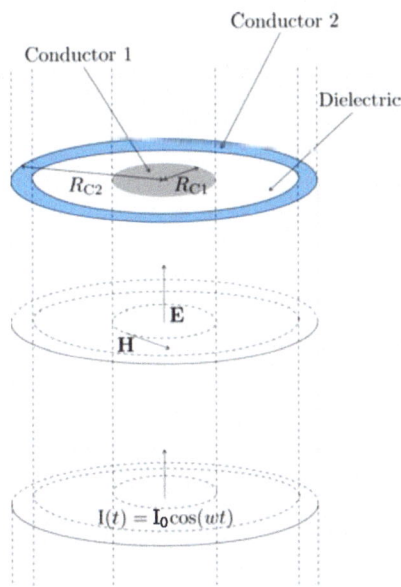

Fig. 16.4. Sketch of a cylindrical domain with two coaxial conductors

$\mathbf{H}(\rho,\theta,z) = H_\theta(\rho)\mathbf{e}_\theta$, with H_θ satisfying the equations

$$i\omega\mu H_\theta(\rho) - \frac{d}{d\rho}\left\{\frac{1}{\sigma_1\rho}\frac{d}{d\rho}[\rho H_\theta(\rho)]\right\} = 0, \quad \rho \in (0, R_{C_1}), \tag{16.4}$$

$$\frac{1}{\rho}\frac{d}{d\rho}[\rho H_\theta(\rho)] = 0, \quad \rho \in (R_{C_1}, R_D), \tag{16.5}$$

$$i\omega\mu H_\theta(\rho) - \frac{d}{d\rho}\left\{\frac{1}{\sigma_2\rho}\frac{d}{d\rho}[d\rho H_\theta(\rho)]\right\} = 0, \quad \rho \in (R_D, R_{C_2}), \tag{16.6}$$

where the second equation is due to the fact that **curl H** = 0 in Ω_D. Moreover, H_θ must be continuous across the interfaces between the different domains and it satisfies the following boundary conditions:

$$|H_\theta(0)| < \infty, \tag{16.7}$$

$$H_\theta(R_{C_1}) = \frac{I_0}{2\pi R_{C_1}}, \tag{16.8}$$

$$H_\theta(R_{C_2}) = \frac{I_0}{2\pi R_{C_2}}. \tag{16.9}$$

The two last conditions are due to the assumption that the current intensity through any horizontal section of the inner conductor is I_0, while it is null through any horizontal section of the outer conductor.

In order to solve this problem, we perform the change of variable $x = \gamma_1 \rho$ in $(0, R_{C_1})$, (respectively, $x = \gamma_2 \rho$ in (R_D, R_{C_2})), where $\gamma_1 = \sqrt{i\omega\mu_1\sigma_1} \in \mathbb{C}$, (respectively, $\gamma_2 = \sqrt{i\omega\mu_2\sigma_2} \in \mathbb{C}$). Then, we obtain the equations

$$x^2\frac{d^2}{dx^2}\widetilde{H}_\theta^1(x) + x\frac{d}{dx}\widetilde{H}_\theta^1(x) - (x^2+1)\widetilde{H}_\theta^1(x) = 0, \quad x \in (0, \gamma_1 R_{C_1}), \tag{16.10}$$

$$H_\theta^D = \frac{C}{r}, \quad r \in (R_{C_1}, R_D), \tag{16.11}$$

$$x^2\frac{d^2}{dx^2}\widetilde{H}_\theta^2(x) + x\frac{d}{dx}\widetilde{H}_\theta^2(x) - (x^2+1)\widetilde{H}_\theta^2(x) = 0, \quad x \in (\gamma_2 R_D, \gamma_2 R_{C_2}), \tag{16.12}$$

where $\widetilde{H}_\theta^1(x) = H_\theta(x/\gamma_1)$, (respectively, $\widetilde{H}_\theta^2(x) = H_\theta(x/\gamma_2)$) and H_θ^D denotes the θ-component of **H** in the dielectric domain. Finally, C is a constant which will be determined by using the continuity conditions between conductors and dielectric.

Equations (16.10) and (16.12) are Bessel's equations with solutions:

$$\widetilde{H}_\theta^1(x) = a_1\mathscr{I}_1(x) + b_1\mathscr{K}_1(x), \tag{16.13}$$

$$\widetilde{H}_\theta^2(x) = a_2\mathscr{I}_1(x) + b_2\mathscr{K}_1(x), \tag{16.14}$$

where $\mathscr{I}_1$ and $\mathscr{K}_1$ are the modified Bessel functions of order 1 of first and second kind, respectively, and $a_1, a_2, b_1, b_2 \in \mathbb{C}$.

Since $\mathscr{I}_1$ is bounded when $x \to 0$ while $\mathscr{K}_1$ is bounded when $x \to \infty$, we deduce that $b_1 = 0$ from condition (16.7). On the other hand, taking into account boundary

16.1 Problems with analytical solution

Table 16.2. Geometrical data and physical parameters used in the solution by MaxFEM

Radius of the inner conducting domain, R_{C_1}:	0.5 m
Radius of the dielectric material, R_D:	1 m
Outer radius of the conducting domain, R_{C_2}:	1.25 m
Relative magnetic permeability of the whole domain, μ_r:	1
Current input intensity in the inner conductor (RMS value and phase):	62000 A, null phase
Frequency, f:	50 Hz

conditions (16.8) and the continuity condition of H_θ at $\rho = R_{C_1}$, we obtain that

$$H_\theta^1(\rho) = \frac{I_0}{2\pi R_{C_1}} \frac{\mathscr{I}_1(\gamma_1 \rho)}{\mathscr{I}_1(\gamma_1 R_{C_1})},$$

and

$$H_\theta^D(\rho) = \frac{I_0}{2\pi\rho}.$$

Finally, by using boundary condition (16.9) and the continuity condition of H_θ at $\rho = R_D$, we obtain the following linear system for the unknowns a_2 and b_2:

$$a_2 \mathscr{I}_1(\gamma_2 R_D) + b_2 \mathscr{K}_1(\gamma_2 R_D) = \frac{I_0}{2\pi R_D}, \quad (16.15)$$

$$a_2 \mathscr{I}_1(\gamma_2 R_{C_2}) + b_2 \mathscr{K}_1(\gamma_2 R_{C_2}) = \frac{I_0}{2\pi R_{C_2}}, \quad (16.16)$$

which can be solved from the geometrical and physical data R_D, R_{C_2}, μ_2, σ_2 and I_0. Once constants a_2 and b_2 are determined, the analytical solution is given by (16.13)–(16.14).

This problem fits into the framework of the 3D time-harmonic model (10.18)–(10.28), and also in that of the 2D eddy currents model (10.79)–(10.82) because **J** is normal to any cross section of the cylindrical domain. Thus, to reduce the computational effort, we will use the 2D model. Let us consider a circle of radius R_{C_2} as the two-dimensional domain $\widehat{\Omega}$. Boundary condition $A_3 = 0$ on the surface of the circle $\widehat{\Omega}$ guarantees that $\mu \mathbf{H} \cdot \mathbf{n} = 0$ and therefore there is no need again to consider air around except if we were interested in computing the magnetic field outside the outer conductor. Table 16.2 shows the data used to compute the numerical solution.

Figure 16.5 shows the modulus of the RMS current density in the inner and outer conductor. Notice that the curent density is higher on the surface of the domains due to the skin effect and it is more important in the inner one. Figure 16.6 shows the modulus of the magnetic field in the whole domain and its variation with respect to the radius which agrees with the analytical solution.

Fig. 16.5. Current density in the whole domain

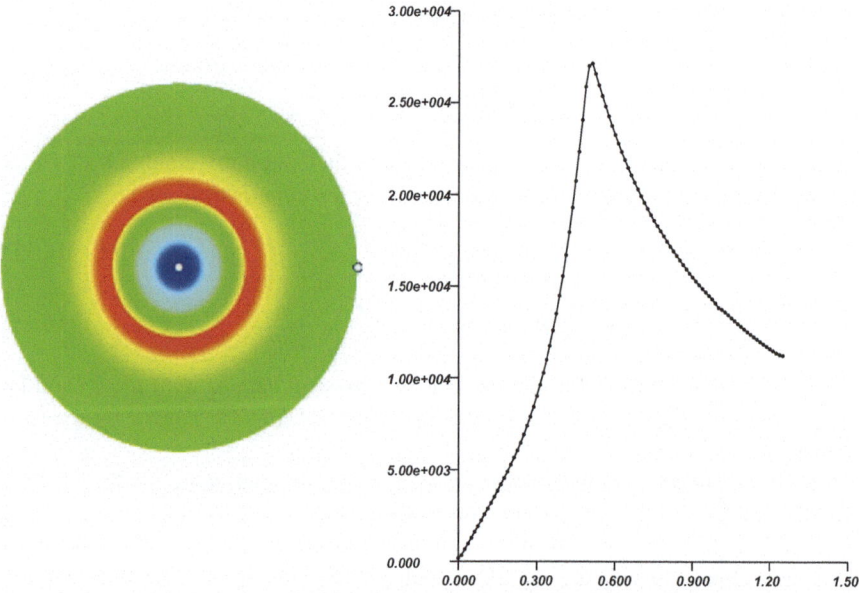

Fig. 16.6. Modulus of the magnetic field in a cross section of the domain (left) and $|\mathbf{H}|$ vs. radius (right)

16.2 Problems arising in physical applications

16.2.1 Metallurgical electrodes of an electric arc furnace

The objective of this example is to compute the induced currents in metallurgical electrodes used in electric furnaces devoted, for instance, to silicon production [18]. A simple sketch of the furnace can be seen in Fig. 16.7. It consists of a cylindrical pot containing charge materials and three electrodes disposed conforming an equilateral triangle. Electrodes are the main components of the reduction furnaces and their purpose is to conduct the electric current to the center of the furnace. The source current,

16.2 Problems arising in physical applications

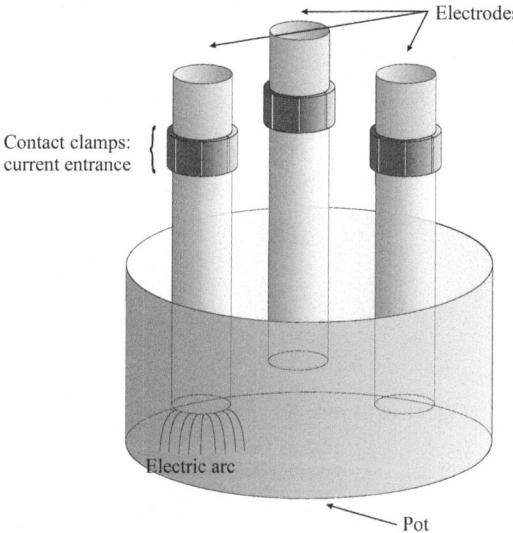

Fig. 16.7. Sketch of a reduction furnace

usually a three-phase alternating current, enters the electrode through the so-called "contact clamps" and goes down crossing the column and generating heat by Joule effect. At the tip of each electrode an electric arc is produced, generating the high temperatures that activate the reduction chemical reaction. The design and control parameters of electrodes play an important role in the performance of the furnace. This is why we can find a lot of publications in the last years devoted to the numerical simulation of different kind of electrodes in order to optimize its behavior (see for instance [6, 18] and references therein). The mathematical modelling of the electrodes requires to consider thermal, electromagnetic and mechanical phenomena, all of them coupled. However, we will only focus here on the electromagnetic problem.

The electromagnetic simulation of the complete furnace would actually require a genuine 3D model including the electrodes and the connection of contact clamps with the electrical network. For instance, in [18] the authors have used a 3D eddy currents model to simulate the behavior of the furnace by simplifying the electrical connection with only one wire per electrode. However, the detailed geometry of the contact clamps and wire connections is complex and would lead to use large meshes. Thus, in this section we will focus on the modelling of a horizontal section of the furnace below the contact clamps by using a two-dimensional eddy currents model. This option is cheaper than a 3D model and allows us to illustrate the proximity effect between the three electrodes [12] and the current distribution in the different materials. It is important to remark that the computation of eddy currents in electrodes with cylindrical geometry is often performed by using an axisymmetric model, neglecting in that case the cited proximity effect (see, for instance, [6]). Indeed, the combination of the two models allows us to analyze the behavior of the electrodes varying its size, the properties of materials, the current intensity, the slipping rate, etc.

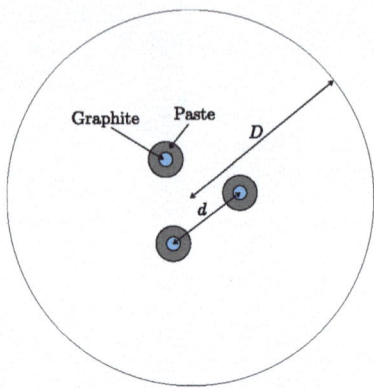

Fig. 16.8. Sketch of two-dimensional domain

Table 16.3. Geometrical data and physical parameters used in the solution by MaxFEM

Radius of the graphite:	0.2 m
Radius of the electrode:	1 m
Edge of the equilateral triangle, d:	2.5 m
Distance from the center of the pot to the outer boundary, D:	20 m
Electrical conductivity of the graphite, σ:	240000 $(\Omega m)^{-1}$
Electrical conductivity of the paste, σ:	10000 $(\Omega m)^{-1}$
Relative magnetic permeability of the whole domain:	1
Current input intensity (RMS value and phase):	70000 A, phase: 0-180-240 degrees
Frequency, f:	50 Hz

Thus, in order to use the 2D eddy currents model (10.79)–(10.82), let us assume that the current density field is orthogonal to the cross-section $\widehat{\Omega}$ depicted in Fig. 16.8. This assumption is valid if we consider a cross-section at low enough level under the contact clamps.

The electrodes considered in this example are known as ELSA electrodes [18]. They consist of a core of graphite and an outer region of paste; the centers of the electrodes are vertices of an equilateral triangle. Table 16.3 specifies the geometrical data and the physical properties of graphite and paste. The latter strongly depend on temperature but we consider here a constant value, for simplicity.

Since we know the current intensity which goes through each electrode, we provide the RMS value of the alternating current and the phase in each electrode (i.e., in the set graphite-paste); in particular, we consider an RMS value of 70000 A with a counterclockwise advance of phase. The frequency is equal to 50 Hz.

Figures 16.9 to 16.11 show some results from the numerical simulations. In particular, Figs. 16.9 and 16.10 show the modulus of the current density in paste and graphite, respectively. Notice that, in both cases, the current is more concentrated on the surface of each material due to the skin effect and its distribution is not axisym-

16.2 Problems arising in physical applications 335

Fig. 16.9. RMS current density in paste

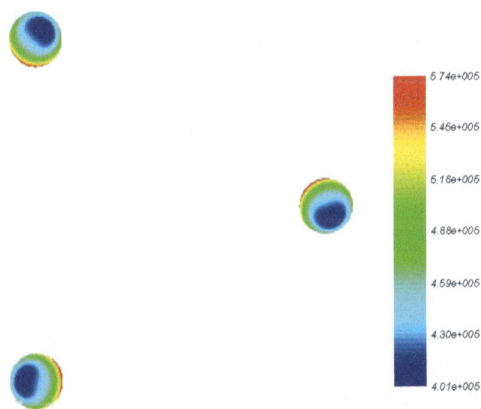

Fig. 16.10. RMS current density in graphite

Fig. 16.11. Active power (Wm^{-3}) in the electrode (graphite and paste)

Table 16.4. Voltage drop per unit length in each electrode

Electrode 1 (RMS value and phase):	10.98 Vm^{-1}	phase: 75.15 degrees
Electrode 2 (RMS value and phase):	10.98 Vm^{-1}	phase: -164.85 degrees
Electrode 3 (RMS value and phase):	10.98 Vm^{-1}	phase: -44.85 degrees

metric due to the proximity effect. Moreover, the current is higher in the graphite because it is a better conductor. Figure 16.11 shows the density of the active power in the electrode which is also greater in the graphite than in the paste.

The numerical simulation also provides the voltage drop per unit length in each electrode which are given in Table 16.4. Notice that the RMS value is the same in all electrodes and they are phase-delayed in a similar way as the current intensities.

We could employ the 3D eddy currents model (10.18)–(10.28) to simulate the complete furnace by imposing the input current intensities at the end of each wire by using a detailed mesh of the corresponding furnace and wires (see, for instance [18], where there are some numerical results in a 3D domain with simplifications in the geometry of the wires).

16.2.2 A two-dimensional example related with non-destructive testing

Eddy currents inspection is widely used as a *Non-Destructive Testing* (NDT) tool due to the variety of inspections and measurements that can be performed, such as detection of small cracks and other defects near the surface of a conducting medium (usually called specimen).

The basis of a standard eddy currents inspection consists of a coil carrying an alternating current which is placed near an electrically conductive specimen. The alternating current in the coil generates a changing magnetic field, which interacts with the test object and induces eddy currents. The presence of any flaws in the specimen will cause a change in the distribution of the magnetic fields. Although usually it is necessary to analyze the variation of the electromagnetic fields under different positions of the defect, here we will consider a fixed position of the defect. Moreover, we will consider a particular device inspired in the geometry presented in [51] in the framework of the so-called *magnetic flux leakage* technique; in particular, the coil is around a ferromagnetic core (steel) and induces currents in the specimen. We notice that the model presented in [51] is more complex because it involves the motion of the specimen.

We solve the problem in a transversal section of the device, because we suppose that the fields do not vary in the normal direction. The geometrical data of the transversal section of the device are specified in Fig. 16.12. The current density in the coil is assumed to be uniformly distributed because it is composed by thin wires. Table 16.5 shows the physical parameters used in the simulation.

16.2 Problems arising in physical applications   337

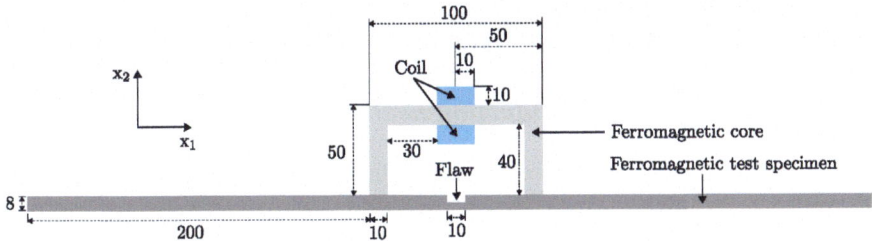

Fig. 16.12. Sketch of the 2D section of the device. Dimensions in mm

Table 16.5. Physical parameters

Electrical conductivity of the copper, σ:	58.e6 $(\Omega m)^{-1}$
Electrical conductivity of the steel, σ:	1.e5 $(\Omega m)^{-1}$
Relative magnetic permeability of copper:	1
Relative magnetic permeability of steel:	3000
RMS Current intensity, I:	4000 A (inner coil)
	−4000 A (outer coil)
Frequency, f:	50 Hz

Fig. 16.13. RMS current density in the magnetic core and the specimen

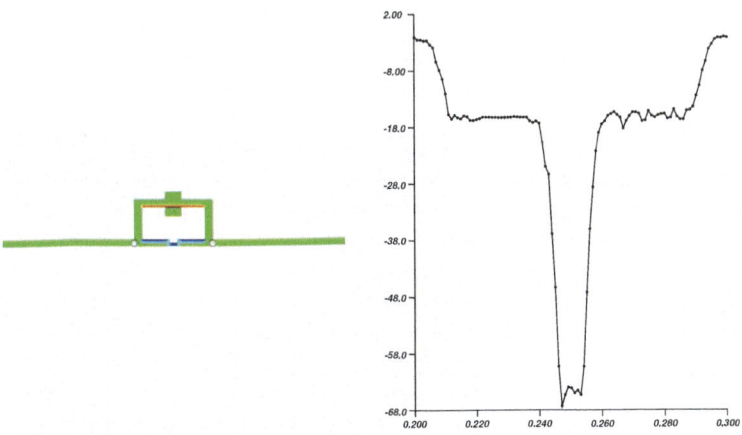

Fig. 16.14. Magnetic induction (x_1-component) and its variation vs. x_1 at the level of the defect and with $0.2 \leq x_1 \leq 0.3$

Figure 16.13 shows the RMS current density in the magnetic core and in the specimen. Figure 16.14 shows the distribution of the x_1-component of the magnetic induction **B** and its variation in the specimen with respect to the length at the level of the defect. We can observe the change in its magnitude when the defect appears.

16.2.3 An axisymmetric induction heating furnace

The aim of this section is to present an application of the eddy currents model in an axisymmetric setting.

The example is based on induction heating, which is a physical process extensively used in the metallurgical industry for different applications such as metal smelting, preheating for operations of welding, purification systems and, in general, in those processes needing a high speed of heating in located zones of a piece of a conductive material.

The main components of an induction heating system are an induction coil connected to a power-supply providing an alternating electric current and a conductive workpiece to be heated, placed inside the coil (see Fig. 16.15). The workpiece is formed by the crucible and the load within, which is the material to melt. The crucible is a cylindrical vessel made of a refractory material with higher temperature resistance than the substances it is designed to hold in. The alternating current traversing the coil generates eddy currents in the workpiece that is subsequently heated by the ohmic losses. When the load melts, the electromagnetic field produced by the coil interacts with the electromagnetic field produced by the induced current. The resulting force causes stirring that helps to homogenize the melt composition and temperature.

Usually, manufactures need to understand the influence on the furnace performance of certain geometrical parameters such as the crucible thickness, its distance to the coil, the number of turns of the coil, or physical parameters such as the electrical conductivity of the refractory materials, the working frequency and the current

Fig. 16.15. Induction furnace off (left) and on (right). Photographs courtesy of Mr. Víctor Valcarcel, Instituto de Cerámica, Universidade de Santiago de Compostela

16.2 Problems arising in physical applications

supplied to the inductor. Numerical simulation proves to be an important tool for this purpose.

The overall process is very complex and involves different physical phenomena: electromagnetic, thermal with phase-change and hydrodynamic in the liquid region; all of them are coupled and it is essential to consider a suitable mathematical model to achieve a realistic simulation. Here we will only deal with the electromagnetic problem. We will give a brief description and refer the reader to reference [8] for further details.

Let us consider an induction furnace where the inductor is a copper helical coil with 12 turns carrying cool water for refrigeration purpose. Inside the coil a crucible is placed, containing the silicon to be melted. The crucible is made of graphite and surrounded by a refractory material to avoid heat losses. For safety reasons, the induction coil is also embedded in the refractory layer. Above this refractory layer there is a layer of another refractory material. Finally, the induction furnace rests on a base of concrete, which will be also considered in our computational domain.

In order to use the axisymmetric eddy currents model (10.99)–(10.102), the coil is replaced by several superimposed rings with toroidal geometry. Thus, the computational domain $\widehat{\Omega}$ is a meridian section of the furnace (load, crucible, refractory materials and coil), the cooling water and the surrounding air, as depicted in Fig. 16.16. In particular, the boundary of $\widehat{\Omega}$ is far enough of the furnace in order to be allowed to consider the approximated Dirichlet boundary condition $A_\theta = 0$ on the whole boundary.

The goal is to compute the distribution of the current density and active power in the conducting parts under different operating frequencies.

It is well known that the high frequencies used in induction heating applications give rise to skin effect. This skin effect forces the alternating current to flow in a thin

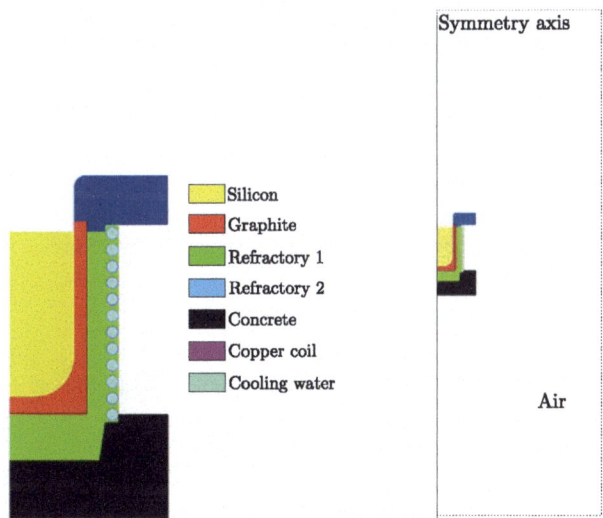

Fig. 16.16. Induction furnace. Distribution of materials (left) and computational domain (right)

layer close to the surface of the workpiece, at an average depth called the skin depth. Thus, the ohmic losses are concentrated in the external part of the workpiece in such a way that the higher the frequency the thinner the skin depth. It is crucial to control the distribution of these ohmic losses, since they could cause very high temperatures in the crucible thus reducing its lifetime.

We have performed two numerical simulations of the furnace with the same value of the current intensity and two different values of the frequency, 200 Hz and 3500 Hz, to see how this parameter affects to the distribution of ohmic losses.

The geometrical data and operating parameters for the numerical tests have been summarized in Tables 16.6 and 16.7 below. In this case, and according to the mathematical model (10.99)–(10.102), the current intensity is provided in each section of the coil while the voltage drop in the workpiece is null. On the other hand, all the materials have a linear magnetic behavior with relative magnetic permeability equal to one; the electrical conductivity is also included in Table 16.8.

Figures 16.17 to 16.19 illustrate some of the results of the numerical simulations. Thus, in Fig. 16.17 we have represented the modulus of the current density on the workpiece. As expected, the skin effect is greater when working at high frequency and fields concentrate on the surface. In Fig. 16.18 we can also observe the difference in the modulus of **J** when changing from silicon to graphite. Similarly, in Fig. 16.19 we have represented the Lorentz's force in the silicon, which is very important on its surface due to the skin effect, specially in the case of 3500 Hz.

Table 16.6. Relevant geometrical data for the numerical test

Height of silicon:	0.410 m
Inner radius of crucible:	0.125 m
Outer radius of crucible:	0.225 m
Crucible height:	0.480 m
Crucible width:	0.200 m
Crucible thickness in the sides:	0.035 m
Crucible thickness in the bottom:	0.045 m
Alumina layer width:	0.050 m
Radius of the coil section:	0.016 m
Radius of the refrigeration tube:	0.014 m
Distance between coil and crucible:	0.025 m
Distance between the turns:	0.012 m
Number of coil turns:	12

Table 16.7. Operating parameters for each simulation

Simulation	*Frequency* (Hz)	*RMS coil current* (A)
1	3500	1074
2	200	1074

16.2 Problems arising in physical applications

Table 16.8. Electrical conductivity in $(\Omega m)^{-1}$ of the materials

Silicon	1234568
Graphite	240000
Refractory 1	1.e-12
Refractory 2	1.e-12
Concrete	1.e-12
Copper	58.e6
Cooling water	1

Fig. 16.17. RMS current density (modulus) in the set graphite-silicon. Simulation at 3500Hz (left) and 200 Hz (right)

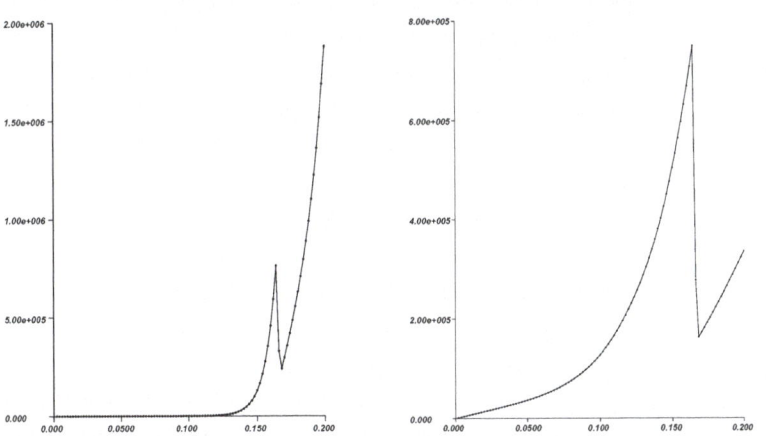

Fig. 16.18. RMS current density vs. the radius in the set graphite-silicon. Simulation at 3500Hz (left) and 200 Hz (right)

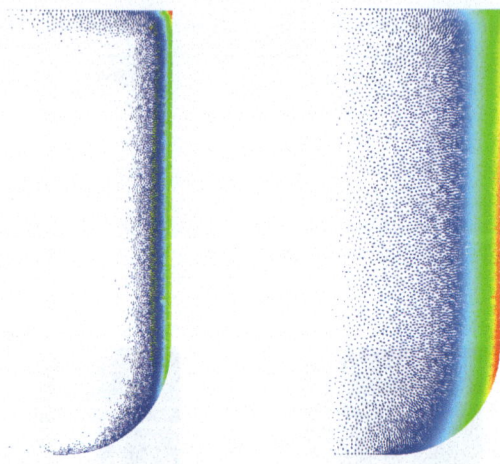

Fig. 16.19. Lorentz's force in the silicon. Simulation at 3500Hz (left) and 200 Hz (right)

Finally, Table 16.9 summarizes the active power dissipated in each conducting domain depending on the operating frequency. The coils are numerated from bottom to top. We can observe that the dissipated power is always greater in the graphite crucible than in the silicon. On the other hand, the values computed on each subdomain are greater when working at high frequency.

Table 16.9. Active power (W) in conducting domains as a function of frequency

Domain	$f = 200$ (Hz)	$f = 3500$ (Hz)
Silicon	2958.380	824.1343
Graphite	5209.158	70106.60
Coil 1	360.9742	1356.862
Coil 2	289.9980	1206.121
Coil 3	282.5386	1212.149
Coil 4	290.6266	1235.677
Coil 5	298.0990	1252.501
Coil 6	301.3538	1261.210
Coil 7	301.3406	1260.697
Coil 8	298.7895	1247.213
Coil 9	295.3134	1227.342
Coil 10	293.6752	1213.962
Coil 11	303.9626	1244.372
Coil 12	369.4638	1465.781

16.2.4 An induction furnace surrounded by an helical coil

The previous example illustrates a typical procedure to model an induction heating furnace with a cylindrical workpiece. However, as we have remarked above, to achieve a formulation in an axisymmetrical setting the helical coil has been replaced by several rings. In this way, the axisymmetric model does not include the detailed geometry of the coil but it is cheaper than a genuine 3D model. Its accuracy can be enough depending on the applications. In this section, we are going to consider a genuine 3D helical coil and compute the induced currents in a cylindrical workpiece by using the 3D eddy currents model (10.18)–(10.28).

We consider a copper coil which carries an alternating current surrounding a cylindrical workpiece made of graphite. We will compute the current density and active power in the conducting part.

The coil and workpiece are shown in Fig. 16.20 which also shows the limits of the box Ω surrounding the conductor domain. The artificial boundary of Ω is supposed to be far enough as to assume boundary condition $\mu \mathbf{H} \cdot \mathbf{n} = 0$ there. The current intensity entering the coil has an RMS value equal to 40000 A and null phase; the current frequency is equal to 200 Hz. The input current intensity is provided through one of the end points of the coil, while it goes out through the other end. Table 16.10 summarizes the main data of the problem.

Figure 16.21 shows the RMS current density in the coil and in the workpiece.

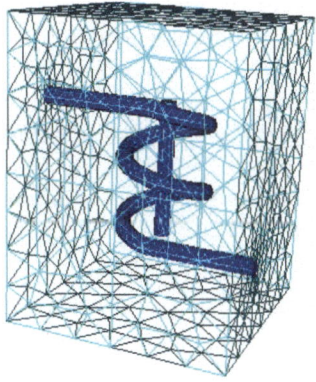

Fig. 16.20. Sketch of the domain

Table 16.10. Geometrical data and physical parameters used to simulate the set coil-cylinder

Electrical conductivity of the copper:	58.e6 $(\Omega m)^{-1}$
Electrical conductivity of the graphite:	240000 $(\Omega m)^{-1}$
Relative magnetic permeability of all materials:	1
Current input intensity (RMS value and phase):	40000 A, phase: 0 degrees
Frequency, f:	200 Hz

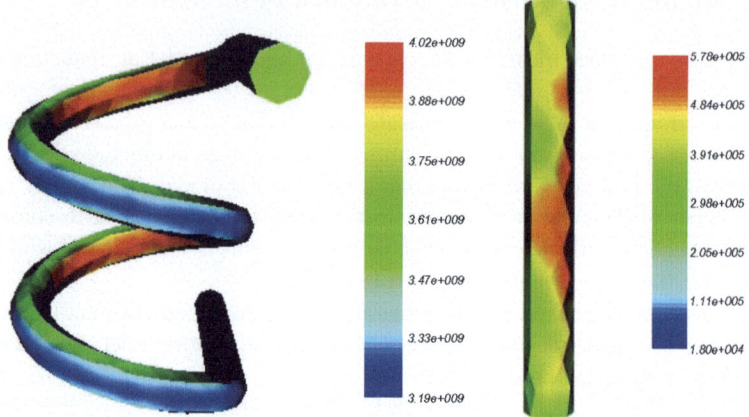

Fig. 16.21. RMS current density in coil (left) and workpiece (right)

16.2.5 A plate over a coil

The objective of this example is to compute the induced currents produced in a plate which is placed near a coil carrying an alternating current. This configuration appears for instance in electromagnetic forming processes where the source is a genuine transient current. Here, we will compute the induced currents in the set coil-plate in the case of a time-harmonic source. The coil and the plate are presented in Fig. 16.22, which also shows a mesh of the conducting domain. The domain has been chosen as a sufficiently large box surrounding the conductor. Table 16.11 shows the physical properties used in the simulation. The material of the workpiece is a magnesium alloy while the coil is made of copper. The input current intensity is provided through one of the end points of the coil.

Figure 16.23 shows the RMS current density in the coil and the workpiece. Since the frequency is quite high, the skin effect is very important.

Table 16.11. Geometrical data and physical parameters used to simulate the set coil-plate

Electrical conductivity of the copper:	58.e6 $(\Omega m)^{-1}$
Electrical conductivity of the magnesium alloy:	11.e6 $(\Omega m)^{-1}$
Relative magnetic permeability of all materials:	1
Current input intensity (RMS value and phase):	40000 A, phase: 0 degrees
Frequency, f:	35000 Hz

16.2 Problems arising in physical applications 345

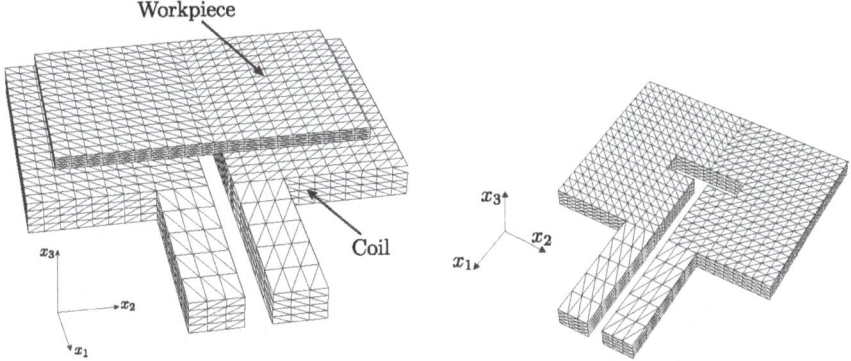

Fig. 16.22. Mesh of the conducting domain (left) and mesh of the coil (right)

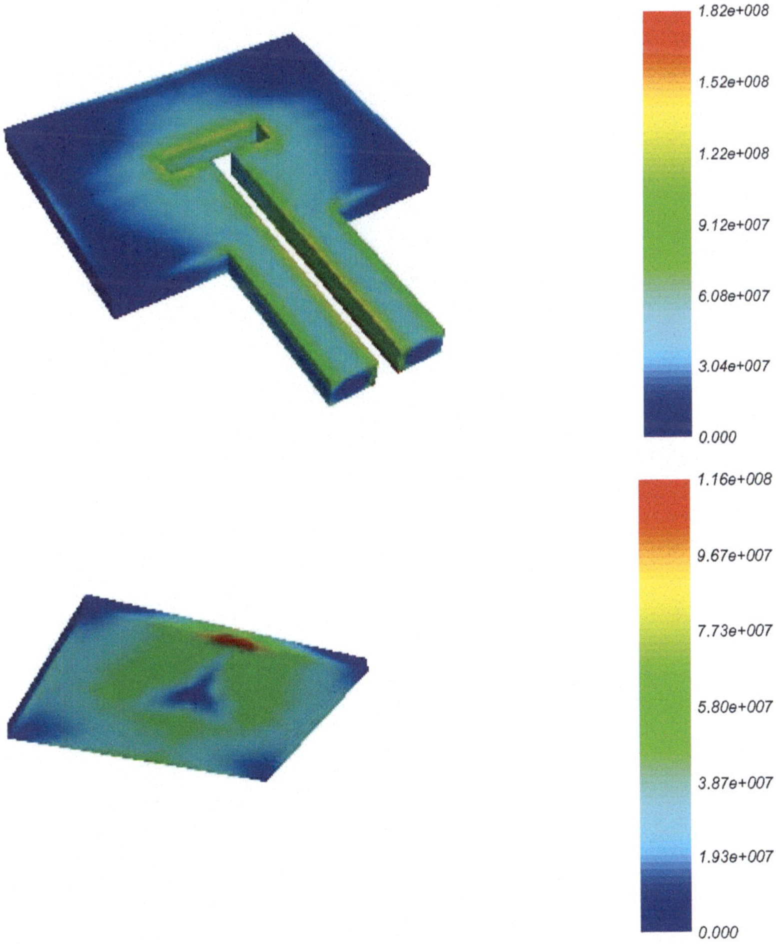

Fig. 16.23. RMS current density in the coil (top) and on the bottom of the plate (bottom)

Appendix A
Elements of graph theory

A.1 Definitions

Definition A.1. A *directed graph* (or *digraph*), $\mathfrak{G}$, is a couple of finite sets (V,A), being $A \subset V \times V$. The elements of V are the *nodes* and those of A the *edges* of the graph. If $r = (a,b)$ is an edge, then a is called the *initial node* and b the *terminal node* of r. Let N be the number of nodes and E the number of edges of $\mathfrak{G}$.

For the sake of simplicity, in what follows all graphs will be directed graphs.

Definition A.2. A set of edges $r_1, \ldots, r_m$ of a graph is a *path* between nodes a and b if:
- Two consecutive edges of the set r_i, r_{i+1} always have a common node.
- There is not any node of the graph belonging to more than two edges of the set.
- Node a is a node of only one edge of the set and the same is true for b.

Definition A.3. Two nodes a y b of a graph $\mathfrak{G}$ are said *connected* if either $a = b$ or there exists a path between them.

Connectivity is an equivalence relation whose equivalence classes are called *connected components* of $\mathfrak{G}$.

Definition A.4. A graph is said *connected* if it only has one connected component.

Definition A.5. A subgraph $\mathfrak{S}$ of a graph $\mathfrak{G}$ is said a *cycle* if the following conditions hold:
- $\mathfrak{S}$ is connected.
- Each node of $\mathfrak{S}$ belongs to exactly two edges of $\mathfrak{S}$.

Definition A.6. A subgraph $\mathfrak{S}$ of a connected graph $\mathfrak{G}$ is a *tree* if:
- $\mathfrak{S}$ is connected.
- $\mathfrak{S}$ contains all nodes of $\mathfrak{G}$.
- $\mathfrak{S}$ has not any cycles.

Definition A.7. If $\mathfrak{T}$ is a tree of $\mathfrak{G}$ their edges are called the *branches* of the tree. Given a tree $\mathfrak{T}$, the set of edges of $\mathfrak{G}$ not belonging to $\mathfrak{T}$ are called *bonds* or *cords* and form the so-called *cotree* of $\mathfrak{T}$.

A.2 Incidence matrix. Connected components

Definition A.8. The *incidence matrix* of $\mathfrak{G}$, $\mathscr{A}$, is the $N \times E$ matrix defined as follows:
For each edge $j \in \{1,\ldots,E\}$:

- $a_{ij} = 0$, if node i does not belong to j.
- $a_{ij} = -1$, if node i is the initial node of j.
- $a_{ij} = 1$, if node i is the terminal node of j.

Lemma A.1. *Let $\mathscr{A}$ be the incidence matrix of a graph $\mathfrak{G}$ with N nodes, then*

$$\mathrm{rank}(\mathscr{A}) \leq N-1. \tag{A.1}$$

Proof. Since each column of $\mathscr{A}$ contains one -1, one 1 and the rest of elements are null, the sum of their rows is the null vector. Therefore, they are linearly dependent which proves the result. □

In what follows we will prove that if $\mathfrak{G}$ is connected then $\mathrm{rank}(\mathscr{A}) = N-1$.

Theorem A.1. *Let $\mathfrak{G}$ a graph with N nodes.*

1. *If $\mathfrak{G}$ is connected any set of $N-1$ rows of $\mathscr{A}$ is linearly independent.*
2. *If $\mathscr{A}$ has a set of $N-1$ linearly independent rows then $\mathfrak{G}$ is connected.*

Proof.

1. Since we can always renumber the nodes it is enough to prove that the first $N-1$ rows of $\mathscr{A}$ are linearly independent. Let $\boldsymbol{\alpha}_i = (a_{i1},\ldots,a_{iE})$ be the i-th row of $\mathscr{A}$. We will prove that vectors $\boldsymbol{\alpha}_1,\ldots,\boldsymbol{\alpha}_{N-1}$ are linearly independent, by contradiction.
Let us suppose there exist numbers $s_1,\ldots,s_{N-1}$, not all null such that

$$s_1\boldsymbol{\alpha}_1 + \cdots + s_{N-1}\boldsymbol{\alpha}_{N-1} = \mathbf{0}. \tag{A.2}$$

Without loss of generality, by permutating the rows of $\mathscr{A}$, we may assume that there exists a number M, $1 \leq M \leq N-1$, such that $s_1,\ldots,s_M$ are all null and $s_{M+1},\ldots,s_{N-1}$ are all non-null. Then we write matrix $\mathscr{A}$ in the form

$$\mathscr{A} = \begin{pmatrix} \mathscr{A}_1 \\ \mathscr{A}_2 \end{pmatrix}, \tag{A.3}$$

where $\mathscr{A}_1$ contains the first M rows of $\mathscr{A}$, and $\mathscr{A}_2$ the remaining $N-M$. Let us notice that each block has at least one row because $M \geq 1$ and moreover, the N-th row belongs to $\mathscr{A}_2$.

A.2 Incidence matrix. Connected components 349

From (A.2) we deduce that there is not any column of $\mathscr{A}_1$ having one single non-null element. Hence, only one of the three following cases may occur:

a. All columns of $\mathscr{A}_1$ are null. Then none of the first M nodes is connected to another one and, consequently, $\mathfrak{G}$ is not connected.
b. Each column of $\mathscr{A}_1$ has a 1 and a -1. Then matrix $\mathscr{A}_2$ is null and $\mathfrak{G}$ is not connected.
c. Some columns of $\mathscr{A}_1$ have a 1 and a -1, and some others are null. In this case we can renumber the columns in such a way that matrix $\mathscr{A}$ can be written as follows:
$$\mathscr{A} = \begin{pmatrix} \mathscr{A}_1 \\ \mathscr{A}_2 \end{pmatrix} = \begin{pmatrix} 0 & (\pm 1, 0) \\ (\pm 1, 0) & 0 \end{pmatrix}, \quad (A.4)$$
which shows that there is not any path between, for instance, nodes 1 and N.

In the three cases above we get a contradiction.

2. We will prove that if the graph is not connected then there is not any set of $N-1$ linearly independent rows of $\mathscr{A}$.

Indeed, let us suppose the graph is not connected. Then it will have M connected components with $M \geq 2$. Let n_i be the number of nodes of the i-th connected component; we have $\sum_{i=1}^{M} n_i = N$. By using a suitable numbering of nodes, the incidence matrix of $\mathfrak{G}$ will have the following structure:

$$\mathscr{A} = \begin{pmatrix} \mathscr{A}_1 & & & \\ & \mathscr{A}_2 & & \\ & & \ddots & \\ & & & \mathscr{A}_M \end{pmatrix} \quad (A.5)$$

being $\mathscr{A}_i$ the incidence matrix of the i-th connected component of $\mathfrak{G}$, $i = 1, \ldots, M$. We notice that $\mathscr{A}_i$ has n_i rows.

Moreover, in virtue of Lemma A.1 and the results already shown in the first part of this proof, we have
$$\text{rank}(\mathscr{A}_i) = n_i - 1.$$

Now, from the structure of $\mathscr{A}$ we deduce,
$$\text{rank}(\mathscr{A}) = \sum_{i=1}^{M} \text{rank}(\mathscr{A}_i) = \sum_{i=1}^{M} (n_i - 1) = N - M < N - 1,$$

since $M > 1$. Therefore, $\mathscr{A}$ has not any set of $N - 1$ linearly independent rows, which finishes the proof. □

Corollary A.1. *The graph $\mathfrak{G}$ has exactly M connected components if and only if*
$$\text{rank}(\mathscr{A}) = N - M$$

Corollary A.2. *Let $\mathfrak{G}$ be a connected graph with N nodes and $\mathscr{A}$ its incidence matrix. Then*
$$\text{rank}(\mathscr{A}) = N - 1.$$

Remark A.1. From the above theorem we deduce that, if $\mathfrak{G}$ is a connected graph with N nodes and E edges then $N - 1 \leq E$

Proposition A.1. *Let $\mathfrak{G}$ be a connected graph with N nodes. Then any tree of $\mathfrak{G}$ has exactly $N - 1$ edges.*

Proof. Let $\mathfrak{T}$ be a tree of $\mathfrak{G}$ with N nodes and $\mathscr{E}$ its incidence matrix. Assuming that $\mathfrak{T}$ has M edges, we will prove that $M = N - 1$.

Indeed, if $M < N - 1$ then $\mathscr{E}$, which is a $N \times M$ matrix, cannot have rank $N - 1$. Theorem A.1 shows that $\mathfrak{T}$ cannot be connected and hence $\mathfrak{T}$ cannot be a tree.

Moreover, let us suppose $M > N - 1$. From the previous theorem $\text{rank}(\mathscr{E}) = N - 1$ and hence $\mathfrak{T}$ has a subgraph $\mathfrak{S}$ with $N - 1$ edges corresponding to $N - 1$ linearly independent columns of $\mathscr{E}$; then from the previous theorem $\mathfrak{S}$ will be connected and hence a tree of $\mathfrak{G}$; $\mathfrak{T}$ would be obtained from $\mathfrak{S}$ by adding $M - N + 1 > 0$ edges. But, if we add an edge to a tree we generate a cycle so $\mathfrak{T}$ will have cycles and then it cannot be a tree. □

We notice that the above proposition does not prove the existence of trees in a graph. Such a result is a corollary of the following theorem.

Theorem A.2. *Let $\mathfrak{G}$ be a connected graph with N nodes. Then $N - 1$ columns of $\mathscr{A}$ are linearly independent if and only if they correspond to the edges of a tree.*

Proof. Let $\mathscr{E} \in M_{N \times (N-1)}$ be a submatrix of the incidence matrix of $\mathfrak{G}$ having $N - 1$ linearly independent columns of $\mathscr{A}$. Let us show that the corresponding edges form a tree. For this purpose let us consider the subgraph $\mathfrak{S}$ of $\mathfrak{G}$ which incidence matrix is $\mathscr{E}$. Then $\mathfrak{S}$ has N nodes and $N - 1$ edges. Since $\mathscr{E}$ has $N - 1$ linearly independent columns, it will also have $N - 1$ linearly independent rows. Hence, $\mathfrak{S}$ that has N nodes and $N - 1$ edges will be connected by the previous lemma. Moreover $\mathfrak{S}$ has not cycles because, otherwise, one could remove at least one edge from $\mathfrak{S}$ keeping connectedness. In this way we would have a tree with less than $N - 1$ edges, in contradiction with Proposition A.1.

Let us suppose $\mathfrak{T}$ is a tree of a connected graph $\mathfrak{G}$. According to the previous proposition $\mathfrak{T}$ has $N - 1$ edges. Without loss of generality, since we can renumber the edges of $\mathfrak{G}$ arbitrarily, we suppose that the first $N - 1$ columns of $\mathscr{A}$ correspond to the edges of $\mathfrak{T}$. Let $\mathscr{E}$ be the submatrix of $\mathscr{A}$ of these first $N - 1$ columns, then $\mathscr{E}$ will be the incidence matrix of subgraph $\mathfrak{T}$. Since $\mathfrak{T}$ is connected, from the previous theorem any $N - 1$ rows of $\mathscr{E}$ are linearly independent and hence its $N - 1$ columns too. □

Corollary A.3. *Let $\mathfrak{G}$ be a connected graph with N nodes and E edges, and $\mathscr{A}$ its incidence matrix. Then, any set of $N - 1$ linearly independent columns of $\mathscr{A}$ determine the edges of a tree and the remaining $E - N + 1$ ones the bonds of its cotree.*

We notice that, according to Theorem A.1, the incidence matrix of any connected graph always has, at least, a set of $N-1$ linearly independent columns which proves the following existence result:

Corollary A.4. *Any connected graph has, at least, a subgraph that is tree.*

A.3 Cycle matrix. Fundamental cycles

The results below are related with the cycles of a graph and with the characterization of $\ker(\mathscr{A})$. In particular, we build a basis of this kernel that can be used in the reduction process shown in Remark 3.1.4.

Definition A.9. Let $\mathfrak{G}$ be a graph with N nodes and E edges. Let L be the number of its cycles which have been provided with an orientation. The *cycle matrix* $\mathscr{B} \in M_{E \times L}$ is defined as follows: for each $l \in \{1,\ldots,L\}$:

- $b_{jl} = 0$, if edge j does belong to cycle l.
- $b_{jl} = -1$, if edge j belongs to cycle l but they have opposed orientations.
- $b_{jl} = 1$, if edge j belongs to cycle l and they have the same orientation.

Let us recall that given any tree, by adding one of the $E - N + 1$ bonds of its cotree a cycle is generated called *fundamental cycle* of the graph. We have the following result:

Proposition A.2. *Given any tree of a connected graph $\mathfrak{G}$ with N nodes and E edges, the $E - N + 1$ columns of $\mathscr{B}$ corresponding to the fundamental cycles associated with this tree are linearly independent.*

Proof. Without loss of generality we can assume that the orientation of each fundamental cycle is the one of the bond of its cotree originating the cycle when added to the tree. Thus, the rows corresponding to these bonds have a 1 at the column corresponding to the fundamental cycle they belong to, and 0 at the rest of the columns. Hence, the submatrix of $\mathscr{B}$ obtained by extracting the rows corresponding to the bonds of the cotree and the columns corresponding to the associated fundamental cycles is a permutation of the identity matrix of order $E - N + 1$. Therefore, the corresponding columns of matrix $\mathscr{B}$ are linearly independent. □

Definition A.10. Any submatrix of $\mathscr{B}$ built with columns associated to a cotree will be called *matrix of fundamental cycles*.

Proposition A.3. *We have*
$$\mathscr{A}\mathscr{B} = 0. \tag{A.6}$$

Proof. Let $\mathscr{G} = \mathscr{A}\mathscr{B}$. Then,
$$g_{il} = \sum_{j=1}^{E} a_{ij} b_{jl}, \quad i = 1,\ldots,N, \ l = 1,\ldots,L.$$

In order to compute this sum of products we proceed by increasing the value of j from 1 to E. The first non-null product appears when some edge j has node i as one of its nodes and belongs to cycle l. Then

$$a_{ij}b_{jl} = \pm 1.$$

But, if edge j belongs to cycle l, node i also belongs to this cycle. Hence, there will be exactly one edge, let say k-th, in cycle l having node i as one of its nodes. By considering all possible cases, we can see that the corresponding term of the above sum, that will be $+1$ or -1, will have an opposite sign to the previous one, namely,

$$a_{ik}b_{kl} = \mp 1.$$

Summarizing,

$$g_{il} = \sum_{j=1}^{E} a_{ij}b_{jl} = 0 + \cdots \pm 1 + 0 + \cdots \mp 1 + 0 + \cdots + 0 = 0. \qquad \square$$

Corollary A.5. *Let $\mathfrak{G}$ be a connected graph with N nodes and E edges. Then,*

$$\operatorname{rank}(\mathscr{B}) = E - N + 1.$$

Proof. The previous proposition shows that the columns of $\mathscr{B}$ belong to $\ker(\mathscr{A})$ which has dimension $E - N + 1$. $\qquad \square$

Corollary A.6. *Let $\mathfrak{G}$ be a connected graph with N nodes and E edges.*

1. *We have,*
$$\ker(\mathscr{A}) = \operatorname{im}(\mathscr{B}).$$

2. *The columns of $\mathscr{B}$ corresponding to a set of fundamental cycles form a basis of $\ker(\mathscr{A})$.*

Definition A.11. We call *reduced incidence matrix* of a connected graph with N nodes to any submatrix of the incidence matrix having $N - 1$ rows.

In what follows we will show how a matrix of fundamental cycles, $\mathscr{B}_F$, can be built from a reduced incidence matrix $\mathscr{A}_R$.

Let $\mathfrak{T}$ be a tree. Without loss of generality we can assume that the edges of the graph are numbered in such a way that those of the cotree appear at the end. Similarly, we will suppose that the fundamental cycles corresponding to the edges of the cotree are also numbered at the end. Then matrix $\mathscr{B}_F$ has the following structure:

$$\mathscr{B}_F = \begin{pmatrix} \mathscr{B}_T \\ \mathscr{I} \end{pmatrix},$$

where $\mathscr{B}_T \in \mathrm{M}_{(N-1) \times (E-N+1)}$ and $\mathscr{I} \in \mathrm{M}_{(E-N+1) \times (E-N+1)}$ is an identity matrix. The rows of $\mathscr{B}_T$ correspond to the branches of tree $\mathfrak{T}$ and its columns to the cycles defined

A.3 Cycle matrix. Fundamental cycles

by the bonds of its cotree. Similarly, an associated reduced incidence matrix writes in the form,

$$\mathscr{A}_R = \begin{pmatrix} \mathscr{A}_T & \mathscr{A}_C \end{pmatrix},$$

with $\mathscr{A}_T \in M_{(N-1)\times(N-1)}$ and $\mathscr{A}_C \in M_{(N-1)\times(E-N+1)}$; the columns of $\mathscr{A}_T$ correspond to the edges of tree $\mathfrak{T}$. From equality (A.6) we deduce

$$\mathscr{A}_R \mathscr{B}_F = \mathscr{A}_T \mathscr{B}_T + \mathscr{A}_C = 0,$$

and then

$$\mathscr{B}_T = -\mathscr{A}_T^{-1} \mathscr{A}_C.$$

Let us recall that $E - N + 1$ columns of $\mathscr{B}_F$ constitute a basis of ker($\mathscr{A}$) and define the fundamental cycles.

Appendix B
Vector calculus

B.1 Vector and tensor algebra

B.1.1 Vector space. Basis

A real *vector space* is an algebraic structure consisting of a set $\mathscr{V}$ endowed with two operations. One of them is an internal operation denoted by $+$ and satisfying the following properties:

i) *Associativity*:
$$\mathbf{a} + (\mathbf{b} + \mathbf{c}) = (\mathbf{a} + \mathbf{b}) + \mathbf{c} \quad \forall \, \mathbf{a}, \, \mathbf{b}, \, \mathbf{c}.$$

ii) *Existence of a neutral element* $\mathbf{0}$ such that:
$$\mathbf{a} + \mathbf{0} = \mathbf{0} + \mathbf{a} = \mathbf{a} \quad \forall \, \mathbf{a}.$$

iii) *Existence of symmetric element*: for each $\mathbf{a}$ there exist $\mathbf{b}$ such that:
$$\mathbf{a} + \mathbf{b} = \mathbf{b} + \mathbf{a} = \mathbf{0}.$$

iv) *Commutativity*:
$$\mathbf{a} + \mathbf{b} = \mathbf{b} + \mathbf{a} \quad \forall \, \mathbf{a}, \, \mathbf{b}.$$

Thus, $\mathscr{V}$ with the $+$ operation is a commutative group.

The second operation is an external one, the product by real numbers, satisfying the following properties:

v) $\lambda(\mathbf{a} + \mathbf{b}) = \lambda \mathbf{a} + \lambda \mathbf{b}.$
vi) $\lambda(\mu \mathbf{a}) = (\lambda \mu)\mathbf{a}.$
vii) $(\lambda + \mu)\mathbf{a} = \lambda \mathbf{a} + \mu \mathbf{a}.$
viii) $1\mathbf{a} = \mathbf{a}.$

Elements of $\mathscr{V}$ are called *vectors*.

A set of vectors $\{\mathbf{e}_i\}_{i=1}^{N}$ is called:

- *linearly independent* if

$$\sum_{i=1}^{N} \lambda_i \mathbf{e}_i = \mathbf{0} \Rightarrow \lambda_i = 0 \ \forall i \in \{1,\ldots,N\};$$

- a *system of generators* if

$$\forall \mathbf{v} \in \mathscr{V} \ \exists \lambda_i, \ i=1,\ldots,N, \text{ such that } \mathbf{v} = \sum_{i=1}^{N} \lambda_i \mathbf{e}_i;$$

- a *basis* if it is both linearly independent and a system of generators.

If $B = \{\mathbf{e}_1, \ldots, \mathbf{e}_N\}$ is a basis of $\mathscr{V}$, then each $\mathbf{v} \in \mathscr{V}$ can be written as a linear combination,

$$\mathbf{v} = \sum_{i=1}^{N} v_i \mathbf{e}_i,$$

in a unique way. The $v_i \in \mathbb{R}$, $i=1,\ldots,N$, are called *coordinates* of $\mathbf{v}$ with respect to basis B.

Let us consider two bases in $\mathscr{V}$, $\{\mathbf{e}_i\}_{i=1}^{N}$ and $\{\mathbf{E}_i\}_{i=1}^{N}$. In particular, vectors $\mathbf{E}_i$ can be written as a linear combination of vectors $\{\mathbf{e}_i\}_{i=1}^{N}$, namely,

$$\mathbf{E}_l = \sum_{i=1}^{N} C_{li} \mathbf{e}_i. \tag{B.1}$$

Now, let $\mathbf{v}$ be any vector in $\mathscr{V}$. Let us denote by $v_i^{\mathbf{e}}$ and $v_i^{\mathbf{E}}$, $i=1,\ldots,N$, the coordinates of $\mathbf{v}$ with respect to the bases $\{\mathbf{e}_i\}_{i=1}^{N}$ and $\{\mathbf{E}_i\}_{i=1}^{N}$, respectively. If we denote by $v^{\mathbf{e}}$ the column vector of the former and by $v^{\mathbf{E}}$ the column vector of the latter, it is easy to see that

$$v^{\mathbf{e}} = [C]^t v^{\mathbf{E}},$$

where $[C]$ is the matrix $[C]_{ij} = C_{ij}$ with C_{ij} given in (B.1).

B.1.2 Inner product

A mapping $\varphi: \mathscr{V} \times \mathscr{V} \longrightarrow \mathbb{R}$ is called *inner product* if the following properties hold:

1) φ is bilinear:

   1.1) $\varphi(\lambda_1 \mathbf{v}_1 + \lambda_2 \mathbf{v}_2, \mathbf{w}) = \lambda_1 \varphi(\mathbf{v}_1, \mathbf{w}) + \lambda_2 \varphi(\mathbf{v}_2, \mathbf{w})$,
   1.2) $\varphi(\mathbf{v}, \lambda_1 \mathbf{w}_1 + \lambda_2 \mathbf{w}_2) = \lambda_1 \varphi(\mathbf{v}, \mathbf{w}_1) + \lambda_2 \varphi(\mathbf{v}, \mathbf{w}_2)$.

2) φ is symmetric:

$$\varphi(\mathbf{v}, \mathbf{w}) = \varphi(\mathbf{w}, \mathbf{v}).$$

3) φ is positive definite:

$$\varphi(\mathbf{v}, \mathbf{v}) > 0 \ \forall \ \mathbf{v} \neq \mathbf{0}.$$

Finite dimensional vector spaces with inner product are called *Euclidean vector spaces*. For them we can define a *norm* by,

$$|\mathbf{v}| = \varphi(\mathbf{v},\mathbf{v})^{\frac{1}{2}}.$$

In what follows we write $\mathbf{v} \cdot \mathbf{w}$ instead of $\varphi(\mathbf{v},\mathbf{w})$.

Two vectors $\mathbf{v}$ and $\mathbf{w}$ are said to be *orthogonal* if $\mathbf{v} \cdot \mathbf{w} = 0$.

A basis $B = \{\mathbf{e}_1, \ldots, \mathbf{e}_N\}$ is said to be *orthogonal* if $\mathbf{e}_i \cdot \mathbf{e}_j = 0 \; \forall \; i \neq j$ and is said to be *orthonormal* if

$$\mathbf{e}_i \cdot \mathbf{e}_j = \delta_{ij} = \begin{cases} 1 & \text{if } i = j, \\ 0 & \text{if } i \neq j. \end{cases}$$

If B is orthonormal, then the i-th coordinate of a vector $\mathbf{v}$ is given by,

$$v_i = \mathbf{v} \cdot \mathbf{e}_i. \tag{B.2}$$

Moreover, the matrix to change coordinates with respect to two orthonormal basis is orthogonal, namely,

$$[C][C]^t = [I],$$

where $[I]$ is the identity matrix.

B.1.3 Tensors

The endomorphisms of $\mathscr{V}$, that is the linear mappings from $\mathscr{V}$ in $\mathscr{V}$ are called (second order) *tensors*. We denote by Lin the set of tensors on a vector space $\mathscr{V}$. In Lin we can define a vector space structure by introducing the following operations:

i) *Sum*:
$$(\mathbf{S} + \mathbf{T})(\mathbf{v}) = \mathbf{S}\mathbf{v} + \mathbf{T}\mathbf{v}.$$

ii) *Product by scalars*:
$$(\lambda \mathbf{S})(\mathbf{v}) = \lambda \mathbf{S}\mathbf{v}.$$

The dimension of this vector space is N^2.

Let $\mathbf{a}$ and $\mathbf{b}$ two vectors in $\mathscr{V}$. The tensor,

$$(\mathbf{a} \otimes \mathbf{b})\mathbf{v} := \mathbf{v} \cdot \mathbf{b} \, \mathbf{a},$$

is called the *tensor product* of $\mathbf{a}$ and $\mathbf{b}$. We notice that $\otimes$ is a bilinear mapping from $\mathscr{V} \times \mathscr{V}$ in Lin.

If $\mathbf{e}$ is a unit vector then

$$(\mathbf{e} \otimes \mathbf{e})\mathbf{v} = (\mathbf{v} \cdot \mathbf{e})\mathbf{e},$$

is the *orthogonal projection* of $\mathbf{v}$ on the straight line generated by $\mathbf{e}$. Moreover,

$$(\mathbf{I} - \mathbf{e} \otimes \mathbf{e})\mathbf{v} = \mathbf{v} - (\mathbf{v} \cdot \mathbf{e})\mathbf{e},$$

is the *projection* of $\mathbf{v}$ on the orthogonal plane to $\mathbf{e}$.

Let $B = \{\mathbf{e}_1,\ldots,\mathbf{e}_N\}$ an orthonormal basis in $\mathscr{V}$. The set of tensors $\{\mathbf{e}_i \otimes \mathbf{e}_j : 1 \leq i, j \leq N\}$ is a basis of Lin. More precisely, we have

$$\mathbf{S} = \sum_{i,j=1}^{N} S_{ij}\, \mathbf{e}_i \otimes \mathbf{e}_j,$$

where

$$S_{ij} = \mathbf{S}\,\mathbf{e}_j \cdot \mathbf{e}_i \tag{B.3}$$

are the *coordinates* of $\mathbf{S}$ with respect to basis B.

We notice that the coordinates of a tensor $\mathbf{S}$ can be arranged as a matrix,

$$[\mathbf{S}] = \begin{pmatrix} S_{11} & \cdots & S_{1N} \\ \vdots & \ddots & \vdots \\ S_{N1} & \cdots & S_{NN} \end{pmatrix}.$$

Actually, this matrix is the standard matrix associated with the endomorphism $\mathbf{S}$ for the basis B.

For example, the coordinate matrix of tensor $\mathbf{a} \otimes \mathbf{b}$ with respect to basis B is the matrix product

$$\begin{pmatrix} a_1 \\ a_2 \\ \vdots \\ a_N \end{pmatrix} (b_1\ b_2\ \ldots\ b_N),$$

where $\mathbf{a} = \sum_{i=1}^{N} a_i \mathbf{e}_i$ and $\mathbf{b} = \sum_{i=1}^{N} b_i \mathbf{e}_i$.

Let us see how tensor coordinates change when we change basis in the vector space. Let $\mathbf{S} \in \text{Lin}$ a tensor. Let us denote by $S^{\mathbf{e}}_{ij}$ and $S^{\mathbf{E}}_{ij}$, $i,j = 1,\ldots,n$, its coordinates with respect to the orthonormal bases $\{\mathbf{e}_i\}_{i=1}^{N}$ and $\{\mathbf{E}_i\}_{i=1}^{N}$, respectively. Then it is not difficult to see that the following relation holds:

$$[\mathbf{S}^{\mathbf{E}}] = [C][\mathbf{S}^{\mathbf{e}}][C]^t.$$

In vector space Lin we can introduce another internal operation which is the mapping *composition*, namely, given two tensors $\mathbf{S}$ and $\mathbf{T}$, $\mathbf{ST}$ is the tensor defined by

$$(\mathbf{ST})\mathbf{v} = \mathbf{S}(\mathbf{T}\mathbf{v}).$$

It is easy to see that the matrix of coordinates of $\mathbf{ST}$ is the product of those corresponding to $\mathbf{S}$ and $\mathbf{T}$.

The *transpose tensor* of $\mathbf{S}$ is the unique tensor $\mathbf{S}^t$ satisfying,

$$\mathbf{S}\mathbf{a} \cdot \mathbf{b} = \mathbf{a} \cdot \mathbf{S}^t \mathbf{b} \quad \forall\, \mathbf{a},\mathbf{b} \in \mathscr{V}.$$

A tensor is called *symmetric* if $\mathbf{S} = \mathbf{S}^t$ and *skew* if $\mathbf{S} = -\mathbf{S}^t$. The subspace of symmetric tensors will be denoted by Sym and that of skew tensors by Skw.

We can also define an *inner product* of tensors by,

$$\mathbf{S} \cdot \mathbf{T} = \mathrm{tr}(\mathbf{S}^t \mathbf{T}),$$

where tr denotes the *trace* operator which is the unique linear operator from Lin to $\mathbb{R}$ satisfying,

$$\mathrm{tr}(\mathbf{a} \otimes \mathbf{b}) = \mathbf{a} \cdot \mathbf{b}.$$

The associated norm is defined by

$$|\mathbf{S}| = (\mathbf{S} \cdot \mathbf{S})^{1/2}.$$

We have,

$$\mathrm{tr}(\mathbf{S}) = \mathrm{tr}\left(\sum_{i,j=1}^{N} S_{ij} \mathbf{e}_i \otimes \mathbf{e}_j \right) = \sum_{i,j=1}^{N} S_{ij} \mathbf{e}_i \cdot \mathbf{e}_j = \sum_{i,j=1}^{N} S_{ij} \delta_{ij} = \sum_{i=1}^{N} S_{ii},$$

and hence

$$\mathbf{S} \cdot \mathbf{T} = \sum_{i,j=1}^{N} S_{ij} T_{ij}.$$

If $\mathbf{T} \in \mathrm{Sym}$ and $\mathbf{W} \in \mathrm{Skw}$ it is easy to show that $\mathbf{T} \cdot \mathbf{W} = 0$.

Any tensor $\mathbf{S}$ is the sum of one symmetric tensor $\mathbf{E}$ and another one skew $\mathbf{W}$. More precisely, $\mathbf{E}$ and $\mathbf{W}$ are defined by

$$\mathbf{E} = \frac{1}{2}(\mathbf{S} + \mathbf{S}^t), \tag{B.4}$$

$$\mathbf{W} = \frac{1}{2}(\mathbf{S} - \mathbf{S}^t). \tag{B.5}$$

Tensor $\mathbf{E}$ is called the *symmetric part* of $\mathbf{S}$ and $\mathbf{W}$ is called the *skew part* of $\mathbf{S}$. Furthermore, this decomposition is unique, more precisely,

$$\mathrm{Lin} = \mathrm{Sym} \overset{\perp}{\oplus} \mathrm{Skw}$$

The following equalities can be easily proved:

$(\mathbf{a} \otimes \mathbf{b})^t = (\mathbf{b} \otimes \mathbf{a}).$

$(\mathbf{a} \otimes \mathbf{b})(\mathbf{c} \otimes \mathbf{d}) = (\mathbf{b} \cdot \mathbf{c}) \mathbf{a} \otimes \mathbf{d}.$

$\mathbf{R} \cdot (\mathbf{ST}) = (\mathbf{S}^t \mathbf{R}) \cdot \mathbf{T} = (\mathbf{RT}^t) \cdot \mathbf{S} \quad \forall \, \mathbf{R}, \mathbf{S}, \mathbf{T} \in \mathrm{Lin}.$ \hfill (B.6)

$\mathbf{I} \cdot \mathbf{S} = \mathrm{tr}(\mathbf{S}).$

$\mathbf{S} \cdot (\mathbf{a} \otimes \mathbf{b}) = \mathbf{a} \cdot \mathbf{Sb}.$

$(\mathbf{a} \otimes \mathbf{b}) \cdot (\mathbf{c} \otimes \mathbf{d}) = (\mathbf{a} \cdot \mathbf{c})(\mathbf{b} \cdot \mathbf{d}).$

We define the *determinant* of a tensor **S** to be the determinant of the matrix [**S**]:

$$\det \mathbf{S} := \det[\mathbf{S}].$$

One can prove that this definition is independent of the basis $\{\mathbf{e}_i\}$.

A tensor **S** is *invertible* if there exists a tensor $\mathbf{S}^{-1}$, called the inverse of **S**, such that

$$\mathbf{S}\mathbf{S}^{-1} = \mathbf{S}^{-1}\mathbf{S} = \mathbf{I}.$$

A tensor **Q** is *orthogonal* if it preserves the inner products:

$$\mathbf{Q}\mathbf{u} \cdot \mathbf{Q}\mathbf{v} = \mathbf{u} \cdot \mathbf{v}.$$

A necessary and sufficient condition for **Q** to be orthogonal is that

$$\mathbf{Q}\mathbf{Q}^t = \mathbf{Q}^t\mathbf{Q} = \mathbf{I},$$

which is equivalent to

$$\mathbf{Q}^t = \mathbf{Q}^{-1}.$$

In particular, $\det \mathbf{Q}$ must be 1 or -1, where $\det \mathbf{Q} := \det[\mathbf{Q}]$.

A *rotation* is an orthogonal tensor with positive determinant. The set of rotations is a subset of Lin which will be denoted by Orth$^+$.

A tensor **S** is *positive definite* if

$$\mathbf{v} \cdot \mathbf{S}\mathbf{v} > 0 \qquad \forall \mathbf{v} \neq \mathbf{0}.$$

A *real* scalar λ is an *eigenvalue* of **S** if there exists a non-null vector **u** such that

$$\mathbf{S}\mathbf{u} = \lambda \mathbf{u}.$$

In this case **u** is an *eigenvector* associated to λ. The set of vectors

$$\mathscr{E}_\lambda = \{\mathbf{u} \in \mathscr{V} : \mathbf{S}\mathbf{u} = \lambda \mathbf{u}\}$$

is a vector subspace of $\mathscr{V}$ called *characteristic space* or *eigenspace* of **S** corresponding to λ. If this space has dimension $M \leq N$, then M is called the *geometric multiplicity* of λ. The *spectrum* of **S** is the set of its eigenvalues.

It is immediate to see that $\lambda \in \mathbb{R}$ is an eigenvalue of tensor **S** if and only if it satisfies the *characteristic equation*

$$\det(\mathbf{S} - \lambda \mathbf{I}) = 0. \tag{B.7}$$

The multiplicity of λ as a root of this algebraic equation is called the *algebraic multiplicity* of λ. The algebraic multiplicity of any eigenvalue is greater than or equal to the geometric one.

Now, let us assume that $N=3$ for the rest of this paragraph. The *principal invariants* of **S** are the three numbers

$$\iota_1(\mathbf{S}) = \mathrm{tr}(\mathbf{S}),$$
$$\iota_2(\mathbf{S}) = \frac{1}{2}[(\mathrm{tr}(\mathbf{S}))^2 - \mathrm{tr}(\mathbf{S}^2)],$$
$$\iota_3(\mathbf{S}) = \det(\mathbf{S}) = \frac{1}{6}[(\mathrm{tr}(\mathbf{S}))^3 + 2(\mathrm{tr}(\mathbf{S}^3)) - 3\mathrm{tr}(\mathbf{S}^2)\mathrm{tr}(\mathbf{S})].$$

The set of principal invariants of **S** will be denoted by $\mathscr{I}_\mathbf{S}$. Moreover, the characteristic equation can be written as

$$\det(\mathbf{S} - \lambda \mathbf{I}) = -\lambda^3 + \iota_1(\mathbf{S})\lambda^2 - \iota_2(\mathbf{S})\lambda + \iota_3(\mathbf{S}).$$

We have the following important results:

Proposition B.1.

1. *The eigenvalues of a positive definite tensor are strictly positive.*
2. *Each symmetric tensor has exactly three eigenvalues accounting for their multiplicities as roots of the characteristic equation* (B.7) *(called algebraic multiplicities).*
3. *The characteristic spaces of a symmetric tensor are mutually orthogonal.*

Theorem B.1. *(Spectral theorem) Let* **S** *be a symmetric tensor. Then there is an orthonormal basis in* $\mathscr{V}$ *consisting of eigenvectors of* **S**. *Furthermore:*

1. **S** *has three distinct eigenvalues if and only it has three eigenspaces which are orthogonal lines through* **0**.
2. **S** *has two distinct eigenvalues if and only if it has two eigenspaces which are a line through* **0** *and its orthogonal plane through* **0**.
3. **S** *has one distinct eigenvalue if and only if it has one eigenspace (which is equal to* $\mathscr{V}$*).*

Moreover, when **S** is symmetric all the roots of the characteristic equation are real and the *Vieta relations* lead to the following equalities:

$$\iota_1(\mathbf{S}) = \lambda_1 + \lambda_2 + \lambda_3,$$
$$\iota_2(\mathbf{S}) = \lambda_1\lambda_2 + \lambda_2\lambda_3 + \lambda_1\lambda_3,$$
$$\iota_3(\mathbf{S}) = \lambda_1\lambda_2\lambda_3.$$

Let us assume that $\{\mathbf{e}_1, \mathbf{e}_2, \mathbf{e}_3\}$ is a positively oriented orthonormal basis in $\mathscr{V}$. We define the *vector product* of two vectors **a** and **b** by,

$$\mathbf{a} \times \mathbf{b} = \sum_{i,j,k=1}^{3} \varepsilon_{ijk}\, a_j b_k\, \mathbf{e}_i.$$

where

$$\varepsilon_{ijk} = \begin{cases} 1 & \text{if } ijk \text{ is an even permutation of 123,} \\ -1 & \text{if } ijk \text{ is an odd permutation of 123,} \\ 0 & \text{otherwise.} \end{cases}$$

Exercise B.1. Check the following equalities:

$$\sum_{i=1}^{3} \varepsilon_{ijk}\varepsilon_{ilm} = \delta_{jl}\delta_{km} - \delta_{jm}\delta_{kl}. \tag{B.8}$$

$$\sum_{i,j=1}^{3} \varepsilon_{ijl}\varepsilon_{ijm} = 2\delta_{lm}. \tag{B.9}$$

$$\sum_{i,j,k=1}^{3} \varepsilon_{ijk}\varepsilon_{ijk} = 6. \tag{B.10}$$

One can prove the following equality:

$$\mathbf{a} \times (\mathbf{b} \times \mathbf{c}) = (\mathbf{a} \cdot \mathbf{c})\mathbf{b} - (\mathbf{a} \cdot \mathbf{b})\mathbf{c}$$

showing, in particular, that the *cross product* is not *associative*.

Another useful property is

$$\mathbf{S}^t\mathbf{a} \times \mathbf{S}^t\mathbf{b} = \det(\mathbf{S})\mathbf{S}^{-1}(\mathbf{a} \times \mathbf{b}). \tag{B.11}$$

There is an isomorphism between vector space $\mathscr{V}$ and the linear space of skew tensors. It is defined by,

$$\mathscr{V} \longrightarrow \text{Skw}$$
$$\mathbf{w} \longrightarrow \mathbf{W}$$

with $\mathbf{W}\mathbf{v} = \mathbf{w} \times \mathbf{v}\ \forall \mathbf{v} \in \mathscr{V}$. Vector $\mathbf{w}$ is called the *axial vector* associated to $\mathbf{W}$.

In an orthonormal basis, the coordinates of $\mathbf{W}$ are

$$W_{ij} = \sum_{l=1}^{3} \varepsilon_{iljw_l}. \tag{B.12}$$

Indeed,

$$W_{ij} = \mathbf{W}\mathbf{e}_j \cdot \mathbf{e}_i = \mathbf{w} \times \mathbf{e}_j \cdot \mathbf{e}_i = \sum_{n,l,m=1}^{3} \varepsilon_{nlm}w_l\delta_{jm}\mathbf{e}_n \cdot \mathbf{e}_i = \sum_{l=1}^{3} \varepsilon_{ilj}w_l.$$

With another notation, if the coordinates of $\mathbf{w}$ are α, β, γ, then

$$[\mathbf{W}] = \begin{pmatrix} 0 & -\gamma & \beta \\ \gamma & 0 & -\alpha \\ -\beta & \alpha & 0 \end{pmatrix}.$$

Multiplying (B.12) by ε_{ijm} and adding with respect to i and j we obtain,

$$\sum_{i,j=1}^{3} \varepsilon_{ijm} W_{ij} = \sum_{i,j,l=1}^{3} \varepsilon_{ijm}\varepsilon_{ilj} w_l = -\sum_{i,j,l=1}^{3} \varepsilon_{ijm}\varepsilon_{ijl} w_l$$

$$= -2\sum_{l=1}^{3} \delta_{lm} w_l = -2 w_m,$$

where we have used (B.9). Hence,

$$w_m = -\frac{1}{2} \sum_{i,j=1}^{3} \varepsilon_{mij} W_{ij}. \tag{B.13}$$

B.1.4 The affine space

Let $\mathscr{V}$ be a vector space. A N-dimensional *affine space* is a triple $(\mathscr{E}, \mathscr{V}, +)$ consisting of a set $\mathscr{E}$, a N-dimensional vector space $\mathscr{V}$ and a mapping

$$\mathscr{E} \times \mathscr{V} \xrightarrow{+} \mathscr{E}$$
$$(x, \mathbf{v}) \longrightarrow x + \mathbf{v}$$

satisfying the following properties:

1. *For any $x, y \in \mathscr{E}$ there exists a unique $\mathbf{v} \in \mathscr{V}$ such that $x + \mathbf{v} = y$.*

   This vector $\mathbf{v}$ is usually denoted by $\overrightarrow{xy}$ or $y - x$.

2. *The following equality holds:*

$$(x + \mathbf{v}) + \mathbf{w} = x + (\mathbf{v} + \mathbf{w}) \quad \forall x \in \mathscr{E} \quad \forall \mathbf{v}, \mathbf{w} \in \mathscr{V}.$$

If $\mathscr{V}$ is an Euclidean vector space then the corresponding affine space is called *Euclidean affine space*.

In an Euclidean affine space we can define a *distance* by,

$$\mathscr{E} \times \mathscr{E} \xrightarrow{d} \mathbb{R}$$
$$(x, y) \longrightarrow d(x, y) = |\overrightarrow{xy}|.$$

Thus $(\mathscr{E}, d)$ is a metric space and hence a topological space.

In an Euclidean affine space, a *cartesian coordinate frame* is a couple $(o, \{\mathbf{e}_i, i = 1, \ldots, N\})$ where o is any fixed point in $\mathscr{E}$ called the *origin* and $\{\mathbf{e}_i, i = 1, \ldots, N\}$ is an orthonormal basis of the Euclidean vector space $\mathscr{V}$.
For any point $x \in \mathscr{E}$ its coordinates with respect to the above cartesian frame are the

numbers x_i, $i = 1, \ldots, N$ such that

$$\vec{ox} = \sum_{i=1}^{N} x_i \mathbf{e}_i.$$

B.2 Vector and tensor analysis

A *field* is any mapping defined in a subset of $\mathscr{E}$ and valued in $\mathbb{R}$, $\mathscr{V}$ or Lin, in which case it is called scalar, vector or tensor field, respectively.

B.2.1 Differential operators

Let W be a normed vector space (in practice W = $\mathbb{R}$, $\mathscr{V}$, or Lin). A mapping $f : \mathscr{R} \subset \mathscr{E} \longrightarrow W$ defined in an open set $\mathscr{R}$ of $\mathscr{E}$ is said *differentiable* at point $x \in \mathscr{R}$ if there exists a linear mapping $\mathscr{X} : \mathscr{V} \longrightarrow W$ such that,

$$f(x+\mathbf{h}) - f(x) - \mathscr{X}\mathbf{h} = o(\mathbf{h}) \text{ as } \mathbf{h} \to \mathbf{0},$$

which means

$$\lim_{\mathbf{h} \to \mathbf{0}} \frac{\|f(x+\mathbf{h}) - f(x) - \mathscr{X}\mathbf{h}\|_W}{\|\mathbf{h}\|_{\mathscr{V}}} = 0.$$

If f is differentiable at x then linear mapping $\mathscr{X}$ is unique and is called the *differential* of f at x. Usually, it is denoted by $Df(x)$.

The above definition can be extended to affine space-valued mappings. In this case the differential $Df(x)$ is a linear mapping from $\mathscr{V}$ into itself (i.e. $Df(x) \in$ Lin) satisfying

$$\lim_{\mathbf{h} \to \mathbf{0}} \frac{\|f(x+\mathbf{h}) - f(x) - Df(x)\mathbf{h}\|_{\mathscr{V}}}{\|\mathbf{h}\|_{\mathscr{V}}} = 0.$$

We notice that the expression in the numerator makes sense because $f(x+\mathbf{h}) - f(x) \in \mathscr{V}$.

From the above definition it can be easily deduced that, for any vector $\mathbf{u} \in \mathscr{V}$ we have

$$Df(x)\mathbf{u} = \lim_{\alpha \to 0} \frac{f(x+\alpha\mathbf{u}) - f(x)}{\alpha}.$$

Let $\mathbf{e} \in \mathscr{V}$ be a unit vector. The *directional derivative* of f in direction $\mathbf{e}$ at point x is defined by

$$D_\mathbf{e} f(x) := Df(x)\mathbf{e} = \lim_{\alpha \to 0} \frac{f(x+\alpha\mathbf{e}) - f(x)}{\alpha}. \tag{B.14}$$

Notice that if f is W-valued then $Df(x)\mathbf{u} \in W$, while $Df(x)\mathbf{u} \in \mathscr{V}$ if f is affine space-valued.

If $\{\mathbf{e}_i\}$ is a basis of unit vectors, the corresponding directional derivatives, $D_{\mathbf{e}_i} f(x)$ are called *partial derivatives* of f at point x. They are usually denoted by $\frac{\partial f}{\partial x_i}(x)$ or $f_{,i}(x)$.

Examples:

1) Let φ be a scalar field which is differentiable at x. Then

$$D\varphi(x): \mathscr{V} \longrightarrow \mathbb{R}.$$

and the unique vector $\operatorname{grad}\varphi(x) \in \mathscr{V}$ such that

$$D\varphi(x)\mathbf{a} = \operatorname{grad}\varphi(x) \cdot \mathbf{a} \quad \forall \mathbf{a} \in \mathscr{V},$$

is called the *gradient of φ at x*.
If φ is differentiable at all points in $\mathscr{R}$ we can define the vector field,

$$\begin{aligned} \mathscr{R} \subset \mathscr{E} &\longrightarrow \mathscr{V} \\ x &\longrightarrow \operatorname{grad}\varphi(x) \end{aligned}$$

which is called the *gradient* of φ.
Let us compute the coordinates of vector $\operatorname{grad}\varphi(x)$ with respect to the orthonormal basis $\{\mathbf{e}_i\}$. According to (B.2) we have

$$(\operatorname{grad}\varphi(x))_i = \operatorname{grad}\varphi(x) \cdot \mathbf{e}_i = D\varphi(x)\mathbf{e}_i = \varphi_{,i}(x)$$

and then

$$\operatorname{grad}\varphi(x) = \sum_{i=1}^{N} \varphi_{,i}(x)\mathbf{e}_i.$$

2) Let us consider the case where $W = \mathscr{V}$ or $W = \mathscr{E}$. Then $D\mathbf{u}(x): \mathscr{V} \longrightarrow \mathscr{V}$.
If $\mathbf{u}$ is differentiable at all points in $\mathscr{R}$, we can define the tensor field,

$$\begin{aligned} \mathscr{R} \subset \mathscr{E} &\longrightarrow \operatorname{Lin} \\ x &\longrightarrow D\mathbf{u}(x) \end{aligned}$$

which is called the *gradient of vector field* $\mathbf{u}$.

We will use the notation $\operatorname{\mathbf{grad}} \mathbf{u}$ instead of $D\mathbf{u}$, to denote this differential operator.

From (B.3) we can deduce the coordinates of $\operatorname{\mathbf{grad}} \mathbf{u}(x)$ with respect to the cartesian basis $\{\mathbf{e}_i\}$. Let $\mathbf{u}(x) = \sum_{i=1}^{N} u_i(x)\mathbf{e}_i$. We have,

$$(\operatorname{\mathbf{grad}}\mathbf{u}(x))_{ij} = \operatorname{\mathbf{grad}}\mathbf{u}(x)\mathbf{e}_j \cdot \mathbf{e}_i = \mathbf{u}_{,j}(x) \cdot \mathbf{e}_i$$

$$= \lim_{\alpha \to 0} \frac{\mathbf{u}(x+\mathbf{h}) - \mathbf{u}(x)}{\alpha} \cdot \mathbf{e}_i = \lim_{\alpha \to 0} \frac{u_i(x+\mathbf{h}) - u_i(x)}{\alpha} = u_{i,j}(x).$$

Hence,

$$\operatorname{\mathbf{grad}}\mathbf{u}(x) = \sum_{i,j=1}^{N} u_{i,j}(x)\mathbf{e}_i \otimes \mathbf{e}_j.$$

Given a differentiable vector field **u**, the scalar field

$$\text{div } \mathbf{u} : \mathscr{R} \subset \mathscr{E} \longrightarrow \mathbb{R}$$
$$x \longrightarrow \text{div } \mathbf{u}(x) := \text{tr}(\mathbf{grad}\, \mathbf{u}(x))$$

is called the *divergence* of **u**.

In coordinates with respect to the orthonormal basis $\{\mathbf{e}_i\}$,

$$\text{div } \mathbf{u}(x) = \text{tr}(\mathbf{grad}\, \mathbf{u}(x)) = \text{tr}\left(\sum_{i,j=1}^{N} u_{i,j}(x)\mathbf{e}_i \otimes \mathbf{e}_j\right)$$
$$= \sum_{i,j=1}^{N} u_{i,j}(x)\text{tr}(\mathbf{e}_i \otimes \mathbf{e}_j) = \sum_{i,j=1}^{N} u_{i,j}(x)\delta_{ij}$$
$$= \sum_{i=1}^{N} u_{i,i}(x).$$

Now, let $\mathbf{S} : \mathscr{R} \subset \mathscr{E} \longrightarrow \text{Lin}$ be a tensor field. We define the vector field div **S**, called *divergence* of **S** as follows:

For any $x \in \mathscr{R}$, div $\mathbf{S}(x)$ is the unique vector such that,

$$\text{div } \mathbf{S}(x) \cdot \mathbf{a} = \text{div}(\mathbf{S}^t \mathbf{a})(x) \quad \forall\, \mathbf{a} \in \mathscr{V}. \tag{B.15}$$

Let us assume that field **S** is expressed in the basis $\{\mathbf{e}_i\}$ as,

$$\mathbf{S}(x) = \sum_{i,j=1}^{N} S_{ij}(x)\mathbf{e}_i \otimes \mathbf{e}_j.$$

Let us write

$$\text{div } \mathbf{S}(x) = \sum_{l=1}^{N} d_l(x)\mathbf{e}_l.$$

By taking $\mathbf{a} = \mathbf{e}_k$ in (B.15) we have

$$d_k(x) = \text{div } \mathbf{S}(x) \cdot \mathbf{e}_k = \text{div}(\mathbf{S}^t \mathbf{e}_k)(x)$$
$$= \text{div}\left(\sum_{i,j=1}^{N} S_{ij}\mathbf{e}_j \otimes \mathbf{e}_i \mathbf{e}_k\right)(x) = \text{div}\left(\sum_{j=1}^{N} S_{kj}\mathbf{e}_j\right)(x) = \sum_{j=1}^{N} S_{kj,j}(x)$$

For a scalar field $\varphi : \mathscr{R} \subset \mathscr{E} \longrightarrow \mathbb{R}$, its *Laplacian* is the scalar field

$$\Delta \varphi(x) = \text{div}(\text{grad}\, \varphi)(x).$$

In coordinates,

$$\Delta \varphi(x) = \sum_{i=1}^{N} \varphi_{,ii}.$$

For a vector field $\mathbf{u} : \mathscr{R} \subset \mathscr{E} \longrightarrow \mathscr{V}$, its *Laplacian* is the vector field

$$\Delta \mathbf{u}(x) = \text{div}(\mathbf{grad}\, \mathbf{u})(x).$$

Let **u** given in coordinates by $\mathbf{u}(x) = \sum_{i=1}^{N} u_i(x)\mathbf{e}_i$. Then

$$\Delta \mathbf{u}(x) = \operatorname{div}\left(\sum_{i,j=1}^{N} u_{i,j}\mathbf{e}_i \otimes \mathbf{e}_j\right)(x) = \sum_{i,j=1}^{N} u_{i,jj}(x)\mathbf{e}_i.$$

We notice that

$$\Delta \mathbf{u}(x) = \sum_{i=1}^{N} \Delta u_i(x)\mathbf{e}_i,$$

but this equality is no longer true for curvilinear coordinates.

Now we assume that $N = 3$. Let $\mathbf{u} : \mathscr{R} \subset \mathscr{E} \longrightarrow \mathscr{V}$ be a smooth vector field. The **curl** of **u** at x is the unique vector, $\operatorname{\mathbf{curl}}\mathbf{u}(x)$, such that,

$$\operatorname{\mathbf{curl}}\mathbf{u}(x) \times \mathbf{a} = (\operatorname{\mathbf{grad}}\mathbf{u}(x) - \operatorname{\mathbf{grad}}\mathbf{u}^t(x))\mathbf{a} \quad \forall \, \mathbf{a} \in \mathscr{V}. \tag{B.16}$$

In other words, $\operatorname{\mathbf{curl}}\mathbf{u}(x)$ is the axial vector of the skew-symmetric tensor $\mathbf{W} = (\operatorname{\mathbf{grad}}\mathbf{u}(x) - \operatorname{\mathbf{grad}}\mathbf{u}^t(x))$. In order to write the **curl** operator in coordinates for an orthonormal basis, we use (B.13) to get

$$\operatorname{\mathbf{curl}}\mathbf{u}(x) = -\frac{1}{2}\sum_{i,j,m=1}^{3} \varepsilon_{mij}(u_{i,j}(x) - u_{j,i}(x))\mathbf{e}_m = \frac{1}{2}\sum_{i,j,m=1}^{3} \varepsilon_{mji} u_{i,j}(x)\mathbf{e}_m$$

$$+ \frac{1}{2}\sum_{i,j,m=1}^{3} \varepsilon_{mij} u_{j,i}(x)\mathbf{e}_m = \sum_{i,j,m=1}^{3} \varepsilon_{mji} u_{i,j}(x)\mathbf{e}_m.$$

It is useful to recall the following nemotechnical rule:

$$\operatorname{\mathbf{curl}}\mathbf{u} = \begin{vmatrix} \mathbf{e}_1 & \mathbf{e}_2 & \mathbf{e}_2 \\ \frac{\partial}{\partial x_1} & \frac{\partial}{\partial x_2} & \frac{\partial}{\partial x_3} \\ u_1 & u_2 & u_3 \end{vmatrix}.$$

The following *product rule* is very useful to prove the equalities below.

Proposition B.2. *Let $\mathscr{R}$ be an open set in the affine space $\mathscr{E}$. Let* f *and* g *be two differentiable fields defined in $\mathscr{R}$ with values in two finite-dimensional normed spaces W and U, respectively. Let π a bilinear map from $W \times U$ into another normed space Y. Let us define the field:*

$$\mathrm{h} : x \in \mathscr{R} \subset \mathscr{E} \to \mathrm{h}(x) := \pi(\mathrm{f}(x), \mathrm{g}(x)) \in Y.$$

Then h *is differentiable in $\mathscr{R}$ and we have*

$$D\mathrm{h}(x)\mathbf{u} = \pi(D\mathrm{f}(x)\mathbf{u}, \mathrm{g}(x)) + \pi(\mathrm{f}(x), D\mathrm{g}(x)\mathbf{u}) \quad \forall \mathbf{u} \in \mathscr{V}. \tag{B.17}$$

Corollary B.1. *Let us choose any $x_0 \in \mathscr{R}$ and define the mappings*

$$\mathrm{h}_1(x) := \pi(\mathrm{f}(x_0), \mathrm{g}(x)), \quad \mathrm{h}_2(x) := \pi(\mathrm{f}(x), \mathrm{g}(x_0)).$$

Then

$$D\mathrm{h}(x_0) = D\mathrm{h}_1(x_0) + D\mathrm{h}_2(x_0).$$

Exercise B.2. Prove the following equalities:

$$\operatorname{grad}(\varphi \mathbf{v}) = \varphi \operatorname{grad} \mathbf{v} + \mathbf{v} \otimes \operatorname{grad} \varphi, \tag{B.18}$$

$$\operatorname{div}(\varphi \mathbf{v}) = \varphi \operatorname{div} \mathbf{v} + \mathbf{v} \cdot \operatorname{grad} \varphi, \tag{B.19}$$

$$\operatorname{grad}(\mathbf{v} \cdot \mathbf{w}) = (\mathbf{grad}\,\mathbf{w})^t \mathbf{v} + (\mathbf{grad}\,\mathbf{v})^t \mathbf{w}, \tag{B.20}$$

$$\operatorname{div}(\mathbf{v} \otimes \mathbf{w}) = \mathbf{v} \operatorname{div} \mathbf{w} + (\mathbf{grad}\,\mathbf{v})\mathbf{w}, \tag{B.21}$$

$$\operatorname{div}(\mathbf{S}^t \mathbf{v}) = \mathbf{S} \cdot \mathbf{grad}\,\mathbf{v} + \mathbf{v} \cdot \operatorname{div} \mathbf{S}, \tag{B.22}$$

$$\operatorname{div}(\varphi \mathbf{S}) = \varphi \operatorname{div} \mathbf{S} + \mathbf{S}\operatorname{grad}\varphi, \tag{B.23}$$

$$\operatorname{div}(\mathbf{v} \times \mathbf{w}) = \mathbf{w} \cdot \mathbf{curl}\,\mathbf{v} - \mathbf{v} \cdot \mathbf{curl}\,\mathbf{w}, \tag{B.24}$$

$$\operatorname{div}\operatorname{grad}\mathbf{v}^t = \operatorname{grad}(\operatorname{div}\mathbf{v}) \tag{B.25}$$

$$\mathbf{curl}(\varphi \mathbf{v}) = \operatorname{grad}\varphi \times \mathbf{v} + \varphi\,\mathbf{curl}\,\mathbf{v} \tag{B.26}$$

$$\mathbf{curl}(\mathbf{v} \times \mathbf{w}) = \mathbf{grad}\,\mathbf{vw} - \mathbf{w}\operatorname{div}\mathbf{v} + \mathbf{v}\operatorname{div}\mathbf{w} - \mathbf{grad}\,\mathbf{wv}, \tag{B.27}$$

$$\operatorname{div}\mathbf{curl}\,\mathbf{v} = 0, \tag{B.28}$$

$$\mathbf{curl}\operatorname{grad}\varphi = 0, \tag{B.29}$$

$$\Delta \mathbf{v} = \operatorname{grad}(\operatorname{div}\mathbf{v}) - \mathbf{curl}(\mathbf{curl}\,\mathbf{v}), \tag{B.30}$$

$$\Delta(\operatorname{grad}\varphi) = \operatorname{grad}(\Delta\varphi), \tag{B.31}$$

$$\operatorname{div}(\Delta\mathbf{v}) = \Delta(\operatorname{div}\mathbf{v}), \tag{B.32}$$

$$\mathbf{curl}(\Delta\mathbf{v}) = \Delta(\mathbf{curl}\,\mathbf{v}), \tag{B.33}$$

$$\Delta(\varphi\psi) = \varphi\Delta\psi + \psi\Delta\varphi + 2\operatorname{grad}\varphi \cdot \operatorname{grad}\psi. \tag{B.34}$$

B.2.2 Curves and curvilinear integrals

A (regular) curve c in $\mathscr{R}$ is a class C^1 mapping,

$$c : [0,1] \longrightarrow \mathscr{R},$$

such that $\dot{c}(s) \neq 0\ \forall s \in [0,1]$, where the derivative $\dot{c}(s) \in \mathscr{V}$ is given by

$$\dot{c}(s) = \lim_{h \to 0} \frac{c(s+h) - c(s)}{h}.$$

Recall that vector $\dot{c}(s)$ is a tangent vector to c at point $c(s)$. Hence, the vector

$$\mathbf{t}(c(s)) = \frac{\dot{c}(s)}{|\dot{c}(s)|}$$

is a unit tangent vector to c at the same point.

B.2 Vector and tensor analysis

The curve is *closed* if $c(0) = c(1)$.

If $x = c(s)$, let us write
$$dl(x) = |\dot{c}(s)|\,ds.$$

The *length* of c is the number
$$\text{length}(c) := \int_c dl(x) = \int_0^1 |\dot{c}(s)|\,ds.$$

Similarly, let us introduce
$$d\mathbf{l}(c(s)) = \mathbf{t}(c(s))dl(c(s)) = \frac{\dot{c}(s)}{|\dot{c}(s)|}|\dot{c}(s)|\,ds = \dot{c}(s)\,ds. \tag{B.35}$$

Given $\mathbf{v} : \mathscr{R} \subset \mathscr{E} \longrightarrow \mathscr{V}$, a continuous *vector field*, the *curvilinear integral* of $\mathbf{v}$ along c is defined by
$$\int_c \mathbf{v}(x) \cdot d\mathbf{l}(x) := \int_c \mathbf{v}(x) \cdot \mathbf{t}(x)dl(x) = \int_0^1 \mathbf{v}(c(s)) \cdot \dot{c}(s)\,ds.$$

Similarly, given $\mathbf{S} : \mathscr{R} \subset \mathscr{E} \longrightarrow \text{Lin}$ a continuous *tensor field*. The *curvilinear integral* of $\mathbf{S}$ along c is the vector defined by
$$\int_c \mathbf{S}(x)d\mathbf{l}(x) := \int_0^1 \mathbf{S}(c(s))\dot{c}(s)\,ds.$$

If $\mathbf{v}$ is the gradient of a scalar field ϕ, then
$$\int_c \mathbf{v}(x) \cdot d\mathbf{l}(x) = \int_c \text{grad}\,\phi(x) \cdot d\mathbf{l}(x) = \int_0^1 \text{grad}\,\phi(c(s)) \cdot \dot{c}(s)\,ds$$
$$= \int_0^1 \frac{d}{ds}(\phi \circ c)(s)\,ds = \phi(c(1)) - \phi(c(0)).$$

In particular $\int_c \text{grad}\,\phi(x) \cdot dx = 0$ whenever c is closed.

A subset $\mathscr{R} \subset \mathscr{E}$ is *simply connected* if any closed curve in $\mathscr{R}$ can be continuously deformed (more precisely, is homotopic) to a point without leaving $\mathscr{R}$.

An *open region* (or *domain*) is any open connected subset of $\mathscr{E}$. The closure of an open region will be called *closed region*. We call *regular region* all closed region with smooth boundary $\partial\mathscr{R}$.

Let $\mathscr{R}$ be a closed region and Φ a field defined in $\mathscr{R}$. We say that Φ is $C^1(\mathscr{R})$ if it is continuously differentiable in the interior of $\mathscr{R}$ and Φ and $\text{grad}\,\Phi$ have continuous extensions to all of $\mathscr{R}$.

Appendix B Vector calculus

We end this section on vector calculus by recalling some fundamental results.

Potential Theorem. *Let* **v** *be a smooth vector field on an open or closed simply connected region $\mathscr{R}$ and assume that,*

$$\mathbf{curl\,v} = 0.$$

Then there is a C^2 scalar field $\phi : \mathscr{R} \longrightarrow \mathbb{R}$ such that,

$$\mathbf{v} = \mathrm{grad}\,\phi.$$

B.2.3 Gauss' and Green's formulas. Stokes' Theorem

Gauss' formulas. Let $\mathscr{R}$ be a bounded regular region and let $\phi : \mathscr{R} \longrightarrow \mathbb{R}$, $\mathbf{v} : \mathscr{R} \longrightarrow \mathscr{V}$ and $\mathbf{S} : \mathscr{R} \longrightarrow \mathrm{Lin}$ be smooth fields. Then,

1. $\int_{\partial \mathscr{R}} \varphi \mathbf{n}\, dA = \int_{\mathscr{R}} \mathrm{grad}\,\varphi\, dV$,
2. $\int_{\partial \mathscr{R}} \mathbf{v} \otimes \mathbf{n}\, dA = \int_{\mathscr{R}} \mathbf{grad\,v}\, dV$,
3. $\int_{\partial \mathscr{R}} \mathbf{v} \cdot \mathbf{n}\, dA = \int_{\mathscr{R}} \mathrm{div}\,\mathbf{v}\, dV$,
4. $\int_{\partial \mathscr{R}} \mathbf{n} \times \mathbf{v}\, dA = \int_{\mathscr{R}} \mathbf{curl\,v}\, dV$,
5. $\int_{\partial \mathscr{R}} \mathbf{Sn}\, dA = \int_{\mathscr{R}} \mathrm{div}\,\mathbf{S}\, dV$,

where $\partial \mathscr{R}$ denotes the boundary of $\mathscr{R}$ and $\mathbf{n}$ is an outward unit normal vector to $\partial \mathscr{R}$.

The following formulas can be obtained from the above Gauss' formulas and product rules in Exercise B.2.

Green's formulas. Let $\mathscr{R}$ be a bounded regular region and let $\phi, \psi : \mathscr{R} \longrightarrow \mathbb{R}$, $\mathbf{v}, \mathbf{w} : \mathscr{R} \longrightarrow \mathscr{V}$ and $\mathbf{S} : \mathscr{R} \longrightarrow \mathrm{Lin}$ be smooth fields. Then,

1. $\int_{\partial \mathscr{R}} \dfrac{\partial \phi}{\partial \mathbf{n}} \psi\, dA = \int_{\mathscr{R}} \mathrm{grad}\,\phi \cdot \mathrm{grad}\,\psi\, dV + \int_{\mathscr{R}} \Delta \phi\, \psi\, dV$,

2. $\int_{\partial \mathscr{R}} \mathbf{Sn} \cdot \mathbf{v}\, dA = \int_{\mathscr{R}} \mathbf{S} \cdot \mathbf{grad\,v}\, dV + \int_{\mathscr{R}} \mathbf{v} \cdot \mathrm{div}\,\mathbf{S}\, dV$,

3. $\int_{\partial \mathscr{R}} (\mathbf{Sn}) \otimes \mathbf{v}\, dA = \int_{\mathscr{R}} \mathbf{S}\,\mathrm{grad}\,\mathbf{v}^t\, dV + \int_{\mathscr{R}} \mathrm{div}\,\mathbf{S} \otimes \mathbf{v}\, dV$,

4. $\int_{\partial \mathscr{R}} \mathbf{v} \cdot \mathbf{n} \psi\, dA = \int_{\mathscr{R}} \mathrm{div}\,\mathbf{v}\, \psi\, dV + \int_{\mathscr{R}} \mathbf{v} \cdot \mathrm{grad}\,\psi\, dV$,

5. $\int_{\partial \mathscr{R}} \mathbf{w} \cdot \mathbf{n} \mathbf{v}\, dA = \int_{\mathscr{R}} \mathrm{div}\,\mathbf{w}\, \mathbf{v}\, dV + \int_{\mathscr{R}} \mathrm{grad}\,\mathbf{v}\mathbf{w}\, dV$,

6. $\int_{\partial \mathscr{R}} \mathbf{v} \times \mathbf{w} \cdot \mathbf{n}\, dA = \int_{\mathscr{R}} \mathbf{w} \cdot \mathbf{curl\,v}\, dV - \int_{\mathscr{R}} \mathbf{v} \cdot \mathbf{curl\,w}\, dV$.

Stokes' Theorem. *Let* **v** *be a smooth vector field on an open set* $\mathscr{R} \subset \mathscr{E}$. *Let S be a smooth surface in* $\mathscr{R}$ *bounded by* c *supposed to be a closed curve. At each point of S we choose a unit normal vector* **n** *consistent with the direction of traversing* c *in that it causes a right-handed screw to advance along* **n**. *Then,*

$$\int_S \mathbf{curl\,v} \cdot \mathbf{n}\, dA = \int_c \mathbf{v} \cdot d\mathbf{l}.$$

B.2.4 The area enclosed by a plane curve

We consider closed curves that are not necessarily simple. Let us suppose that c is a plane closed curve which intersects itself at most at a finite number of points. Let $c(s)$, $s \in [a,b]$ be a parametrization of c. Then this curve is the union of a finite collection of simple closed curves c_i, $i = 1,\ldots,N$. Each of these curves is walked in only one sense. Thus we can associate a sign s_i to curve c_i which is $+1$ if it is walked counter-clockwise or -1, otherwise.

We denote by $\mathscr{R}$ the open region enclosed by c and by $\mathscr{R}_i$ the region enclosed by c_i, $i = 1,\ldots,N$. The *net area* of $\mathscr{R}$ is defined as

$$\text{net area}(\mathscr{R}) = \sum_{i=1}^N s_i\, \text{area}(\mathscr{R}_i).$$

Then we have the following formulas for the *net area* of $\mathscr{R}$.

Proposition B.3. *We have*

$$\text{net area}(\mathscr{R}) = \int_a^b c_1(s)\frac{dc_2}{ds}(s)\,ds = -\int_a^b c_2(s)\frac{dc_1}{ds}(s)\,ds.$$

Proof.- It is an immediate consequence of the Stokes' theorem (in fact, for plane curves it is the Green's theorem). For a smooth plane vector field **w** defined in $\mathscr{R}$ it says that

$$\int_{\mathscr{R}_i} \mathbf{curl\,w} \cdot s_i \mathbf{e}_3\, dA = \int_{\partial \mathscr{R}_i} \mathbf{w} \cdot d\mathbf{l}. \tag{B.36}$$

Let us take

$$\mathbf{w}(x) = -x_2 \mathbf{e}_1 + x_1 \mathbf{e}_2.$$

Then $\mathbf{curl\,w} = 2\mathbf{e}_3$ and (B.36) yields

$$2\, s_i \text{area}(\mathscr{R}_i) = \int_{\partial \mathscr{R}_i} \mathbf{w} \cdot d\mathbf{x}, \quad i = 1,\ldots,N.$$

Adding the previous equalities we obtain

$$\text{net area}(\mathscr{R}) = \frac{1}{2}\sum_{i=1}^N \int_{\partial \mathscr{R}_i} \mathbf{w} \cdot d\mathbf{x} = \frac{1}{2}\int_c \mathbf{w} \cdot d\mathbf{x}$$

$$= \frac{1}{2}\int_a^b \left(-c_2(s)\frac{dc_1}{ds}(s) + c_1(s)\frac{dc_2}{ds}(s)\right) ds. \tag{B.37}$$

Moreover, by integration by parts

$$\int_a^b -c_2(s)\frac{dc_1}{ds}(s)\,ds = -(c_1(b)c_2(b) - c_1(a)c_2(a)) + \int_a^b c_1(s)\frac{dc_2}{ds}(s)\,ds$$
$$= \int_a^b c_1(s)\frac{dc_2}{ds}(s)\,ds, \tag{B.38}$$

because c is closed. The result follows immediately by using (B.38) in (B.37). □

B.2.5 Change of variable in integrals

Theorem B.2. *Let* $\mathbf{f}$ *be an injective smooth mapping,* $\mathbf{f} : \mathscr{B} \subset \mathscr{E} \to \mathscr{E}$, *where $\mathscr{B}$ is an open subset of the affine space $\mathscr{E}$, and* $\phi : \mathbf{f}(\mathscr{B}) \to \mathbb{R}$ *a smooth scalar field. Let* c *(resp. S) be a smooth curve (resp. surface) in $\mathscr{B}$ and $\mathscr{P}$ an open subset of $\mathscr{B}$. Then we have*

1. $\int_{\mathbf{f}(c)} \phi \mathbf{t}\,dl(x) = \int_c \phi \circ \mathbf{f}\mathbf{F}\mathbf{k}\,dl(p)$,
2. $\int_{\mathbf{f}(S)} \phi \mathbf{n}\,dA(x) = \int_S \phi \circ \mathbf{f}\det\mathbf{F}\,\mathbf{F}^{-t}\mathbf{m}\,dA(p)$,
3. $\int_{\mathbf{f}(c)} \phi\,dl(x) = \int_c \phi \circ \mathbf{f}|\mathbf{F}\mathbf{k}|\,dl(p)$,
4. $\int_{\mathbf{f}(S)} \phi\,dA(x) = \int_{\mathscr{S}} \phi \circ \mathbf{f}\det\mathbf{F}\,\mathbf{F}^{-t}\mathbf{m}\,dA(p)$,
5. $\int_{\mathbf{f}(\mathscr{P})} \phi\,dV(x) = \int_{\mathscr{P}} \phi \circ \mathbf{f}\det\mathbf{F}\,dV(p)$,

where $\mathbf{F}$ denotes the gradient of $\mathbf{f}$, which is supposed to have a positive determinant, $\mathbf{k}$ *(resp.* $\mathbf{t}$*) denotes a unit tangent vector to curve* c *(resp.* $\mathbf{f}(c)$*) and* $\mathbf{m}$ *(resp.* $\mathbf{n}$*) denotes a unit normal vector to surface S (resp.* $\mathbf{f}(S)$*).*

B.2.6 Kinematics and transport theorems

B.2.6.1 The motion of a body

Let $\mathscr{E}$ be an affine space on a three-dimensional Euclidean vector space $\mathscr{V}$. Let us denote by Lin the vector space of endomorphisms of $\mathscr{V}$ (which is isomorphic to the space of second order tensors) and by Sym the subspace of those which are symmetric.

Definition B.1. A *body* $\mathscr{B}$ is a regular region of the Euclidean space $\mathscr{E}$. Sometimes we will refer to $\mathscr{B}$ as the *reference configuration*. Elements in $\mathscr{B}$ are called *material points*.

Definition B.2. A *deformation* of $\mathscr{B}$ is a smooth one-to-one mapping $\mathbf{f}$ which maps $\mathscr{B}$ onto a closed region in $\mathscr{E}$ and satisfies

$$\det(\mathrm{grad}\,\mathbf{f}) > 0.$$

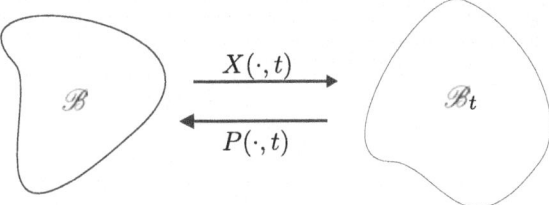

Fig. B.1. Motion and reference map

Definition B.3. A *motion* of $\mathscr{B}$ is a class C^3 mapping

$$X : \mathscr{B} \times \mathbb{R} \to \mathscr{E}$$

with $X(\cdot,t)$, a deformation of $\mathscr{B}$ for each fixed t.

We refer to $x = X(p,t)$ as the *place* occupied by the material point p at time t while

$$\mathscr{B}_t = X(\mathscr{B},t)$$

is the space occupied by the body at time t (see Fig. B.1).

Definition B.4. We call *trajectory of the material point p* the set in the affine space,

$$\mathscr{T}_p = \{X(p,t),\ t \in \mathbb{R}\}.$$

Definition B.5. We call *trajectory of the motion* the set

$$\mathscr{T} = \{(x,t) \,:\, x \in \mathscr{B}_t,\ t \in \mathbb{R}\}.$$

The vector $\mathbf{u}(p,t) = X(p,t) - p$ represents the *displacement* of p at time t while the *gradient of the motion* is the tensor field defined by

$$\mathbf{F}(p,t) := \nabla X(p,t),$$

so that we have $\mathbf{F}(p,t) = \mathbf{I} + \nabla \mathbf{u}(p,t)$. The gradient gives information on the local deformation of the body.

Let $P(\cdot,t) : \mathscr{B}_t \to \mathscr{B}$ be the inverse mapping of $X(\cdot,t)$. Then the mapping

$$P : \mathscr{T} \to \mathscr{B}$$

gives the material point which occupies the place x at time t. It is called the *reference map of the motion*.

We call $\frac{\partial X}{\partial t}$ the *velocity* and $\frac{\partial^2 X}{\partial t^2}$ the *acceleration* of the material point p at time t.

B.2.6.2 Material and spatial fields. Notations

Fields defined in $\mathscr{T}$ are called *spatial* (or *Eulerian*) *fields* while those defined in $\mathscr{B} \times \mathbb{R}$ are called *material* (or *Lagrangian*) *fields*.

Let $\boldsymbol{\Psi} : \mathscr{B} \times \mathbb{R} \to \mathscr{T}$ be the mapping given by

$$\boldsymbol{\Psi}(p,t) := (X(p,t),t),$$

the inverse of which is $\boldsymbol{\Psi}^{-1}(x,t) = (P(x,t),t)$. By using these mappings any spatial field can be transformed into a material field and vice versa. Let θ be a spatial field. Its material description is defined by

$$\theta_m := \theta \circ \boldsymbol{\Psi}.$$

Similarly, let ϕ be a material field. Its material description is defined by

$$\phi_s := \phi \circ \boldsymbol{\Psi}^{-1}.$$

In particular, the *spatial description of the velocity* is $\mathbf{v} = (\frac{\partial X}{\partial t})_s$, i.e.,

$$\mathbf{v}(x,t) := \frac{\partial X}{\partial t}(P(x,t),t).$$

Throughout this book, the following notations for differential operators are used:

- *Material fields:*

$$\dot{\theta} = \frac{\partial \theta}{\partial t}, \nabla \theta = \text{gradient}_p \theta,$$
$$\text{Div}\,\boldsymbol{\theta} = \text{divergence}_p \boldsymbol{\theta}, \text{Curl}\,\boldsymbol{\theta} = \text{Curl}_p \boldsymbol{\theta}.$$

- *Spatial fields:*

$$\phi' = \frac{\partial \phi}{\partial t}, \text{grad}\,\phi = \text{gradient}_x \phi,$$
$$\text{div}\,\boldsymbol{\phi} = \text{divergence}_x \boldsymbol{\phi}, \text{curl}\,\boldsymbol{\phi} = \text{curl}_x \boldsymbol{\phi},$$
$$\dot{\phi} : \text{material time derivative (see below)}.$$

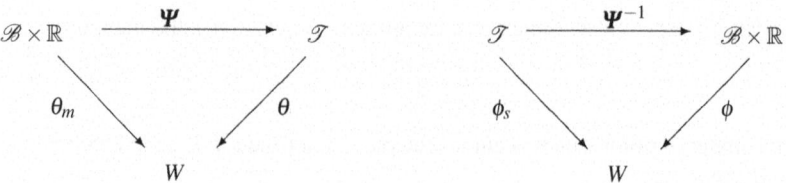

Fig. B.2. Material and spatial descriptions

B.2.6.3 Material time derivative of a spatial field

Definition B.6. Given a *spatial scalar field* ϕ, the *material time derivative* is defined by
$$\dot{\phi} =: (\frac{\partial}{\partial t}(\phi_m))_s.$$

Similarly, given a *spatial vector field* **w**, the *material time derivative* is defined by
$$\dot{\mathbf{w}} =: (\frac{\partial}{\partial t}(\mathbf{w}_m))_s.$$

The next proposition allows us to compute the material time derivatives.

Proposition B.4. *We have*
$$\dot{\phi} = \frac{\partial \phi}{\partial t} + \mathbf{v} \cdot \operatorname{grad} \phi,$$
$$\dot{\mathbf{w}} = \frac{\partial \mathbf{w}}{\partial t} + \operatorname{grad} \mathbf{w} \, \mathbf{v}.$$

Definition B.7. The position vector **r** with respect to an origin $o \in \mathscr{B}$ is the spatial vector field defined by
$$\mathbf{r}: (x,t) \in \mathscr{T} \longrightarrow \mathbf{r}(x,t) = x - o \in \mathscr{V}. \tag{B.39}$$

It is easy to see that $\dot{\mathbf{r}} = \mathbf{v}$.

B.2.6.4 Transport theorems

Let X be a motion of a body $\mathscr{B}$. Let c, S and $\mathscr{P}$ be a curve, a surface and a part of $\mathscr{B}$, respectively. Let $\phi: \mathscr{T} \to \mathbb{R}$ and $\mathbf{w}: \mathscr{T} \to \mathscr{V}$ be a scalar field and a vector field, respectively. Let c_t, S_t and $\mathscr{P}_t$ be defined by
$$c_t = X(c,t)$$
$$S_t = X(S,t)$$
$$\mathscr{P}_t = X(\mathscr{P},t).$$

Then we have

1. $\dfrac{d}{dt} \displaystyle\int_{c_t} \phi \mathbf{t} \, dl = \int_{c_t} (\dot{\phi}\mathbf{t} + \phi L\mathbf{t}) \, dl,$

2. $\dfrac{d}{dt} \displaystyle\int_{S_t} \phi \mathbf{n} \, dA = \int_{S_t} (\dot{\phi}\mathbf{n} + \phi \operatorname{div}\mathbf{v}\,\mathbf{n} - \phi L^t \mathbf{n}) \, dA,$

3. $\dfrac{d}{dt} \displaystyle\int_{\mathscr{P}_t} \phi \, dV = \int_{\mathscr{P}_t} (\dot{\phi} + \phi \operatorname{div}\mathbf{v}) \, dV = \int_{\mathscr{P}_t} (\phi' + \operatorname{div}(\phi \mathbf{v})) \, dV$
$$= \int_{\mathscr{P}_t} \phi' \, dV + \int_{\partial \mathscr{P}_t} \phi \mathbf{v} \cdot \mathbf{n} \, dA,$$

4. $\dfrac{\mathrm{d}}{\mathrm{d}t}\int_{c_t}\mathbf{w}\otimes\mathbf{t}\,\mathrm{d}l = \int_{c_t}\left(\dot{\mathbf{w}}\otimes\mathbf{t}+\mathbf{w}\otimes L\mathbf{t}\right)\mathrm{d}l,$

5. $\dfrac{\mathrm{d}}{\mathrm{d}t}\int_{c_t}\mathbf{w}\cdot\mathbf{t}\,\mathrm{d}l = \int_{c_t}\left(\dot{\mathbf{w}}\cdot\mathbf{t}+L^t\mathbf{w}\cdot\mathbf{t}\right)\mathrm{d}l_x,$

6. $\dfrac{\mathrm{d}}{\mathrm{d}t}\int_{S_t}\mathbf{w}\otimes\mathbf{n}\,\mathrm{d}A = \int_{S_t}\left(\dot{\mathbf{w}}\otimes\mathbf{n}+\mathrm{div}\,\mathbf{v}\,\mathbf{w}\otimes\mathbf{n}-\mathbf{w}\otimes L^t\mathbf{n}\right)\mathrm{d}A,$

7. $\dfrac{\mathrm{d}}{\mathrm{d}t}\int_{S_t}\mathbf{w}\cdot\mathbf{n}\,\mathrm{d}A = \int_{S_t}\left(\dot{\mathbf{w}}\cdot\mathbf{n}+\mathrm{div}\,\mathbf{v}\,\mathbf{w}\cdot\mathbf{n}-\mathrm{grad}\,\mathbf{v}\mathbf{w}\cdot\mathbf{n}\right)\mathrm{d}A,$
$\qquad = \int_{S_t}\left(\mathbf{w}'\cdot\mathbf{n}+\mathrm{div}\,\mathbf{w}\,\mathbf{v}\cdot\mathbf{n}+\mathrm{curl}(\mathbf{w}\times\mathbf{v})\cdot\mathbf{n}\right)\mathrm{d}A,$

8. $\dfrac{\mathrm{d}}{\mathrm{d}t}\int_{\mathscr{P}_t}\mathbf{w}\,\mathrm{d}V = \int_{\mathscr{P}_t}\left(\dot{\mathbf{w}}+\mathrm{div}\,\mathbf{v}\,\mathbf{w}\right)\mathrm{d}V = \int_{\mathscr{P}_t}\left(\mathbf{w}'+\mathrm{div}(\mathbf{w}\otimes\mathbf{v})\right)\mathrm{d}V$
$\qquad = \int_{\mathscr{P}_t}\mathbf{w}'\,\mathrm{d}V + \int_{\partial\mathscr{P}_t}\mathbf{w}\mathbf{v}\cdot\mathbf{n}\,\mathrm{d}A,$

where $\mathbf{t}$ denotes a unit tangent vector to c_t and $\mathbf{n}$ is a unit normal vector to S_t.

B.2.7 Localization theorem

Theorem B.3. *Let Φ be a continuous scalar mapping defined in an open set $\mathscr{R}\subset\mathscr{E}$. If*
$$\int_B \Phi\,\mathrm{d}V = 0,$$
for every closed ball $B\subset\mathscr{R}$, then $\Phi = 0$ in $\mathscr{R}$. Moreover, if Φ is scalar and
$$\int_B \Phi\,\mathrm{d}V \geq 0,$$
for every closed ball $B\subset\mathscr{R}$, then $\Phi \geq 0$ in $\mathscr{R}$.

B.2.8 Curvilinear coordinates

Sometimes, local curvilinear coordinates rather than fixed cartesian ones are used to simplify equations. This can be done when the problem exhibits some symmetries. By writing the equations in some particular system of coordinates these symmetries allows us reduce the number of unknowns and/or the number of independent variables. The inconvenient lies in that expressions of differential operators in curvilinear coordinates are more complicated than in fixed cartesian ones. For cylindrical and spherical coordinates we adopt the notations included in the norm ISO 80000-2:2009.

B.2.8.1 Cartesian Coordinates

Definition

$$f : \Omega = \mathbb{R}^3 \to \mathbb{R}^3$$
$$f(u_1, u_2, u_3) = (u_1, u_2, u_3)$$

Basis

- *Contravariant:* $\quad \mathbf{g}_1 = \mathbf{e}_1, \quad \mathbf{g}_2 = \mathbf{e}_2, \quad \mathbf{g}_3 = \mathbf{e}_3.$
- *Covariant:* $\quad \mathbf{g}^1 = \mathbf{e}_1, \quad \mathbf{g}^2 = \mathbf{e}_2, \quad \mathbf{g}^3 = \mathbf{e}_3.$
- *Physical:* $\quad \mathbf{e}_1, \quad \mathbf{e}_2, \quad \mathbf{e}_3.$

Christoffel symbols of the second kind

$$\Gamma^1_{jk} = \begin{pmatrix} 0 & 0 & 0 \\ 0 & 0 & 0 \\ 0 & 0 & 0 \end{pmatrix}$$

$$\Gamma^2_{jk} = \begin{pmatrix} 0 & 0 & 0 \\ 0 & 0 & 0 \\ 0 & 0 & 0 \end{pmatrix}$$

$$\Gamma^3_{jk} = \begin{pmatrix} 0 & 0 & 0 \\ 0 & 0 & 0 \\ 0 & 0 & 0 \end{pmatrix}$$

Differential operators in the physical basis

- *Gradient of scalar field:*

$$\text{grad}\,\varphi = \frac{\partial \varphi}{\partial x_1} \mathbf{e}_1 + \frac{\partial \varphi}{\partial x_2} \mathbf{e}_2 + \frac{\partial \varphi}{\partial x_3} \mathbf{e}_3. \tag{B.40}$$

- *Gradient of a vector field:*

$$\begin{aligned}\mathbf{grad\,w} &= \frac{\partial w_1}{\partial x_1} \mathbf{e}_1 \otimes \mathbf{e}_1 + \frac{\partial w_1}{\partial x_2} \mathbf{e}_1 \otimes \mathbf{e}_2 + \frac{\partial w_1}{\partial x_3} \mathbf{e}_1 \otimes \mathbf{e}_3 \\ &+ \frac{\partial w_2}{\partial x_1} \mathbf{e}_2 \otimes \mathbf{e}_1 + \frac{\partial w_2}{\partial x_2} \mathbf{e}_2 \otimes \mathbf{e}_2 + \frac{\partial w_2}{\partial x_3} \mathbf{e}_2 \otimes \mathbf{e}_3 \\ &+ \frac{\partial w_3}{\partial x_1} \mathbf{e}_3 \otimes \mathbf{e}_1 + \frac{\partial w_3}{\partial x_2} \mathbf{e}_3 \otimes \mathbf{e}_2 + \frac{\partial w_3}{\partial x_3} \mathbf{e}_3 \otimes \mathbf{e}_3. \end{aligned} \tag{B.41}$$

- *Convective term:*

$$(\mathbf{grad\,w})\mathbf{w} = \left\{w_1\frac{\partial w_1}{\partial x_1} + w_2\frac{\partial w_1}{\partial x_2} + w_3\frac{\partial w_1}{\partial x_3}\right\}\mathbf{e}_1$$
$$+ \left\{w_1\frac{\partial w_2}{\partial x_1} + w_2\frac{\partial w_2}{\partial x_2} + w_3\frac{\partial w_2}{\partial x_3}\right\}\mathbf{e}_2$$
$$+ \left\{w_1\frac{\partial w_3}{\partial x_1} + w_2\frac{\partial w_3}{\partial x_2} + w_3\frac{\partial w_3}{\partial x_3}\right\}\mathbf{e}_3.$$

- *Curl of a vector field:*

$$\mathbf{curl\,w} = \left(\frac{\partial w_3}{\partial x_2} - \frac{\partial w_2}{\partial x_3}\right)\mathbf{e}_1$$
$$+ \left(\frac{\partial w_1}{\partial x_3} - \frac{\partial w_3}{\partial x_1}\right)\mathbf{e}_2$$
$$+ \left(\frac{\partial w_2}{\partial x_1} - \frac{\partial w_1}{\partial x_2}\right)\mathbf{e}_3.$$

- *Divergence of a vector field:*

$$\mathrm{div}\,\mathbf{w} = \frac{\partial w_1}{\partial x_1} + \frac{\partial w_2}{\partial x_2} + \frac{\partial w_3}{\partial x_3}.$$

- *Divergence of a tensor field:*

$$\mathrm{div}\,\mathbf{S} = \left[\frac{\partial S_{11}}{\partial x_1} + \frac{\partial S_{12}}{\partial x_2} + \frac{\partial S_{13}}{\partial x_3}\right]\mathbf{e}_1$$
$$+ \left[\frac{\partial S_{21}}{\partial x_1} + \frac{\partial S_{22}}{\partial x_2} + \frac{\partial S_{23}}{\partial x_3}\right]\mathbf{e}_2$$
$$+ \left[\frac{\partial S_{31}}{\partial x_1} + \frac{\partial S_{32}}{\partial x_2} + \frac{\partial S_{33}}{\partial x_3}\right]\mathbf{e}_3. \qquad (\text{B.42})$$

- *Laplacian of a scalar field:*

$$\Delta\varphi = \frac{\partial^2\varphi}{\partial x_1^2} + \frac{\partial^2\varphi}{\partial x_2^2} + \frac{\partial^2\varphi}{\partial x_3^2}.$$

- *Laplacian of a vector field:*

$$\Delta\mathbf{w} = \Delta w_1\mathbf{e}_1 + \Delta w_2\mathbf{e}_2 + \Delta w_3\mathbf{e}_3. \qquad (\text{B.43})$$

Measure elements for integration

- *Line:*
$$dx_1, \; dx_2, \; dx_3. \tag{B.44}$$
$$dl(x) = e_1 dx_1 + e_2 dx_2 + e_3 dx_3. \tag{B.45}$$

- *Surface:*
$$dx_2 \, dx_3, \; dx_1 \, dx_3, \; dx_1 \, dx_2. \tag{B.46}$$

- *Volume:*
$$dx_1 \, dx_2 \, dx_3. \tag{B.47}$$

B.2.8.2 Cylindrical Coordinates

Definition
$$f : \Omega = (0, \infty) \times (0, 2\pi) \times (-\infty, \infty) \to \mathbb{R}^3$$
$$f(\rho, \theta, z) = (\rho \cos \theta, \rho \sin \theta, z).$$

Basis

- *Contravariant:*
$$\begin{cases} g_1(\rho, \theta, z) = \cos \theta \, e_1 + \sin \theta \, e_2, \\ g_2(\rho, \theta, z) = -\rho \sin \theta \, e_1 + \rho \cos \theta \, e_2, \\ g_3(\rho, \theta, z) = e_3. \end{cases}$$

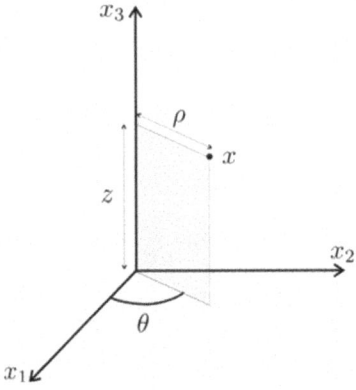

Fig. B.3. Cylindrical coordinates

Appendix B Vector calculus

- *Covariant:*

$$\begin{cases} \mathbf{g}^1(\rho,\theta,z) = \cos\theta\,\mathbf{e}_1 + \sin\theta\,\mathbf{e}_2, \\ \mathbf{g}^2(\rho,\theta,z) = -\frac{1}{\rho}\sin\theta\,\mathbf{e}_1 + \frac{1}{\rho}\cos\theta\,\mathbf{e}_2, \\ \mathbf{g}^3(\rho,\theta,z) = \mathbf{e}_3. \end{cases}$$

- *Physical:*

$$\begin{cases} \mathbf{e}_\rho(\rho,\theta,z) = \cos\theta\,\mathbf{e}_1 + \sin\theta\,\mathbf{e}_2, \\ \mathbf{e}_\theta(\rho,\theta,z) = -\sin\theta\,\mathbf{e}_1 + \cos\theta\,\mathbf{e}_2, \\ \mathbf{e}_z(\rho,\theta,z) = \mathbf{e}_3. \end{cases}$$

Christoffel symbols of the second kind

$$\Gamma^1_{jk} = \begin{pmatrix} 0 & 0 & 0 \\ 0 & -\rho & 0 \\ 0 & 0 & 0 \end{pmatrix},$$

$$\Gamma^2_{jk} = \begin{pmatrix} 0 & \frac{1}{\rho} & 0 \\ \frac{1}{\rho} & 0 & 0 \\ 0 & 0 & 0 \end{pmatrix},$$

$$\Gamma^3_{jk} = \begin{pmatrix} 0 & 0 & 0 \\ 0 & 0 & 0 \\ 0 & 0 & 0 \end{pmatrix}.$$

Differential operators in the physical basis

- *Gradient of scalar field:*

$$\operatorname{grad}\theta = \frac{\partial\theta}{\partial\rho}\mathbf{e}_\rho + \frac{1}{\rho}\frac{\partial\theta}{\partial\theta}\mathbf{e}_\theta + \frac{\partial\theta}{\partial z}\mathbf{e}_z.$$

- *Gradient of a vector field:*

$$\begin{aligned}\operatorname{grad}\mathbf{w} =\ & \frac{\partial w_\rho}{\partial\rho}\mathbf{e}_\rho\otimes\mathbf{e}_\rho + \left[\frac{1}{\rho}\frac{\partial w_\rho}{\partial\theta} - \frac{1}{\rho}w_\theta\right]\mathbf{e}_\rho\otimes\mathbf{e}_\theta + \frac{\partial w_\rho}{\partial z}\mathbf{e}_\rho\otimes\mathbf{e}_z \\ & + \frac{\partial w_\theta}{\partial\rho}\mathbf{e}_\theta\otimes\mathbf{e}_\rho + \left[\frac{1}{\rho}\frac{\partial w_\theta}{\partial\theta} + \frac{1}{\rho}w_\rho\right]\mathbf{e}_\theta\otimes\mathbf{e}_\theta + \frac{\partial w_\theta}{\partial z}\mathbf{e}_\theta\otimes\mathbf{e}_z \\ & + \frac{\partial w_z}{\partial\rho}\mathbf{e}_z\otimes\mathbf{e}_\rho + \frac{1}{\rho}\frac{\partial w_z}{\partial\theta}\mathbf{e}_z\otimes\mathbf{e}_\theta + \frac{\partial w_z}{\partial z}\mathbf{e}_z\otimes\mathbf{e}_z. \end{aligned} \quad (B.48)$$

- *Convective term:*

$$\begin{aligned}(\operatorname{grad}\mathbf{w})\mathbf{w} =\ & \left\{w_\rho\frac{\partial w_\rho}{\partial\rho} + \frac{1}{\rho}w_\theta\frac{\partial w_\rho}{\partial\theta} + w_z\frac{\partial w_\rho}{\partial z} - \frac{(w_\theta)^2}{\rho}\right\}\mathbf{e}_\rho \\ & + \left\{w_\rho\frac{\partial w_\theta}{\partial\rho} + \frac{1}{\rho}w_\theta\frac{\partial w_\theta}{\partial\theta} + w_z\frac{\partial w_\theta}{\partial z} + \frac{w_\rho w_\theta}{\rho}\right\}\mathbf{e}_\theta \\ & + \left\{w_\rho\frac{\partial w_z}{\partial\rho} + \frac{1}{\rho}w_\theta\frac{\partial w_z}{\partial\theta} + w_z\frac{\partial w_z}{\partial z}\right\}\mathbf{e}_z. \end{aligned}$$

B.2 Vector and tensor analysis

- Curl of a vector field:

$$\mathbf{curl\,w} = \left(\frac{1}{\rho}\frac{\partial w_z}{\partial \theta} - \frac{\partial w_\theta}{\partial z}\right)\mathbf{e}_\rho$$
$$+ \left(-\frac{\partial w_z}{\partial \rho} + \frac{\partial w_\rho}{\partial z}\right)\mathbf{e}_\theta$$
$$+ \left(\frac{1}{\rho}\frac{\partial}{\partial \rho}(\rho w_\theta) - \frac{1}{\rho}\frac{\partial w_\rho}{\partial \theta}\right)\mathbf{e}_z. \tag{B.49}$$

- Divergence of a vector field:

$$\mathrm{div}\,\mathbf{w} = \frac{1}{\rho}\frac{\partial}{\partial \rho}(\rho w_\rho) + \frac{1}{\rho}\frac{\partial w_\theta}{\partial \theta} + \frac{\partial w_z}{\partial z}.$$

- Divergence of a tensor field:

$$\mathrm{div}\,\mathbf{S} = \frac{1}{\rho}\left[\frac{\partial}{\partial \rho}(\rho S_{\rho\rho}) + \frac{\partial S_{\rho\theta}}{\partial \theta} + \rho\frac{\partial S_{\rho z}}{\partial z} - S_{\theta\theta}\right]\mathbf{e}_\rho$$
$$+ \left[\frac{\partial S_{\theta\rho}}{\partial \rho} + \frac{1}{\rho}\frac{\partial S_{\theta\theta}}{\partial \theta} + \frac{\partial S_{\theta z}}{\partial z} + \frac{1}{\rho}S_{\rho\theta} + \frac{1}{\rho}S_{\theta\rho}\right]\mathbf{e}_\theta$$
$$+ \frac{1}{\rho}\left[\frac{\partial}{\partial \rho}(\rho S_{z\rho}) + \frac{\partial S_{z\theta}}{\partial \theta} + \rho\frac{\partial S_{zz}}{\partial z}\right]\mathbf{e}_z. \tag{B.50}$$

- Laplacian of a scalar field:

$$\Delta\theta = \frac{1}{\rho}\left[\frac{\partial}{\partial \rho}\left(\rho\frac{\partial\theta}{\partial \rho}\right) + \frac{\partial}{\partial \theta}\left(\frac{1}{\rho}\frac{\partial\theta}{\partial \theta}\right) + \frac{\partial}{\partial z}\left(\rho\frac{\partial\theta}{\partial z}\right)\right]$$
$$= \frac{1}{\rho}\frac{\partial}{\partial \rho}\left(\rho\frac{\partial\theta}{\partial \rho}\right) + \frac{1}{\rho^2}\frac{\partial^2\theta}{\partial \theta^2} + \frac{\partial^2\theta}{\partial z^2}.$$

- Laplacian of a vector field:

$$\Delta\mathbf{w} = \left[\Delta w_\rho - \frac{1}{\rho^2}w_\rho - \frac{2}{\rho^2}\frac{\partial w_\theta}{\partial \theta}\right]\mathbf{e}_\rho + \left[\Delta w_\theta - \frac{1}{\rho^2}w_\theta + \frac{2}{\rho^2}\frac{\partial w_\rho}{\partial \theta}\right]\mathbf{e}_\theta + \Delta w_z\mathbf{e}_z. \tag{B.51}$$

Measure elements for integration

- Line:
$$d\rho,\ \rho\,d\theta,\ dz,$$
$$\mathbf{dl} = \mathbf{e}_\rho d\rho + \rho\mathbf{e}_\theta d\theta + \mathbf{e}_z dz. \tag{B.52}$$

- Surface:
$$\rho\,d\theta dz,\ d\rho dz,\ \rho\,d\rho d\theta. \tag{B.53}$$

- Volume:
$$\rho\,d\rho d\theta dz. \tag{B.54}$$

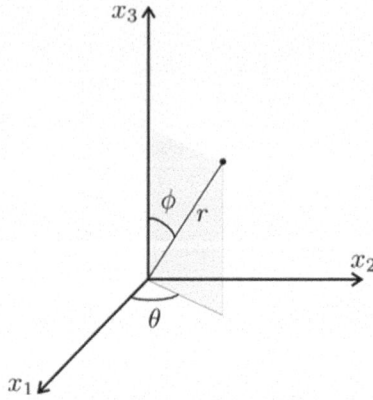

Fig. B.4. Spherical coordinates

B.2.8.3 Spherical Coordinates

Definition
$$f : \Omega = (0,\infty) \times (0,\pi) \times (0,2\pi) \to \mathbb{R}^3$$
$$f(r,\phi,\theta) = (r\sin\phi\cos\theta, r\sin\phi\sin\theta, r\cos\phi).$$

Basis

- *Contravariant:*
$$\begin{cases} \mathbf{g}_1(r,\phi,\theta) = \sin\phi\cos\theta\,\mathbf{e}_1 + \sin\phi\sin\theta\,\mathbf{e}_2 + \cos\phi\,\mathbf{e}_3, \\ \mathbf{g}_2(r,\phi,\theta) = r\cos\phi\cos\theta\,\mathbf{e}_1 + r\cos\phi\sin\theta\,\mathbf{e}_2 - r\sin\phi\,\mathbf{e}_3, \\ \mathbf{g}_3(r,\phi,\theta) = -r\sin\phi\sin\theta\,\mathbf{e}_1 + r\sin\phi\cos\theta\,\mathbf{e}_2. \end{cases}$$

- *Covariant:*
$$\begin{cases} \mathbf{g}^1(r,\phi,\theta) = \sin\phi\cos\theta\,\mathbf{e}_1 + \sin\phi\sin\theta\,\mathbf{e}_2 + \cos\phi\,\mathbf{e}_3, \\ \mathbf{g}^2(r,\phi,\theta) = \frac{1}{r}\cos\phi\cos\theta\,\mathbf{e}_1 + \frac{1}{r}\cos\phi\sin\theta\,\mathbf{e}_2 - \frac{1}{r}\sin\phi\,\mathbf{e}_3, \\ \mathbf{g}^3(r,\phi,\theta) = -\frac{1}{r\sin\phi}\sin\theta\,\mathbf{e}_1 + \frac{1}{r\sin\phi}\cos\theta\,\mathbf{e}_2. \end{cases}$$

- *Physical:*
$$\begin{cases} \mathbf{e}_r(r,\phi,\theta) = \sin\phi\cos\theta\,\mathbf{e}_1 + \sin\phi\sin\theta\,\mathbf{e}_2 + \cos\phi\,\mathbf{e}_3, \\ \mathbf{e}_\phi(r,\phi,\theta) = \cos\phi\cos\theta\,\mathbf{e}_1 + \cos\phi\sin\theta\,\mathbf{e}_2 - \sin\phi\,\mathbf{e}_3, \\ \mathbf{e}_\theta(r,\phi,\theta) = -\sin\theta\,\mathbf{e}_1 + \cos\theta\,\mathbf{e}_2. \end{cases}$$

Christoffel symbols of the second kind

$$\Gamma^1_{jk} = \begin{pmatrix} 0 & 0 & 0 \\ 0 & -r & 0 \\ 0 & 0 & -r\sin^2\phi \end{pmatrix},$$

$$\Gamma_{jk}^2 = \begin{pmatrix} 0 & \frac{1}{r} & 0 \\ \frac{1}{r} & 0 & 0 \\ 0 & 0 & -\frac{1}{2}\sin 2\phi \end{pmatrix},$$

$$\Gamma_{jk}^3 = \begin{pmatrix} 0 & 0 & \frac{1}{r} \\ 0 & 0 & \cot\phi \\ \frac{1}{r} & \cot\phi & 0 \end{pmatrix}.$$

Differential operators

- *Gradient of a scalar field:*

$$\operatorname{grad}\phi = \frac{\partial\phi}{\partial r}\mathbf{g}^1 + \frac{\partial\phi}{\partial\phi}\mathbf{g}^2 + \frac{\partial\phi}{\partial\theta}\mathbf{g}^3$$
$$= \frac{\partial\phi}{\partial r}\mathbf{e}_r + \frac{1}{r}\frac{\partial\phi}{\partial\phi}\mathbf{e}_\phi + \frac{1}{r\sin\phi}\frac{\partial\phi}{\partial\theta}\mathbf{e}_\theta.$$

- *Gradient of a vector field:*

$$\begin{aligned}
\operatorname{grad}\mathbf{w} &= \frac{\partial w_r}{\partial r}\mathbf{e}_r\otimes\mathbf{e}_r + \left[\frac{1}{r}\frac{\partial w_r}{\partial\phi} - \frac{1}{r}w_\phi\right]\mathbf{e}_r\otimes\mathbf{e}_\phi \\
&+ \left[\frac{1}{r\sin\phi}\frac{\partial w_r}{\partial\theta} - \frac{1}{r}w_\theta\right]\mathbf{e}_r\otimes\mathbf{e}_\theta \\
&+ \frac{\partial w_\phi}{\partial r}\mathbf{e}_\phi\otimes\mathbf{e}_r + \left[\frac{1}{r}\frac{\partial w_\phi}{\partial\phi} + \frac{1}{r}w_r\right]\mathbf{e}_\phi\otimes\mathbf{e}_\phi \\
&+ \left[\frac{1}{r\sin\phi}\frac{\partial w_\phi}{\partial\theta} - \frac{1}{r}\cot\phi\, w_\theta\right]\mathbf{e}_\phi\otimes\mathbf{e}_\theta \\
&+ \frac{\partial w_\theta}{\partial r}\mathbf{e}_\theta\otimes\mathbf{e}_r + \frac{1}{r}\frac{\partial w_\theta}{\partial\phi}\mathbf{e}_\theta\otimes\mathbf{e}_\phi \\
&+ \left[\frac{1}{r\sin\phi}\frac{\partial w_\theta}{\partial\theta} + \frac{1}{r}w_r + \frac{1}{r}\cot\phi\, w_\phi\right]\mathbf{e}_\theta\otimes\mathbf{e}_\theta.
\end{aligned} \quad (B.55)$$

- *Convective term:*

$$\begin{aligned}
(\operatorname{grad}\mathbf{w})\mathbf{w} &= \left\{w_r\frac{\partial w_r}{\partial r} + \frac{1}{r}w_\phi\frac{\partial w_r}{\partial\phi} + \frac{w_\theta}{r\sin\phi}\frac{\partial w_r}{\partial\theta} - \frac{w_\phi^2 + w_\theta^2}{r}\right\}\mathbf{e}_r \\
&+ \left\{w_r\frac{\partial w_\phi}{\partial r} + \frac{1}{r}w_\phi\frac{\partial w_\phi}{\partial\phi} + \frac{w_\theta}{r\sin\phi}\frac{\partial w_\phi}{\partial\theta} + \frac{w_r w_\phi}{r} - \frac{w_\theta^2\cot\phi}{r}\right\}\mathbf{e}_\phi \\
&+ \left\{w_r\frac{\partial w_\theta}{\partial r} + \frac{1}{r}w_\phi\frac{\partial w_\theta}{\partial\phi} + \frac{w_\theta}{r\sin\phi}\frac{\partial w_\theta}{\partial\theta} + \frac{w_r w_\theta}{r} + \frac{w_\phi w_\theta\cot\phi}{r}\right\}\mathbf{e}_\theta.
\end{aligned}$$

Appendix B Vector calculus

- *Curl of a vector field:*

$$\mathbf{curl\,w} = \frac{1}{r\sin\phi}\left(\frac{\partial}{\partial\phi}(w_\theta \sin\phi) - \frac{\partial w_\phi}{\partial\theta}\right)\mathbf{e}_r$$
$$+ \left(\frac{1}{r\sin\phi}\frac{\partial w_r}{\partial\theta} - \frac{1}{r}\frac{\partial}{\partial r}(rw_\theta)\right)\mathbf{e}_\phi$$
$$+ \frac{1}{r}\left(\frac{\partial}{\partial r}(rw_\phi) - \frac{\partial w_r}{\partial\phi}\right)\mathbf{e}_\theta.$$

- *Divergence of a vector field:*

$$\mathrm{div}\,\mathbf{w} = \frac{1}{r^2}\frac{\partial}{\partial r}(r^2 w_r) + \frac{1}{r\sin\phi}\frac{\partial}{\partial\phi}(\sin\phi\, w_\phi) + \frac{1}{r\sin\phi}\frac{\partial w_\theta}{\partial\theta}.$$

- *Divergence of a tensor field*

$$\mathrm{div}\,\mathbf{S} = \left[\frac{\partial S_{rr}}{\partial r} + \frac{1}{r}\frac{\partial S_{r\phi}}{\partial\phi} + \frac{1}{r\sin\phi}\frac{\partial S_{\theta r}}{\partial\theta} + \frac{1}{r}\left(2S_{rr} - S_{\phi\phi} - S_{\theta\theta} + S_{r\phi}\cot\phi\right)\right]\mathbf{e}_r$$
$$+ \left[\frac{\partial S_{r\phi}}{\partial r} + \frac{1}{r}\frac{\partial S_{\phi\phi}}{\partial\phi} + \frac{1}{r\sin\phi}\frac{\partial S_{\phi\theta}}{\partial\theta} + \frac{1}{r}\left((S_{\phi\phi} - S_{\theta\theta})\cot\phi + 3S_{r\phi}\right)\right]\mathbf{e}_\phi$$
$$+ \left[\frac{\partial S_{\theta r}}{\partial r} + \frac{1}{r}\frac{\partial S_{\phi\theta}}{\partial\phi} + \frac{1}{r\sin\phi}\frac{\partial S_{\theta\theta}}{\partial\theta}\right]\mathbf{e}_\theta. \qquad (B.56)$$

- *Laplacian of a scalar field:*

$$\Delta\phi = \frac{1}{r^2 \sin\phi}\left[\frac{\partial}{\partial r}\left(r^2 \sin\phi\,\frac{\partial\phi}{\partial r}\right) + \frac{\partial}{\partial\phi}\left(\sin\phi\,\frac{\partial\phi}{\partial\phi}\right) + \frac{\partial}{\partial\phi}\left(\frac{1}{\sin\phi}\frac{\partial\phi}{\partial\theta}\right)\right]$$
$$= \frac{1}{r^2}\frac{\partial}{\partial r}\left(r^2 \frac{\partial\phi}{\partial r}\right) + \frac{1}{r^2 \sin\phi}\frac{\partial}{\partial\phi}\left(\sin\phi\,\frac{\partial\phi}{\partial\phi}\right) + \frac{1}{r^2 \sin^2\phi}\frac{\partial^2\phi}{\partial\theta^2}. \qquad (B.57)$$

- *Laplacian of a vector field:*

$$\Delta\mathbf{w} = \left[\Delta w_r - \frac{2}{r^2}w_r - \frac{2}{r^2}\frac{\partial w_\phi}{\partial\phi} - \frac{2}{r^2}w_\phi \cot\phi - \frac{2}{r^2 \sin\phi}\frac{\partial w_\theta}{\partial\theta}\right]\mathbf{e}_r$$
$$+ \left[\Delta w_\phi + \frac{2}{r^2}\frac{\partial w_r}{\partial\phi} - \frac{1}{r^2 \sin^2\phi}w_\phi - \frac{2\cos\phi}{r^2 \sin^2\phi}\frac{\partial w_\theta}{\partial\theta}\right]\mathbf{e}_\phi$$
$$+ \left[\Delta w_\theta - \frac{1}{r^2 \sin^2\phi}w_\theta + \frac{2}{r^2 \sin\phi}\frac{\partial w_r}{\partial\theta} + \frac{2\cos\phi}{r^2 \sin^2\phi}\frac{\partial w_\phi}{\partial\theta}\right]\mathbf{e}_\theta, \qquad (B.58)$$

where the Laplacian operators in the right-hand side are given by (B.57).

Measure elements for integration

- *Line:*
$$dr,\ rd\phi,\ r\sin\phi\, d\theta$$
$$d\mathbf{l} = \mathbf{e}_r dr + r\mathbf{e}_\phi d\phi + r\sin\phi\, \mathbf{e}_\theta d\theta. \tag{B.59}$$

- *Surface:*
$$r^2 \sin\phi\, d\phi d\theta,\ r\sin\phi\, drd\theta,\ rdrd\phi. \tag{B.60}$$

- *Volume:*
$$r^2 \sin\phi\, drd\phi d\theta. \tag{B.61}$$

Appendix C

Function spaces for electromagnetism

Point charges, surface currents, electric dipoles, all appear in electromagnetism but cannot be represented by classical mathematical functions, mostly because they are supported on zero measure sets. This is the reason why electromagnetism needs tools from the mathematical theory of distributions. In this section we just recall some essential elements of this theory. In particular, the weak formulation of many electromagnetic problems involves functional spaces of distributions which are also recalled below. Further details and the proofs of the results can be found, for instance, in [37].

C.1 Scalar, vector and tensor distributions. Differential operators

C.1.1 Introduction

We motivate the need of extending the concept of function through a simple example. Let us consider a unit charge uniformly distributed in a ball $B(o, \varepsilon)$ of the affine space. The corresponding charge density is define by the function:

$$\rho_\varepsilon(x) = \begin{cases} \frac{3}{4\pi\varepsilon^3} & \text{if } |x-o| \leq \varepsilon \\ 0 & \text{otherwise.} \end{cases}$$

Now, we reduce the radius of the ball while keeping the same unit total charge inside and we ask for the following question: what is the limit of functions ρ_ε? Formally, this limit would be the "function"

$$\rho_0(x) = \begin{cases} \infty & \text{if } x = o \\ 0 & \text{otherwise,} \end{cases}$$

but this is not a function. Moreover, we should require

$$\int \rho_0(x)\, dV = 1.$$

A. Bermúdez, D. Gómez, P. Salgado: *Mathematical Models and Numerical Simulation in Electromagnetism.* UNITEXT – La Matematica per il 3+2 74
DOI 10.1007/978-3-319-02949-8_C, © Springer International Publishing Switzerland 2014

"Function" ρ_0 satisfying the above requirements has been defined by physicists as the Dirac's delta function. Later on, mathematicians introduce a framework where this limit makes a sense: the Theory of Distributions. The idea is the following: associated to each ρ_ε, let us introduce the linear operator

$$T_{\rho_\varepsilon}(\varphi) := \int_{\mathscr{E}} \rho_\varepsilon(x)\varphi(x)\,dV.$$

Notice that this makes sense as far as, for instance, φ is a continuous function. The interesting feature is that the right-hand side of the above equality has a limit as $\varepsilon \to 0$. Indeed, we have

$$\lim_{\varepsilon \to 0} \int_{\mathscr{E}} \rho_\varepsilon(x)\varphi(x)\,dV = \lim_{\varepsilon \to 0} \frac{3}{4\pi\varepsilon^3} \int_{B(o,\varepsilon)} \varphi(x)\,dV. \tag{C.1}$$

From the mean-value theorem of integration there exist a point $x_\varepsilon \in B(o,\varepsilon)$ such that

$$\int_{B(o,\varepsilon)} \varphi(x)\,dV = \varphi(x_\varepsilon)\frac{4}{3}\pi\varepsilon^3$$

By replacing this equality in (C.1) we easily get

$$\lim_{\varepsilon \to 0} \int_{\mathscr{E}} \rho_\varepsilon(x)\varphi(x)\,dV = \varphi(o),$$

simply by using the continuity of φ.

According to the previous computations, we are led to define the limit of operators T_{ρ_ε} as the one defined by

$$T(\varphi) := \varphi(o).$$

This operator T is the mathematical definition of the *Dirac's delta*. In what follows we will make more precise and general the above ideas.

C.1.2 The space of distributions

Let Ω denote an open subset of the N-dimensional affine space $\mathscr{E}$ with boundary Γ. We call $\mathscr{D}(\Omega)$ the linear space of infinitely differentiable scalar fields, with compact support in Ω. This means that, in particular, any function in $\mathscr{D}(\Omega)$ and its partial derivatives are null on the boundary of Ω. We also set,

$$\mathscr{D}(\overline{\Omega}) = \{\phi|_{\overline{\Omega}} : \phi \in \mathscr{D}(\mathscr{E})\}. \tag{C.2}$$

The space of scalar distributions on Ω is defined as the topological dual space of $\mathscr{D}(\Omega)$ endowed with the *inductive limit topology of L. Schwartz*. It will be denoted by $\mathscr{D}'(\Omega)$. While defining this topology is beyond the scope of this notes, convergence of sequences is easy to express: a sequence $\{\varphi_n\}$ converges to φ in $\mathscr{D}(\Omega)$ if and only if there exist a compact set $K \subset \Omega$ and a positive integer M such that support$(\varphi_n) \subset K\ \forall n > M$ and $\{D^\alpha \varphi_n\}$ converges uniformly to $D^\alpha \varphi$ in K, where

C.1 Scalar, vector and tensor distributions. Differential operators

$\alpha = (\alpha_1, \ldots, \alpha_N)$ is a multi-index and

$$D^\alpha \varphi = \frac{\partial^{|\alpha|}}{\partial x_1^{\alpha_1} \cdots \partial x_N^{\alpha_N}} \varphi,$$

being $|\alpha| = \alpha_1 + \cdots + \alpha_N$.

Notation $< \cdot, \cdot >$ will be used for the duality pairing between $\mathscr{D}'(\Omega)$ and $\mathscr{D}(\Omega)$. One can show that a linear mapping $S : \mathscr{D}(\Omega) \to \mathbb{R}$ is a distribution (i.e., is continuous) if and only if

$$\lim_{n \to \infty} \langle S, \varphi_n \rangle = \langle S, \varphi \rangle$$

as far as $\lim_{n \to \infty} \varphi_n = \varphi$ in $\mathscr{D}(\Omega)$, in the above sense.

Similar definitions are made in the case of vector or tensor fields. The respective spaces are denoted by $\vec{\mathscr{D}}(\Omega)$, $\vec{\mathscr{D}}(\overline{\Omega})$ and $\vec{\mathscr{D}}'(\Omega)$ for vector fields and by $\mathfrak{D}(\Omega)$, $\mathfrak{D}(\overline{\Omega})$, and $\mathfrak{D}'(\Omega)$ for tensor fields. For instance, $\mathfrak{D}(\Omega)$ denotes the linear space of infinitely differentiable tensor fields with compact support in Ω, and $\mathfrak{D}'(\Omega)$, the space of tensor distributions in Ω, is the linear space of continuous linear forms in $\mathfrak{D}(\Omega)$.

We will consider the weak* topology in the distribution spaces. Then a sequence of distributions $\{T_n\}$ converges to a distribution T if and only if

$$\lim_{n \to \infty} \langle T_n, \varphi \rangle = \langle T, \varphi \rangle \quad \forall \varphi \in \mathscr{D}(\Omega).$$

Let $L^1_{loc}(\Omega)$ (respectively, $\mathbf{L}^1_{loc}(\Omega)$) be the space of (class of)[1] locally integrable scalar fields (respectively, vector fields) in Ω, i.e., such that,

$$\int_K |f(x)| \, dV < \infty,$$

for all compact set $K \subset \Omega$. Similarly, $\mathfrak{L}^1_{loc}(\Omega)$ denotes the space of (class of) locally integrable tensor fields in Ω, i.e., such that,

$$\int_K |\mathbf{S}(x)| \, dV < \infty.$$

Any $f \in L^1_{loc}(\Omega)$ can be identified to a *scalar distribution* T_f defined by

$$< T_f, \varphi > = \int_\Omega f(x) \varphi(x) \, dV \quad \forall \varphi \in \mathscr{D}(\Omega).$$

Similarly, any $\mathbf{f} \in \mathbf{L}^1_{loc}(\Omega)$ can be identified to the *vector distribution* $T_\mathbf{f}$ defined by

$$< T_\mathbf{f}, \varphi > = \int_\Omega \mathbf{f}(x) \cdot \varphi(x) \, dV \quad \forall \varphi \in \mathscr{D}(\Omega),$$

---

[1] Two measurable functions are equivalent if and only if they differ in a zero-measure set.

and if $\mathbf{S} \in \mathfrak{L}^1_{\mathrm{loc}}(\Omega)$ it can be identified to the *tensor distribution* $\mathbf{T_S}$ defined by

$$< \mathbf{T_S}, \boldsymbol{\Phi} >= \int_\Omega \mathbf{S}(x) \cdot \boldsymbol{\Phi}(x)\, \mathrm{d}V \quad \forall \boldsymbol{\Phi} \in \mathfrak{D}(\Omega).$$

One can introduce derivatives of any order for distributions :

$$\langle D^\alpha \mathrm{T}, \varphi \rangle := (-1)^{|\alpha|} \langle \mathrm{T}, D^\alpha \varphi \rangle.$$

In particular

$$\langle \frac{\partial \mathrm{T}}{\partial x_i}, \varphi \rangle := -\langle \mathrm{T}, \frac{\partial \varphi}{\partial x_i} \rangle.$$

This definition is consistent with the partial derivative of a differentiable function: let $f \in C^1$, then

$$\frac{\partial \mathrm{T}_f}{\partial x_i} = \mathrm{T}_{\frac{\partial f}{\partial x_i}}.$$

Indeed, by using a Green's formula and the fact that $\varphi|_\Gamma = 0\ \forall \varphi \in \mathscr{D}(\Omega)$, we have

$$\int_\Omega \frac{\partial f}{\partial x_i} \varphi\, \mathrm{d}V = -\int_\Omega f \frac{\partial \varphi}{\partial x_i}\, \mathrm{d}V.$$

Next, we recall the definition of the usual differential operators: if $\mathrm{T} \in \mathscr{D}'(\Omega)$, the gradient of T, grad T, is the *vector distribution* defined by

$$< \operatorname{grad} \mathrm{T}, \varphi >:= - < \mathrm{T}, \operatorname{div} \varphi > \quad \forall\, \varphi \in \vec{\mathscr{D}}(\Omega). \tag{C.3}$$

For vector distributions we have an analogous definition. If $\mathbf{T} \in \vec{\mathscr{D}}'(\Omega)$, its gradient $\mathbf{grad\,T} \in \mathfrak{D}'(\Omega)$ is the *tensor distribution* defined by

$$< \mathbf{grad\,T}, \boldsymbol{\Phi} >:= - < \mathbf{T}, \operatorname{div} \boldsymbol{\Phi} > \quad \forall\, \boldsymbol{\Phi} \in \mathfrak{D}(\Omega). \tag{C.4}$$

In a similar way we can extend the other classical differential operators to the spaces of distributions.

Let $\mathrm{T} \in \mathscr{D}'(\Omega)$. The *Laplacian* of T is the *scalar distribution* $\Delta \mathrm{T}$ satisfying

$$< \Delta \mathrm{T}, \varphi >:=< \mathrm{T}, \Delta \varphi > \quad \forall\, \varphi \in \mathscr{D}(\Omega). \tag{C.5}$$

Let $\mathbf{T} \in \vec{\mathscr{D}}'(\Omega)$. The *divergence* of $\mathbf{T}$ is the *vector distribution* defined by

$$< \operatorname{div} \mathbf{T}, \varphi >:= - < \mathbf{T}, \operatorname{grad} \varphi > \quad \forall\, \varphi \in \mathscr{D}(\Omega), \tag{C.6}$$

while the **curl** of $\mathbf{T}$ is the *vector distribution* given by

$$< \mathbf{curl\,T}, \varphi >:=< \mathbf{T}, \mathbf{curl}\, \varphi > \quad \forall\, \varphi \in \vec{\mathscr{D}}(\Omega), \tag{C.7}$$

and the *Laplacian* of $\mathbf{T}$ is the *vector distribution* $\Delta \mathbf{T}$ satisfying

$$< \Delta \mathbf{T}, \varphi >:=< \mathbf{T}, \Delta \varphi > \quad \forall\, \varphi \in \vec{\mathscr{D}}(\Omega). \tag{C.8}$$

C.2 Lebesgue and Sobolev spaces. Trace theorems

The duality pairing $<\cdot,\cdot>$ can be considered as an extension of the scalar product in the Lebesgue space $L^2(\Omega)$ (respectively $\mathbf{L}^2(\Omega)$ or $\mathcal{L}^2(\Omega)$), which is the linear space of (equivalence classes of) measurable scalar fields (respectively, vector or tensor fields) such that,

$$\int_\Omega |f(x)|^2 \, dV < \infty,$$

(respectively,

$$\int_\Omega |\mathbf{f}(x)|^2 \, dV < \infty \text{ or } \int_\Omega |\boldsymbol{\Phi}(x)|^2 \, dV < \infty, \mathbf{f} \in \mathbf{L}^2(\Omega), \boldsymbol{\Phi} \in \mathcal{L}^2(\Omega)).$$

Endowed with the scalar product,

$$(f, g)_{L^2(\Omega)} = \int_\Omega f(x) g(x) \, dV, \tag{C.9}$$

(respectively, $(\mathbf{f}, \mathbf{g})_{\mathbf{L}^2(\Omega)} = \int_\Omega \mathbf{f}(x) \cdot \mathbf{g}(x) \, dV$ or $(\mathbf{S}, \boldsymbol{\Phi})_{\mathcal{L}^2(\Omega)} = \int_\Omega \mathbf{S}(x) \cdot \boldsymbol{\Phi}(x) \, dV$), they are Hilbert spaces. Their corresponding norms are

$$\|f\|_{L^2(\Omega)} = \left(\int_\Omega (f(x))^2 \, dV \right)^{1/2}, \tag{C.10}$$

$$\|\mathbf{f}\|_{\mathbf{L}^2(\Omega)} = \left(\int_\Omega |\mathbf{f}(x)|^2 \, dV \right)^{1/2}, \tag{C.11}$$

and

$$\|\boldsymbol{\Phi}\|_{\mathcal{L}^2(\Omega)} = \left(\int_\Omega |\boldsymbol{\Phi}(x)|^2 \, dV \right)^{1/2}. \tag{C.12}$$

Since they are included in their respective locally integrable spaces, both are (identifiable to) spaces of distributions which, for the sake simplicity in notations, will be also denoted by $L^2(\Omega)$, $\mathbf{L}^2(\Omega)$ and $\mathcal{L}^2(\Omega)$.

Given a scalar distribution $u \in L^2(\Omega)$, suppose the vector distribution $\operatorname{grad} u$ belongs to (i.e., is identifiable to an element of) $\mathbf{L}^2(\Omega)$. Then we say that u belongs to the *Sobolev space* $H^1(\Omega)$. Precisely, the first order Sobolev space is defined by

$$H^1(\Omega) = \{ u \in L^2(\Omega) : \operatorname{grad} u \in \mathbf{L}^2(\Omega) \}, \tag{C.13}$$

which is a Hilbert space with the following scalar product:

$$(u, v)_{H^1(\Omega)} = \int_\Omega uv \, dV + \int_\Omega \operatorname{grad} u \cdot \operatorname{grad} v \, dV. \tag{C.14}$$

Since $\mathscr{D}(\Omega) \subset H^1(\Omega)$, we can define the closed subspace

$$H_0^1(\Omega) = \overline{\mathscr{D}(\Omega)}^{H_1(\Omega)},$$

When Ω is a proper subset of $\mathscr{E}$ then $H_0^1(\Omega)$ is generally a proper subspace of $H^1(\Omega)$.

On the other hand, $\mathscr{D}(\Omega)$ is dense in $L^2(\Omega)$, i.e.

$$\overline{\mathscr{D}(\Omega)}^{L_2(\Omega)} = L^2(\Omega).$$

We denote by $H^{-1}(\Omega)$ the dual space of $H_0^1(\Omega)$, that is the linear space of continuous linear forms in $H_0^1(\Omega)$. It is also a Hilbert space with the norm

$$\|f\|_{H_{-1}(\Omega)} = \sup_{\substack{v \in H_0^1(\Omega) \\ v \neq 0}} \frac{<f,v>}{\|v\|_{H_0^1(\Omega)}}.$$

Since $\mathscr{D}(\Omega)$ is dense in $H_0^1(\Omega)$, $H^{-1}(\Omega)$ is a subspace of $\mathscr{D}'(\Omega)$ characterized as follows: a distribution f belongs to $H^{-1}(\Omega)$ if and only if there exist a vector distribution $\mathbf{g} \in \mathbf{L}^2(\Omega)$ such that

$$f = \text{div}\,\mathbf{g}.$$

In what follows, we assume that Ω has a Lipschitz-continuous boundary. Then $\mathscr{D}(\overline{\Omega})$ is dense in $H^1(\Omega)$.

Let us consider the linear mapping

$$u \in \mathscr{D}(\overline{\Omega}) \longrightarrow u|_\Gamma \in L^2(\Gamma).$$

Since it is continuous when $\mathscr{D}(\overline{\Omega})$ is endowed with the norm of $H^1(\Omega)$, it can be extended to the whole $H^1(\Omega)$ by a density argument. Let us call this extension the *trace operator* which will be denoted by γ_0. We have,

$$\ker(\gamma_0) := \{u \in H^1(\Omega) : \gamma_0 u = 0\} = H_0^1(\Omega).$$

This mapping is not surjective. We denote by $H^{1/2}(\Gamma)$ its image set, i.e.,

$$H^{1/2}(\Gamma) = \gamma_0(H^1(\Omega)). \tag{C.15}$$

One can show that, with the norm

$$\|f\|_{H^{1/2}(\Gamma)} = \inf\{\|v\|_{H^1(\Omega)} : \gamma_0 v = f\}, \tag{C.16}$$

$H^{1/2}(\Gamma)$ is a Hilbert space. We denote by $H^{-1/2}(\Gamma)$ its dual space.

Similar spaces and definitions can be introduced for vector fields. In this case spaces and their elements are denoted with bold face symbols, namely, $\mathbf{L}^2(\Omega)$, $\mathbf{H}^1(\Omega)$, $\mathbf{H}^{1/2}(\Gamma)$, etc.

C.3 The spaces $\mathbf{H}(\mathrm{div},\Omega)$ and $\mathbf{H}(\mathbf{curl},\Omega)$

Let us introduce the space of vector fields in Ω

$$\mathbf{H}(\mathrm{div},\Omega) = \{\mathbf{v} \in \mathbf{L}^2(\Omega) : \mathrm{div}\,\mathbf{v} \in L^2(\Omega)\}. \tag{C.17}$$

It is a Hilbert space for the scalar product

$$\langle \mathbf{u},\mathbf{v}\rangle_{\mathbf{H}(\mathrm{div},\Omega)} = \int_\Omega \mathbf{u}\cdot\mathbf{v}\,dV + \int_\Omega \mathrm{div}\,\mathbf{u}\,\mathrm{div}\,\mathbf{v}\,dV. \tag{C.18}$$

We also introduce its closed subspace

$$\mathbf{H}_0(\mathrm{div},\Omega) = \overline{\mathscr{D}(\Omega)}^{\mathbf{H}(\mathrm{div},\Omega)}.$$

Moreover, one can show that $\mathscr{D}(\overline{\Omega})$ is dense in $\mathbf{H}(\mathrm{div},\Omega)$.

This fact allows us to define the normal component of boundary values of functions of $\mathbf{H}(\mathrm{div},\Omega)$. Indeed, one can show that the mapping

$$\gamma_n : \mathbf{v} \longrightarrow \mathbf{v}\cdot\mathbf{n}|_\Gamma,$$

defined on $\mathscr{D}(\overline{\Omega})$, can be extended by continuity to a linear continuous mapping, still denoted by γ_n, from $\mathbf{H}(\mathrm{div},\Omega)$ to $H^{-1/2}(\Gamma)$. Furthermore $\gamma_n(\mathbf{H}(\mathrm{div},\Omega)) = H^{-1/2}(\Gamma)$ and we also have

$$\ker(\gamma_n) := \{\mathbf{v} \in \mathbf{H}(\mathrm{div},\Omega)/\gamma_n\mathbf{v} = 0\} = \mathbf{H}_0(\mathrm{div},\Omega).$$

Another important function space in electromagnetism is

$$\mathbf{H}(\mathbf{curl},\Omega) = \{\mathbf{v} \in \mathbf{L}^2(\Omega) : \mathbf{curl}\,\mathbf{v} \in \mathbf{L}^2(\Omega)\}. \tag{C.19}$$

Endowed with the scalar product,

$$(\mathbf{u},\mathbf{v})_{\mathbf{H}(\mathbf{curl},\Omega)} = \int_\Omega \mathbf{u}\cdot\mathbf{v}\,dV + \int_\Omega \mathbf{curl}\,\mathbf{u}\cdot\mathbf{curl}\,\mathbf{v}\,dV, \tag{C.20}$$

it is a Hilbert space.

Let us define its closed subspace

$$\mathbf{H}_0(\mathbf{curl},\Omega) = \overline{\mathscr{D}(\Omega)}^{\mathbf{H}(\mathbf{curl},\Omega)}.$$

One can prove that $\mathscr{D}(\overline{\Omega})$ is dense in $\mathbf{H}(\mathbf{curl},\Omega)$.

This fact allows us to define the tangential component of boundary values of fields in $\mathbf{H}(\mathbf{curl},\Omega)$. More precisely, one can show that the mapping

$$\gamma_\tau : \mathbf{v} \longrightarrow \mathbf{v}\times\mathbf{n}|_\Gamma$$

defined, in principle, in $\mathscr{D}(\overline{\Omega})$, can be extended by continuity and density to a linear continuous mapping, still denoted by γ_τ, from $\mathbf{H}(\mathbf{curl},\Omega)$ into $\mathbf{H}^{-1/2}(\Gamma)$.

Moreover,

$$\ker(\gamma_\tau) := \{\mathbf{v} \in \mathbf{H}(\mathbf{curl},\Omega) / \gamma_\tau \mathbf{v} = 0\} = \mathbf{H}_0(\mathbf{curl},\Omega).$$

Remark C.1. The tangential component on Γ of a vector field $\mathbf{v}$ is formally defined by,

$$\mathbf{v}_T = \mathbf{v} - (\mathbf{v}\cdot\mathbf{n})\mathbf{n}.$$

Since we have,

$$\mathbf{a}\times(\mathbf{b}\times\mathbf{c}) = (\mathbf{c}\cdot\mathbf{a})\mathbf{b} - (\mathbf{b}\cdot\mathbf{a})\mathbf{c},$$

then

$$\mathbf{n}\times(\mathbf{v}\times\mathbf{n}) = (\mathbf{n}\cdot\mathbf{n})\mathbf{v} - (\mathbf{v}\cdot\mathbf{n})\mathbf{n} = \mathbf{v} - (\mathbf{v}\cdot\mathbf{n})\mathbf{n},$$

and therefore

$$\mathbf{v}_T = \mathbf{n}\times(\mathbf{v}\times\mathbf{n}).$$

In particular, $\mathbf{v}_T = 0$ if and only if $\mathbf{v}\times\mathbf{n} = 0$. We notice that, like $\mathbf{v}_T$, vector $\mathbf{v}\times\mathbf{n}$ is on the tangent plane to Γ.

Finally we have the following equality both algebraically and topologically:

$$\mathbf{H}_0^1(\Omega) = \mathbf{H}_0(\mathrm{div},\Omega) \cap \mathbf{H}_0(\mathbf{curl},\Omega),$$

but, in general, we can only affirm that

$$\mathbf{H}^1(\Omega) \subset \mathbf{H}(\mathrm{div},\Omega) \cap \mathbf{H}(\mathbf{curl},\Omega).$$

C.4 The spaces $\mathrm{H}_{00}^{1/2}(S)$ and $\mathrm{H}_{00}^{-1/2}(S)$

Let S be a Lipschitz-continuous subset of Γ. We introduce the space

$$\mathrm{H}_{00}^{1/2}(S) = \{u \in \mathrm{H}^{1/2}(S) \,/\, \tilde{u} \in \mathrm{H}^{1/2}(\Gamma)\}, \tag{C.21}$$

where $\tilde{u}$ denotes the extension of u to Γ by 0 outside S. We call $\mathrm{H}_{00}^{-1/2}(S)$ its dual space. Similar spaces can be introduced for vector fields.

We have:

- If $\mathbf{v} \in \mathbf{H}(\mathrm{div},\Omega)$ then its normal trace on S belongs to $\mathrm{H}_{00}^{-1/2}(S)$.
- If $\mathbf{v} \in \mathbf{H}(\mathbf{curl},\Omega)$, then its tangential trace on S belongs to $\mathrm{H}_{00}^{-1/2}(S)$.

C.5 Green's formulas

Let us suppose, for the moment, that $\mathbf{v}$ and ϕ are smooth fields,

$$\mathbf{v} : \Omega \longrightarrow \mathscr{V},$$
$$\phi : \Omega \longrightarrow \mathbb{R}.$$

Then we have

$$\int_\Omega \operatorname{div}(\mathbf{v}\phi)\, dV = \int_\Omega \phi \operatorname{div} \mathbf{v}\, dV + \int_\Omega \mathbf{v} \cdot \operatorname{grad} \phi\, dV,$$

and using the Gauss' theorem in the integral of the left-hand side we deduce the Green's formula

$$\int_\Gamma \phi \mathbf{v} \cdot \mathbf{n} dA = \int_\Omega \phi \operatorname{div} \mathbf{v}\, dV + \int_\Omega \mathbf{v} \cdot \operatorname{grad} \phi\, dV. \tag{C.22}$$

Now, let us suppose $\mathbf{v} \in \mathbf{H}(\operatorname{div}, \Omega)$ and $\phi \in H^1(\Omega)$. Then $\operatorname{grad} \phi \in \mathbf{L}^2(\Omega)$, $\operatorname{div} \mathbf{v} \in L^2(\Omega)$, $\gamma_0 \phi \in H^{1/2}(\Gamma)$ and $\gamma_n \mathbf{v} \in H^{-1/2}(\Gamma)$. By using continuity and density arguments one can show that the above formula is still valid. More precisely,

$$\langle \gamma_n \mathbf{v}, \phi \rangle_{H^{-1/2}(\Gamma), H^{1/2}(\Gamma)} = \int_\Omega \phi \operatorname{div} \mathbf{v}\, dV + \int_\Omega \mathbf{v} \cdot \operatorname{grad} \phi\, dV \tag{C.23}$$
$$\forall \mathbf{v} \in \mathbf{H}(\operatorname{div}, \Omega)\ \forall \phi \in H^1(\Omega).$$

Let u be a scalar field in $H^1(\Omega)$ such that $\Delta u \in L^2(\Omega)$. Then the vector field $\operatorname{grad} u \in \mathbf{L}^2(\Omega)$ (because $u \in H^1(\Omega)$) and, furthermore,

$$\operatorname{div} \operatorname{grad} u = \Delta u \in L^2(\Omega).$$

Thus $\operatorname{grad} u \in \mathbf{H}(\operatorname{div}, \Omega)$ and we can take $\mathbf{v} = \operatorname{grad} u$ in Green's formula (C.23) to get,

$$\langle \gamma_n(\operatorname{grad} u), \phi \rangle_{H^{-1/2}(\Gamma), H^{1/2}(\Gamma)} = \int_\Omega \Delta u \phi\, dV + \int_\Omega \operatorname{grad} u \cdot \operatorname{grad} \phi\, dV. \tag{C.24}$$

We notice that, if $\gamma_n(\operatorname{grad} u)$ is more regular, let say $\gamma_n(\operatorname{grad} u) \in L^2(\Gamma)$, then

$$\langle \gamma_n(\operatorname{grad} u), \phi \rangle_{H^{-1/2}(\Gamma), H^{1/2}(\Gamma)} = \int_\Gamma \frac{\partial u}{\partial n} \phi dA.$$

Now let us consider two smooth vector fields,

$$\mathbf{u}, \mathbf{v} : \Omega \longrightarrow \mathscr{V}.$$

We have,

$$\operatorname{div}(\mathbf{u} \times \mathbf{v}) = \mathbf{v} \cdot \operatorname{\mathbf{curl}} \mathbf{u} - \mathbf{u} \cdot \operatorname{\mathbf{curl}} \mathbf{v}.$$

By integrating this equality in Ω we get,

$$\int_\Omega \mathrm{div}(\mathbf{u} \times \mathbf{v})\, dV = \int_\Omega \mathbf{v} \cdot \mathbf{curl}\,\mathbf{u}\, dV - \int_\Omega \mathbf{u} \cdot \mathbf{curl}\,\mathbf{v}\, dV.$$

By using the Gauss' theorem to transform the term on the left-hand side we get the Green's formula

$$\int_\Gamma \mathbf{u} \times \mathbf{v} \cdot \mathbf{n}\, dA = \int_\Omega \mathbf{v} \cdot \mathbf{curl}\,\mathbf{u}\, dV - \int_\Omega \mathbf{u} \cdot \mathbf{curl}\,\mathbf{v}\, dV,$$

which can also be written as

$$-\int_\Gamma \mathbf{u} \times \mathbf{n} \cdot \mathbf{v}\, dA = \int_\Omega \mathbf{v} \cdot \mathbf{curl}\,\mathbf{u}\, dV - \int_\Omega \mathbf{u} \cdot \mathbf{curl}\,\mathbf{v}\, dV.$$

Let us assume $\mathbf{u} \in \mathbf{H}(\mathbf{curl}, \Omega)$ and $\mathbf{v} \in \mathbf{H}^1(\Omega)$. By continuity and density arguments we can show that the previous formula is still valid. More precisely,

$$\langle \gamma_\tau \mathbf{u}, \mathbf{v} \rangle_{\mathbf{H}^{-1/2}(\Gamma), \mathbf{H}^{1/2}(\Gamma)} = \int_\Omega \mathbf{u} \cdot \mathbf{curl}\,\mathbf{v}\, dV - \int_\Omega \mathbf{v} \cdot \mathbf{curl}\,\mathbf{u}\, dV \quad (C.25)$$
$$\forall \mathbf{u} \in \mathbf{H}(\mathbf{curl}, \Omega)\ \forall \mathbf{v} \in (\mathbf{H}^1(\Omega))^3.$$

Remark C.2. A sense for the normal trace $\mathbf{u} \times \mathbf{v} \cdot \mathbf{n}|_\Gamma$ when $\mathbf{u}, \mathbf{v} \in \mathbf{H}(\mathbf{curl}, \Omega)$. Since we have,

$$\mathrm{div}(\mathbf{u} \times \mathbf{v}) = \mathbf{v} \cdot \mathbf{curl}\,\mathbf{u} - \mathbf{u} \cdot \mathbf{curl}\,\mathbf{v},$$

then

$$\mathrm{div}(\mathbf{u} \times \mathbf{v}) \in \mathrm{L}^1(\Omega).$$

Moreover $\mathbf{u} \times \mathbf{v} \in \mathbf{L}^1(\Omega)$ too and hence,

$$\mathbf{u} \times \mathbf{v} \in \mathbf{W}^1(\mathrm{div}, \Omega) = \{ \mathbf{w} \in \mathbf{L}^1(\Omega) : \mathrm{div}\,\mathbf{w} \in \mathrm{L}^1(\Omega) \}.$$

One can show that, if Ω is bounded and its boundary Γ is Lipschitz-continuous, the trace mapping,

$$\mathbf{w} \longrightarrow \mathbf{w} \cdot \mathbf{n}|_\Gamma$$

is well defined and continuous from $\mathbf{W}^1(\mathrm{div}, \Omega)$ into the dual space of Lipschitz-continuous functions on Γ (see Cessenat [4] page 40).

The proof of most of the results enounced above can be seen in [37].

C.6 Weighted Sobolev spaces

In this section we define appropriate weighted Sobolev spaces that are used in the weak formulations of axisymmetric problems and establish some of their properties. Further details can be found in [39, 53].

Let us consider a bounded three-dimensional domain Ω which has cylindrical symmetry in a cylindrical coordinate system (ρ, θ, z). Let us denote by $\mathbf{e}_\rho$, $\mathbf{e}_\theta$ and $\mathbf{e}_z$ the orthonormal vectors of the local basis associated to this coordinate system. Let $\widehat{\Omega}$ be a Lipschitz bounded connected two-dimensional open set which is a meridional section (θ = constant) of the three-dimensional domain Ω. Let Γ_A be the intersection between $\partial\widehat{\Omega}$ and the axis of revolution ($\rho = 0$). Given a scalar function $u(\rho, z)$ defined in this domain we will denote the partial derivatives of u by ∂_ρ and ∂_z.

Let $L^2_\rho(\widehat{\Omega})$ be the weighted Lebesgue space of all measurable functions u defined in $\widehat{\Omega}$ for which

$$\|u\|_{L^2_\rho(\widehat{\Omega})} := \int_{\widehat{\Omega}} |u|^2 \rho \, d\rho dz < \infty.$$

The weighted Sobolev space $H^k_\rho(\widehat{\Omega})$ consists of all functions in $L^2_\rho(\widehat{\Omega})$ whose derivatives up to the order k are also in $L^2_\rho(\widehat{\Omega})$. We define the norms and semi-norms in the standard way; in particular,

$$|u|^2_{H^1_\rho(\widehat{\Omega})} := \int_{\widehat{\Omega}} \left(|\partial_\rho u|^2 + |\partial_z u|^2 \right) \rho \, d\rho dz.$$

Let $\widetilde{H}^1_\rho(\widehat{\Omega}) := H^1_\rho(\widehat{\Omega}) \cap L^2_{1/\rho}(\widehat{\Omega})$, where $L^2_{1/\rho}(\widehat{\Omega})$ denotes the set of all measurable functions u defined in $\widehat{\Omega}$ for which

$$\|u\|^2_{L^2_{1/\rho}(\widehat{\Omega})} := \int_{\widehat{\Omega}} \frac{|u|^2}{\rho} \, d\rho dz < \infty.$$

$\widetilde{H}^1_\rho(\widehat{\Omega})$ is a Hilbert space with the norm

$$\|u\|_{\widetilde{H}^1_\rho(\widehat{\Omega})} := \left(\|u\|^2_{H^1_\rho(\widehat{\Omega})} + \|u\|^2_{L^2_{1/\rho}(\widehat{\Omega})} \right)^{1/2}.$$

The set of $\mathscr{C}^\infty(\overline{\widehat{\Omega}})$ functions vanishing in a neighborhood of Γ_A is dense in $\widetilde{H}^1_\rho(\widehat{\Omega})$. On the other hand, functions in $\widetilde{H}^1_\rho(\widehat{\Omega})$ have traces on Γ_A ($\rho = 0$). Moreover, since the set of functions in $\mathscr{C}^\infty(\overline{\widehat{\Omega}})$ vanishing in a neighborhood of Γ_A is dense in $\widetilde{H}^1_\rho(\widehat{\Omega})$, the functions in $\widetilde{H}^1_\rho(\widehat{\Omega})$ have vanishing traces on Γ_A.

C.7 Complements

This section includes some results related to distributions and functional spaces. In particular we consider distributions supported on a surface and study the properties of some radial distributions appearing in 3D potential theory.

C.7.1 Distributions supported on a surface. Examples

Sometimes, electric charges or currents are concentrated at points, curves or surfaces in the space. In these cases, they cannot be represented by locally integrable fields but by more general distributions.

As an example, let us consider a charge distributed on a surface S. Let ρ_S denote its surface density and suppose $\rho_S \in L^1_{\text{loc}}(S)$. Then this charge can be represented by the distribution, T_{ρ_S}, defined by

$$T_{\rho_S} : \mathscr{D}(\Omega) \longrightarrow \mathbb{R} \qquad (C.26)$$
$$\varphi \rightsquigarrow \int_S \rho_S \varphi \, dA.$$

We notice that this distribution is not defined by a function in $L^1_{\text{loc}}(\Omega)$. In particular, if $\mathbf{u} \in \mathbf{L}^2(\Omega)$ is such that $\text{div}\,\mathbf{u} = T_{\rho_S}$ then $\mathbf{u} \notin \mathbf{H}(\text{div},\Omega)$.

According to the definition of the divergence of vector distributions (see C.6), $\text{div}\,\mathbf{u} = \rho_S$ means

$$-\int_\Omega \mathbf{u} \cdot \text{grad}\,\varphi \, dV = \int_S \rho_S \varphi \, dA \quad \forall \varphi \in \mathscr{D}(\Omega). \qquad (C.27)$$

If $\varphi \in \mathscr{D}(\Omega \setminus S)$ then this equality yields

$$-\int_\Omega \mathbf{u} \cdot \text{grad}\,\varphi \, dV = 0 \qquad (C.28)$$

Thus, the distribution $\text{div}\,\mathbf{u} \in \mathscr{D}'(\Omega \setminus S)$ can be identified to the (locally integrable) null function:

$$\langle \text{div}\,\mathbf{u}, \varphi \rangle = -\int_\Omega \mathbf{u} \cdot \text{grad}\,\varphi \, dV = 0 \quad \forall \varphi \in \mathscr{D}(\Omega \setminus S). \qquad (C.29)$$

Since $\mathbf{u} \in \mathbf{H}(\text{div}, A)$, for any open subset of $\Omega \setminus S$ then $\mathbf{u}$ has normal traces on both sides of surface S, belonging to the space $H^{-1/2}_{00}(S)$ (see Fig. C.1), namely,

$$\mathbf{u}^+ \cdot \mathbf{n}^+|_S$$
$$\mathbf{u}^- \cdot \mathbf{n}^-|_S \in H^{-1/2}_{00}(S),$$

where $\mathbf{n}^+$ and $\mathbf{n}^-$ denotes the two unit vectors normal to S.

We are going to prove that

$$[\mathbf{u} \cdot \mathbf{n}] := \mathbf{u}^+ \cdot \mathbf{n}^+ + \mathbf{u}^- \cdot \mathbf{n}^- = -\rho_S,$$

on S.

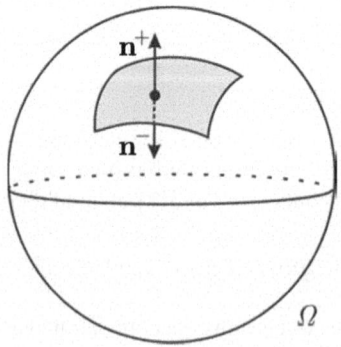

Fig. C.1. Normal traces on both sides of surface S

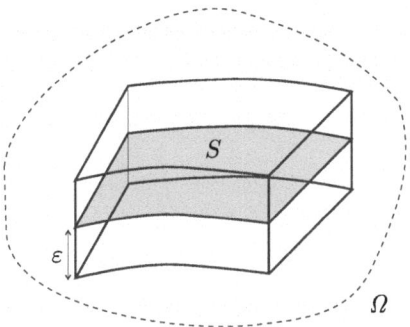

Fig. C.2. The set B

For this purpose, let us consider the set (see Fig. C.2)
$$B = \{x \in \mathscr{E} / x = x_0 + \theta \mathbf{n}^+(x_0) \ \forall \theta \in [-\varepsilon, \varepsilon] \ \forall x_0 \in S\},$$
with ε chosen in such a way that $B \subset \Omega$.

Let us take $\varphi \in \mathscr{D}(\overset{\circ}{B})$ in (C.27), $\overset{\circ}{B}$ being the interior of B. We get
$$-\int_{\overset{\circ}{B}} \mathbf{u} \cdot \mathrm{grad}\, \varphi \, dV = \int_S \rho_S \varphi \, dA \ \forall \varphi \in \mathscr{D}(\overset{\circ}{B}),$$
and by continuity and density
$$-\int_{\overset{\circ}{B}} \mathbf{u} \cdot \mathrm{grad}\, \varphi \, dV = \langle \rho_S, \varphi \rangle_{H_{00}^{-1/2}(S), H_{00}^{1/2}(S)} \ \forall \varphi \in H_0^1(\overset{\circ}{B}) \quad (\text{C.30})$$
(if $\rho_S \in H_{00}^{-1/2}(S)$).

Moreover, since $\mathbf{u} \in \mathbf{L}^2(\Omega)$ and $\mathrm{div}\, \mathbf{u} = 0$ in $\Omega \setminus S$, we have
$$-\int_{\overset{\circ}{R}} \mathbf{u} \cdot \mathrm{grad}\, \varphi \, dV = -\int_{B^+} \mathbf{u} \cdot \mathrm{grad}\, \varphi \, dV - \int_{D^-} \mathbf{u} \cdot \mathrm{grad}\, \varphi \, dV$$
$$= -\langle \mathbf{u}^+ \cdot \mathbf{n}^+, \varphi \rangle_{H_{00}^{-1/2}(S), H_{00}^{1/2}(S)} + \int_{B^+} \mathrm{div}\, \mathbf{u}\, \varphi \, dV$$
$$- \langle \mathbf{u}^- \cdot \mathbf{n}^-, \varphi \rangle_{H_{00}^{-1/2}(S), H_{00}^{1/2}(S)} + \int_{B^-} \mathrm{div}\, \mathbf{u}\, \varphi \, dV$$
$$= -\langle \mathbf{u}^+ \cdot \mathbf{n}^+ + \mathbf{u}^- \cdot \mathbf{n}^-, \varphi \rangle_{H_{00}^{-1/2}(S), H_{00}^{1/2}(S)} \ \forall \varphi \in H_0^1(\overset{\circ}{B}),$$
with
$$B^+ = \mathrm{int}\{x \in B / x = x_0 + \theta \mathbf{n}^+(x_0) \ \forall \theta \in (0, \varepsilon) \ \forall x_0 \in S\},$$
$$B^- = \mathrm{int}\{x \in B / x = x_0 + \theta \mathbf{n}^+(x_0) \ \forall \theta \in (-\varepsilon, 0) \ \forall x_0 \in S\},$$
where we have used that open sets B^+ and B^+ are subsets of $\Omega \setminus S$ and (C.29).

Finally, by using (C.30) we get
$$-\left(\mathbf{u}^+ \cdot \mathbf{n}^+ + \mathbf{u}^- \cdot \mathbf{n}^-\right) = \rho_S \text{ in } H_{00}^{-1/2}(S).$$

A similar situation arises for surface currents. Let $\mathbf{J}_S$ be a surface current density on surface S. Then it can be represented by the vector distribution $T_{\mathbf{J}_S}$ defined by

$$T_{\mathbf{J}_S} : \mathscr{D}(\Omega) \longrightarrow \mathbb{R}$$

$$\varphi \rightsquigarrow \int_S \varphi \cdot \mathbf{J}_S dA,$$

as far as $\mathbf{J}_S \in \mathbf{L}^1_{\text{loc}}(S)$.

Let $\mathbf{u} \in \mathbf{L}^2(\Omega)$ be a vector field such that,

$$\mathbf{curl}\,\mathbf{u} = T_{\mathbf{J}_S},$$

in the sense of distributions. This means,

$$\int_\Omega \mathbf{u} \cdot \mathbf{curl}\,\varphi \, dV = \int_S \mathbf{J}_S \cdot \varphi \, dA \quad \forall \varphi \in \mathscr{D}(\Omega). \tag{C.31}$$

Since $T_{\mathbf{J}_S} \notin \mathbf{L}^1_{\text{loc}}(\Omega)$, (i.e., there is no $\mathbf{f} \in \mathbf{L}^1_{\text{loc}}(\Omega)$ such that

$$\int_\Omega \mathbf{f} \cdot \varphi \, dV = \int_S \mathbf{J} \cdot \varphi \, dA$$

$\forall \varphi \in \mathscr{D}(\Omega))$, then $\mathbf{u} \notin \mathbf{H}(\mathbf{curl},\Omega)$.

Moreover, by taking $\varphi \in \mathscr{D}(\Omega\setminus S)$ in (C.31) we get

$$\int_{\Omega\setminus S} \mathbf{u} \cdot \mathbf{curl}\,\varphi \, dV = 0 \quad \forall \varphi \in \mathscr{D}(\Omega\setminus S), \tag{C.32}$$

which means that $\mathbf{curl}\,\mathbf{u} = 0$ in $\mathscr{D}'(\Omega\setminus S)$.

Since $\mathbf{u} \in \mathbf{H}(\mathbf{curl},A)$, for any open subset of $A \subset \Omega \setminus S$, it has tangential traces on both sides of surface S belonging to the space $\mathbf{H}_{00}^{-1/2}(S)$, namely,

$$\begin{matrix} \mathbf{u}^+ \times \mathbf{n}^+ \\ \mathbf{u}^- \times \mathbf{n}^- \end{matrix} \in \mathbf{H}_{00}^{-1/2}(S).$$

We are going to prove that

$$\mathbf{u}^+ \times \mathbf{n}^+ + \mathbf{u}^- \times \mathbf{n}^- = \mathbf{J}_S \text{ on } S.$$

For this we consider the set B defined above and take $\varphi \in \mathscr{D}(\overset{\circ}{B})$ in (C.31). We get

$$\int_B \mathbf{u} \cdot \mathbf{curl}\,\varphi \, dV = \int_S \mathbf{J}_S \cdot \varphi \, dA \quad \forall \varphi \in \mathscr{D}(\overset{\circ}{B}) \tag{C.33}$$

and, by continuity and density, if $\mathbf{J}_S \in \mathbf{H}_{00}^{-1/2}(S)$ then

$$\int_{\overset{\circ}{B}} \mathbf{u} \cdot \mathbf{curl}\, \varphi \, dV = \langle \mathbf{J}_S, \varphi \rangle_{\mathbf{H}_{00}^{-1/2}(S), \mathbf{H}_{00}^{1/2}(S)} \quad \forall \varphi \in \mathbf{H}_0^1(\overset{\circ}{B}).$$

Moreover, since $\mathbf{u} \in \mathbf{L}^2(\Omega)$ and $\mathbf{u} \in \mathbf{H}(\mathbf{curl}, A)$ for any open subset of $A \subset \Omega \setminus S$, we have

$$\int_{\overset{\circ}{B}} \mathbf{u} \cdot \mathbf{curl}\, \varphi \, dV = \int_{B+} \mathbf{u} \cdot \mathbf{curl}\, \varphi \, dV + \int_{B-} \mathbf{u} \cdot \mathbf{curl}\, \varphi \, dV$$

$$= \langle \mathbf{u}^+ \times \mathbf{n}^+, \varphi \rangle_{\mathbf{H}_{00}^{-1/2}(S), \mathbf{H}_{00}^{1/2}(S)} + \int_{B+} \mathbf{curl}\, \mathbf{u} \cdot \varphi \, dV$$

$$+ \langle \mathbf{u}^- \times \mathbf{n}^-, \varphi \rangle_{\mathbf{H}_{00}^{-1/2}(S), \mathbf{H}_{00}^{1/2}(S)} + \int_{B-} \mathbf{curl}\, \mathbf{u} \times \varphi \, dV$$

$$= \langle \mathbf{u}^+ \times \mathbf{n}^+ + \mathbf{u}^- \times \mathbf{n}^-, \varphi \rangle_{\mathbf{H}_{00}^{-1/2}(S), \mathbf{H}_{00}^{1/2}(S)}.$$

Finally, by using (C.33), we get

$$\mathbf{u}^+ \times \mathbf{n}^+ + \mathbf{u}^- \times \mathbf{n}^- = \mathbf{J}_S \text{ in } \mathbf{H}_{00}^{-1/2}(S).$$

C.8 Radial functions and distributions

Recall that the physical space is modelled as a three-dimensional Euclidean affine space $\mathscr{E}$. We consider a cartesian fixed coordinate system $(o, \{\mathbf{e}_i\})$, where o is the origin and $\{\mathbf{e}_i\}$ is an orthonormal basis in vector space $\mathscr{V}$. Thus, given any $x \in \mathscr{E}$ there exist unique coordinates x_i, $i = 1, 2, 3$, such that

$$x = o + \sum_{i=1}^{3} x_i \mathbf{e}_i.$$

The *position vector* is the vector field defined by

$$x \in \mathscr{E} \longrightarrow \mathbf{r}(x) = x - o \in \mathscr{V}.$$

We denote $r(x) = |\mathbf{r}(x)|$. We have the following results:

i) $\operatorname{grad} r(x) = \dfrac{\mathbf{r}(x)}{r(x)} \quad \forall x \neq o.$

ii) $\operatorname{grad} \dfrac{1}{r(x)} = -\dfrac{\mathbf{r}(x)}{r^3(x)} \quad \forall x \neq o.$

iii) $r, \dfrac{1}{r} \in W_{\text{loc}}^{1,1}(\mathscr{E})$. Let us prove that $\dfrac{1}{r} \in W_{\text{loc}}^{1,1}(\mathscr{E})$.

Firstly,

$$\int_{B(o,\alpha)} \frac{1}{r(x)} dV = \int_0^\alpha dr \int_0^\pi d\theta \int_0^{2\pi} d\phi \frac{1}{r} r^2 \sin\theta$$

$$= 4\pi \frac{r^2}{2}\Big|_0^\alpha = 2\pi\alpha^2 < \infty.$$

Secondly,

$$\int_{B(o,\alpha)} |\operatorname{grad} \frac{1}{r}| dV = \int_{B(o,\alpha)} \frac{1}{r^2(x)} dV = \int_0^\alpha dr \int_0^\pi d\theta \int_0^{2\pi} d\phi \frac{1}{r^2} r^2 \sin\theta$$

$$= 4\pi r\Big|_0^\alpha = 4\pi\alpha < \infty.$$

iv) $\operatorname{grad} r(x) = \dfrac{\mathbf{r}(x)}{r(x)}$ in $\mathscr{D}'(\mathscr{E})$.

v) $\operatorname{grad} \dfrac{1}{r(x)} = -\dfrac{\mathbf{r}(x)}{r^3(x)}$ in $\mathscr{D}'(\mathscr{E})$.

vi) $\operatorname{div} \left(\dfrac{\mathbf{r}}{r^3}\right) = 4\pi\delta_o$ in $\mathscr{D}'(\mathscr{E})$.

Indeed, let $\mathbf{g} = \dfrac{\mathbf{r}}{r^3} = -\operatorname{grad} f$, with $f(x) = \dfrac{1}{r(x)}$. Then, from iii), $\mathbf{g} \in \mathbf{L}^1_{\text{loc}}(\mathscr{E})$ and defines the distribution

$$\langle T_\mathbf{g}, \varphi \rangle = \int_\mathscr{E} \mathbf{g} \cdot \varphi \, dV \quad \forall \varphi \in \mathscr{D}(\mathscr{E}),$$

the divergence of which is given by

$$\langle \operatorname{div} T_\mathbf{g}, \psi \rangle = -\langle T_\mathbf{g}, \operatorname{grad} \psi \rangle$$
$$= -\int_\mathscr{E} \mathbf{g} \cdot \operatorname{grad} \psi \, dV.$$

Since $|\mathbf{g}(x)| \to \infty$ as $x \to o$, this integral is improper but it converges as we will show below.

Indeed, let R be such that $\operatorname{supp} \psi \subset B(o,R)$. We have

$$\int_{\{x\in\mathscr{E}/r(x)>\alpha\}} \mathbf{g} \cdot \operatorname{grad} \psi \, dV = \int_\alpha^R dr \int_0^\pi d\theta \int_0^{2\pi} d\phi \frac{\mathbf{r}}{r^3} \cdot \frac{\partial \psi}{\partial r} \frac{\mathbf{r}}{r} r^2 \sin\theta$$

$$= \int_0^\pi d\theta \sin\theta \left(\int_0^{2\pi} d\phi \left(\int_\alpha^R \frac{\partial \psi}{\partial r} dr\right)\right)$$

$$= \int_0^\pi d\theta \sin\theta \int_0^{2\pi} d\phi \, (-\psi(\alpha, \theta, \phi)).$$

Taking the limit as $\alpha \to 0$ we get,

$$\langle T_{\mathbf{g}}, \operatorname{grad} \psi \rangle = \lim_{\alpha \to 0} \int_{\{x \in \mathscr{E}/r(x) > \alpha\}} \mathbf{g} \cdot \operatorname{grad} \psi \, dV$$

$$= \lim_{\alpha \to 0} \int_0^\pi d\theta \sin\theta \left(\int_0^{2\pi} -\psi(\alpha, \theta, \phi) d\phi \right)$$

$$= -\psi(o) \left(\int_0^\pi \sin\theta d\theta \right) \left(\int_0^{2\pi} d\phi \right)$$

$$= -4\pi \psi(o) = -4\pi \langle \delta_o, \psi \rangle.$$

vii) $\Delta(\frac{1}{r}) = -4\pi \delta_o$ in the sense of distributions, that is, $v(x) = -\frac{1}{4\pi r(x)}$ is a fundamental solution of the Laplace's equation. Actually, it is the unique fundamental solution such that $\lim_{r(x) \to \infty} v(x) = 0$ (see, for instance, Dautray-Lions [33]).

The result is an immediate consequence of the fact that $\Delta = \operatorname{div} \operatorname{grad}$, v) and vi). However, we give a direct proof.

Firstly, the $L^1_{\text{loc}}(\mathscr{E})$ function $f(x) = \frac{1}{r(x)}$ defines the distribution

$$T_f : \mathscr{D}(\mathscr{E}) \longrightarrow \mathbb{R}$$

$$\psi \rightsquigarrow \langle T_f, \psi \rangle = \int_{\mathscr{E}} \frac{1}{r(x)} \psi(x) \, dV.$$

Secondly, the Laplacian of T_f is defined by

$$\langle \Delta T_f, \psi \rangle = \langle T_f, \Delta \psi \rangle = \int_{\mathscr{E}} \frac{1}{r(x)} \Delta \psi(x) \, dV \quad \forall \psi \in \mathscr{D}(\mathscr{E}).$$

This is an improper integral because $\lim_{x \to 0} \frac{1}{r(x)} = \infty$, but it is convergent. Indeed, since $\frac{1}{r}$ is a radial function, it is enough to take ψ to be also a radial function, i.e., $\psi(x) = \widetilde{\psi}(r(x))$. Then,

$$\int_{\{x \in \mathscr{E}/r(x) > \alpha\}} \frac{1}{r(x)} \Delta \psi(x) \, dV = \int_\alpha^R dr \int_0^\pi d\theta \int_0^{2\pi} d\phi \frac{1}{r} \frac{1}{r^2} \frac{d}{dr}\left(r^2 \frac{d\widetilde{\psi}}{dr}\right) r^2 \sin\theta$$

$$= \int_\alpha^R dr \int_0^\pi d\theta \int_0^{2\pi} d\phi \frac{1}{r} \left[2r \frac{d\widetilde{\psi}}{dr} + r^2 \frac{d^2\widetilde{\psi}}{dr^2}\right] \sin\theta$$

$$= 4\pi \int_\alpha^R \left[2 \frac{d\widetilde{\psi}}{dr} + r \frac{d^2\widetilde{\psi}}{dr^2}\right] dr = 4\pi \left[-\alpha \frac{d\widetilde{\psi}}{dr}(\alpha) - \widetilde{\psi}(\alpha)\right].$$

Passing to the limit as $\alpha \to 0$ we get

$$\lim_{\alpha \to 0} \int_{\{x \in \mathscr{E}/r(x) > \alpha\}} \frac{1}{r(x)} \Delta \psi(x) \, dV = -4\pi \widetilde{\psi}(o) = -4\pi \langle \delta_o, \psi \rangle.$$

404 Appendix C Function spaces for electromagnetism

viii) $\mathbf{curl}\left(\dfrac{\mathbf{r}}{r^3}\right) = \mathbf{0}$ in the sense of distributions, i.e.,

$$\int_{\mathscr{E}} \frac{\mathbf{r}}{r^3} \cdot \mathbf{curl}\,\Psi\,dV = 0 \quad \forall \Psi \in \mathscr{D}(\mathscr{E}).$$

Indeed,

$$\int_{\mathscr{E}} \frac{\mathbf{r}}{r^3} \cdot \mathbf{curl}\,\Psi\,dV = \int_{\mathscr{E}} \frac{\mathbf{r}}{r^3} \cdot \frac{1}{r\sin\theta}\left(\frac{\partial}{\partial\theta}(\psi_\phi \sin\theta) - \frac{\partial \psi_\theta}{\partial \phi}\right)\mathbf{e}_r\,dV$$

$$= \int_{\mathscr{E}} \frac{1}{r^3 \sin\theta}\left[\frac{\partial}{\partial\theta}(\psi_\phi \sin\theta) - \frac{\partial \psi_\theta}{\partial \phi}\right]dV.$$

The right-hand side is an improper integral. We have

$$l \int_{\{x \in \mathscr{E}\,/\,r(x) > \alpha\}} \frac{1}{r^3 \sin\theta}\left[\frac{\partial}{\partial\theta}(\psi_\phi \sin\theta) - \frac{\partial \psi_\theta}{\partial \phi}\right]dV$$

$$= \int_\alpha^R dr \int_0^\pi d\theta \int_0^{2\pi} d\phi \frac{1}{r^3 \sin\theta}\left[\frac{\partial}{\partial\theta}(\psi_\phi \sin\theta) - \frac{\partial \psi_\theta}{\partial \phi}\right]r^2 \sin\theta$$

$$= \int_\alpha^R \frac{1}{r}\left[\int_0^{2\pi} \psi_\phi \sin\theta\big|_0^\pi d\phi - \int_0^\pi \psi_\theta\big|_0^{2\pi} d\theta\right]dr = 0,$$

which implies the result.

Appendix D
Harmonic regime: average values

Lemma D.1. *Let $\mathscr{X}$ and $\mathscr{Y}$ be two complex linear spaces each of them endowed with an anti-linear mapping denoted by "¯" ($\bar{\mathbf{A}}$ will be called the conjugate of $\mathbf{A}$, for $\mathbf{A}$ in $\mathscr{X}$ or $\mathscr{Y}$). Let us introduce the "real part" operator, Re, defined by*

$$\mathrm{Re}(\mathbf{A}) = \frac{1}{2}(\mathbf{A} + \bar{\mathbf{A}}).$$

Let $f : \mathscr{X} \times \mathscr{X} \longrightarrow \mathscr{Y}$ be a bilinear mapping such that

$$f(\bar{\mathbf{A}}, \bar{\mathbf{B}}) = \overline{f(\mathbf{A}, \mathbf{B})} \;\; \forall \mathbf{A}, \mathbf{B} \in \mathscr{X}.$$

Let

$$\mathscr{A}, \mathscr{B} : \mathbb{R} \longrightarrow \mathscr{X}$$

be the two mappings defined by

$$\mathscr{A}(t) = \mathrm{Re}(e^{i\omega t}\mathbf{A}), \;\; \mathscr{B}(t) = \mathrm{Re}(e^{i\omega t}\mathbf{B}),$$

for a real number ω. Then we have

$$\frac{1}{T}\int_0^T f(\mathscr{A}(t), \mathscr{B}(t))\, dt = \frac{1}{2}\mathrm{Re}\left(f(\mathbf{A}, \bar{\mathbf{B}})\right),$$

where $T = \frac{2\pi}{\omega}$.

Proof.

$$\frac{1}{T}\int_0^T f(\mathscr{A}(t), \mathscr{B}(t))\, dt = \frac{1}{T}\int_0^T f(\mathrm{Re}(e^{i\omega t}\mathbf{A}), \mathrm{Re}(e^{i\omega t}\mathbf{B}))\, dt$$

$$= \frac{1}{4}\frac{1}{T}\int_0^T f\left(e^{i\omega t}\mathbf{A} + e^{-i\omega t}\bar{\mathbf{A}}, e^{i\omega t}\mathbf{B} + e^{-i\omega t}\bar{\mathbf{B}}\right) dt$$

$$= \frac{1}{4}\frac{1}{T}\left(f(\mathbf{A},\mathbf{B}) \int_0^T e^{2i\omega t}\, dt + f(\mathbf{A},\bar{\mathbf{B}}) \int_0^T dt + f(\bar{\mathbf{A}},\mathbf{B}) \int_0^T dt \right.$$

$$\left. + f(\bar{\mathbf{A}},\bar{\mathbf{B}}) \int_0^T e^{-2i\omega t}\, dt \right) = \frac{1}{4}\left(f(\mathbf{A},\bar{\mathbf{B}}) + f(\bar{\mathbf{A}},\mathbf{B}) \right)$$

$$= \frac{1}{4}\left(f(\mathbf{A},\bar{\mathbf{B}}) + \overline{f(\mathbf{A},\bar{\mathbf{B}})} \right) = \frac{1}{2}\mathrm{Re}\left(f(\mathbf{A},\bar{\mathbf{B}}) \right). \qquad \square$$

Appendix E
Linear nodal and edge finite elements

In this appendix we present a brief description of the finite element spaces frequently used for the discretization of the weak problems proposed along the book. To simplify the presentation we will focus on the case where Ω is a bounded domain of $\mathbb{R}^3$ and a simple remark will be added for the particular cases of two-dimensional and axisymmetric domains.

We recall that the functional spaces appearing in the weak formulations and which must be approximated by finite element spaces are essentially $H^1(\Omega)$, $\mathbf{H}(\mathbf{curl}, \Omega)$ and subsets of them. Thus, we will recall the nodal piecewise linear continuous finite elements to approximate $H^1(\Omega)$ and the lowest-order Nédélec edge finite elements to approximate $\mathbf{H}(\mathbf{curl}, \Omega)$. We will restrict ourselves to the lowest order finite elements in both cases due to its simplicity and that they are the used ones in the numerical codes implemented in MaxFEM. We refer the reader to [27, 60, 55] for a more detailed presentation and the description of nodal and edge elements of higher order.

The finite element approximation requires three basic tools: a mesh of domain Ω, a finite-dimensional space consisting of piecewise-polynomials functions and a suitable basis of this space. Thus, we will start by defining a mesh of Ω composed by tetrahedra. Let $\Omega \in \mathbb{R}^3$ be a Lipschitz polyhedral domain and let us consider a finite decomposition of Ω given by

$$\overline{\Omega} = \cup_{K \in \mathscr{T}_h} K,$$

where:

- Each K is a tetrahedron with positive volume.
- $\overset{\circ}{K_1} \cap \overset{\circ}{K_2} = \emptyset$ for each distinct pair of elements $K_1, K_2 \in \mathscr{T}_h$.
- If K_1 and K_2 are distinct elements in $\mathscr{T}_h$ and $F = K_1 \cap K_2 \neq \emptyset$, then F is a common face, side, or vertex of K_1 and K_2.
- The diameter of the smallest sphere containing K (so-called diameter of K) is less or equal than h for each $K \in \mathscr{T}_h$.

Under these conditions, $\mathscr{T}_h$ is called a *triangulation* of Ω composed by tetrahedra or a tetrahedral mesh of the domain. The parameter h is usually known as the *mesh-size*.

A. Bermúdez, D. Gómez, P. Salgado: *Mathematical Models and Numerical Simulation in Electromagnetism.* UNITEXT – La Matematica per il 3+2 74
DOI 10.1007/978-3-319-02949-8_E, © Springer International Publishing Switzerland 2014

The next sections deal with approximations of $H^1(\Omega)$ and $\mathbf{H}(\mathbf{curl},\Omega)$ based on polynomials of degree one on each tetrahedron. For each case, we will describe the finite-dimensional space and a suitable basis of polynomial functions. To attain this goal, in the sequel we denote by $\mathscr{P}_1$ the space of polynomials of degree less than or equal to one in the variables x_1, x_2, x_3.

E.1 Linear nodal finite elements

Here, we introduce the classical discrete space of piecewise continuous linear nodal elements suitable to approximate $H^1(\Omega)$. This space will be denoted by $\mathscr{L}_h$ and is defined as

$$\mathscr{L}_h := \{V_h \in C(\Omega) : V_h|_K \in \mathscr{P}_1 \ \forall K \in \mathscr{T}_h\},$$

where $\mathscr{T}_h$ is a tetrahedral mesh of the domain. The proof that $\mathscr{L}_h$ is a subspace of $H^1(\Omega)$ can be found, for instance, in [60, Proposition 3.2.1]. We refer the reader to the same book to see convergence results by using this approximation.

To identify a basis for space $\mathscr{L}_h$ it is necessary to choose a set of degrees of freedom on each element K. The degrees of freedom are the parameters which allow us to uniquely determine a polynomial in K belonging to $\mathscr{P}_1$. In this case, we have to choose four degrees of freedom on each K. The simplest choice for the degrees of freedom are the values of V_h at vertices v_i of each tetrahedron K, i.e., $V_h(v_i), i = 1, \ldots, 4$. Furthermore, this choice guarantees that V_h will be globally continuous in Ω.

Now, we can construct a basis of $\mathscr{L}_h$. By denoting by $v_j, j = 1, \ldots, N_h$ the global set of vertices in $\overline{\Omega}$, it is enough to choose those functions $\lambda_i \in \mathscr{L}_h$ such that

$$\lambda_i(v_j) = \delta_{ij} \quad i, j = 1, \ldots, N_h,$$

δ_{ij} being the Kronecker symbol. The basis functions are also known as shape functions.

Remark E.1. The weak formulations of 2D and axisymmetric problems presented in the book involve functional spaces $H^1(\widehat{\Omega})$ and $\widetilde{H}^1_\rho(\widehat{\Omega})$ respectively. In these cases, the space of continuous linear nodal elements defined in a triangular mesh of the domain is also suitable for the corresponding discretization and guarantees convergence of the finite element method. The shape functions are similar to the ones presented in the 3D case and the degrees of freedom are again the values of the unknown at the vertices of the mesh.

E.2 Nédélec edge elements of first order

Now, in order to approximate the function space $\mathbf{H}(\mathbf{curl},\Omega)$ we recall the lowest-order finite elements of the family introduced by Nédélec in [49] and commonly

E.2 Nédélec edge elements of first order

known as *edge* elements; namely, the finite-dimensional space

$$\mathcal{N}_h(\Omega) := \{\mathbf{G}_h \in \mathbf{H}(\mathbf{curl}, \Omega) : \mathbf{G}_h|_K \in \mathcal{N}(K) \ \forall K \in \mathcal{T}_h\},$$

where

$$\mathcal{N}(K) := \{\mathbf{G}_h \in \mathscr{P}_1^3 : \mathbf{G}_h(x) = \mathbf{a} \times \mathbf{r}(x) + \mathbf{b}, \ \mathbf{a}, \mathbf{b} \in \mathbb{R}^3, \ x \in K\}.$$

The components of vectors $\mathbf{a}$ and $\mathbf{b}$ are uniquely determined by the values

$$\left\{\int_e \mathbf{G}_h \cdot \mathbf{t}_e \, dl \text{ for the six edges } e \text{ of } K\right\}, \tag{E.1}$$

where $\mathbf{t}_e$ denotes a unit tangential vector along edge e. Actually, one can prove that by choosing these six values as degrees of freedom to identify a piecewise polynomial function that locally belongs to $\mathcal{N}(K)$, we obtain an element of $\mathbf{H}(\mathbf{curl}, \Omega)$ and hence an element of $\mathcal{N}_h(\Omega)$.

Notice that the edges have to be endowed with an orientation before an evaluation of (E.1). The elements belonging to $\mathcal{N}_h(\Omega)$ are piecewise linear vector fields with tangential traces that are continuous through the edges of the mesh.

The term *edge elements* to denote the finite elements defined by $\mathcal{N}_h(\Omega)$ is due to the definition of the degrees of freedom. Actually, this term has been extended to the higher order elements introduced by Nédélec. Moreover, the elements of $\mathcal{N}_h(\Omega)$ also belong to the family of the so-called Whitney elements described for instance in [23].

Next, we will describe a basis function of space $\mathcal{N}_h(\Omega)$. For this purpose, we choose an orientation for each edge e of the mesh $\mathcal{T}_h$ and denote by p_i and p_j its initial and end points, respectively. Let $\mathbf{t}_e$ be a unit tangent vector to edge e, namely,

$$\mathbf{t}_e = \frac{\overrightarrow{p_i p_j}}{\|\overrightarrow{p_i p_j}\|}.$$

Then, a basis for $\mathcal{N}_h(\Omega)$ is the set

$$\{\mathbf{w}_e\}_{e \in \mathcal{T}_h}$$

of functions of $\mathcal{N}_h(\Omega)$ such that

$$\int_e \mathbf{w}_e \cdot \mathbf{t}_{e'} \, dl = \begin{cases} 1 & \text{if } e' = e \\ 0 & \text{otherwise.} \end{cases} \tag{E.2}$$

Notice that the dimension of $\mathcal{N}_h(\Omega)$ is the number of edges of the mesh of domain Ω. Moreover, in order to implement a discrete problem based on the lowest-order edge element it is very convenient to use the following expression of the basis functions

$$\mathbf{w}_e|_K = \lambda_i^K \operatorname{grad} \lambda_j^K - \lambda_j^K \operatorname{grad} \lambda_i^K, \tag{E.3}$$

where λ_i^K denotes the *shape function* of the continuous piecewise linear finite element space corresponding to vertex p_i of tetrahedron K, namely,

$$\lambda_i^K(p_m) = \delta_{im}. \tag{E.4}$$

We recall that, actually, $\lambda_i^K(x)$ is nothing but the i-th barycentric coordinate of x with respect to the vertices of tetrahedron K.

Along an edge e joining vertices p_i and p_j we have

$$\operatorname{grad} \lambda_i^K \cdot \overrightarrow{p_i p_j} = -\operatorname{grad} \lambda_j^K \cdot \overrightarrow{p_i p_j}$$

and

$$\lambda_i^K + \lambda_j^K = 1.$$

Hence,

$$\int_e \mathbf{w}_e \cdot \mathbf{t}_e \, dl = \int_e (\lambda_i^K \operatorname{grad} \lambda_j^K - \lambda_j^K \operatorname{grad} \lambda_i^K) \cdot \mathbf{t}_e \, dl$$

$$= \int_e (\lambda_i^K + \lambda_j^K) \operatorname{grad} \lambda_j^K \cdot \mathbf{t}_e \, dl$$

$$= \operatorname{grad} \lambda_j^K \cdot \overrightarrow{p_i p_j} \int_e \frac{(\lambda_i^K + \lambda_j^K)}{\|\overrightarrow{p_i p_j}\|} \, dl = 1.$$

A similar calculus allows us to conclude that $\int_{e'} \mathbf{w}_e \cdot \mathbf{t}_{e'} \, dl = 0$ along any other edge e'.

On the other hand, from the computational point of view it is also interesting to take into account the following expression (see property (B.26) in Appendix B)

$$\operatorname{\mathbf{curl}} \mathbf{w}_e |_K = \operatorname{grad} \lambda_i^K \times \operatorname{grad} \lambda_j^K + \lambda_i^K \operatorname{\mathbf{curl}} \operatorname{grad} \lambda_j^K$$

$$- \operatorname{grad} \lambda_j^K \times \operatorname{grad} \lambda_i^K - \lambda_j^K \operatorname{\mathbf{curl}} \operatorname{grad} \lambda_i^K$$

$$= 2 \operatorname{grad} \lambda_i^K \times \operatorname{grad} \lambda_j^K.$$

Moreover, it is easy to see that $\operatorname{div} \mathbf{w}_e$ is null in each tetrahedron K. Indeed, by using property (B.19) of Appendix B,

$$\operatorname{div} \mathbf{w}_e |_K = \operatorname{div}(\lambda_i^K \operatorname{grad} \lambda_j^K) - \operatorname{div}(\lambda_j^K \operatorname{grad} \lambda_i^K)$$

$$= \operatorname{grad} \lambda_i^K \cdot \operatorname{grad} \lambda_j^K + \lambda_i^K \operatorname{div}(\operatorname{grad} \lambda_j^K)$$

$$- \operatorname{grad} \lambda_j \cdot \operatorname{grad} \lambda_i^K - \lambda_j^K \operatorname{div}(\operatorname{grad} \lambda_i^K).$$

Notice that the barycentric coordinates functions are linear functions so their gradients are constant in each tetrahedron and hence the second and last term of the previous identity are null. On the other hand, the first and third term are identical. Thus, $\operatorname{div} \mathbf{w}_e = 0$ in each tetrahedron. However, this does not imply that it be globally divergence-free because the normal components of the basis funcions are not continuous through the faces of the mesh.

Finally, we refer the reader to [55] to see theoretical convergence results based on the approximation of $\mathbf{H}(\mathbf{curl}, \Omega)$ by means of linear edge elements.

Appendix F

Maxwell's equations in Lagrangian coordinates

Some applications of electromagnetism as electromagnetic forming are modelled by using the Maxwell's equations in Lagrangian coordinates. In this appendix we obtain these equations following [48].

F.1 Preliminary results and notation

Our goal is to write Maxwell's equations for a body subjected to a motion, in Lagrangian (or material) coordinates.

Let us call $\mathscr{B}$ the reference configuration and X the motion. Thus

$$X : \mathscr{B} \times \mathbb{R} \longrightarrow \mathscr{E}.$$

For a material point $p \in \mathscr{B}$, its *position* at time t is given by $x = X(p,t)$. The gradient of the motion is the tensor field

$$\mathbf{F}(p,t) := \nabla X(p,t).$$

The *velocity* of material point p at time t is the vector field $\dot{X}(p,t)$ and the spatial description of the velocity is $\mathbf{v}(x,t) = \dot{X}(P(x,t),t)$, where P is the so-called reference mapping of the motion, which is defined by

$$p = P(x,t) \Leftrightarrow x = X(p,t).$$

Let us denote by Ω a part of $\mathscr{B}$. The position of Ω at time t is the set $\Omega_t = X(\Omega,t)$. We assume regularity of X so that $\partial \Omega_t = X(\partial \Omega, t)$.

Since

$$p = P(X(p,t),t),$$

then

$$0 = \text{grad} P(x,t) \dot{X}(p,t) + P'(x,t)$$

A. Bermúdez, D. Gómez, P. Salgado: *Mathematical Models and Numerical Simulation in Electromagnetism.* UNITEXT – La Matematica per il 3+2 74
DOI 10.1007/978-3-319-02949-8_F, © Springer International Publishing Switzerland 2014

Appendix F Maxwell's equations in Lagrangian coordinates

from which follows that
$$\dot{X} = -FP'_m.$$
The field $P'_m(p,t) = P'(X(p,t),t)$ is called the *matter flow field*.

Moreover, let $c : [0,1] \to \Omega$ be a material curve. The position of the points of this material curve at time t is the curve
$$c_t(s) = X(c(s),t).$$
If S is a material surface having the curve c as bord, then the bord of the spatial surface $S_t = X(S,t)$ is the spatial curve c_t.

Lemma F.1. *Let $\mathbf{w}$ be a spatial vector field. We have the following transport theorem*
$$\frac{d}{dt}\int_{S_t} \mathbf{w} \cdot \mathbf{n}\, dA_x = \int_{S_t} (\dot{\mathbf{w}} \cdot \mathbf{n} + \operatorname{div} \mathbf{v}\, \mathbf{w} \cdot \mathbf{n} - \operatorname{\mathbf{grad}} \mathbf{v}\mathbf{w} \cdot \mathbf{n})\, dA_x$$
$$= \int_{S_t} \left[\mathbf{w}' \cdot \mathbf{n} + \operatorname{div} \mathbf{w}\, \mathbf{v} \cdot \mathbf{n} + \operatorname{\mathbf{curl}}(\mathbf{w} \times \mathbf{v}) \cdot \mathbf{n} \right] dA_x.$$

Proof. We make use of several results that can be found, for instance, in Gurtin [41] (see also Theorem B.2)::

i) $\int_{S_t} \mathbf{u} \cdot \mathbf{n}\, dA_x = \int_S \mathbf{u}_m \cdot \det \mathbf{F}\mathbf{F}^{-t}\mathbf{m}\, dA_p.$

ii) $\dot{\mathbf{F}} = \mathbf{L}_m \mathbf{F},$ where $\mathbf{L} = \operatorname{\mathbf{grad}} \mathbf{v}.$

iii) $(\det \mathbf{F})\dot{} = \det \mathbf{F}\, \operatorname{tr}(\dot{\mathbf{F}}\mathbf{F}^{-1}) = \det \mathbf{F}\, \operatorname{tr}(\mathbf{L}_m) = \det \mathbf{F}(\operatorname{div} \mathbf{v}).$

We have
$$\frac{d}{dt}\int_{S_t} \mathbf{w} \cdot \mathbf{n}\, dA_x = \frac{d}{dt}\int_S \mathbf{w}_m \cdot \det \mathbf{F}\mathbf{F}^{-t}\mathbf{m}\, dA_p = \int_S \frac{\partial}{\partial t}(\mathbf{w}_m \cdot \det \mathbf{F}\mathbf{F}^{-t}\mathbf{m})\, dA_p$$
$$= \int_S \dot{\mathbf{w}}_m \cdot \det \mathbf{F}\mathbf{F}^{-t}\mathbf{m}\, dA_p + \int_S \mathbf{w}_m \cdot (\det \mathbf{F})\dot{}\, \mathbf{F}^{-t}\mathbf{m}\, dA_p$$
$$+ \int_S \mathbf{w}_m \cdot (\det \mathbf{F})(\mathbf{F}^{-t})\dot{}\, \mathbf{m} A_p.$$

Moreover $(\mathbf{F}^{-t})\dot{} = -\mathbf{F}^{-t}\dot{\mathbf{F}}^t \mathbf{F}^{-t},$ then
$$\frac{d}{dt}\int_{S_t} \mathbf{w} \cdot \mathbf{n}\, dA_x = \int_{S_t} \dot{\mathbf{w}} \cdot \mathbf{n}\, dA_x + \int_S \mathbf{w}_m \cdot \det \mathbf{F}\, \operatorname{tr}(\dot{\mathbf{F}}\mathbf{F}^{-1})\mathbf{F}^{-t}\mathbf{m}\, dA_p$$
$$- \int_S \mathbf{w}_m \cdot \det\mathbf{F}(\dot{\mathbf{F}}\mathbf{F}^{-1})^t \mathbf{F}^{-t}\mathbf{m}\, dA_p$$
$$= \int_{S_t} \dot{\mathbf{w}} \cdot \mathbf{n}\, dA_x + \int_{S_t} \dot{\mathbf{w}} \cdot \operatorname{div} \mathbf{v}\, \mathbf{n}\, dA_x - \int_{S_t} \mathbf{w} \cdot (\operatorname{\mathbf{grad}} \mathbf{v})^t \mathbf{n}\, dA_x,$$

from which the first equality follows.

In order to prove the second one observe that

$$\dot{\mathbf{w}} = \mathbf{w}' + \operatorname{grad}\mathbf{w}\mathbf{v}$$

and

$$\operatorname{grad}\mathbf{w}\mathbf{v} - \operatorname{grad}\mathbf{v}\mathbf{w} = \operatorname{curl}(\mathbf{w} \times \mathbf{v}).$$

□

F.2 Transforming the Maxwell's equations

Firstly, let us recall that Maxwell's equations in spatial coordinates and differential form are (see Chap. 6)

$$\frac{\partial \mathbf{D}}{\partial t} - \operatorname{curl}\mathbf{H} = -\mathbf{J},$$

$$\frac{\partial \mathbf{B}}{\partial t} + \operatorname{curl}\mathbf{E} = \mathbf{0},$$

$$\operatorname{div}\mathbf{D} = q,$$

$$\operatorname{div}\mathbf{B} = 0.$$

The current density consists of a conduction current $\mathbf{J}^c$ and a convection current $q\mathbf{v}$ such that

$$\mathbf{J} = \mathbf{J}^c + q\mathbf{v}. \tag{F.1}$$

In order to transform these equations to material form we first consider their integral forms.

Integrating the first one over the spatial surface S_t we have

$$\int_{S_t} \left(\frac{\partial \mathbf{D}}{\partial t} \cdot \mathbf{n} - \operatorname{curl}\mathbf{H} \cdot \mathbf{n} + \mathbf{J} \cdot \mathbf{n} \right) dA_x = 0. \tag{F.2}$$

By using the transport theorem for surfaces we obtain

$$\frac{d}{dt} \int_{S_t} \mathbf{D} \cdot \mathbf{n}\, dA_x = \int_{S_t} \left(\frac{\partial \mathbf{D}}{\partial t} \cdot \mathbf{n} + \operatorname{div}\mathbf{D}\mathbf{v} \cdot \mathbf{n} + \operatorname{curl}(\mathbf{D} \times \mathbf{v}) \right) dA_x$$

$$= \int_{S_t} \frac{\partial \mathbf{D}}{\partial t} \cdot \mathbf{n}\, dA_x + \int_{S_t} q\mathbf{v} \cdot \mathbf{n}\, dA_x + \int_{c_t} (\mathbf{D} \times \mathbf{v}) \cdot \mathbf{t}\, dl_x,$$

by using Stokes' theorem.

From this equality and (F.2) we get

$$\frac{d}{dt} \int_{S_t} \mathbf{D} \cdot \mathbf{n}\, dA_x - \int_{S_t} q\mathbf{v} \cdot \mathbf{n}\, dA_x - \int_{c_t} \mathbf{D} \times \mathbf{v} \cdot \mathbf{t}\, dl_x - \int_{c_t} \mathbf{H} \cdot \mathbf{t}\, dl_x + \int_{S_t} \mathbf{J} \cdot \mathbf{n}\, dA_x = 0.$$

By changing variables in the integrals from spatial to material coordinates we deduce, after using (F.1),

$$\frac{d}{dt}\int_S \mathbf{D}_m \cdot \det \mathbf{F} \mathbf{F}^{-t} \mathbf{m}\, dA_p - \int_c \mathbf{D}_m \times \dot{X} \cdot \mathbf{F}\mathbf{k}\, dl_p$$

$$- \int_c \mathbf{H}_m \cdot \mathbf{F}\mathbf{k}\, dl_p + \int_S \mathbf{J}_m^c \cdot \det \mathbf{F}\mathbf{F}^{-t} \mathbf{m}\, dA_p = 0$$

and then

$$\int_S \frac{d}{dt}(\det \mathbf{F}\mathbf{F}^{-1} \mathbf{D}_m) \cdot \mathbf{m}\, dA_p - \int_S \mathbf{curl}\left[\mathbf{F}^t(\mathbf{D}_m \times \dot{X} + \mathbf{H}_m)\right] \cdot \mathbf{m}\, dA_p$$

$$+ \int_S \det \mathbf{F}\mathbf{F}^{-1} \mathbf{J}_m^c \cdot \mathbf{m}\, dA_p = 0.$$

Let us introduce the following transformed fields

$$\mathscr{H}(p,t) := \mathbf{F}^t(p,t)\left[\mathbf{H}_m(X(p,t),t) + \mathbf{D}_m(X(p,t),t) \times \dot{X}(p,t)\right]$$

$$= \mathbf{F}^t(p,t)\left[\mathbf{H}_m(X(p,t),t) - \dot{X}(p,t) \times \mathbf{D}_m(X(p,t),t)\right]$$

$$\mathscr{D}(p,t) := \det \mathbf{F}(p,t)\mathbf{F}^{-1}(p,t)\mathbf{D}_m(X(p,t),t)$$

$$\mathscr{J}^c(p,t) := \det \mathbf{F}(p,t)\mathbf{F}^{-1}(p,t)\mathbf{J}_m^c(X(p,t),t)$$

We have the following result:

Proposition F.1.

$$\mathbf{H}_m = \mathbf{F}^{-t}(\mathscr{H} - \mathbf{P}'_m \times \mathscr{D}).$$

Proof. From the definition of $\mathscr{H}$ we get

$$\mathbf{H}_m = \mathbf{F}^{-t}\mathscr{H} + \dot{X} \times \mathbf{D}_m = \mathbf{F}^{-t}\left[\mathscr{H} + \mathbf{F}^t(\dot{X} \times \mathbf{D}_m)\right]$$

$$= \mathbf{F}^{-t}\left[\mathscr{H} - \mathbf{P}'_m \times (\det \mathbf{F}\mathbf{F}^{-1}\mathbf{D}_m)\right] = \mathbf{F}^{-t}\left[\mathscr{H} - \mathbf{P}'_m \times \mathscr{D}\right]$$

by using (B.11) for $\mathbf{S} = \mathbf{F}^{-1}$, $\mathbf{a} = \dot{X}$ and $\mathbf{b} = \mathbf{D}_m$, and the definition of $\mathscr{D}$. □

Then, from the above calculations we finally get

$$\frac{\partial \mathscr{D}}{\partial t} - \mathbf{Curl}\,\mathscr{H} = -\mathscr{J}^c.$$

Similarly, by integrating the Faraday's law on the spatial surface S_t we get

$$\int_{S_t} \frac{\partial \mathbf{B}}{\partial t} \cdot \mathbf{n}\, dA_x + \int_{S_t} \mathbf{curl}\,\mathbf{E} \cdot \mathbf{n}\, dA_x = 0.$$

From the transport theorem we can replace the first term with

$$\int_{S_t} \frac{\partial \mathbf{B}}{\partial t} \cdot \mathbf{n}\, dA_x = \frac{d}{dt}\int_{S_t} \mathbf{B} \cdot \mathbf{n}\, dA_x - \int_{S_t} \mathrm{div}\,\mathbf{B}\mathbf{v} \cdot \mathbf{n}\, dA_x - \int_{S_t} \mathbf{curl}(\mathbf{B} \times \mathbf{v}) \cdot \mathbf{n}\, dA_x$$

and, from the Stokes' theorem, the second one with $\int_{C_t} \mathbf{E} \cdot \mathbf{t} \, dl_x$.
Thus we obtain

$$\frac{d}{dt} \int_{S_t} \mathbf{B} \cdot \mathbf{n} \, dA_x - \int_{C_t} (\mathbf{B} \times \mathbf{v} - \mathbf{E}) \cdot \mathbf{t} \, dl_x = 0.$$

Transformation of the integration variables to the material frame yields

$$\frac{d}{dt} \int_{S} \mathbf{B}_m \cdot \det \mathbf{F} \mathbf{F}^{-t} \mathbf{m} \, dA_p - \int_C (\mathbf{B}_m \times \dot{X} - \mathbf{E}_m) \cdot \mathbf{F} \mathbf{k} \, dl_p = 0$$

and then

$$\int_S \frac{d}{dt} [\det \mathbf{F} \mathbf{F}^{-1} \mathbf{B}_m] \cdot \mathbf{m} \, dA_p + \int_S \mathbf{curl} \left[\mathbf{F}^t (\mathbf{E}_m + \dot{X} \times \mathbf{B}_m) \right] \cdot \mathbf{m} \, dA_p = 0. \quad (\text{F.3})$$

This leads us to define the transformed fields

$$\mathscr{E} = \mathbf{F}^t \left[\mathbf{E}_m + \dot{X} \times \mathbf{B}_m \right],$$
$$\mathscr{B} = \det \mathbf{F} \mathbf{F}^{-1} \mathbf{B}_m. \quad (\text{F.4})$$

We have the following result

Proposition F.2.

$$\mathbf{E}_m = \mathbf{F}^{-t} (\mathscr{E} + P'_m \times \mathscr{B}).$$

Proof. From the definition of $\mathscr{E}$ we get

$$\mathbf{E}_m = \mathbf{F}^{-t} \mathscr{E} - \dot{X} \times \mathbf{B}_m = \mathbf{F}^{-t} \left[\mathscr{E} - \mathbf{F}^t (\dot{X} \times \mathbf{B}_m) \right]$$
$$= \mathbf{F}^{-t} \left[\mathscr{E} + P'_m \times (\det \mathbf{F} \mathbf{F}^{-1} \mathbf{B}_m) \right] = \mathbf{F}^{-t} \left[\mathscr{E} + P'_m \times \mathscr{B} \right])$$

by using (B.11) for $\mathbf{S} = \mathbf{F}^{-1}$, $\mathbf{b} = \dot{X}$ and $\mathbf{c} = \mathbf{B}_m$, and the definition of $\mathscr{B}$. $\square$

Then (F.3) yields

$$\frac{\partial \mathscr{B}}{\partial t} + \mathbf{Curl} \, \mathscr{E} = 0.$$

Now, let us consider the Gauss' law $\text{div} \, \mathbf{D} = q$. By integrating in a part Ω_t we obtain

$$\int_{\Omega_t} \text{div} \, \mathbf{D} \, dV_x = \int_{\Omega_t} q \, dV_x$$

and using the Gauss' theorem

$$\int_{\partial \Omega_t} \mathbf{D} \cdot \mathbf{n} \, dA_x = \int_{\Omega_t} q \, dV_x.$$

By changing to material variables we get

$$\int_{\partial \Omega} \mathbf{D}_m \det \mathbf{F} \mathbf{F}^{-t} \mathbf{m} \, dA_p = \int_\Omega q_m \det \mathbf{F} \, dV_p$$

and using the Gauss' theorem

$$\int_\Omega \text{Div}(\det \mathbf{F}\mathbf{F}^{-1}\mathbf{D}_m)\,dV_p = \int_\Omega q_m \det \mathbf{F}\,dV_p. \qquad (F.5)$$

By introducing

$$\mathscr{D}(p,t) = q_m(X(p,t),t)\det \mathbf{F}(p,t)$$

and using the definition of $\mathscr{D}$, (F.5) yields

$$\text{Div}\,\mathscr{D} = \mathscr{Q}.$$

In a similar way we obtain

$$\text{Div}\,\mathscr{B} = 0.$$

F.3 Transforming the constitutive laws

In spatial coordinates the constitutive laws are

$$\mathbf{D} = \varepsilon_0 \mathbf{E} + \mathbf{P}, \qquad (F.6)$$

$$\mathbf{H} = \frac{\mathbf{B}}{\mu_0} - \mathbf{M} - \mathbf{P} \times \mathbf{v}, \qquad (F.7)$$

$$\mathbf{J}^c = \sigma(\mathbf{E} + \mathbf{v} \times \mathbf{B}), \qquad (F.8)$$

where $\mathbf{M}$ and $\mathbf{P}$ are the magnetization and polarization fields, respectively. For linear materials

$$\mathbf{M} = \chi_B \mu_0^{-1} \mathbf{B}, \qquad (F.9)$$

$$\mathbf{P} = \chi_E \varepsilon_0 \mathbf{E}, \qquad (F.10)$$

where χ_E and χ_B are the electric and the magnetic *susceptibilities*, respectively.

In order to write these constitutive laws in material coordinates we define

$$\mathscr{P} = \det \mathbf{F}\mathbf{F}^{-1}\mathbf{P}_m,$$

$$\mathscr{M} = \mathbf{F}^t \mathbf{M}_m.$$

We have the following results.

Proposition F.3.

$$\mathscr{H} = \frac{1}{\mu_0 \det \mathbf{F}} \mathbf{C}\mathscr{B} + \varepsilon_0 \det \mathbf{F}\,P'_m \times \left[\mathbf{C}^{-1}(\mathscr{E} + P'_m \times \mathscr{B})\right] - \mathscr{M}. \qquad (F.11)$$

Proof. From Proposition F.1 we deduce

$$\mathcal{H} = \mathbf{F}^t \mathbf{H}_m + \mathbf{P}'_m \times \mathcal{D} = \mathbf{F}^t \left(\frac{\mathbf{B}_m}{\mu_0} - \mathbf{M}_m - \mathbf{P}_m \times \dot{\mathbf{X}} \right) + \mathbf{P}'_m \times \mathcal{D}$$

$$= \frac{1}{\mu_0 \det \mathbf{F}} \mathbf{F}^t \mathbf{F} \mathcal{B} - \mathcal{M} - \mathbf{F}^t (\mathbf{P}_m \times \dot{\mathbf{X}}) + \mathbf{P}'_m \times \mathcal{D}$$

$$= \frac{1}{\mu_0 \det \mathbf{F}} \mathbf{C} \mathcal{B} - \mathcal{M} - \det \mathbf{F} (\mathbf{F}^{-1} \mathbf{P}_m \times \mathbf{F}^{-1} \dot{\mathbf{X}}) + \mathbf{P}'_m \times \mathcal{D}$$

$$= \frac{1}{\mu_0 \det \mathbf{F}} \mathbf{C} \mathcal{B} - \mathcal{M} + \det \mathbf{F} (\mathbf{F}^{-1} \mathbf{P}_m \times \mathbf{P}'_m + \mathbf{P}'_m \times \mathbf{F}^{-1} \mathbf{D}_m)$$

$$= \frac{1}{\mu_0 \det \mathbf{F}} \mathbf{C} \mathcal{B} - \mathcal{M} + \det \mathbf{F} (\mathbf{P}'_m \times \mathbf{F}^{-1} \varepsilon_0 \mathbf{E}_m)$$

$$= \frac{1}{\mu_0 \det \mathbf{F}} \mathbf{C} \mathcal{B} - \mathcal{M} + \varepsilon_0 \det \mathbf{F} (\mathbf{P}'_m \times \mathbf{F}^{-1} \mathbf{F}^{-t} (\mathcal{E} + \mathbf{P}'_m \times \mathcal{B}))$$

$$= \frac{1}{\mu_0 \det \mathbf{F}} \mathbf{C} \mathcal{B} - \mathcal{M} + \varepsilon_0 \det \mathbf{F} (\mathbf{P}'_m \times \mathbf{C}^{-1} (\mathcal{E} + \mathbf{P}'_m \times \mathcal{B})). \quad \square$$

Proposition F.4.
$$\mathcal{D} = \varepsilon_0 \det \mathbf{F} \mathbf{C}^{-1} (\mathcal{E} + \mathbf{P}'_m \times \mathcal{B}) + \mathcal{P}. \quad \text{(F.12)}$$

Proof. By definition of $\mathcal{D}$

$$\mathcal{D} = \det \mathbf{F} \mathbf{F}^{-1} \mathbf{D}_m = \det \mathbf{F} \mathbf{F}^{-1} (\varepsilon_0 \mathbf{E}_m + \mathbf{P}_m)$$

$$= \varepsilon_0 \det \mathbf{F} \mathbf{F}^{-1} \mathbf{E}_m + \det \mathbf{F} \mathbf{F}^{-1} \mathbf{P}_m$$

$$= \varepsilon_0 \det \mathbf{F} \mathbf{F}^{-1} \mathbf{F}^{-t} (\mathcal{E} + \mathbf{P}'_m \times \mathcal{B}) + \mathcal{P}$$

$$= \varepsilon_0 \det \mathbf{F} \mathbf{C}^{-1} (\mathcal{E} + \mathbf{P}'_m \times \mathcal{B}) + \mathcal{P}. \quad \square$$

Corollary F.1.
$$\mathcal{H} = \frac{1}{\mu_0 \det \mathbf{F}} \mathbf{C} \mathcal{B} + \mathbf{P}'_m \times (\mathcal{D} - \mathcal{P}) - \mathcal{M}.$$

Moreover, from the definitions of $\mathcal{P}$ and $\mathcal{M}$ and the constitutive laws (F.9) and (F.10) we deduce the following

Proposition F.5.

a) $\mathcal{P} = \dfrac{\chi_E}{1 + \chi_E} \mathcal{D}.$ (F.13)

b) $\mathcal{M} = \dfrac{\chi_B}{\mu_0 \det \mathbf{F}} \mathbf{C} \mathcal{B}.$ (F.14)

Proof.

a) $\mathcal{P} = \det \mathbf{F} \mathbf{F}^{-1} \chi_E \varepsilon_0 \mathbf{E}_m = \det \mathbf{F} \mathbf{F}^{-1} \chi_E (\mathbf{D}_m - \mathbf{P}_m) = \chi_E (\mathcal{D} - \mathcal{P})$

from which (F.13) follows.

b) $\mathscr{M} = \mathbf{F}^t \mathbf{M}_m = \mathbf{F}^t \chi_B \dfrac{\mathbf{B}_m}{\mu_0} = \dfrac{\chi_B}{\mu_0} \mathbf{F}^t \mathbf{F} \mathbf{F}^{-1} \mathbf{B}_m = \dfrac{\chi_B}{\mu_0 \det \mathbf{F}} \mathbf{C}\mathscr{B}.$ □

By using (F.14) and (F.13) respectively in (F.11) and (F.12), we get the following final forms of the constitutive laws in material coordinates.

Proposition F.6.

a) $\mathscr{H} = \dfrac{1-\chi_B}{\mu_0 \det \mathbf{F}} \mathbf{C}\mathscr{B} + \dfrac{1}{1+\chi_E} \mathbf{P}'_m \times \mathscr{D}.$

b) $\mathscr{E} = \dfrac{1}{(1+\chi_E)\varepsilon_0 \det \mathbf{F}} \mathbf{C}\mathscr{D} - \mathbf{P}'_m \times \mathscr{B}.$

Finally let us write the Ohm's law in material coordinates.

Proposition F.7.

$$\mathscr{J}^c = \det \mathbf{F}\, \sigma\, \mathbf{C}^{-1} \mathscr{E}.$$

Proof. From the definition of $\mathscr{J}^c$ we have

$$\mathscr{J}^c = \det \mathbf{F} \mathbf{F}^{-1} \mathbf{J}_m^c = \det \mathbf{F} \mathbf{F}^{-1} \sigma (\mathbf{E}_m + \dot{\mathbf{X}} \times \mathbf{B}_m)$$

where we have used the Ohm's law in spatial coordinates (F.8).
Then

$$\mathscr{J}^c = \sigma \det \mathbf{F} \mathbf{F}^{-1} \mathbf{F}^{-t} \mathbf{F}^t (\mathbf{E}_m + \dot{\mathbf{X}} \times \mathbf{B}_m) = \sigma \det \mathbf{F}\, \mathbf{C}^{-1} \mathscr{E}$$

simply by using the definition of $\mathscr{E}$ in (F.4). □

Summarizing, the system of Maxwell equations in Lagrangian configuration is

$$\dfrac{\partial \mathscr{D}}{\partial t} - \mathrm{Curl}\, \mathscr{H} = -\mathscr{J}^c,$$

$$\dfrac{\partial \mathscr{B}}{\partial t} - \mathrm{Curl}\, \mathscr{E} = \mathbf{0},$$

$$\mathrm{Div}\, \mathscr{D} = \mathscr{Q},$$

$$\mathrm{Div}\, \mathscr{B} = 0,$$

$$\mathscr{H} = \dfrac{1-\chi_B}{\mu_0 \det \mathbf{F}} \mathbf{C}\mathscr{B} + \dfrac{1}{1+\chi_E} \mathbf{P}'_m \times \mathscr{D},$$

$$\mathscr{E} = \dfrac{1}{(1+\chi_E)\varepsilon_0 \det \mathbf{F}} \mathbf{C}\mathscr{D} - \mathbf{P}'_m \times \mathscr{B},$$

$$\mathscr{J}^c = \det \mathbf{F}\, \sigma\, \mathbf{C}^{-1} \mathscr{E}.$$

References

[1] Alexander, C.K., Sadiku, M.N.O: Fundamentals of Electric Circuits. McGraw-Hill Science/Engineering/Math, London (2012)

[2] Alonso Rodríguez, A.: Formulation via vector potentials of eddy-current problems with voltage or current excitation. In: Communications in Applied and Industrial Mathematics (2011), **2** (1). http://caim.simai.eu/index.php/caim/article/view/369. Cited 12 April 2013

[3] Alonso Rodríguez, A., Hiptmair, R., Valli, A.: Mixed finite element approximation of eddy current problems. IMA J. Numer. Anal. **24**, 255–271 (2004)

[4] Alonso A., Valli, A.: Eddy Current Approximation of Maxwell Equations. Springer–Verlag, Italy (2010)

[5] Amrouche, C., Bernardi, C., Dauge, M., Girault, V.: Vector potentials in three-dimensional non-smooth domains. Math. Methods Appl. Sci. **21**, 823–864 (1998)

[6] Bermúdez, A., Bullón, J., Pena, F., Salgado, P.: A numerical method for transient simulation of metallurgical compound electrodes. Finite Elem. Anal. Des. **39**, 283–299 (2003)

[7] Bermúdez, A., Gómez, D., Muñiz, M.C., Salgado, P., Vázquez, R.: Numerical simulation of a thermo–electromagneto–hydrodynamic problem in an induction heating furnace. Appl. Numer. Math. **59**, 2082–2104 (2009)

[8] Bermúdez, A., Gómez, D., Muñiz, M.C., Salgado, P., Vázquez, R.: Numerical modelling of industrial induction. In: Grundas, S. (eds.) Advances in induction and microwave heating of mineral and organic materials, Intech (2011). http:////www.intechopen.com/books/advances-in-induction-and-microwave-heating-of-mineral-and-organic-materials/numerical-modelling-of-industrial-induction. Cited 12 April 2013

[9] Bermúdez, A., Gómez, D., Salgado, P.: On the eddy current losses in laminated cores and the computation of an equivalent conductivity. IEEE Trans. Magn. **44** (12), 4730–4738 (2008)

[10] Bermúdez, A., López-Rodríguez, B., Rodríguez, R., Salgado, P.: Equivalence between two finite element methods for the eddy current problem. C. R. Acad. Sci. Paris, Ser. I **348**, 769–774 (2010)

[11] Bermúdez, A., López-Rodríguez, B., Rodríguez, R., Salgado, P.: An eddy current problem in terms of a time-primitive of the electric field with non-local source conditions. ESAIM, Math. Model. Numer. Anal. **47**(3), 875–902 (2013)
[12] Bermúdez, A., Muñiz, M. C., Pena, F., Bullón, J.: Numerical computation of the electromagnetic field in the electrodes of a three-phasearc furnace. Int. J. Numer. Meth. Engng. **46**, 649–658 (1999)
[13] Bermúdez, A., Muñiz, M. C., Quintela, P.: Numerical solution of a three-dimensional thermoelectric problem taking place in an aluminium electrolytic cell. Comput. Methods Appl. Mech. Engrg. **106** (1-2), 129–142 (1993)
[14] Bermúdez, A., Reales, C., Rodríguez, R., Salgado, P.: Numerical analysis of a finite element method for the axisymmetric eddy current model of an induction furnace. IMA J. Numer. Anal. **30**, 654–676 (2010)
[15] Bermúdez, A., Rodríguez, R., Salgado, P.: A finite element method with Lagrange multipliers for low-frequency harmonic Maxwell equations. SIAM J. Numer. Anal. **40**, 1823–1849 (2002)
[16] Bermúdez, A., Rodríguez, R., Salgado, P.: Numerical treatment of realistic boundary conditions for the eddy current problem in an electrode via Lagrange multipliers, Math. Comp. **74**, 123–151 (2005)
[17] Bermúdez, A., Rodríguez, R., Salgado, P.: Numerical solution of eddy current problems in bounded domains using realistic boundary conditions. Comput. Methods Appl. Mech. Engrg. **194**, 411–426 (2005)
[18] Bermúdez, A., Rodríguez, R., Salgado, P.: FEM for 3D eddy current problems in bounded domains subject to realistic boundary conditions. An application to metallurgical electrodes. Arch. Comput. Methods Engrg. **12**, 67-114 (2005)
[19] Bermúdez, A., Rodríguez, R., Salgado, P.: A finite element method for the magnetostatics problem in terms of scalar potentials. SIAM J. Numer. Anal. **46**, 1338–1363 (2008)
[20] Bermúdez, A., Salgado, P. A domain decomposition/finite element method for the numerical simulation of electrolytic cells. Comput. Methods Appl. Mech. Engrg. **188**, 391–412 (2000)
[21] Bertotti, G.: Hysteresis in Magnetism. Academic Press, New York (1998)
[22] Bossavit, A.: Eddy currents in dimension 2: voltage drops. In: Mathis, W., Schindler, T. (eds.) Proc. ISTET'99, pp. 103–107. University Otto-von-Guericke, Magdeburg, Germany (1999)
[23] Bossavit, A.: Computational Electromagnetism. Variational Formulations, Complementarity, Edge Elements. Academic Press, San Diego, CA (1998)
[24] Bossavit, A.: Most general "non-local" boundary conditions for the Maxwell equation in a bounded region, COMPEL. **19**, 3239–3245 (2000)
[25] Cadwell, J., Douglas, K.S.Ng.: Mathematical Modelling. Case Studies and Projects. Kluwer, Boston (2004)
[26] Casas, E.: L^2 estimates for the finite element method for the Dirichlet problem with singular data. Numer. Math., **47**, 627–632 (1985)
[27] Ciarlet, P.G.: The Finite Element Method for Elliptic Problems. North-Holland, Amsterdam (1978)

[28] Lipnikov, K., Manzini, G., Brezzi, F., Buffa, A.: The mimetic finite difference method for 3D magnetostatics fields problems. J. Comput. Phys., **20**(2), 305–328 (2011)
[29] Cessenat, M.: Mathematical Methods in Electromagnetism: linear theory and applicarions. World Scientific, Singapore (1996)
[30] Chua, L. O., Desoer, Ch.A., Kuh, E.S.: Linear and Nonlinear Circuits. Mc Graw-Hill, New York (1987)
[31] Chari, M.V.K., Silvester, P.P.: Finite elements in electrical and magnetic field problems. John Wiley & Sons, Chichester (1980)
[32] Consiglieri, L., Muñiz, M.C.: Existence of a solution for a free boundary problem in the thermoelectrical modelling of an aluminium electrolytic cell. European J. Appl. Math. **14**(2), 201–216 (2003)
[33] Dautray, R., Lions, J. L.: Mathematical Analysis and Numerical Methods for Science and Technology. Springer-Verlag, Berlin Heidelberg New York (1990)
[34] Dyer, S, A.: Wiley Survey of Instrumentation and Measurement. John Wiley & Sons (2004)
[35] Fernandes, P., Gilardi, G.: Magnetostatic and electrostatic problems in inhomogeneous anisotropic media with irregular boundary and mixed boundary conditions. Math. Models Meth. Appl. Sci. **7**, 957–991 (1997)
[36] Fernandes, P., Perugia, I.: Vector potential formulation for magnetostatic and modelling of permanent magnets. IMA J. Appl. Math. **66**, 293–318 (2001)
[37] Girault, V., Raviart, P.A.: The finite element method for Navier-Stokes equations. Springer-Verlag, New York (1986)
[38] Griffiths, D.J.: Introduction to electrodynamics. Prentice-Hall Inc., Upper Saddle River, New Jersey (1999)
[39] Gopalakrishnan, J., Pasciak, J.: The convergence of V-cycle multigrid algorithms for axisymmetric Laplace and Maxwell equations. Math. Comp. **75**, 1697–1719 (2006)
[40] Grjotheim, K., Kvande, H.: Understanding the Hall-Héroult Process for Production of Aluminium. Aluminium Verlag, Dusseldorf (1986)
[41] Gurtin, M.E.: An Introduction to Continuum Mechanics. Academic Press, New York (1981)
[42] Helrich, C.S.: The Classical Theory of Fields. Springer-Verlag, Berlin Heidelberg (2012)
[43] Honig, B., Nicholls, A.: Classical electrostatics in biology and chemistry. Science **268**(5214), 1144–1149 (1995)
[44] Kikuchi, F.: Mixed formulations for finite element analysis of magnetostatic and electrostatic problems. Japan. J. Appl. Math. **6**, 209–221 (1989)
[45] Krasnoselskii, M.A. and Pokrovskii, A.V.: System with Hysteresis. Springer-Verlag, Berlin (1989)
[46] Jin, J.: The finite element mehtod in electromagnetics. John Wiley & Sons Inc., New York (2002)
[47] Johnk, C. T. A.: Engineering Electromagnetic Fields and Waves. John Wiley & Sons Inc., New York (1988)

[48] Lax, M., Nelson, D.F.: Maxwell equations in material form. Phys. Rev. B **13**, 1777–1784 (1976)
[49] Nédélec, J.-C.: Mixed finite elements in $\mathbb{R}^3$. Numer. Math. **35**, 315–341 (1980)
[50] Lorrain, P., Corson, D. L.: Electromagnetism: principles and applications. W. H. Freeman and Company, San Francisco (1978)
[51] Li, Y., Tian, G. Y., Ward, S.: Numerical simulation on magnetic flux leakage evaluation at high speed. NDT &E International **39**, 367–373 (2006)
[52] Mayergoyz, I.D.: Mathematical models of hysteresis and their applications. Academic Press, New York (2003)
[53] Mercier, B., Raugel, G.: Resolution d'un problème aux limites dans un ouvert axisymétrique par éléments finis en r, z et séries de Fourier en θ, RAIRO, Anal. Numér. **16**, 405–461 (1982)
[54] Moore A.D. (ed.): Electrostatics and its applications. John Wiley & Sons Inc., (1973)
[55] Monk, P.: Finite Element Methods for Maxwell's Equations. Clarendon Press, Oxford (2003)
[56] Nagle, R.K., Saff, E.B., Snider, A.D.: Fundamentals of Differential Equations and Boundary Value Problems. Pearson Education (2012)
[57] Nédélec, J. C.: Acoustic and Electromagnetic Equations. Springer-Verlag, New York (2001)
[58] ON Semiconductor n. 4 (2011). Power Factor Correction Handbook. http://www.onsemi.com/pub_link/Collateral/HBD853_D.PDF. Cited 21 July 2013
[59] OpenStax College, College Physics. OpenStax College. http://cnx.org/content/col11406/latest/. Cited 12 April 2013
[60] Quarteroni, A., Valli, A.: Numerical Approximation of Partial Differential Equations. Springer-Verlag, Berlin Heidelberg (1994)
[61] Popović, Z., Popović, B.D.: Introductory Electromagnetics. Prentice-Hall, Inc., Upper Saddle River, New Jersey (2000)
[62] Preisach, F.: Über die magnetische Nachwirkung, Zeitschrift für Physik, **94**, 277–302 (1935).
[63] Sadiku, M.N.O.: Elements of Electromagnetics. Oxford University Press Inc., New York (2001)
[64] Silvester, P.P., Ferrari, R.L.: Finite Elements for Electrical Engineers. Cambridge University Press, Cambridge (1996)
[65] Steinmetz, C.P.: On the law of hyesteresis, Trans. Ame. Inst. Elect. Engrs. **9**, 3–51 (1892). Reprinted in Proc. of the IEEE, **72**(2) (1984)
[66] Venegas, P.: Contribution to the mathematical and numerical analysis of some electromagnetic problems. PhD Thesis. Universidad de Concepción (Chile) (2013)
[67] Visintin, A.: Differential models of hysteresis. Springer-Verlag, Berlin Heidelberg New York (1994)
[68] Ansys. Electromagnetics, Circuit & Systems Solutions. http://www.ansys.com/Products/Simulation+Technology/Electromagnetics

[69] Flux 2D/3D. Electromagnetis and thermal finite element analysis.
http://www.cedrat.com/en/software/flux.html
[70] COMSOL Multiphysics.
http://www.comsol.com/
[71] Magnet 2D/3D Electromagnetic Field Simulation Software.
http://www.infolytica.com/en/products/magnet/
[72] MaxFEM. Software for electromagnetism simulation.
https://sourceforge.net/projects/maxfem/
[73] Opera Simulation Software.
http://operafea.com/

Index

acceleration, 373
affine space, 363
air-gap, 318
Ampére's law, 58
ampere, 57
amplification factor, 12
amplitude, 5, 12, 56
- complex, 5, 325
attenuation constant, 79, 107
average value, 405
axial vector, 362

backward formula, 38
basis
- of a vector space, 356
- orthogonal, 357
- orthonormal, 357
Bessel's equation, 210, 330
Bessel's functions, 210
Biot-Savart law, 154
body, 372
busbar, 198

capacitance, 21, 35
- reduced, 125, 128
capacitor, 3, 21, 23, 35, 125, 292
- circular, 294
- planar, 295
cartesian coordinate frame, 363
charge, 22
- point, 56, 115, 119
charge conservation, 85, 93

charge density
- linear, 119
- surface, 119
- volume, 119
circuit
- in parallel, 147
- in series, 147
- magnetically coupled, 35, 243
- problem, 36
- resonant, 24
- RLC, 3, 21
comfort criterion, 16
congruency property, 231
connectivity matrix, 33
constitutive laws, 35, 418
- in material coordinates, 420
contactor, 318
convection current, 415
coordinate system
- cartesian, 377
- cylindrical, 379
- spherical, 382
Coulomb's gauge, 153
Coulomb's law, 68
critical damping coefficient, 7
critical resistance, 7
critically damped, 7, 8
curl
- in cartesian system, 378
- in cylindrical system, 381
- in spherical system, 384
- of a vector distribution, 390
- of a vector field, 367

current
- density, 57, 86, 87, 91, 92, 108, 151, 159, 174, 175, 191, 200, 212, 415
- surface density, 95, 400
current intensity, 22, 57
curve, 368
- closed, 369
- length, 369
curvilinear integral
- of a tensor field, 369
- of a vector field, 369

d'Alembert solution, 54
damping, 79
- factor, 7
- mechanism, 6
dashpot, 3
deformation, 372
depth of penetration, 107
dielectrics
- lossless, 111
- lossy, 111
differentiable mapping, 364
digraph, 33, 347
direct currents, 94
Dirichlet boundary condition, 118
displacement, 373
displacement current density, 108
dissipation, 77
distance, 363
distributed models, 53, 151, 213
distributions, 397, 401
- derivatives of, 390
- scalar, 389
- space of, 388
- tensor, 390
- vector, 389
divergence
- of a vector field, 366
- in cartesian system, 378
- in cylindrical system, 381
- in spherical system, 384
- of a scalar distribution, 390
- of a tensor field, 366

electric
- susceptibility, 84
- arc furnace, 332
- dipole, 81, 83

- displacement, 83
- energy, 97
- field, 57
- field intensity, 57
- permittivity, 58, 84, 109
- potential, 67
- relative permittivity, 84
electrical conductivity, 80
electrode, 332
electrolytic cell, 300
electromagnet, 319
endomorphisms, 372
energy
- dissipated electrostatic, 23
- elastic potential, 9
- electric, 97
- electromagnetic, 96
- electrostatic, 23, 132
- kinetic, 9
- magnetic, 23, 97
energy balance, 8, 23, 96, 222
Euclidean
- affine space, 363
- vector space, 357
Eulerian field, 374
Everett function, 233

farad, 21
Faraday's law, 58
field
- coercive, 219
- remanent, 219
- scalar, 364
- tensor, 364
- vector, 364
forced oscillations, 10
frequency, 4, 7, 63
- angular, 56
- natural, 7, 12
- resonance, 24
- spatial, 56
fringing fields, 292
fundamental solution, 116

Gauss'
- electric law, 58, 84
- formulas, 370
- magnetic law, 58, 91

gradient
- in cartesian system, 377
- in cylindrical system, 380
- in spherical system, 383
- of a scalar field, 365
- of a vector field, 365
- of the motion, 373

graph, 33, 172, 221, 244
- connected, 347
- connected components, 347
- connected nodes, 347
- cotree, 347
- directed, 33, 172, 347
- edge, 347
- incidence matrix, 348
- node, 347
- path, 347
- reduced incidence matrix, 352
- tree, 347

Green's formulas, 370, 395

harmonic, 23
- oscillator, 3, 5, 21

Heaviside function, 30
Helmholtz's equation, 56, 79
henry, 21
hertz, 4
Hooke's law, 3
hysteresis
- cycle, 218
- loop, 218
- losses, 99
- major loop, 100, 219
- minor loop, 219
- operator, 224
- vector operator, 239

impedance, 24, 26
- boundary condition, 112
- electrical, 21
- intrinsic, 78
- intrinsic wave, 64, 108, 109
- matrix, 195
- mechanical, 11
- nodal, 45
- reduced, 196
- reduced matrix, 45, 46
- surface, 95, 112
- surface boundary condition, 95

inductance, 21, 35
- mutual, 37, 196, 243
- self-, 196

inductance matrix, 35, 244, 254
induction furnace, 343
inductor, 3, 21, 185, 339, 343
- uncoupled, 35

inner product, 356
integration measure elements
- in cartesian system, 379
- in cylindrical system, 381
- in spherical system, 385

intensity of magnetization, 91
internal resistance, 35, 39, 41, 146, 147

Joule effect, 97, 98

Kirchhoff's law, 22, 35, 37
Kronecker symbol, 408

Lagrange multipliers, 203
Lagrangian coordinates, 413
Lagrangian field, 374
Lambert's law, 112
Lamé-Navier equations, 53
Laplacian
- in cartesian system, 378
- in cylindrical system, 381
- in spherical system, 384
- of a scalar distribution, 390
- of a scalar field, 366
- of a vector distribution, 390
- of a vector field, 366

Lebesgue space, 391
Leontovich boundary condition, 95
Localization theorem, 376
Lorentz's force law, 57

losses
- classical eddy current, 240
- dynamic, 240
- eddy current, 98
- excess eddy current, 240
- hysteresis, 98, 240

lumped models, 3, 33, 139, 151, 183, 213

magnetic
- dipole, 85
- energy, 97
- flux density, 57

- flux leakage, 336
- force, 57
- intensity field, 63, 91
- permeability, 58, 91
- relative permeability, 91
- saturation, 219
- susceptibility, 91
- vector potential, 87, 152, 187, 201

magnetization
- curve, 223
- vector, 91

magnetomotive force, 164, 172

magnetron, 112

material
- anisotropic, 113
- conductor, 80
- demagnetized, 218
- diamagnetic, 217
- dielectric, 80
- ferromagnetic, 217
- hard magnetic, 219
- homogeneous, 113
- inhomogeneous, 113
- insulator, 80
- isotropic, 100
- linear, 96
- magnetic, 217
- soft magnetic, 219, 223

material field, 374

material point, 372

material time derivative, 375

matrix
- admittance, 195
- capacitance, 125
- conductance, 144
- connectivity , 33
- impedance, 195
- inductance, 35, 244, 254
- reduced admittance, 198
- reduced impedance, 196–199
- reluctance, 165
- resistance, 144

matter flow field, 414

Maxwell's equations
- boundary conditions, 94
- in free space, 53, 58, 67
- in Lagrangian coordinates, 413
- in material regions, 77

- interface boundary conditions, 94
- time harmonic, 61, 111, 181

Maxwell's stress tensor, 101

mechanical
- impedance, 11
- reactance, 6, 11

mesh-size, 407

momentum, 81

motion, 373

Nédélec edge elements, 156, 157, 407, 409

net area, 371

Neumann boundary condition, 118

Newton's second law, 3

nodal
- admittance, 45
- finite elements, 408

Non-Destructive Testing, 336

nonlocal memory, 221

norm, 357

ohm, 21

Ohm's law, 80
- for circuits, 144
- in material coordinates, 420

overdamped, 8

perfect conductor condition, 112

period, 4, 7

permeance, 167

phase, 5, 26, 56, 64
- constant, 63, 79, 107
- three-phase system, 250
- velocity, 63, 107

phasor, 5, 44

point charge, 117

polarization, 64
- current, 84
- current density, 85
- vector, 83

port, 213

position of material point, 413

position vector, 401

potential
- reduced, 198
- reduced scalar magnetic, 159
- total scalar, 161

Potential theorem, 370

power, 194
- active, 21, 25, 194
- apparent, 26
- factor, 26
- reactive, 21, 25, 194
- real, 25
Poynting vector, 98
Preisach
- classical model, 217, 225
- dynamic model, 240
- function, 225
- operator, 231
- triangle, 225
pressure waves, 54
primary coil, 243
product rule, 367
propagation constant, 79, 106

rate independent, 221, 240
reactance, 24, 195
- capacitive, 24
- inductive, 24
reduced current intensity, 198
reference configuration, 372
reference map, 373
reflection, 77
reflection coefficients, 78
relay operator, 224
reluctance, 163, 167, 172
remanence, 219
resistance, 3, 21, 24, 35, 195
- approximate, 146
- critical, 7
- internal, 39, 41, 146, 147
resistor, 21, 35, 147
resonance, 13, 21
reversal
- branch, 221
- point, 221
rigid media, 101
RLC circuit, 253

Schur complement, 40
secondary coil, 243
self-inductances, 37
SI, 4
siemens, 80
Silver-Müller conditions, 61
simply connected set, 369

skin depth, 95, 107
Sobolev space, 391
- weighted, 396
spatial field, 374
speed sound, 54
spring, 3
stability criterion, 16
stationary solution, 10, 12, 25
Steinmetz coefficient, 223
stiffness, 4
Stokes' theorem, 371

tensor, 357
- coordinates, 358
- eigenvalue of a, 360
- inner product, 359
- invertible, 360
- orthogonal, 360
- positive definite, 360
- principal invariants of a, 361
- product, 357
- skew part of, 359
- spectrum of a, 360
- symmetric, 358
- symmetric part of, 359
- transpose, 358
tesla, 57
trace operator, 392
trajectory
- of the material point, 373
- of the motion, 373
transformer, 37, 243
- step down, 244
- step up, 244, 246
transient solution, 10
transmission coefficients, 78
transmission conditions, 78
Transport theorems, 375, 414
trapezoidal rule, 38
tuning, 31

undamped, 8
underdamped, 6, 8
unit of
- apparent power, 26
- capacitance, 21
- capacitive reactance, 24
- charge, 22, 57
- current density, 57

- current intensity, 22, 57
- electric field intensity, 57
- electric permittivity, 58
- electrical conductivity, 80
- frequency, 4
- inductance, 21
- inductive reactance, 24
- magnetic flux density, 57
- magnetic permeability, 58
- magnetization density, 90
- mechanical impedance, 11
- mechanical resistance, 6
- reactive power, 26
- real power, 26
- resistance, 21
- voltage, 21

vector space, 355
vectors, 355
- coordinates, 356
- linearly indenpendent, 356
- orthogonal, 357
- system of generators, 356
- vector product of, 361
velocity, 373
- of material point, 413
Vieta relations, 361
volt, 21
voltage, 21, 35

wave
- dispersive, 107
- equations, 53, 56, 59, 60, 77
- linearly polarized, 64
- number, 79
- progressive, 56
- reflected, 78
- regressive, 56
- transmitted, 78
- transversal, 62
- vector, 61
wave number, 56
wavelength, 63, 107
weak formulation, 120, 123, 124, 140, 141, 143, 155, 158, 159, 161, 162, 165, 175, 188, 191, 193, 203, 204, 208, 213

Collana Unitext – La Matematica per il 3+2

Series Editors:
A. Quarteroni (Editor-in-Chief)
L. Ambrosio
P. Biscari
C. Ciliberto
G. van der Geer
G. Rinaldi
W.J. Runggaldier

Editor at Springer:
F. Bonadei
francesca.bonadei@springer.com

As of 2004, the books published in the series have been given a volume number. Titles in grey indicate editions out of print.
As of 2011, the series also publishes books in English.

A. Bernasconi, B. Codenotti
Introduzione alla complessità computazionale
1998, X+260 pp, ISBN 88-470-0020-3

A. Bernasconi, B. Codenotti, G. Resta
Metodi matematici in complessità computazionale
1999, X+364 pp, ISBN 88-470-0060-2

E. Salinelli, F. Tomarelli
Modelli dinamici discreti
2002, XII+354 pp, ISBN 88-470-0187-0

S. Bosch
Algebra
2003, VIII+380 pp, ISBN 88-470-0221-4

S. Graffi, M. Degli Esposti
Fisica matematica discreta
2003, X+248 pp, ISBN 88-470-0212-5

S. Margarita, E. Salinelli
MultiMath – Matematica Multimediale per l'Università
2004, XX+270 pp, ISBN 88-470-0228-1

A. Quarteroni, R. Sacco, F.Saleri
Matematica numerica (2a Ed.)
2000, XIV+448 pp, ISBN 88-470-0077-7
2002, 2004 ristampa riveduta e corretta
(1a edizione 1998, ISBN 88-470-0010-6)

13. A. Quarteroni, F. Saleri
    Introduzione al Calcolo Scientifico (2a Ed.)
    2004, X+262 pp, ISBN 88-470-0256-7
    (1a edizione 2002, ISBN 88-470-0149-8)

14. S. Salsa
    Equazioni a derivate parziali - Metodi, modelli e applicazioni
    2004, XII+426 pp, ISBN 88-470-0259-1

15. G. Riccardi
    Calcolo differenziale ed integrale
    2004, XII+314 pp, ISBN 88-470-0285-0

16. M. Impedovo
    Matematica generale con il calcolatore
    2005, X+526 pp, ISBN 88-470-0258-3

17. L. Formaggia, F. Saleri, A. Veneziani
    Applicazioni ed esercizi di modellistica numerica
    per problemi differenziali
    2005, VIII+396 pp, ISBN 88-470-0257-5

18. S. Salsa, G. Verzini
    Equazioni a derivate parziali – Complementi ed esercizi
    2005, VIII+406 pp, ISBN 88-470-0260-5
    2007, ristampa con modifiche

19. C. Canuto, A. Tabacco
    Analisi Matematica I (2a Ed.)
    2005, XII+448 pp, ISBN 88-470-0337-7
    (1a edizione, 2003, XII+376 pp, ISBN 88-470-0220-6)

20. F. Biagini, M. Campanino
    Elementi di Probabilità e Statistica
    2006, XII+236 pp, ISBN 88-470-0330-X

21. S. Leonesi, C. Toffalori
    Numeri e Crittografia
    2006, VIII+178 pp, ISBN 88-470-0331-8

22. A. Quarteroni, F. Saleri
    Introduzione al Calcolo Scientifico (3a Ed.)
    2006, X+306 pp, ISBN 88-470-0480-2

23. S. Leonesi, C. Toffalori
    Un invito all'Algebra
    2006, XVII+432 pp, ISBN 88-470-0313-X

24. W.M. Baldoni, C. Ciliberto, G.M. Piacentini Cattaneo
    Aritmetica, Crittografia e Codici
    2006, XVI+518 pp, ISBN 88-470-0455-1

25. A. Quarteroni
    Modellistica numerica per problemi differenziali (3a Ed.)
    2006, XIV+452 pp, ISBN 88-470-0493-4
    (1a edizione 2000, ISBN 88-470-0108-0)
    (2a edizione 2003, ISBN 88-470-0203-6)

26. M. Abate, F. Tovena
    Curve e superfici
    2006, XIV+394 pp, ISBN 88-470-0535-3

27. L. Giuzzi
    Codici correttori
    2006, XVI+402 pp, ISBN 88-470-0539-6

28. L. Robbiano
    Algebra lineare
    2007, XVI+210 pp, ISBN 88-470-0446-2

29. E. Rosazza Gianin, C. Sgarra
    Esercizi di finanza matematica
    2007, X+184 pp, ISBN 978-88-470-0610-2

30. A. Machì
    Gruppi – Una introduzione a idee e metodi della Teoria dei Gruppi
    2007, XII+350 pp, ISBN 978-88-470-0622-5
    2010, ristampa con modifiche

31. Y. Biollay, A. Chaabouni, J. Stubbe
    Matematica si parte!
    A cura di A. Quarteroni
    2007, XII+196 pp, ISBN 978-88-470-0675-1

32. M. Manetti
    Topologia
    2008, XII+298 pp, ISBN 978-88-470-0756-7

33. A. Pascucci
    Calcolo stocastico per la finanza
    2008, XVI+518 pp, ISBN 978-88-470-0600-3

34. A. Quarteroni, R. Sacco, F. Saleri
    Matematica numerica (3a Ed.)
    2008, XVI+510 pp, ISBN 978-88-470-0782-6

35. P. Cannarsa, T. D'Aprile
    Introduzione alla teoria della misura e all'analisi funzionale
    2008, XII+268 pp, ISBN 978-88-470-0701-7

36. A. Quarteroni, F. Saleri
    Calcolo scientifico (4a Ed.)
    2008, XIV+358 pp, ISBN 978-88-470-0837-3

37. C. Canuto, A. Tabacco
    Analisi Matematica I (3a Ed.)
    2008, XIV+452 pp, ISBN 978-88-470-0871-3

38. S. Gabelli
    Teoria delle Equazioni e Teoria di Galois
    2008, XVI+410 pp, ISBN 978-88-470-0618-8

39. A. Quarteroni
    Modellistica numerica per problemi differenziali (4a Ed.)
    2008, XVI+560 pp, ISBN 978-88-470-0841-0

40. C. Canuto, A. Tabacco
    Analisi Matematica II
    2008, XVI+536 pp, ISBN 978-88-470-0873-1
    2010, ristampa con modifiche

41. E. Salinelli, F. Tomarelli
    Modelli Dinamici Discreti (2a Ed.)
    2009, XIV+382 pp, ISBN 978-88-470-1075-8

42. S. Salsa, F.M.G. Vegni, A. Zaretti, P. Zunino
    Invito alle equazioni a derivate parziali
    2009, XIV+440 pp, ISBN 978-88-470-1179-3

43. S. Dulli, S. Furini, E. Peron
    Data mining
    2009, XIV+178 pp, ISBN 978-88-470-1162-5

44. A. Pascucci, W.J. Runggaldier
    Finanza Matematica
    2009, X+264 pp, ISBN 978-88-470-1441-1

45. S. Salsa
    Equazioni a derivate parziali – Metodi, modelli e applicazioni (2a Ed.)
    2010, XVI+614 pp, ISBN 978-88-470-1645-3

46. C. D'Angelo, A. Quarteroni
    Matematica Numerica – Esercizi, Laboratori e Progetti
    2010, VIII+374 pp, ISBN 978-88-470-1639-2

47. V. Moretti
    Teoria Spettrale e Meccanica Quantistica – Operatori in spazi di Hilbert
    2010, XVI+704 pp, ISBN 978-88-470-1610-1

48. C. Parenti, A. Parmeggiani
    Algebra lineare ed equazioni differenziali ordinarie
    2010, VIII+208 pp, ISBN 978-88-470-1787-0

49. B. Korte, J. Vygen
    Ottimizzazione Combinatoria. Teoria e Algoritmi
    2010, XVI+662 pp, ISBN 978-88-470-1522-7

50. D. Mundici
    Logica: Metodo Breve
    2011, XII+126 pp, ISBN 978-88-470-1883-9

51. E. Fortuna, R. Frigerio, R. Pardini
    Geometria proiettiva. Problemi risolti e richiami di teoria
    2011, VIII+274 pp, ISBN 978-88-470-1746-7

52. C. Presilla
    Elementi di Analisi Complessa. Funzioni di una variabile
    2011, XII+324 pp, ISBN 978-88-470-1829-7

53. L. Grippo, M. Sciandrone
    Metodi di ottimizzazione non vincolata
    2011, XIV+614 pp, ISBN 978-88-470-1793-1

54. M. Abate, F. Tovena
    Geometria Differenziale
    2011, XIV+466 pp, ISBN 978-88-470-1919-5

55. M. Abate, F. Tovena
    Curves and Surfaces
    2011, XIV+390 pp, ISBN 978-88-470-1940-9

56. A. Ambrosetti
    Appunti sulle equazioni differenziali ordinarie
    2011, X+114 pp, ISBN 978-88-470-2393-2

57. L. Formaggia, F. Saleri, A. Veneziani
    Solving Numerical PDEs: Problems, Applications, Exercises
    2011, X+434 pp, ISBN 978-88-470-2411-3

58. A. Machì
    Groups. An Introduction to Ideas and Methods of the Theory of Groups
    2011, XIV+372 pp, ISBN 978-88-470-2420-5

59. A. Pascucci, W.J. Runggaldier
    Financial Mathematics. Theory and Problems for Multi-period Models
    2011, X+288 pp, ISBN 978-88-470-2537-0

60. D. Mundici
    Logic: a Brief Course
    2012, XII+124 pp, ISBN 978-88-470-2360-4

61. A. Machì
    Algebra for Symbolic Computation
    2012, VIII+174 pp, ISBN 978-88-470-2396-3

62. A. Quarteroni, F. Saleri, P. Gervasio
    Calcolo Scientifico (5a Ed.)
    2012, XVIII+450 pp, ISBN 978-88-470-2744-2

63. A. Quarteroni
    Modellistica Numerica per Problemi Differenziali (5a Ed.)
    2012, XVIII+628 pp, ISBN 978-88-470-2747-3

64. V. Moretti
   Spectral Theory and Quantum Mechanics
   With an Introduction to the Algebraic Formulation
   2013, XVI+728 pp, ISBN 978-88-470-2834-0

65. S. Salsa, F.M.G. Vegni, A. Zaretti, P. Zunino
   A Primer on PDEs. Models, Methods, Simulations
   2013, XIV+482 pp, ISBN 978-88-470-2861-6

66. V.I. Arnold
   Real Algebraic Geometry
   2013, X+110 pp, ISBN 978-3-642-36242-2

67. F. Caravenna, P. Dai Pra
   Probabilità. Un'introduzione attraverso modelli e applicazioni
   2013, X+396 pp, ISBN 978-88-470-2594-3

68. A. de Luca, F. D'Alessandro
   Teoria degli Automi Finiti
   2013, XII+316 pp, ISBN 978-88-470-5473-8

69. P. Biscari, T. Ruggeri, G. Saccomandi, M. Vianello
   Meccanica Razionale
   2013, XII+352 pp, ISBN 978-88-470-5696-3

70. E. Rosazza Gianin, C. Sgarra
   Mathematical Finance: Theory Review and Exercises. From Binomial
   Model to Risk Measures
   2013, X+278pp, ISBN 978-3-319-01356-5

71. E. Salinelli, F. Tomarelli
   Modelli Dinamici Discreti (3a Ed.)
   2014, XVI+394pp, ISBN 978-88-470-5503-2

72. C. Presilla
   Elementi di Analisi Complessa. Funzioni di una variabile (2a Ed.)
   2014, XII+360pp, ISBN 978-88-470-5500-1

73. S. Ahmad, A. Ambrosetti
   A Textbook on Ordinary Differential Equations
   2014, XIV+324pp, ISBN 978-3-319-02128-7

74. A. Bermúdez, D. Gómez, P. Salgado
   Mathematical Models and Numerical Simulation in Electromagnetism
   2014, XVIII+432pp, ISBN 978-3-319-02948-1

The online version of the books published in this series is available at SpringerLink.
For further information, please visit the following link:
http://www.springer.com/series/5418

The manufacturer's authorised representative in the EU is Springer Nature Customer Service Centre GmbH, Europaplatz 3, 69115 Heidelberg, Germany. If you have any concerns regarding our products, please contact ProductSafety@springernature.com

Printed and bound by CPI Group (UK) Ltd, Croydon, CR0 4YY
23/03/2026
02076665-0002